Fruit and Vegetable Phytochemicals

Chemistry, Nutritional Value, and Stability

Fruit and Vegetable Phytochemicals

Chemistry, Nutritional Value, and Stability

Laura A. de la Rosa
Emilio Alvarez-Parrilla
Gustavo A. González-Aguilar

WILEY-BLACKWELL
A John Wiley & Sons, Inc., Publication

Edition first published 2010

Blackwell Publishing was acquired by John Wiley & Sons in February 2007. Blackwell's publishing program has been merged with Wiley's global Scientific, Technical, and Medical business to form Wiley-Blackwell.

Editorial Office
2121 State Avenue, Ames, Iowa 50014-8300, USA

For details of our global editorial offices, for customer services, and for information about how to apply for permission to reuse the copyright material in this book, please see our website at www.wiley.com/wiley-blackwell.

Library of Congress Cataloging-in-Publication Data

de la Rosa, Laura A.
Fruit and vegetable phytochemicals : chemistry, nutritional value and stability /
Laura A. de la Rosa, Emilio Alvarez-Parrilla, Gustavo A. González-Aguilar. – 1. ed.
p. cm.
Includes bibliographical references and index.
ISBN 978-0-8138-0320-3 (hardback : alk. paper)
1. Phytochemicals. 2. Polyphenols. 3. Carotenoids. 4. Fruit–Analysis. 5. Vegetables–Analysis. I. Alvarez-Parrilla, Emilio. II. González-Aguilar, Gustavo A. III. Title.
QK898.P764R67 2009
615'.321–dc22

2009031846

A catalog record for this book is available from the U.S. Library of Congress.

Set in 10/12pt Times by Aptara® Inc., New Delhi, India
Printed in Singapore

1 2010

Contents

Contributors

Emilio Alvarez-Parrilla (Chapter 11)
Departamento de Ciencias Básicas
Instituto de Ciencias Biomédicas
Universidad Autónoma de Ciudad Juárez
Ciudad Juárez, Chihuahua
México

Cristina Andres-Lacueva (Chapter 2)
Nutrition and Food Science Department
Pharmacy School
University of Barcelona
Av Joan XXIII s/n
Barcelona 08028
Spain

Sofía Arvizu-Medrano (Chapter 12)
Departamento de Investigación y Posgrado en Alimentos Facultad de Química
Universidad Autónoma de Querétaro Programa de Posgrado en Alimentos del Centro de la República Centro Universitario Cerro de las Campanas s/n
Col. Centro. Querétaro Querétaro, Mexico

J. Fernando Ayala-Zavala (Chapter 11)
Coordinación de Tecnología de Alimentos de Origen Vegetal
Centro de Investigación en Alimentación y Desarrollo, A.C.
Apdo. Postal 1735
Hermosillo, Sonora 83000
Mexico

Adriana Verónica Bolaños-Villar (Chapter 6)
Department of Human Nutrition
Centro de Investigación en Alimentación y Desarrollo A.C.
Carr. a la Victoria Km. 0.6
C.P. 83000
Hermosillo, Sonora
Mexico

Gemma Chiva-Blanch (Chapter 2)
Nutrition and Food Science Department
Pharmacy School
University of Barcelona
Av Joan XXIII s/n
Barcelona 08028
Spain

Saul Ruiz Cruz (Chapter 12)
Instituto Tecnológico de Sonora
Departamento de Biotecnología y Ciencias Alimentarias
5 de Febrero 818 Sur
Colonia centro
Ciudad Obregón, Sonora 85000
México

Laura A. de la Rosa (Chapters 6 and 11)
Departamento de Ciencias Básicas
Instituto de Ciencias Biomédicas
Universidad Autónoma de Ciudad Juárez

Ciudad Juárez, Chihuahua
México

Felix M. C. Gamarra (Chapter 9)
Department of Agricultural
Food and Nutritional Science
University of Alberta
Edmonton, Alberta T6G 2P5
Canada

Francisco García-Carmona (Chapter 4)
Department of Biochemistry and
Molecular Biology-A
Faculty of Biology
University of Murcia
Campus de Espinardo
30071 Murcia
Spain

Isabel Goñi (Chapter 8)
Nutrition and Gastrointestinal Health
Unit (UCM/CSIC)
Department of Nutrition I
Universidad Complutense de Madrid
(UCM)

Gustavo A. González-Aguilar (Chapters 6
and 11)
Coordinación de Tecnología de
Alimentos de Origen Vegetal
Centro de Investigación en
Alimentación y Desarrollo, A.C.
Apdo. Postal 1735
Hermosillo, Sonora 83000
Mexico

Nuria Grigelmo-Miguel (Chapter 10)
Department of Food Technology
UTPV-CeRTA
University of Lleida
Av. Alcalde Rovira Roure 191
25198 Lleida
Spain

Nasiruddin Khan (Chapter 2)
Nutrition and Food Science Department
Pharmacy School
University of Barcelona
Av Joan XXIII s/n
Barcelona 08028
Spain

Rosa M. Lamuela-Raventós (Chapter 2)
Nutrition and Food Science Department
Pharmacy School
University of Barcelona
Av Joan XXIII s/n
Barcelona 08028
Spain

Rafael Llorach (Chapter 2)
Nutrition and Food Science Department
Pharmacy School
University of Barcelona
Av Joan XXIII s/n
Barcelona 08028
Spain

José Manuel López-Nicolás (Chapter 4)
Department of Biochemistry and
Molecular Biology-A
Faculty of Biology
University of Murcia
Campus de Espinardo
30071 Murcia
Spain

Olga Martín-Belloso (Chapter 10)
Department of Food Technology
UTPV-CeRTA
University of Lleida
Av. Alcalde Rovira Roure 191
25198 Lleida
Spain

Alex Medina-Remon (Chapter 2)
Nutrition and Food Science Department
Pharmacy School
University of Barcelona
Av Joan XXIII s/n
Barcelona 08028
Spain

Jason McCallum (Chapter 5)
Regis and Joan Duffy Research Centre
UPEI, Agriculture & Agri-Food Canada
440 University Avenue
Charlottetown
Prince Edward Island C1A 4N6
Canada

José De Jesús Ornelas-Paz (Chapter 7)
Centro de Investigación en Alimentación y Desarrollo A.C.
Unidad Cuauhtémoc
Av. Río Conchos S/N Parque Industrial
Apdo. Postal 781
C.P. 31570 Cd.
Cuauhtémoc, Chihuahua
México

Jara Pérez-Jiménez (Chapter 8)
Department of Metabolism and Nutrition
ICTAN-IF
Consejo Superior de Investigaciones Cientificas (CSIC)

Alma E. Robles-Sardin (Chapter 6)
Department of Human Nutrition
Centro de Investigación en Alimentación y Desarrollo A.C.
Carr. a la Victoria Km. 0.6
C.P. 83000
Hermosillo, Sonora
Mexico

M[a] Alejandra Rojas-Graü (Chapter 10)
Department of Food Technology
UTPV-CeRTA
University of Lleida
Av. Alcalde Rovira Roure 191
25198 Lleida
Spain

Maria Rotchcs-Ribalta (Chapter 2)
Nutrition and Food Science Department
Pharmacy School
University of Barcelona
Av Joan XXIII s/n
Barcelona 08028
Spain

Marleny D. A. Saldaña (Chapter 9)
Department of Agricultural
Food and Nutritional Science
University of Alberta
Edmonton, Alberta T6G 2P5
Canada

Mikal E. Saltveit (Chapter 3)
Mann Laboratory, Department of Plant Sciences
University of California
One Shields Ave.
Davis, CA 95616-8631
USA

Fulgencio Saura-Calixto (Chapter 8)
Department of Metabolism and Nutrition
ICTAN-IF
Consejo Superior de Investigaciones Cientificas (CSIC)

Rodrigo M. P. Siloto (Chapter 9)
Department of Agricultural
Food and Nutritional Science
University of Alberta
Edmonton, Alberta T6G 2P5
Canada

Robert Soliva-Fortuny (Chapter 10)
Department of Food Technology
UTPV-CeRTA
University of Lleida
Av. Alcalde Rovira Roure 191
25198 Lleida
Spain

Rong Tsao (Chapter 5)
Guelph Food Research Centre
Agriculture & Agri-Food Canada
93 Stone Road West
Guelph, Ontario N1G 5C9
Canada

Mireia Urpi-Sarda (Chapter 2)
Nutrition and Food Science Department
Pharmacy School
University of Barcelona
Av Joan XXIII s/n
Barcelona 08028
Spain

Elhadi M. Yahia (Chapters 1 and 7)
Facultad de Ciencias Naturales
Universidad Autónoma de Querétaro
Avenida de las Ciencias s/n, Juriquilla
Querétaro, 76230, Qro.
México

Raul Zamora-Ros (Chapter 2)
Nutrition and Food Science Department
Pharmacy School
University of Barcelona
Av Joan XXIII s/n
Barcelona 08028
Spain

Preface

There has been increasing interest in the nutritional properties of fruits and vegetables as sources of health-promoting phytochemical compounds, due to epidemiological data that correlate a high consumption of these compounds with a lower risk of several cardiovascular and degenerative diseases. Phytochemicals are a heterogeneous group of chemical compounds with numerous biological actions. The aim of this book is to provide scientists in the areas of food technology and nutrition with accessible and up-to-date information about the chemical nature, classification, and analysis of the main phytochemicals present in fruits and vegetables: polyphenols and carotenoids. Special care is taken to analyze the health benefits of these compounds and their interaction with fiber, antioxidants, and other biological activities, as well as the degradation processes that occur after harvest and minimal processing.

This book was possible thanks to the valuable collaboration of many scientists who agreed to dedicate part of their time to participate in this project, with valuable contributions from their different fields of expertise. To them we are sincerely thankful.

Dr. Laura A. de la Rosa
Dr. Emilio Alvarez-Parrilla
Dr. Gustavo A. González-Aguilar

Fruit and Vegetable Phytochemicals

Chemistry, Nutritional Value, and Stability

1 The Contribution of Fruit and Vegetable Consumption to Human Health

Elhadi M. Yahia

Introduction

Increasing incidences of some chronic diseases, including cancer and cardiovascular disease, especially in industrial countries, have raised awareness regarding the importance of diet (Erbersdobler 2003). It is estimated that one-third of the cancer cases and up to half of cardiovascular disease cases are thought to be diet related (Goldberg 1994). Numerous epidemiological studies have shown an inverse association between fruit and vegetable consumption and chronic diseases, including different types of cancer and cardiovascular disease (Hirayama 1990; Block and others 1992; Howe and others 1992; Steinmetz and Potter 1991, 1996; World Cancer Research Fund 1997; Joshipura and others 2001; Bazzano and others 2002; Kris-Etherton and others 2002). These studies have shown mounting evidence that people who avoid fruit and vegetables completely, or who consume very little, are indeed at increased risk of these diseases. Therefore, interest in the health benefits of fruit and vegetable consumption is increasing, and the interest in understanding the type, number, and mode of action of the different components in fruits and vegetables that confer health benefits is also increasing.

Fruits and vegetables have historically been considered rich sources of some essential dietary micronutrients and fibers, and more recently they have been recognized as important sources for a wide array of phytochemicals that individually, or in combination, may benefit health (Stavric 1994; Rechkemmer 2001). Therefore, some people have conferred on fruits and vegetables the status of "functional foods." There are many biologically plausible reasons for this potentially protective association, including the fact that many of the phytochemicals act as antioxidants.

Phytochemicals present in fruits and vegetables are very diverse, such as ascorbic acid, carotenoids, and phenolic compounds (Liu 2004; Percival and others 2006; Syngletary and others 2005; Yahia and others 2001a, 2001b). Plant polyphenols are ubiquitous in the diet, with rich sources being tea, wine, fruits, and vegetables; they demonstrate considerable antioxidative activity *in vitro*, which can have important implications for health (Duthie and others 2000).

Naturally occurring compounds such as phytochemicals, which possess anticarcinogenic and other beneficial properties, are referred to as chemopreventers. One of the predominant mechanisms of their protective action is due to their antioxidant activity and the capacity to scavenge free radicals. Among the most investigated chemopreventers are some vitamins, plant polyphenols, and pigments such as carotenoids, chlorophylls, flavonoids, and betalains. Resolution of the potential protective roles of specific antioxidants and other constituents of fruits and vegetables deserves major attention.

Evidence indicates that for the effect of fruit and vegetable consumption on health, the whole may be more than the sum of the parts, individual components appear to act synergistically where the influence of at least some of these is additive.

Consumption of a diet high in fruits and vegetables increases antioxidant concentration in blood and body tissues and potentially protects against oxidative damage to cells and tissues. Olmedilla and others (2001) described blood concentration of carotenoids, tocopherols, ascorbic acid, and retinol in well-defined groups of healthy nonsmokers, aged 25–45 years; the study included 175 men and 174 women from five European countries (France, Northern Ireland, Republic of Ireland, the Netherlands, and Spain). Analysis was centralized and performed within 18 months. Within-gender, vitamin C showed no significant differences between countries. Females in France, Republic of Ireland, and Spain had significantly higher plasma vitamin C concentration than their male counterparts. Serum retinol and α-tocopherol levels were similar, but γ-tocopherol showed great variability, being lowest in Spain and France and highest in the Netherlands. The ratio of provitamin A to non–provitamin A carotenoids was similar among countries, whereas the ratio of xanthophylls (lutein, zeaxanthin, and β-cryptoxanthin) to carotenes (α-carotene, β-carotene, and lycopene) was double in southern areas (Spain) compared to northern areas (Northern Ireland and Republic of Ireland). Serum concentration of lutein and zeaxanthin were highest in France and Spain and β-cryptoxanthin was highest in Spain and the Netherlands. *trans*-Lycopene tended to be highest in Irish males and lowest in Spanish males, whereas α-carotene and β-carotene were higher in the French volunteers. Because of the study design, concentration of carotenoids and vitamins A, C, and E represent physiological ranges achievable by dietary means and may be considered as "reference values" in serum of healthy, nonsmoking middle-aged subjects from the five European countries. Results suggest that lutein (and zeaxanthin), β-cryptoxanthin, total xanthophylls, γ-tocopherol, and the ratio of β-tocopherol to γ-tocopherol may be important markers related to the healthy or protective effects of a Mediterranean-like diet.

The epidemiological evidence indicates that avoidance of smoking, increased consumption of fruits and vegetables, and control of infections can have a major effect on reducing rates of several chronic diseases including cardiovascular disease and different types of cancer (Ames and others 1995; Graham and Mettlin 1981; Giovannucci 1999; Liu 2004; Syngletary and others 2005; Percival and others 2006). It is argued that increasing intake of 400 to 800 g/day of fruits and vegetables is a public health strategy of considerable importance for individuals and communities worldwide. For this reason the World Health Organization (WHO) recommends a daily intake of more than 400 g per person daily, and health authorities worldwide promote high consumption of fruits and vegetables. Many of the putative chemoprotective phytochemicals in fruits and vegetables are colored (due to various pigments). The guidelines are based on selecting one daily serving of fruits and vegetables from each of seven color classes (red, yellow-green, red-purple, orange, orange-yellow, green, white-green) so that a variety of phytochemicals is consumed. A study by Johnston and others (2000) during 1994–1996, a "continuing survey of food intakes by individuals," was used to examine the types of fruits and vegetables consumed in the US. The sample populations consisted of 4,806 men and women (25–75 years old) who completed two nonconsecutive 24-hr recalls, consumed 3.6–2.3 servings of vegetables and 1.6 ± 2.0 servings of fruit daily. Iceberg lettuce, tomatoes, French fried potatoes, bananas, and orange juice were the most commonly consumed fruits and vegetables, accounting for nearly 30% of all fruits and vegetables consumed. The most popular items, lettuce and tomatoes, were consumed by 39–42% of the sample population during the reporting period. Fewer respondents (16–24%) consumed French fried potatoes, bananas, or orange juice. Only 3% of the sample consumed broccoli during the reporting period. White potato consumption averaged 1.1 servings daily, with French fried potatoes representing 0.4 serving. Tomato product consumption averaged 0.5 serving daily, dark green vegetable consumption averaged 0.2 serving daily, and citrus, berries, or melon consumption amounted to nearly 0.8 serving daily. These data have indicated that people in the US are consuming more fruits and vegetables compared to previous years but that dark green and cruciferous vegetable intake is low. Many studies suggest that consumption of fruit and vegetables is still low in many countries (Naska and others 2000; Agudo and others 2002; USDA 2002; Blanck and others 2008), and efforts are still needed to increase it. This chapter will highlight the potential health benefits of fruit and vegetable consumption on several diseases, as well as the nutritional and health importance of some fruits and vegetables.

Effect of Consumption of Fruit and Vegetables on Some Diseases

Cancer

Epidemiological evidence of cancer-protective effects of fruits and vegetables, as well as the basic mechanisms by which phytochemicals in fruits and vegetables can protect against cancer development, has been previously surveyed by Wargovich (2000). Sometimes it was difficult to associate total fruit and vegetable consumption and cancer prevention, but it could be associated with some specific families or types of fruits and vegetables (Steinmetz and Potter 1996; Voorips and others 2000). For example, a high consumption of tomato or tomato-based products is consistently associated with lower risk of different cancer types as shown by meta-analysis, with the highest evidence found for lung, prostate, and stomach cancer (Giovannucci 1999). The metabolism of chemical carcinogens has been shown to be influenced by dietary constituents (Wattenberg 1975). Naturally occurring inducers of increased activity of the microsomal mixed-function oxidase system are present in plants; cruciferous vegetables are particularly potent in this regard. From Brussels sprouts, cabbage, and cauliflower, three indoles with inducing activity have been identified: indole-3-acetonitrile, indole-3-carbinol, and 3,3′-diindolylmethane. A second type of dietary constituent that affects the microsomal mixed-function oxidase system is added phenolic antioxidants, butylated hydroxyanisole (BHA), and butylated hydroxytoluene. The feeding of BHA has resulted in microsomal changes in the liver. Spectral characteristics of cytochrome P450 were changed, and the aryl hydrocarbon hydroxylasc system of these microsomes demonstrated increased sensitivity to inhibition by α-naphthoflavone. A decrease in the binding of metabolites of benzo[*a*]pyrene to DNA was noted upon incubation of these microsomes with benzo[*a*]pyrene. BHA and butylated hydroxytoluene exert a protective effect against chemical carcinogens. Therefore, it seems that the constituents of the diet could be of consequence in the neoplastic response to exposure to carcinogens in the environment.

Several *in vitro* studies have shown that phenolic compounds in fruits and vegetables have antiproliferative effect in different cancer cell lines (Eberhardt and others 2000; Chu and others 2002; Sun and others 2002; Liu and others 2005; Mertens-Talcott and others 2005; Percival and others 2006).

Quercetin, a flavonoid found in many fruits and vegetables, has been shown to affect the metastatic potential of mouse tumor cells (Suzuki and others 1991). Mutagenicity of quercetin was examined by means of DNA fingerprint analysis using the Pc-1 probe that efficiently detects mutations due to recombination. Treatment of BMT-11 and FM3A tumor cells with quercetin resulted in gain and loss of bands in the fingerprints in both cell lines. The frequencies of the clones having undergone mutation were 3/11 and 6/26, respectively. This suggests that quercetin is mutagenic and induces recombination, and the results seem to provide a molecular basis for the phenotypic variations of BMT-11 tumor cells induced by quercetin. It is important to mention, however, that although *in vitro* quercetin has consistently tested positive for mutagenic activity in prokaryotic and eukaryotic cells, *in vivo* experiments have not confirmed these findings, supporting the toxicological safety of quercetin (Harwood and others 2007). Other flavonoids also show mutagenic activity (Takahashi and others 1979).

Chlorophyllin (CHL), a food-grade derivative of the ubiquitous fruit and vegetable pigment chlorophyll, has been shown to be a potent, dose-responsive inhibitor of aflatoxin B_1 DNA adduction and hepatocarcinogenesis in the rainbow trout model when fed with carcinogen (Breinholt and others 1995). Chlorophyllins are derivatives of chlorophyll in which the central magnesium atom is replaced by other metals, such as cobalt, copper, or iron. The relative efficacy of chlorophyll and chlorophyllin has been shown to modify the genotoxic effects of various known toxicants (Sarkar and others 1994). CHL neither promoted nor suppressed carcinogenesis with chronic postinitiation feeding. By molecular dosimetry analysis, reduced aflatoxin B_1-DNA adduction accounted quantitatively for reduced tumor response up to 2000 ppm dietary CHL, but an additional protective mechanism was operative at 4000 ppm CHL. The finding of potent inhibition (up to 77%) at CHL levels well within the chlorophyll content of some green fruits and vegetables may have important implications in intervention and dietary management of human cancer risks.

Monoterpenes are natural plant products found in the essential oils of many commonly consumed fruits and vegetables and have been widely used as flavor and fragrance additives in food and beverages. Monoterpenes have been shown to possess antitumorigenic activities (Kelloff and others 1996). Limonene, the simplest monocyclic monoterpene, which is found in some citrus fruits, and perrillyl alcohol, a hydroxylated limonene analog, have demonstrated chemopreventive and chemotherapeutic activity against mammary, skin, lung, pancreas, and colon tumors in rodent models (Crowell and Gould 1994; Wattenburg and Coccia 1991; Stark and others 1995).

Experimental studies (Sugie and others 1993; Dashwood and others 1989; Tanaka and others 1990) have shown that indoles and isothiocyanates (found in *Brassica* vegetables) given to animals after a carcinogen insult reduced tumor incidence and multiplicity at a number of sites including the liver, mammary gland, and colon. A possible inhibitory activity of isothiocyanates and indoles against tumorigenesis apparently stems from their ability to influence Phase I and Phase II biotransformation enzyme activities (Zhang and Talalay 1994; Boone and others 1990; McDannell and McLean 1988). Sulforaphane, which is present in broccoli, is a potent inducer of the Phase II detoxification enzymes quinine reductase and glutathione transferase and an inhibitor of the carcinogen-activating cytochrome P450 2E1 (Zhang and others 1992; Barcelo and others 1996).

Numerous studies (Rungapamestry and others 2007) have indicated that the hydrolytic products of at least three glucosinolates, 4-methylsulfinylbutyl (glucoraphanin), 2-phenylethyl (gluconasturtiin), and 3-indolylmethyl (glucobrassicin), have anticarcinogenic activity. Indole-3-carbinol, a metabolite of glucobrassicin, has shown inhibitory effects in studies of human breast and ovarian cancers. *S*-Methylcysteine sulfoxide and its metabolite methylmethane thiosulfinate were shown to inhibit chemically induced genotoxicity in mice. Thus, the cancer chemopreventive effects of *Brassica* vegetables that have been shown in human and animal studies may be due to the presence of both types of sulfur-containing phytochemicals (i.e., certain glucosinolates and *S*-methylcysteine sulfoxide).

Stomach

Reports on an inverse relationship between the consumption of fresh vegetables and human gastrointestinal cancer have been followed by screening for the protective activity of a large number of plant extracts, including leafy vegetables (Botterweck and others 1998; Larsson and others 2006).

Protection for all sites of digestive-tract cancers (oral cavity and pharynx, esophagus, stomach, colon, rectum) was associated with an increased intake of tomato-based foods and an increased supply of lycopene (Franceschi and others 1994). People who ate at least one serving of tomato-based product per day had 50% less chance of developing digestive tract cancer than those who did not eat tomatoes (Franceschi and others 1994). The intake of lycopene has also been associated with a reduced risk of cancers of sites other than the digestive tract, such as the pancreas and the bladder (Gerster 1997). Older subjects who regularly ate tomatoes were found to be less likely to develop all forms of cancer (Colditz and others 1985). A significant trend in risk reduction of gastric cancers by high tomato consumption was also observed in a study that estimated dietary intake in low-risk versus high-risk areas in Italy (Buiatti and others 1990). A similar regional impact on stomach cancer was also observed in Japan (Tsugani and others 1992), where out of several micronutrients including vitamins A, C, and D and β-carotene, only lycopene was strongly inversely associated with stomach cancer. Franceschi and others (1994) have shown a consistent pattern of protection for many sites of digestive tract cancer associated with an increased intake of fresh tomatoes.

Habitual consumption of garlic has been reported to correlate with a reduction in gastric nitrite content and reduction in gastric cancer mortality (Mei and others 1982; You and others 1989). *Allium* compounds present in garlic have been reported to inhibit the synthesis of *N*-nitroso compounds (Mei and others 1989). In a human study, 5 g of fresh garlic consumption has shown to markedly suppress urinary excretion of *N*-nitrosoproline in individuals given supplemental nitrate and proline (Mei and others 1989). In addition, two ecological studies (World Cancer Research Fund Food 1997) showed that in areas where garlic or onion production is very high, mortality rates from stomach cancer are very low.

Colon

Consumption of fruits and vegetables, and the associated vitamin C, carotene, and fiber, has been reported to reduce risk of colon cancer (Ziegler and others 1981). Since plant sterols are plentiful in vegetarian diets, the effect of β-sitosterol on colon tumor formation in rats treated with the carcinogen *N*-methyl-*N*-nitrosourea was studied by Raicht and others (1980). They have demonstrated that β-sitosterol nullified in part the effect of this direct-acting carcinogen on the colon. They suggest that plant sterols may have a protective dietary action to retard colon tumor formation, and therefore the beneficial effects of vegetarian diets may be enhanced because of the presence of these compounds.

An increased risk of colon cancer has been associated with decreases in the frequency with which vegetables were eaten in a study of 214 females with cancer of the colon, and 182 females with cancer of the rectum yielded similar results (Graham and others 1978). The decrease in risk was found to be associated with frequent ingestion of vegetables, especially cabbage, Brussels sprouts, and broccoli, and it is consistent with the decreased numbers of tumors observed in animals challenged with carcinogens and fed compounds found in these same vegetables.

Associations between fruit, vegetable, and dietary fiber consumption and colorectal cancer risk were investigated in a population that consumes relatively low amounts of fruit and vegetables and high amounts of cereals (Terry and others 2001). Data were examined from a food-frequency questionnaire used in a population-based prospective mammography screening study of women in central Sweden. Women with a diagnosis of colorectal cancer were identified by linkage to regional cancer registries, and Cox proportional hazards models were used to estimate relative risks; all statistical tests were two-sided. During an average 9.6 years of follow-up of 61,463 women, 460 incident cases of colorectal cancer were observed. In the entire studied population, total fruit and vegetable consumption was inversely associated with colorectal cancer risk, but subanalyses showed that this association was largely due to fruit consumption. The association was stronger and the dose–response effect was more evident among individuals who consumed the lowest amounts of fruit and vegetables; individuals who consumed less than 1.5 servings of fruit and vegetables/day had a higher relative risk of developing colorectal cancer compared with individuals who consumed greater than 2.5 servings.

Diets containing citrus fiber have been reported to reduce the risk of intestinal cancer. The effect of dietary dehydrated citrus fiber on carcinogenesis of the colon and small intestine was studied in male F344 rats by Reddy and others (1981). Weanling rats were fed semipurified diets containing 5% fat and 15% citrus fiber; at 7 weeks of age, all animals, except vehicle-treated controls, received weekly injections of 8 mg azoxymethane (AOM)/kg body weight for 10 weeks, and the AOM- or vehicle-treated groups were necropsied 20 weeks after the last injection of AOM. The animals fed the citrus fiber diet and treated with AOM had a lower incidence (number of animals with tumors) and multiplicity (number of tumors/tumor-bearing animal) of colon tumors and tumors of the small intestine than did those fed the control diet and treated with

AOM. The number of adenomas but not the number of adenocarcinomas was reduced in rats fed on the citrus pulp diet.

The effect of a diet containing 10–40% lyophilized cabbage or broccoli as cruciferous vegetable or 10–40% lyophilized potato as noncruciferous vegetable fed for 14 days on the colon mucosal glutathione (GSH) level was studied in male rats (Chen and others 1995). The GSH levels of the duodenum mucosa and the liver were also measured. Cabbage and broccoli enhanced the colon and duodenum mucosal GSH levels in a dose-related manner, but potato had no effect. All three vegetables had no effect on the liver GSH level. The effect of GSH on colon tumorigenesis induced by 1,2-dimethylhydrazine (DMH) was also examined in rats. Male Sprague-Dawley rats were injected weekly with DMH (20 mg/kg body wt) for 20 weeks (Chen and others 1995). DMH lowered the colon mucosal GSH level. GSH (100 mg/day/rat) dissolved in the drinking water and given to rats during and after DMH injections had little or no effect on tumor incidence and total number of colon tumors. Tumors were larger in rats that received GSH than in those that received water. Therefore, it seems likely that the colon mucosal GSH level can be enhanced by feeding rats a diet high in cabbage or broccoli and that GSH added to the drinking water did not affect DMH-induced colon tumorigenesis under the experimental conditions used.

Steinmetz and others (1994) have shown that risk of cancer in the distal colon was 50% lower in women with the highest consumption of garlic than in women who did not consume garlic.

However, some large cohort studies (Michels and others 2000; Voorips and others 2000) showed no appreciable association between fruit and vegetable intake and colon and rectal cancer.

Breast

A meta-analysis of 26 prospective and retrospective studies (Gandini and others 2000) confirmed the reduction of the risk of breast cancer with enhanced intake of fruit and vegetables, in contrast to a pooled analysis of eight prospective studies that indicated that fruit and vegetable consumption did not significantly reduce the risk of breast cancer (Smith-Warner and others 2001). A Europe-wide prospective EPIC study (European Prospective Investigation into Cancer and Nutrition) with 285,526 women has demonstrated that a daily intake of 370 g fruit or 246 g vegetables is not associated with a reduced breast cancer risk (van Gils and others 2005). Shannon and others (2003) reported a protective effect of vegetable and fruit intake at the highest quartile of consumption, suggesting a threshold effect in reducing breast cancer risk. However, reports indicated that 100 g of fresh apples have an antioxidant activity equivalent to 1700 mg of vitamin C and that whole apple extracts prevent breast cancer in a rat model in a dose-dependent manner at doses comparable to human consumption of one, three, and six apples a day (Liu and others 2005). Whole apple extracts were reported to effectively inhibit mammary cancer growth in the rat model; thus consumption of apples was suggested as an effective means of cancer protection. Female Sprague-Dawley rats treated with the carcinogen 7,12-dimethylbenz[*a*]anthracene (DMBA) at 50 days of age developed mammary tumors with 71% tumor incidence during a 24-week study, but a dose-dependent inhibition of mammary carcinogenesis by whole apple extracts was observed, where application of low, medium, and high doses of whole apple extracts, comparable to 3.3, 10, and 20 g of apples/kg of body weight, reduced the tumor incidence by 17, 39 ($p < 0.02$), and 44% ($p < 0.01$), respectively; this is comparable to human consumption of one (200 g/60 kg), three, and six apples per day.

Garcia Solis and others (2008, 2009) studied the antineoplastic properties of some fruits and vegetables using *in vivo* and *in vitro* models. The effect of "Ataulfo" mango fruit consumption was studied on chemically induced mammary carcinogenesis and plasma antioxidant capacity (AC) in rats treated with the carcinogen *N*-methyl-*N*-nitrosourea (MNU) (Garcia-Solis and others 2008). Mango was administered in the drinking water (0.02–0.06 g/mL) during both short-term and long-term periods to rats, and plasma antioxidant capacity was measured by ferric reducing/antioxidant power and total oxyradical scavenging capacity assays. Rats treated with MNU had no differences in mammary carcinogenesis (incidence, latency, and number of tumors), nor differences in plasma antioxidant capacity. On the other hand, they (Garcia-Solis and others 2009) used a methylthiazolydiphenyltetrazolium bromide assay to screen the antiproliferative activity of aqueous extracts of avocado, black sapote, guava, mango, cactus stems (cooked and raw), papaya, pineapple, four different prickly-pear fruit, grapes, and tomato on the breast cancer cell line MCF-7. Only the papaya extract had a significant antiproliferative effect, and there was no relationship between total phenolic content and AC with antiproliferative effect. These results suggested that each plant food has a unique combination in quantity and quality of phytochemicals that could determine its biological activity.

Indole-3-carbinol, 3,3′-diindolylmethane, and indole-3-acetonitrile, three indoles occurring in edible cruciferous vegetables, have been studied for their effects on 7,12-dimethylbenz[*a*]anthracene-induced mammary tumor formation in female Sprague-Dawley rats and on benzo[*a*]pyrene-induced neoplasia of the forestomach in female ICR/Ha mice (Wattenberg and Loub 1978). When given by oral intubation 20 hr prior to 7,12-dimethylbenz[*a*]anthracene administration, indole-3-carbinol and 3,3′-diindolylmethane had an inhibitory effect on mammary tumor formation, but indole-3-acetonitrile was inactive. Indole-3-carbinol, when added to the diet for 8 days prior to challenge with 7,12-dimethylbenz[*a*]anthracene, inhibited mammary tumor formation, whereas indole-3-acetonitrile did not. Dietary administration of all three indoles inhibited benzo[*a*]pyrene-induced neoplasia of the forestomach in ICR/Ha mice.

The consumption of a mixture of phenolic compounds presented in apple or purple grape juice inhibited mammary carcinogenesis in 7,12-dimethylbenzo[*a*]anthracene (DMBA) treated rats (Liu and others 2005; Jung and others 2006). However, the individual antioxidants of these foods studied in clinical trials, including β-carotene, vitamin C, and vitamin E, do not appear to have consistent preventive effects comparable to the observed health benefits of diets rich in fruits and vegetables, suggesting that natural phytochemicals in fresh fruits and vegetables could be more effective than a dietary supplement.

Associations between breast cancer and total and specific fruit and vegetable group intakes were examined using standardized exposure definitions (Smith-Warner and others 2001). Data sources were eight prospective studies that had at least 200 incident breast cancer cases, included assessment of usual dietary intake, and had completed a validation study of the diet assessment method or a closely related instrument.

These researchers studied 7,377 incident invasive breast cancer cases among 351,825 women whose diet was analyzed at baseline. For highest vs. lowest quartiles of intake, weak nonsignificant associations were observed for total fruits, total vegetables, and total fruits and vegetables. There was no apparent additional benefit for the highest and lowest deciles of intake. There were no associations for green leafy vegetables, eight botanical groups, or 17 specific fruits and vegetables. This study concluded that consumption of fruits and vegetables during adulthood is not significantly associated with reduced risk of breast cancer.

Some research has suggested that diets high in mushrooms may modulate aromatase activity and be useful in chemoprevention against breast cancer by reducing the *in situ* production of estrogen. The white button mushroom (*Agaricus bisporus*) suppressed aromatase activity dose dependently (Grube and others 2001). Enzyme kinetics demonstrated mixed inhibition, suggesting the presence of multiple inhibitors or more than one inhibitory mechanism. Aromatase activity and cell proliferation were then measured using MCF-7aro, an aromatase-transfected breast cancer cell line. Phytochemicals in the mushroom aqueous extract inhibited aromatase activity and proliferation of MCF-7aro cells.

Prostate

Some studies have suggested that ingestion of some fruits and vegetables may potentially reduce risk of prostate cancer (Giovannucci and others 2003; Campbell and others 2004; Stram and others 2006). Several epidemiological studies have reported associations between fruit and vegetable intake and reduced risk of prostate cancer, but the findings are inconsistent and data on clinically relevant advanced prostate cancer are limited (Kirsh and others 2007).

A study at the Harvard School of Public Health done on 48,000 men for 4 years reported that men who ate 10 or more servings of tomato products (such as tomatoes, tomato sauce, pizza sauce) per week had up to 34% less chance to develop prostate cancer (Giovannucci and others 1995). They showed that lycopene intake from tomato-based products is related to a low risk of prostate cancer, but consumption of other carotenoids (β-carotene, α-carotene, lutein, β-cryptoxanthin) or retinol was not associated with the risk of prostate cancer.

Activities of various carotenoids present in foods against human prostate cancer cell lines were investigated (Kotake-Nara and others 2001). Effects of 15 carotenoids on the viability of three lines of human prostate cancer cells, PC-3, DU 145, and LNCaP, were evaluated. When cancer cells were cultured in a carotenoid-supplemented medium for 72 hr at 20 mmol/liter, 5,6-monoepoxy carotenoids, namely, neoxanthin from spinach and fucoxanthin from brown algae, significantly reduced cell viability to 10.9 and 14.9% for PC-3, 15.0 and 5.0% for DU 145, and nearly zero and 9.8% for LNCaP, respectively. Acyclic carotenoids such as phytofluene, xi-carotene, and lycopene, all of which are present in tomato, also significantly reduced cell viability. However, phytoene, canthaxanthin, β-cryptoxanthin, and zeaxanthin did not affect the growth of the prostate cancer cells. DNA fragmentation of nuclei in neoxanthin- and fucoxanthin-treated cells was detected by *in situ* TdT-mediated dUTP nick end labeling (TUNEL) assay. Neoxanthin and fucoxanthin reduced cell viability through induction of apoptosis.

Lung

Increased fruit and vegetable consumption may protect against lung cancer, although epidemiologic findings are inconclusive. Dosil-Díaz and others (2008) analyzed the effect of fruit and vegetable intake on lung cancer risk in a population in northwest Spain, using data from a hospital-based case-control study including 295 histologically confirmed cases and 322 controls. Controls were patients attending the hospital for minor surgery. After adjustment for sex, age, tobacco use, and occupation, the researchers found no protective effect of overall consumption of fruit (odds ratio 1.49, 95% confidence interval 0.81–2.73), but green leafy vegetables conferred a protective effect (odds ratio 0.50, 95% confidence interval 0.30–0.83).

The association of fruit and vegetable consumption and lung cancer incidence was evaluated by Linseisen and others (2007) using recent data from the European Prospective Investigation into Cancer and Nutrition (EPIC), applying a refined statistical approach (calibration) to account for measurement error potentially introduced by using food frequency questionnaire (FFQ) data. Between 1992 and 2000, detailed information on diet and lifestyle of 478,590 individuals participating in EPIC was collected, and during a median follow-up of 6.4 years, 1,126 lung cancer cases were observed. In the whole study population, fruit consumption was significantly inversely associated with lung cancer risk, whereas no association was found for vegetable consumption. In current smokers, however, lung cancer risk significantly decreased with higher vegetable consumption, and this association became more pronounced after calibration, the hazard ratio (HR) being 0.78 (95% CI 0.62–0.98) per 100 g increase in daily vegetable consumption. In comparison, the HR per 100 g fruit was 0.92 (0.85–0.99) in the entire cohort and 0.90 (0.81–0.99) in smokers. Cancer incidence decreased with higher consumption of apples and pears (entire cohort) as well as root vegetables (in smokers).

Wright and others (2008) prospectively examined associations between lung cancer risk and intakes of fruit, vegetables, and botanical subgroups in 472,081 participants aged 50–71 years in the National Institutes of Health (NIH)-AARP Diet and Health Study. Diet was assessed at baseline (1995–1996) with a 124-item dietary questionnaire. A total of 6,035 incident lung cancer cases were identified between 1995 and 2003. Total fruit and vegetable intake was unrelated to lung cancer risk in both men and women, but higher consumption of several botanical subgroups, however, was significantly inversely associated with risk only in men. Association between lung cancer risk and fruit and vegetable consumption was also investigated by Feskanich and others (2000) in 77,283 women in a Nurses' Health Study and 47,778 men in a Health Professionals' Follow-up Study. Diet was assessed using a food frequency questionnaire including 15 fruits and 23 vegetables. Relative risk (RR) of lung cancer within each cohort was estimated using logistic regression models; all statistical tests were two-sided. 519 and 274 lung cancer cases were documented among women and men, respectively. Total fruit consumption was associated with a modestly lower risk among women but not among men. RR for highest versus lowest quintile of intake was 0.79 (95% confidence interval, CI = 0.59–1.06) for women and 1.12 (95% CI = 0.74–1.69) for men after adjustment for smoking parameters. Total fruit and vegetable consumption was associated with lower risk of lung cancer among men and women who had never smoked, but the reduction was not statistically significant (RR = 0.63; 95% CI = 0.35–1.12 in the highest tertile). It is suggested that the inverse association among women was confounded by unmeasured smoking characteristics.

Goodman and others concluded in 1992 that some β-carotene rich fruits such as papaya, sweet potato, mango, and yellow orange showed little influence on survival of lung cancer patients, and the intake of β-carotene before diagnosis of lung cancer does not affect the progression of the disease. These authors also concluded that a tomato-rich diet (which contributes high lycopene content but only a little β-carotene) had a strong positive relationship with survival, particularly among women. However, other studies (Le Marchand and others 1989; Steinmetz and others 1993) concluded that lycopene intake was unrelated to lung cancer.

Cardiovascular disease (CVD)

CVD is the number one cause of death in the United States, and prevention is at the top of the public health agenda (Retelny and others 2008). Evidence shows that reducing the incidence of coronary heart disease with diet is possible (Retelny and others 2008).

Numerous epidemiological studies have demonstrated evidence that diet rich in fruits and vegetables may protect against CVD (Ness and Powles 1997; Law and Morris 1998; Ness and others 1999; Joshipura and others 2001; Bazzano and others 2002). The positive effect has been accomplished by three servings of vegetables and fruits, and the relative risk could be minimized to a great extent by enhancing the vegetable and fruit consumption by up to 10.2 servings per day. A study on 2,682 men in Finland has also indicated that high intake of fruits and vegetables correlates with reduced risk of cardiovascular disease (Rissanen and others 2003). The inverse relationship between vegetable intake and cardiovascular disease was more evident with smokers consuming at least 2.5 servings per day in comparison with less than one serving per day (Liu and others 2001). Legume consumption was significantly and inversely associated with cardiovascular disease, lowering the risk by about 11% (Bazzano and others 2001).

The Japan Collaborative Cohort Study for Evaluation of Cancer Risk (Nagura and others 2009), with 25,206 men and 34,279 women aged 40–79 years, whose fruit, vegetable, and bean intakes were assessed by questionnaire at baseline in 1988–1990 and followed for 13 years, concluded that intakes of plant-based foods, particularly fruit intake, were associated with reduced mortality from CVD and all causes among Japanese men and women. However, Nakamura and others (2008) assessed the intake of fruits and vegetables in 13,355 men and 15,724 women in Takayama, Gifu, Japan, using a validated FFQ and found that for women, the highest quartile of vegetable intake compared with the lowest was marginally significant and inversely associated with CVD mortality after adjusting for total energy, age, and nondietary and dietary covariates, but in men, CVD death was not associated with fruit or with vegetable intake.

Radhika and others (2008) examined the relationship between fruit and vegetable intake (g/day) and CVD risk factors in urban south Indians. The study population composed of 983 individuals aged at least 20 years selected from the Chennai Urban Rural Epidemiological Study (CURES), a population-based cross-sectional study on a representative population of Chennai in southern India, and fruit and vegetable intake (g/day) was measured using a validated semiquantitative FFQ. Linear regression analysis revealed that after adjusting for potential confounders such as age, sex, smoking, alcohol, body mass index (BMI), and total energy intake, the highest quartile of fruit and vegetable intake (g/day) showed a significant inverse association with systolic blood pressure, total cholesterol, and low-density lipoprotein (LDL)-cholesterol concentration when compared with the lowest quartile. A higher intake of fruit and vegetables explained 48% of the protective effect against CVD risk factors.

It has been reported that death attributed to cardiovascular and coronary heart diseases showed strong and consistent reductions with increasing nut/peanut butter consumption (Blomhoff and others 2006). Nuts, including peanuts, have been recognized as having the potential to improve the blood lipid profile, and, in cohort studies, nut consumption has been associated with a reduced risk of coronary heart disease (CHD) (Jenkins and others 2008). Data from the Nurses Health Study indicate that frequent nut consumption is associated with a reduced risk of developing cardiovascular disease. Epidemiologic and clinical trial evidence has demonstrated consistent benefits of nut and peanut consumption on coronary heart disease (CHD) risk and associated risk factors (Kris-Etherton and others 2008). A pooled analysis of four US epidemiologic studies showed that subjects in the highest intake group for nut consumption had an approximately 35% reduced risk of CHD incidence, and the reduction in total CHD death was due primarily to a decrease in sudden cardiac death. Clinical studies have evaluated the effects of many different nuts and peanuts on lipids, lipoproteins, and various CHD risk factors, including oxidation, inflammation, and vascular reactivity (Kris-Etherton and others 2008). Evidence from these studies consistently shows a beneficial effect on these CHD risk factors. The LDL cholesterol-lowering response of nut and peanut studies is greater than expected on the basis of blood cholesterol-lowering equations that are derived from changes in the fatty acid profile of the diet. Thus, in addition to a favorable fatty acid profile, nuts and peanuts contain other bioactive compounds that explain their multiple cardiovascular benefits. Other macronutrients include plant protein and fiber; micronutrients including potassium, calcium, magnesium, and tocopherols; and phytochemicals such as phytosterols, phenolic compounds, resveratrol, and arginine. Kris-Etherton and others (2008) have indicated that nuts and peanuts are food sources that are a composite of numerous cardioprotective nutrients and if routinely incorporated in a healthy diet, population risk of CHD would therefore be expected to decrease markedly.

Hung and others (2003) evaluated the association of consumption of fruits and vegetables with peripheral arterial disease in a cohort study of 44,059 men initially free of cardiovascular disease and diabetes, reporting no evidence that fruit and vegetable consumption protects against peripheral arterial disease, although a modest benefit cannot be excluded. In the age-adjusted model, men in the highest quintile had a relative risk of 0.55 (95% confidence interval = 0.38–0.80) for overall fruit and vegetable intake, 0.52 (0.36–0.77) for fruit intake, and 0.54 (0.36–0.81) for vegetable intake, compared with those in the lowest quintile of intake. However, the associations were greatly weakened after adjustment for smoking and other traditional cardiovascular disease risk factors.

Pure β-carotene supplementation (30–50 mg per day) had no depressive effects on cardiovascular disease risk (Hennekens and others 1996; Omenn and others 1996; Lee and others 1999).

Monounsaturated fats in avocados have been shown to reduce blood cholesterol while preserving the level of high-density lipoproteins (Yahia 2009b). An avocado-enriched diet produced a significant reduction in LDL and total cholesterol in patients with high cholesterol levels, whereas diets enriched with soy and sunflower did not change the total cholesterol concentrations (Carranza and others 1997).

Lycopene from tomato fruit was found to prevent the oxidation of LDL cholesterol and to reduce the risk of developing atherosclerosis and coronary heart disease (Agarwal and Rao 1998), and a daily consumption of tomato products providing at least 40 mg of lycopene was reported to be enough to substantially reduce LDL oxidation. Lycopene is recognized as the most efficient singlet oxygen quencher among biological carotenoids (Di Mascio and others 1991, 1989). Lycopene has also been reported to increase gap-junctional communication between cells and to induce the synthesis of connexin-43 (Zhang and others 1992).

Diabetes Mellitus

The association between the intake of fruit and vegetables and the risk of type 2 diabetes is unclear. Hamer and Chida (2007) have suggested that the consumption of three or more daily servings of fruit or vegetables was not associated with a substantial reduction in the risk of type 2 diabetes, but the intake of antioxidants was associated with a 13% reduction in risk, mainly attributed to vitamin E. Five cohort studies of fruit and vegetables intake and the risk of diabetes using 167,128 participants and 4,858 incident cases of type 2 diabetes, with a mean follow-up of 13 years, were reviewed by Hamer and Chida (2007). The relative risk of type 2 diabetes for consuming five or more servings of fruit and vegetables daily was 0.96 (95% CI, 0.79–1.17, $P = 0.96$), 1.01 (0.88–1.15, $P = 0.88$) for three or more servings of fruit, and 0.97 (0.86–1.10, $P = 0.59$) for three or more servings of vegetables. Nine cohort studies of antioxidant intake and the risk of diabetes were also identified, incorporating 139,793 participants and 8,813 incident cases of type 2 diabetes, with a mean follow-up of 13 years (Hamer and Chida 2007) indicated that the pooled relative risk was 0.87 (0.79–0.98, $P = 0.02$) for the highest compared with the lowest antioxidant intake. Villegas and others (2008) examined the associations between fruit and vegetable intake and the incidence of type 2 diabetes (T2D) in a population-based prospective study of 64,191 Chinese women with no history of T2D or other chronic diseases at study recruitment and with valid dietary information. Individual vegetable groups were all inversely and significantly associated with the risk of T2D, but there was no association with fruit intake.

Randomized controlled trials of patients with type 2 diabetes have confirmed the beneficial effects of nuts on blood lipids, also seen in nondiabetic subjects, but the trials have not reported improvement in A1c or other glycated proteins (Jenkins and others 2008). Therefore, Jenkins and others (2008) concluded that there is justification to consider the inclusion of nuts in the diets of individuals with diabetes in view of their potential to reduce CHD risk, even though their ability to influence overall glycemic control remains to be established.

Nettleton and others (2008) characterized dietary patterns and their relation to incident type 2 diabetes in 5,011 participants from the Multi-Ethnic Study of Atherosclerosis (MESA) and found that high intake of whole grains, fruit, nuts/seeds, green leafy vegetables, and low fat dairy was associated with a 15% lower diabetes risk.

Bazzano and others (2008) concluded that consumption of green leafy vegetables and fruit was associated with a lower hazard of diabetes, whereas consumption of fruit juices may be associated with an increased hazard among women. A total of 71,346 female nurses aged 38–63 years who were free of cardiovascular disease, cancer, and diabetes in 1984 were followed for 18 years, and dietary information was collected using a semiquantitative FFQ every 4 years (Bazzano and others 2008). During follow-up, 4,529 cases of diabetes were documented; the cumulative incidence of diabetes was 7.4%, and an increase of three servings/day in total fruit and vegetable consumption was not associated with development of diabetes, whereas the same increase in whole fruit consumption was associated with a lower hazard of diabetes. An increase of 1 serving/day in green leafy vegetable consumption was associated with a modestly lower hazard of diabetes, whereas the same change in fruit juice intake was associated with an increased hazard of diabetes.

Experimental evidence showed that consumption of prickly pear cladodes (nopal) could decrease blood glucose levels (Frati and others 1990a). The intake of broiled *Opuntia* stems for 10 days improved glucose control in a small group of adults with non-insulin-dependent diabetes mellitus (NIDDM) (Frati and others 1990b, 1983). The rise in serum glucose levels that follows the intake of a sugar load (oral glucose tolerance test) was lower with previous ingestion of *Opuntia* stems compared to the ingestion of sugar alone (Frati and others 1990a). In patients with NIDDM, the ingestion of some species of nopal (*Opuntia streptacantha, O. ficus-indica)* in fasting condition is generally followed by a decrease of serum glucose and serum insulin levels (Frati 1992). These positive health effects of *Opuntia* stems might be associated with dietary fiber, since similar results can be achieved by *Plantago psyllium* or other sources of dietary fiber (Frati 1992). Ingestion of raw and cooked *Opuntia ficus-indica* extracts resulted in beneficial effects on total cholesterol, without any secondary effect on glucose and lipoproteins amounts in blood (Medellin and others 1998).

Some flavonoids, such as procyanidins, have antidiabetic properties because they improve altered glucose and oxidative metabolisms of diabetic states (Pinent and others 2004). Extract of grape seed procyanidins (PE) administered orally to streptozotocin-induced diabetic rats resulted in an antihyperglycemic effect, which was significantly increased if PE administration was accompanied by a low insulin dose (Pinent and others 2004). The antihyperglycemic effect of PE may be partially due to the insulinomimetic activity of procyanidins on insulin-sensitive cell lines.

Obesity

Nowadays, obesity is considered a major public health issue, especially in most developed countries, because it is widely spread across population groups and because it contributes to the development of chronic diseases, particularly cardiovascular diseases and diabetes.

Despite the alarming increase in the prevalence of obesity in the world, epidemiologic studies on the relation between fruit and vegetable consumption and weight gain

(WG) are still insufficient. Vioque and others (2008) explored the associations between fruit and vegetable intake and WG over a 10-year period in an adult Mediterranean population of 206 aged 15–80 years at baseline in 1994, who participated in a nutrition survey in Valencia, Spain. They concluded that dietary patterns associated with a high intake of fruits and vegetables in Mediterranean populations may reduce long-term risk of subsequent WG and obesity among adults.

Svendsen and others (2007) assessed the effect of an increased consumption of vegetables and fruit on body weight, risk factors for CVD and antioxidant defense in obese patients with sleep-related breathing disorders (SRBDs). They concluded that targeted dietary advice to increase the intake of vegetables and fruit among subjects with SRBD contributed to weight reduction and reduced systolic and diastolic blood pressure, but had no effect on antioxidant defense measured by ferric-reducing/antioxidant power (FRAP) assay.

He and others (2004) examined the changes in intake of fruits and vegetables in relation to risk of obesity and weight gain among middle-aged women through a prospective cohort study with 12 years of follow-up, conducted in the Nurses' Health Study with a total of 74,063 female nurses aged 38–63 years, who were free of cardiovascular disease, cancer, and diabetes at baseline in 1984. During the 12-year follow-up, participants tended to gain weight with aging, but those with the largest increase in fruit and vegetable intake had a 24% lower risk of becoming obese compared with those who had the largest decrease in intake after adjustment for age, physical activity, smoking, total energy intake, and other lifestyle variables. For major weight gain (≥25 kg), women with the largest increase in intake of fruits and vegetables had a 28% lower risk compared to those in the other extreme group.

Bes-Rastrollo and others (2006) assessed the association of fiber intake and fruit and vegetable consumption with the likelihood of weight gain in a Mediterranean (Spain) population with a cross-sectional analysis of 5,094 men and 6,613 women in a multipurpose prospective cohort (Seguimiento Universidad de Navarra Study). There was a significant inverse association between total fruit/vegetable consumption and weight gain, but only among men, and it was more evident among those with a high intake of total fiber; the benefit of total fiber was more evident among those with a high consumption of fruits and vegetables.

de Carvalho and others (2006) evaluated the dietary fiber intake of adolescents in the metropolitan area of Sao Paulo city and the association between low dietary fiber intake and constipation and overweight. The study included 716 adolescents, and evaluation of fiber intake was based on a 24-hr daily intake record and a frequency questionnaire. Adolescents who did not eat beans on more than 4 days per week presented a higher risk of fiber intake below that recommended, and dietary fiber intake below that recommended was associated with a greater risk of overweight in students attending public schooling.

Pulmonary Health

Several studies have suggested a positive association between fruit and vegetable intake (particularly fruit) and pulmonary function (Walda and others 2002; Watson and others 2002; Celik and Topcu 2006). Fruit and vegetable consumption has been suggested to maintain a healthy pulmonary function in well-adult populations, and improving lung function in those with established pulmonary disorders. Phytochemicals, especially of antioxidant potential, have been suggested to be important in protecting lungs from oxidative stress. In a study of 2,917 men in seven European countries, higher fruit intake was found to be consistently associated with lowered mortality from chronic obstructive pulmonary disease (COPD)-related causes and that the trend was statistically significant when age, country of residence, and smoking were considered (Walda and others 2002). This study has shown that vegetables and the antioxidant nutrients vitamins C and β-carotene did not correlate with COPD mortality, but vitamin E intake did appear to be protective when data were adjusted for age, country, and smoking. Fruit intakes of over 121 g/day and increased vegetable consumption were reported to be associated with significantly reduced COPD (Watson and others 2002; Celik and Topcu 2006).

It has been suggested that reduced antioxidant intake is one critical factor associated with increased susceptibility to asthma (Devereux and Seaton 2005), and therefore fruit and vegetable intake has been suggested to reduce it (Patel and others 2006) by improving ventilatory function and respiratory symptoms (Romieu and others 2006). Fruits and vegetables that were found to be associated with reduced incidence of asthma included green leafy vegetables (intake of >90 g/day; 22% risk reduction), tomato (intake of >28.2 g/day; 15% risk reduction), carrots (intake of >24.9 g/day; 19% risk reduction), and apples (intake of >31.2 g/day; 10% risks reduction). These studies have suggested that high carotenoid content in the food could account for the effect, although results of studies focused on β-carotene have not been consistent. The consumption of fruit and vegetables was reported to be significantly associated with reduced occurrence of wheezing and shortness of breath in 2,103 boys and 2,001 girls aged 6–7 years from Italy (Farchi and others 2003), and in more than 20,000 children, aged 7–11, from 25 areas in Central and Eastern Europe (Antova and others 2003). Kelly and others (2003) suggested an association between fruit and vegetable intake and improved respiratory function in more than 6,000 healthy adults in Scotland. However, contradicting data have also been reported, such as the study of Lewis and others (2005) reporting no association between asthma prevalence and fruit intake in 11,562 children aged 4–6 years and living in the United Kingdom based on data provided by parents of the study subjects and in 598 Dutch children aged 8–13 years (Tabak and others 2006).

Bone Health

The loss of bone mass is a global epidemic associated with osteoporosis (Lanham-New 2006). Fruit and vegetable consumption has been suggested to improve bone status (Lanham-New 2006; Bueline and others 2001; Pryne and others 2006). Higher fruit and vegetable intake was associated with improved markers of bone status in males and females ranging between 16 and 83 years old (Pryne and others 2006). Tylavsky and others (2004) showed that fruit and vegetable intake might be important in bone health in white girls aged 8–13 years. The effect was high with 3 servings/day or more and low with less than 3 servings/day, with 4.0 servings (1.6 fruit/2.4 vegetables) in the high group and 1.7 servings (0.6 fruit/1.1 vegetable) in the low group. Girls in the high fruit and vegetable intake group had significantly larger bone area of the whole body and wrist, and higher mineral content for whole body and at the wrist. A study of 1,407 premenopausal farm women from five rural districts in Japan has concluded that fruit and vegetable intake is positively correlated with bone health (Okabo and others 2006). In a study conducted with 85 boys and 67 girls, aged 8–20 years, in Saskatchewan, Canada (Vatanparast and others 2005), fruit and vegetable intake was reported to be an important independent predictor of accrued total body bone mineral content in boys but not in girls. In a study with adolescents aged 12 and 15 years in Northern Ireland (n = 1,345), 12-year-olds consumed the greatest quantity of fruit, and a positive association has been demonstrated between bone density and fruit intake.

Antioxidants in fruits and vegetables including vitamin C and β-carotene reduce oxidative stress on bone mineral density, in addition to the potential role of some nutrients such as vitamin C and vitamin K that can promote bone cell and structural formation (Lanham-New 2006). Many fruits and vegetables are rich in potassium citrate and generate basic metabolites to help buffer acids and thereby may offset the need for bone dissolution and potentially preserve bone. Potassium intake was significantly and linearly associated with markers of bone turnover and femoral bone mineral density (Macdonald and others 2005).

Lin and others (2003) indicated that high potassium, magnesium, and calcium content in addition to antioxidants, phytochemicals, and lower acidity of fruits and vegetables could be important factors for bone health.

In a study with 40 healthy men and women, average age 63.7 years, who were randomized to either an "alkali" diet (meat plus fruits and vegetables) or an "acid" diet (meat plus cereal grains) (Jajoo and others 2006), altering the renal net acid excretion over a period of 60 days affected several biochemical markers of bone turnover and calcium excretion. The acidity of the diet had a significant effect on increasing NTX, a urinary marker of bone breakdown, and increasing the amount of calcium excreted in the urine.

Cataracts and Eye Health

Oxidative mechanisms have been implicated in the etiology of cataracts in humans, and fruit and vegetables intake has been associated with this problem. A study involving 35,724 healthy professional women over the age of 45 years in the US was conducted to determine the potential association between fruit and vegetable intake and subsequent risk of cataract development over a 10-year follow-up period (Christen and others 2005). Relative risk of developing cataracts during the 10-year study was only slightly reduced in women with the highest intake of fruits and vegetables (10 servings/day) compared to those with the lowest intake (2.6 servings/day). A study of 479 women with an average fruit intake of 2.5 servings/day and average vegetable intake of approximately 4 servings/day also indicated that fruit and vegetable intake did not differ between women with and without nuclear opacities (Moeller and others 2004). The authors concluded that multiple aspects of the diet are more important in reducing the risk of cataracts than emphasizing one particular food group or component over another. In a study with 98 participants, 68% women, ranging in age between 45 and 73 years, macular pigment optical density (MPOD) was used as a marker, to correlate diet and serum carotenoid levels with the amount of molecular pigment in the retina showing that high ($\geq$5 servings/day) intake of fruit and vegetables was associated with significantly higher MPOD compared to measurement in subjects with lower (<3 servings /day) intake (Burke and others 2005).

Arthritis

Dietary antioxidants and anti-inflammatory components in food are thought to be important in reducing the risk or improving the course of rheumatoid arthritis (RA), and therefore fruit and vegetable consumption has been associated with reduced risk of RA (Pattison and others 2004b). A study with 29,368 married women from the US, predominately white, average age 61.4 years, lasting 11 years, indicated that total fruit consumption (>83 servings per month) was associated with reduced risk of RA (Cerhan and others 2003). Oranges were the only individual fruit linked to reduce incidence of RA, and β-cryptoxanthin, a carotenoid found in this fruit, was consistently highly protective. Total vegetable intake was not associated with reduced incidence of RA, but intake of cruciferous vegetables (>11 servings/month), particularly broccoli (>3 servings/month), was associated with a moderate effect on RA. In a study where dietary intake for 73 cases was compared to intake of 146 controls (mean age 60–61 years; 70% women), lower, but not statistically significant, intake of fruits and vegetables was weakly associated with higher incidence of inflammatory polyarthritis (IP) (Pattison and others 2005). Subjects with the lowest intake (<55.7 mg/day) of vitamin C were three times more likely to develop IP than those with the highest intake (>94.9 mg/day). In a related study to determine the relationship with carotenoid intake, the diets of 88 cases were compared to those of 176 controls (mean age 61 years; 69% women); intake of vitamin C and dietary carotenoids, particularly β-cryptoxanthin and to a lesser extent, zeaxanthin, was found to be significantly correlated with reduced risk of IP (Pattison and others 2005).

Birth Defects

The effect of folic acid supplementation on reducing the risk of neural tube defects of the brain and spine, including spina bifida and anencephaly, is well documented (Eichholzer and others 2006). Fruits and vegetables are an important source of dietary folate, and their consumption has been associated with increased plasma levels of folate. Plasma folate concentration increased by 13–27% after short-term feeding experiments with fruits and vegetables (Brevick and others 2005), and red blood cell folate also increased with increasing fruit and vegetable intake (from 1 to 7 servings/day) (Silaste and others 2003).

Diverticulosis

Diverticulosis, or the presence of several diverticula, affects 50% or more of the population over the age of 60 years in several countries (Ye and others 2005). Studies have established an association between low-fiber diets and the presence of diverticulosis (Aldoori and Rayan-Harschman 2002). The intake of fruit and vegetable fiber was inversely associated with risk of diverticulosis in a large prospective study of male health professionals, and therefore a high-fiber diet including fruits and vegetables remains an important aspect of therapy for diverticulosis (American Dietetic Association 2002).

Skin Diseases

Lycopene had a protective effect on the oxidative stress-mediated damage of the human skin after irradiation with UV light (Ribaya-Mercado and others 1995).

Aging and Cognition

Oxidative stress and inflammation are considered significant mediators in healthy aging of the brain and in age-related neurodegenerative diseases such as Alzheimer's and Parkinson's disease (Shukitt-Hale and others 2006). Animal and human studies have suggested that fruits and vegetables have the potential to decrease some age-related processes, primarily due to the antioxidant and anti-inflammatory properties of several phytochemicals. An *in vitro* study has suggested that some classes of phytochemicals also act in cell signaling and thus may protect against aging by mechanisms other than oxidative and inflammatory processes (William and others 2004). Fruit and vegetable extracts have been demonstrated to reverse or retard various age-related cognitive and motor deficits in rats (Lau and others 2005).

Strawberry and spinach extracts attenuated age-related cognitive and neuronal decline in rats over 6–15 months, and blueberry extracts were effective in reversing existing cognitive deficits and improving motor function in aged rats (Joseph and others 2005). Examination of the brain tissue from these animals showed evidence of reduced inflammatory and oxidative processes in the supplemented groups. A transgenic mouse model of Alzheimer's disease fed blueberry extract exhibited cognitive performance equivalent to that of normal nonsupplemented mice and were significantly improved compared to nonsupplemented transgenic mice (Joseph and others 2003). Blueberry supplementation in the transgenic mice increased concentration of cell signaling kinases thought to be involved in converting short-term memory to long-term memory, and increased other aspects of cell signaling including increased muscarinic receptor activity that is also known to be important in cognitive function (Casadesus and others 2004). Aging rats provided with Concord grape juice at low concentration (10%) for 9 weeks improved cognitive performance, while high (50%) concentration improved motor performance (Shukitt-Hale and others 2006). Concord grape and juice contain a variety of flavonoids, and 10% grape juice supplementation was reported to be associated with the most effective increase in muscarinic receptor sensitivity in aging rats.

A study that interviewed 13,388 women living in 11 US states over a period of 10–16 years (Kang and others 2005) showed that baseline cognitive performance was stronger in women who reported the highest intake of cruciferous vegetables compared to those with lower intake.

It is believed that oxidative stress plays a key role in the development of Alzheimer's disease because of the characteristic lesions associated with free radical damage and the attenuation of these processes with supplementation of some antioxidants. To date, there are no clinical trials that specifically address the role of dietary fruits and vegetables, although there are trials to investigate the association between dietary antioxidants in food and risk of Alzheimer's disease. A study involving 5,395 men with an average age of 67.7 years, living in the Netherlands and followed for 6 years (Engelhart and others 2002), reported that baseline dietary intake of vitamin C and E as well as the use of antioxidant supplements was associated with reduced risk of developing Alzheimer's during the follow-up period, with a stronger protective effect in subjects who were smokers, and flavonoid and β-carotene intake was protective in smokers but not in nonsmokers.

Nutritional and Health Importance of Some Fruits and Vegetables

Fruits and vegetables are rich sources of phytochemicals, in addition to other components that may act synergistically with phytochemicals to contribute to the nutritional and health benefits of these food commodities.

Apples and Pears

Apples are a widely consumed, rich source of phytochemicals, including phenolic compounds, pigments, vitamin C, among others. There are six classes of polyphenols in apple fruit (Thompson and others 1972; Lea and Timberlake 1974; Whiting and Coggins 1975a, b; Lea 1978; Oleszek and others 1988; Spanos and others 1990). The anthocyanins and flavonol glycosides are mainly found in the skin. The phenolic acids are mainly chlorogenic and *p*-coumaroylquinic acid. The dihydrochalcones are phloretin glucoside (phloridzin) and xyloglucoside. The main catechin is (–)-epicatechin, and the procyanidins are the 4—β-8-linked epicatechin series with some mixed (+)-catechin/(–)-epicatechin. There is no significant amount of glutathione in apple fruit, and none in apple juice (Jones and other 1992). Fresh apple fruit may contain up to 100 ppm of vitamin C, but during processing into juice this is rapidly lost (Lea 1992). The phytochemical composition of apples varies greatly between different varieties, and it changes during maturation, ripening, storage, and processing (Curry 1997). The accumulation of antioxidants in apple peels before harvest was found to be affected more by ripening and light intensity than by low temperature (Barden and Bramlage 1994). Total antioxidant activity of apples is different depending on the cultivar; in one study, antioxidant capacity of four apple cultivars was in the order "Golden delicious" > "Empire" > "Delicious" > "Cortland" (Ju and Bramlage 1999).

Ascorbate and nonprotein thiol (glutathione) content significantly decreased with increasing maturity of pears (*Pyrus communis* L. cv. Conference; Lentheric and others 1999). Concomitantly, the activity of superoxide dismutase and catalase decreased about fivefold and twofold, respectively, when the fruit was picked more mature. Pears were reported to have lower antioxidant activity compared to pigmented fruits (Prior and Cao 2000). The antioxidant potentials measured by 1,1-diphenyl-2-picrylhydrazyl (DPPH), β-carotene bleaching, and nitric oxide inhibition radical scavenging tests in apple peel and pulp were significantly higher than in pear peel and pulp (Leontowicz and others 2003). The ethanol extract of apple peels showed the strongest inhibition of lipid peroxidation as a function of its concentration, pear pulp had the weakest antioxidant ability, whereas apple pulp and pear peel were equal (Leontowicz and others 2003). Diets supplemented with peels of both fruits exercised a significantly higher positive influence on plasma lipid levels and on plasma antioxidant capacity of rats than diets with fruit pulps (Leontowicz and others 2003).

In the laboratory, apples have been found to have very strong antioxidant activity, inhibit cancer cell proliferation, decrease lipid oxidation, and lower cholesterol (Boyer and Liu 2004). Epidemiological studies have linked the consumption of apples with reduced risk of some cancers, cardiovascular disease, asthma, and diabetes (Liu and others 2005). Apple extracts with 70% acetone tested on the basis of their dry weight showed strong antioxidant activities toward oxidation of methyl linoleate, although apples were reported to be low in total phenolics (Kahkonen and others 1999). In apple juice, vitamin C activity represented a minor fraction of the total antioxidant activity with chlorogenic acid and phloretin glycosides as the major identifiable antioxidants (Miller and Rice-Evans 1997). Chlorogenic acid was found to contribute 27% of the total activity of apple extract to scavenge hydroxyl radicals (Plumb and others 1996b). It has been reported (Eberhardt and others 2000; Lee and others 2003) that the antioxidant and antiproliferative activities of apples are the consequences of synergistic activities of phenolics rather than vitamin C. In pear fruit, the contribution of phenolic compounds to antioxidant activity has also been reported to be much greater than that of vitamin C (Galvis-Sanchez and others 2003). Phenolics in apple skin showed a much higher degree of contribution to the total antioxidant and antiproliferative activities of whole apple than those in apple flesh (Eberhardt and others 2000; Wolfe and others 2003). Some of the important antioxidant phenolics found in apples include chlorogenic acid, epicatechin, procyanidin B_2, phloretin, and quercetin (Burda and others 1990; Mayr and others 1995). The relative antioxidant activity of some of these phenolic compounds in apples were in the order quercetin > epicatechin > procyanidin > phloretin > vitamin C > chlorogenic acid (Lee and others 2003).

Citrus

Citrus fruit and their juices are rich sources of vitamin C, where one 8 fl-oz serving was reported to supply the entire Reference Daily Intake (RDI) amount for vitamin C (Rouseff and Nagy 1994). There is considerable variation in the vitamin C content of juices of different citrus fruits. Grapefruit, tangerine, and lemon generally contain between 20 and 50 mg/100 mL juice. Citrus limonoids (such as limonin and nomilin), one of the two main classes of compounds responsible for the bitter taste in citrus fruits, have certain biological activities that may be used as chemopreventive agents (Lam and others 1994). The main flavonoids in oranges and grapefruits are hesperidin and naringin, respectively, and hesperidin is also found in mandarins, lemons, and limes. The polymethoxyflavones and the glycosylated flavones are only found in citrus fruits, and their pattern is specific for each species; they have been shown to have a broad spectrum of biological activities such as anti-inflammatory, anticancer, and antiatherogenic properties (Li and others 2009). Citrus flavonoids have been reported to possess anticancer activity both *in vitro* and *in vivo* (Middleton and Kandaswami 1994), in addition to antiviral and anti-inflammatory activities and an ability to inhibit human platelet aggregation (Huet 1982; Benavente-Garcia and others 1997). Citrus fruit are a particularly rich source of pectin, which occurs both in the edible portion and in the inedible residues such as peel, rag, and core (Baker 1994). Dietary incorporation of citrus pectins appears to affect several metabolic and digestive processes, such as glucose absorption, and cholesterol levels (Baker 1994).

Orange (*Citrus sinensis*) was found to be more active than pink grapefruit (*Citrus paradisi*) in scavenging peroxyl radicals, whereas grapefruit juice was more active than orange juice, when the oxygen radical antioxidant capacity (ORAC) assay was used

(Wang and others 1996). Grapefruit extracts inhibited ascorbate-iron-induced lipid peroxidation of liver microsomes in a dose-dependent way, but were less effective antioxidant toward an NADH-iron induced system (Plumb and others 1996b). The total antioxidant activity in orange juice was thought to be accounted for by hesperidin and narirutin (Miller and Rice-Evans 1997). Seeds of lemons, sour orange, sweet orange, mandarins, and limes had greater antioxidant activity than the peels (Bocco and others 1998). Antioxidant activity is generally higher in the seeds than in the peels (Bocco and others 1998).

Grapes and Berries

Grapes and berries are rich sources of phytochemicals including phenolic compounds, pigments, and ascorbic acid.

Fresh grapes and grape juices are excellent sources of phenolic antioxidants (Frankel and Meyer 1998). Grapes and other dietary constituents derived from grapes, such as grape juice and wine, have attracted a great deal of attention in recent years, and therefore, the composition and properties of grapes have been extensively investigated. Grapes are considered one of the major sources of phenolic compounds among various fruits (Macheix and others 1990). The phenolic compounds found in *Vitis vinifera* include phenolic acids, stilbenes, and flavonoids, which include flavonols, flavanols, and anthocyanins, and play an important role in the quality of grapes and wines (Downey and others 2006). Anthocyanins are directly responsible for the color of grape fruit and young wines, whereas astringency of wines is related to catechins, proanthocyanidins, flavonols, and low-molecular-weight polyphenols (hydroxycinnamic and hydroxybenzoic acid derivatives); the latter also contribute to bitterness (Hufnagel and Hoffman 2008). The diverse classes and large amounts of phenolic compounds found in grapes (Somers and Ziemelis, 1985; Macheix and others, 1990; Ricardo-da-Silva and others, 1990) were reported to play an important role in human health, such as lowering of low-density lipoprotein (Frankel and others 1993; Tussedre and others 1996). It has been demonstrated that products derived from grapes have high antioxidant capabilities (Alonso and others 2002). In addition, viniferin, a potent antifungal agent, and anthocyanins, which are strong antioxidants that inhibit platelet aggregation, are also present in grapes (Escarpa and Gonzalez 2001). Extracts of fresh grapes inhibited human LDL oxidation from 22% to 60% and commercial grape juice from 68% to 75% when standardized at 10 μM gallic acid equivalent (Frankel and others 1998). The LDL antioxidant activity correlated highly with the concentration of total phenolics for both grape extracts and commercial grape juices; with the level of anthocyanins and flavonols for grape extracts; with the levels of anthocyanins for Concord grape juices; and with the levels of hydroxycinnamates and flavan-3-ols in the white grape juice samples (Frankel and others 1998).

Berries contribute a significant number and amounts of phytochemicals such as ascorbic acid, carotenoids, flavonoids, phenolic acids, and tocopherols. Some wild berries with very high antioxidant activities include crowberry (*Empetrum nigrum*), cloudberry (*Rubus chamaemorus*), whortleberry (*Vaccinium uliginosum*), lingonberry (*V. vitis-idaea*), aronia (*Aronia melanocarpa*), cranberry (*V. oxycoccus*), and rowanberry (*Sorbus aucuparia*), but some of the cultivated berries such as strawberry (*Fragaria ananassa*), red currant (*Ribes rubrum*), black currant (*Ribes nigrum*), and raspberry (*Rubus idaeus*) exerted lower antioxidant activity (Kahkonen and others 1999). Different blueberries (*V. corymbosum*) and bilberries (*V. myrtillus*) were reported to exhibit good antioxidant activity in the ORAC assay (Prior and others 1998; Kalt and others 1999). The antioxidant capacity of blueberries was about threefold higher than either strawberries or raspberries with only a small contribution of ascorbic acid to the total antioxidant capacity compared to total phenolics and anthocyanins (Kalt and others 1999).

Strawberries are good sources of natural antioxidants (Heinonen and others 1998; Wang and others 1996). The extract of strawberries (*F. ananassa*) had the highest total antioxidant activity compared with extracts of plums, orange, red grapes, kiwi fruit, pink grapefruit, white grapes, banana, apple, tomato, pears, and honeydew melons when the ORAC assay was used (Wang and others 1996), but ranked among the least compared to some other berries when lipid oxidation models (methyl linoleate, LDL) assays were used (Heinonen and others 1998; Kahkonen and others 1999). Strawberry extracts exhibited high enzymatic activity for oxygen detoxification (Wang and others 2005) and a high level of antioxidant capacity against free radical species including peroxyl radicals, superoxide radicals, hydrogen peroxide, hydroxyl radicals, and singlet oxygen (Wang and Lin 2000; Wang and Jiao 2000). The major pigments of wild strawberries (*F. vesca*) were pelargonidin-3-monoglucoside and cyanidin-3-monoglucoside, whereas in cultivated strawberries the main pigment was only pelargonidin-3-monoglucoside (Sondheimer and Karash 1956). Strawberry extracts exhibited chemopreventive and chemotherapeutic activities *in vitro* and *in vivo* (Carlton and others 2001; Meyers and others 2003; Wang and others 2005). They also inhibited proliferation of the human lung epithelial cancer cell line A549 and decreased tetradecanoylphorbol-13-acetate (TPA)-induced neoplastic transformation of JB6 P+ mouse epidermal cells (Wang and others 2005).

Avocados

Avocado fruit is a high-fat fruit, contains rare sugars of high carbon number, and is relatively rich in certain vitamins, dietary fiber, minerals, and nitrogenous substances (Yahia 2009b). It has a high oil (3–30%) and low sugar content (about 1%); hence it is recommended as a high-energy food for diabetics (Yahia 2009b). It is a rich source of potassium, containing 1.6 times as much as bananas. A 100-g serving has about 177 calories, contains no cholesterol, and has about 17 g of fat, which is primarily monounsaturated type. Oil content is a key part of the sensory quality of the fruit. Oil quality is very similar to that of olive oil with a high proportion of the oil being approximately 75% monounsaturated, 15% saturated, and 10% polyunsaturated fatty acids. The high mono- and poly-unsaturation and low saturated content make it a "healthy" oil in terms of effect on heart disease. Monounsaturated fats in avocados have been reported to reduce blood cholesterol while preserving the level of high-density lipoproteins. In addition, avocado oil contains a range of other health-promoting compounds such as chlorophyll, carotenoids, α-tocopherol, and β-sitosterol. The edible portion of the fruit is rich in oleic, palmitic, linoleic, and palmitoleic acids, whereas stearic acid is present only in trace amounts. The fatty acid composition of the lipids of avocado fruit and avocado oil differs greatly with cultivar, stage of ripening, anatomical region of the fruit, and geographic location (Itoh and others 1975). However, the major fatty acid is always oleic followed by palmitic and linoleic acids, while the fatty acids present in trace amounts are myristic, stearic, linolenic, and arachidonic (Itoh and others 1975; Gutfinger and Letan 1974; Tango and others 1972; Mazliak 1971, Swisher 1984). The cuticular wax contains C20 to C27 long chain fatty acids (Mazliak 1971). Avocados are rich in vitamin B_6 (3.9–6.1 μg/g pyridoxine) and contain lesser amounts of biotin, folic acid, thiamine, and riboflavin (Hall and others 1955), calciferol (vitamin D), α-tocopherol (vitamin E), and 2-methyl-1,4-naphthoquinone (vitamin K) (Yahia 2009b). Therefore, besides being a source of energy and vitamins, avocado also contains several phytochemicals that are thought to be beneficial for health, and therefore it is considered by some as a "functional food" (Mazza 1998). Some nutraceutical ingredients that have been found in avocado pulp are antioxidants, such as tocopherols (about 4.3 UI/100 g) and glutathione (18 mg/100g). It has also been reported that avocado is a source of lutein (containing up to 248 mg/100g). The amount of β-sitosterol in this fruit is comparable to the level found in soy and olives. An avocado-enriched diet produced a significant reduction in low-density lipoproteins and total cholesterol in patients with high cholesterol levels, whereas diets enriched with soy and sunflower did not change the total cholesterol concentrations (Carranza and others 1997). Pigments are important contributors to the appearance and healthful properties of both avocado fruit and oil. Ashton and others (2006) identified in the skin, flesh, and oil of avocado fruit lutein, α-carotene, β-carotene, neoxanthin, violaxanthin, zeaxanthin, antheraxanthin, a and b chlorophylls, and a and b pheophytins, with the highest concentrations of all pigments in the skin. Chlorophyllides a and b were identified in the skin and flesh tissues only.

Stone fruits

Yellow flesh peaches (*Prunus persica* L.), such as the cv "Fay Alberta," contain 54 RE/100 g of provitamin A carotenoids (USDA 1991), predominantly as β-carotene and β-cryptoxanthin (Gross 1987). The two dominant polyphenols (nonflavonoid) in sweet cherries (*P. avium* L.) are caffeoyltartaric acid and 3-*p*-coumaroylquinic acid (Robards and others 1999). However, sweet cherries are characterized to have anthocyanins as the major phenolic compounds, the aglicon cyanidin bound to the glycosides 3-rutinoside and 3-glucoside being the main compounds, and pelargonidin-3-rutinoside, peonidin-3-rutinoside, and peonidin-3-glucoside as minor contributors (Gao and Mazza 1995; Chaovanalikit and Wrolstad 2004). Ascorbic acid, total phenolic compounds, and total antioxidant activity decreased during the early stages of sweet cherry fruit development, but exponentially increased coinciding with the stage of anthocyanin accumulation and fruit darkening (Serrano and others 2005). Prunes (*Prunus domestica*) and prune juice were antioxidant toward oxidation of human LDL (Donovan and others 1998). Prunes ranked highest with more than twice the level of antioxidants found in other high-scoring fruits, when the ORAC assay was used (Wang and others 1996). The inhibition of LDL oxidation by raw and canned peaches (*Prunus persica*) ranged between 56% and 87%, with oxidation activity mainly attributed to the presence of hydroxycinnamic acids and chlorogenic and neochlorogenic acids, but not to carotenoids such as β-carotene and β-cryptoxanthin. However, Plumb and others (1996b) reported that hydroxycinnamic acids do not contribute to the inhibition of lipid peroxidation of liver and cell microsomes by fruit extracts including plums and peaches, although these fruits had the ability to scavenge hydroxyl radicals.

Tropical Fruits

Tropical fruits are important sources of nutrients (Yahia 2006). Bananas, plantains, and breadfruit are widely used as a source of starch, acerola fruit contains the highest known ascorbic acid content among all fruits (1,000–3,300 mg/100 g fresh weight), and some other good sources of vitamin C include guava, lychee, papaya, and passion fruit (Yahia 2006). Mango and papaya are good sources of vitamin A, breadfruit and cherimoya contain relatively high amounts of niacin and thiamin, and most tropical fruit are good sources of minerals, especially potassium and iron (Yahia 2006). Tropical fruits are thought to contain more carotenoids compared to temperate fruits, which contain more anthocyanins. Mango and papaya are among the tropical fruits rich in carotenoids (Rivera-Pastarna and others 2009; Ornelas-Paz and others 2007, 2008).

Mango fruit is a rich source of vitamin C and carotenoids, some of which function as provitamin A (Siddappa and Bhatia 1954; Thomas 1975, Yahia and others 2006). β-Carotene (all-*trans*), β-cryptoxanthin (all-*trans* and -*cis*), zeaxanthin (all-*trans*), luteoxanthin isomers, violaxanthin (all-*trans* and -*cis*), and neoxanthin (all-*trans* and -*cis*) were identified in several mango cultivars (Mercadante and others 1997; Ornelas-Paz and others 2007, 2008). Mango retinol was found to be highly bioavailable by estimating vitamin A and carotene reserves in the liver and plasma of rats. Information on the tocopherol content in mango is very scarce, but it seems to be low (Burns and others 2003; Ornelas-Paz and others 2007).

Some chemical components of *Carica papaya* fruit pulp have been reported by Oloyede (2005), suggesting that the astringent action of the plant encountered in numerous therapeutic uses is due to the presence of some phytochemicals such as saponins and cardenolides. Liquid chromatography–mass spectrometry analyses of ripe and green papaya showed few candidate phenols, other than catechin conjugates (Mahattanatawe and others 2006), which is consistent with the small number of compounds reported previously in this fruit (Agrawal and Agrawal 1982). Quercetin and kaempferol, previously reported in leaves and shoots of *C. papaya* (Canini and others 2007), were found only in trace amounts in fruit peel extracts (Canini and others 2007; Miean and Mohamed 2001). Corral-Aguayo and others (2008) reported that the antioxidant capacity of papaya extract ranked one of the lowest among eight different fruits.

Dopamine, a strong water-soluble antioxidant, was identified in banana fruit (*Musa cavendishii*) by Kanazawa and Sakakibara (2000). Banana fruit contained high levels in the pulp and peel: 2.5–10 mg/100 g and 80–560 mg/100 g, respectively. A banana water extract was reported to suppress the autoxidation of linoleic acid by 65–70% after a 5-day incubation in an emulsion system, as determined from peroxide value and thiobarbituric acid reactivity (Kanazawa and Sakakibara 2000).

Pineapple fiber showed higher (86.7%) antioxidant activity than orange peel fiber (34.6%), and myricetin was the major polyphenol identified in pineapple fiber (Larrauri and others 1997).

Ma and others (2004) isolated and identified seven phenolic compounds in *Pouteria campechiana*, *P. sapota,* and *P. viridis*, namely, gallic acid, (+)-gallocatechin,

(+)-catechin, (−)-epicatechin, dihydromyricetin, (+)-catechin-3-*O*-gallate, and myricitrin. The highest level of the seven phenolic compounds was found in *P. sapota*, the second highest in *P. viridis*, and the lowest in *P. campechiana*.

Nuts

Several nuts are among the dietary plants with high content of total antioxidants. Of the tree nuts, walnuts, pecans, and chestnuts have the highest contents of antioxidants. Walnuts contain more than 20 mmol antioxidants per 100 g, mostly in the walnut pellicles. Peanuts also contribute significantly to dietary intake of antioxidants (Blomhoff and others 2006). Almonds are rich sources of proteins, dietary fats, fibers, and several minerals (Ren and others 1993; Turnball and others 1993). Nine phenolic compounds have been identified in almonds by Sang and others (2002), of which eight exhibited strong antioxidant activity. Almond skins were found to contain high levels of four different types of flavanol glycosides, which are thought to have powerful effects as antioxidants (Frison and Sporns 2002). Polyphenolic compounds in almonds were reported to be absorbed well in the body and are active in preventing the oxidation of LDL cholesterol; vitamin E in almonds acts in synergy with phenolic compounds to reduce oxidation (Millburry and others 2002). Individuals who replaced half of their daily fat intake with almonds or almond oil for a period of 6 weeks showed reduced total and LDL cholesterol by 4% and 6%, respectively, and HDL cholesterol increased by 6% (Hyson and others 2002). Research has suggested a possible effect of almond consumption on cancer (Davis and Iwahashi 2000), including colon cancer (Davis and others 2003).

Tomatoes

Tomato and tomato-based products are considered healthy foods because they are low in fat and calories, cholesterol-free, and a good source of fiber, vitamins A and C, β-carotene, lycopene, and potassium (Yahia and Brecht 2009). The interest in the nutritional and health benefits of tomato fruit and their products has increased greatly (Geeson and others 1985; Giovannucci and Clinton 1998; Guester 1997; Yahia and Brecht 2009). Vitamin C content in tomato (23 mg/100 g) is not as high as in several other fruits, but its contribution is very important because of the common use of tomato in the diet of many cultures. A 100-g tomato can supply about 20% and 40% of the adult US recommended daily intake of vitamins A and C, respectively. The selection of tomato genotypes that are rich in vitamins A and C has been accomplished, and cultivars with very high vitamin A content have been developed, but their orange color was not highly accepted by consumers. Epidemiological studies indicated that tomato fruit had one of the highest inverse correlations with cancer risk and cardiovascular disease, including stroke (Giovannucci and others 1995). Lycopene, the principal pigment responsible for the characteristic deep-red color of ripe tomato fruit and tomato products, is a natural antioxidant that can prevent cancer and heart disease (Shi and Le Maguer 2000). Although lycopene has no provitamin A activity, as is the case with some other carotenoids, it does exhibit a physical quenching rate constant with singlet oxygen almost twice as high as that of β-carotene.

Increasing clinical evidence supports the role of lycopene as a micronutrient with important health benefits, due to its role in the protection against a broad range of epithelial cancers (Shi and Le Maguer 2000). The serum level of lycopene and the dietary intake of tomatoes have been inversely correlated with the incidence of cancer (Helzlsouer and others 1989; VanEenwyk and others 1991).

Fresh tomato fruit contains about 0.72 to 20 mg of lycopene per 100 g of fresh weight, which accounts for about 30% of the total carotenoids in plasma (Stahl and Sies 1996). In contrast to other pigments such as β-carotene, lutein, violaxanthin, auroxanthin, neoxanthin, and chlorophylls a and b, which accumulate in inner pulp and in the outer region of the pericarp, lycopene appears only at the end of the maturation period and almost exclusively in the external part of the fruit (Laval-Martin and others 1975). Other tomato components that can contribute to health include flavonoids, folic acid, and vitamin E (Dorais and others 2001a,b).

Tomato was reported to exert antioxidant activity in some studies (Vinson and others 1998; Kahkonen and others 1999), whereas it showed no antioxidant activity or even acted as a pro-oxidant in others (Gazzani and others 1998). The antioxidant effect of tomato is most probably due to synergism between several compounds and not due to lycopene content alone, as pure lycopene and several other carotenoids act as pro-oxidants in a lipid environment (Al-Saikhan and others 1995; Haila and others 1996).

Cruciferous Vegetables

Sulfur-containing phytochemicals of two different kinds are present in all *Brassica oleracea* (Cruciferae) vegetables (cabbage, broccoli, cauliflower, Brussels sprouts, kale): (1) glucosinolates (previously called thioglucosides) and (2) *S*-methylcysteine sulfoxide. These compounds, which are derived in plant tissue by amino acid biosynthesis, show quite different toxicological effects and appear to possess anticarcinogenic properties (Stoewsand 1995). Glucosinolates have been extensively studied since the mid-nineteenth century. They are present in plant foods besides *Brassica* vegetables with especially high levels in a number of seed meals fed to livestock. About 100 different kinds of glucosinolates are known to exist in the plant kingdom, but only about 10 are present in *Brassica*. The first toxic effects of isothiocyanates and other hydrolytic products from glucosinolates that were identified were goiter and a general inhibition of iodine uptake by the thyroid. Numerous studies have indicated that the hydrolytic products of at least three glucosinolates, 4-methylsulfinylbutyl (glucoraphanin), 2-phenylethyl (gluconasturtiin), and 3-indolylmethyl (glucobrassicin), have anticarcinogenic activity. Indole-3-carbinol, a metabolite of glucobrassicin, has shown inhibitory effects in studies of human breast and ovarian cancers. *S*-Methylcysteine sulfoxide and its metabolite methylmethane thiosulfinate were shown to inhibit chemically induced genotoxicity in mice. Thus, the cancer chemopreventive effects of *Brassica* vegetables that have been shown in human and animal studies may be due to the presence of both types of sulfur-containing phytochemicals (i.e., certain glucosinolates and *S*-methylcysteine sulfoxide).

The effect of consumption of Brussels sprouts on levels of 8-oxo-7,8-dihydro-2′-deoxyguanosine (8-oxodG) in human urine was investigated in ten healthy, male, nonsmoking volunteers by Verhagen and others (1995). Following a 3-week run-in period, five volunteers continued on a diet free of cruciferous vegetables for a subsequent 3-week intervention period (control group), while the other five (sprouts group) consumed 300 g of cooked Brussels sprouts per day, at the expense of 300 g of a glucosinolate-free vegetable. In the control group there was no difference between the two periods in levels of 8-oxodG ($P = 0.72$), but in contrast, in the sprouts group the levels of 8-oxodG were decreased by 28% during the intervention period ($P = 0.039$), and therefore these results support the results of epidemiologic studies that consumption of cruciferous vegetables may result in a decreased cancer risk.

Extracts from broccoli (*Brassica oleracea* L. cv. Italica), Brussels sprouts (*B. oleracea* L. cv. Rubra), white cabbage (*B. oleracea* L. cv. Alba), and cauliflower (*B. oleracea* L. cv. Botrytis) showed significant antioxidant properties against lipid peroxidation (Plumb and others 1996a). However, it is thought that most of the direct antioxidant action of the cruciferous vegetables is not due to the glucosinolate content, but probably involves the hydroxylated phenols and polyphenol content, as has been identified in broccoli (Plumb and others 1997b). Cabbage, cauliflower, and Brussels sprouts were reported to be pro-oxidants toward lipid peroxidation in microsomes containing specific cytochrome P450s (Plumb and others 1997a). Kale (*B. oleracea* L. cv. Acephala), Brussels sprouts, and broccoli were found to exert higher antioxidant activity than cauliflower and some other vegetables (Al-Saikhan and others 1995; Cao and others 1996; Ramarathnam and others 1997; Vinson and others 1998).

Leafy Vegetables

Some contradictory results have been reported regarding the antioxidant activities of leafy vegetables. For example, among 23 vegetables, spinach (*Spinacia oleracea* L.) ranked 18th and head lettuce (*Lactuca sativa* L. cv. Capita) 22nd when assayed for inhibition of LDL (Vinson and others 1998). However, the ORAC antioxidant activity of spinach was reported to be very high, whereas that of leaf lettuce and iceberg lettuce was poor (Cao and others 1996).

Root and Tuberous Vegetables

Carrots (*Daucus carota*) are excellent sources of β-carotene and vitamin A, although they have been reported to exert low antioxidant activity compared to some other vegetables (Al-Saikhan and others 1995; Cao and others 1996; Ramarathnam and others 1997; Vinson and others 1998; Beom and others 1998). However, boiling carrots for 30 min significantly improved their antioxidant activity toward coupled oxidation of β-carotene and linolenic acid (Gazzani and others 1998).

In addition to being excellent source for some carbohydrates, potatoes (*Solanum tuberosum*) are considered a source for some antioxidants such as ascorbic acid, α-tocopherol and polyphenolic compounds (Al-Saikhan and others 1995; Velioglu and others 1998). Potato peels have been reported to show high antioxidant activity (Rodriguez de Sotillo and others 1994). Purple potatoes and peel have been shown to exhibit greater antioxidant activities than the white and yellow varieties (Velioglu and others 1998), which is due to the presence of anthocyanins such as pelargonidin-3-rutinoside-5-glucoside (Rodriguez-Saona and others 1998). Potato is not considered as a rich source of carotenoids, but some have been identified. Pendlington and others (1965) tentatively identified eight carotenoids (e.g., β,β-carotene, lutein) as most abundant in most cultivars, although Muller (1997) found that violaxanthin is the main potato carotenoid, followed by lutein, antheraxanthin, and others. Iwanzik and others (1983) compared the carotenoid content among 13 potato cultivars in the same habitat and reported violaxanthin as the main carotenoid, followed by lutein, lutein-5,6-epoxide, and neoxanthin. Breithaupt and Bamedi (2002) investigated the carotenoid pattern of four yellow- and four white-fleshed potato cultivars and reported that it was dominated by violaxanthin, antheraxanthin, lutein, and zeaxanthin, whereas neoxanthin, β-cryptoxanthin, and β,β-carotene generally were only minor constituents. The color of orange-fleshed potato was suggested to originate from large amounts of zeaxanthin (Brown and others 1993).

In addition of having some important phytochemicals such as betalains, beetroot (*Beta vulgaris* L.) and sugar beet (*Beta vulgaris esculenta*) peels showed remarkably high antioxidant activities (Kahkonen and others 1999). For example, beet ranked eighth among 23 vegetables assayed for inhibition of LDL oxidation (Vinson and others 1998).

Peppers

Fresh peppers are excellent sources of vitamins A and C, as well as neutral and acidic phenolic compounds (Howard and others 2000). Levels of these can vary by genotype and maturity and are influenced by growing conditions and processing (Mejia and others 1988; Howard and others 1994; Lee and others 1995; Daood and others 1996; Simmone and others 1997; Osuna-Garcia and others 1998; Markus and others 1999; Howard and others 2000). Peppers have been reported to be rich in the provitamin A carotenoids β-carotene, α-carotene, and β-cryptoxanthin (Minguez-Mosquera and Hornero-Mendez 1994; Markus and others 1999), as well as xanthophylls (Davies and others 1970; Markus and others 1999). Bell peppers have been shown to exert low antioxidant activity (Al-Saikhan and others 1995; Cao and others 1996; Vinson and others 1998) or may even act as pro-oxidants (Gazzani and others 1998).

Alliums

Edible alliums, especially onions and garlic, have been an important part of the daily diet of several cultures for hundreds of years. In many cultures alliums were thought to have medicinal properties. *Allium* vegetables, such as onions, garlic, scallions, chives, and leeks, include high concentrations of compounds such as diallyl sulfide and allyl methyl trisulfide (Steinmetz and Potter 1996). These compounds have been shown to inhibit cell proliferation and growth, enhance the immune system, alter carcinogen activation, stimulate detoxification enzymes, and reduce carcinogen-DNA binding (Hatono and others 1996; Lin and others 1994; Lee and others 1994; Amagase and Milner 1993). Hageman and others (1997) studied the effects of supplemental garlic consumption on *ex vivo* production of benzo[*a*]pyrene-DNA adducts in lymphocytes in a nonrandomized pilot study of nine men and observed that isolated lymphocytes from the blood of participants eating garlic (3 g raw garlic per day for 8 days) developed fewer adducts when incubated with benzo[*a*]pyrene.

Selenium has been shown to possess antitumor activity in humans. Efforts to produce Se-enriched vegetables (including garlic and broccoli) with anticarcinogenic activity have been successful. A number of studies (Ip and others 1992; Ip and Lisk 1994) have shown that selenium-enriched garlic is an effective anticarcinogen. Whanger and others (2000) have suggested that Se-enriched ramps (*Allium tricoccum*, a wild leek species) appear to have potential for reduction of cancer in humans. There was approximately 43% reduction in chemically induced mammary tumors when rats were fed a diet with Se-enriched ramps. Bioavailability studies with rats indicated that Se in ramps was 15–28% more available for regeneration of glutathione peroxidase activity than inorganic Se as selenite.

Yellow and red onions (*Allium cepa*) were reported to be poor antioxidants toward oxidation of methyl linoleate (Kahkonen and others 1999) in contrast to their high antioxidant activity toward oxidation of LDL (Vinson and others 1998). Garlic (*Allium sativum* L) was reported to have four times more antioxidant activity than onions when using the ORAC assay (Cao and others 1996).

Prickly Pear Fruit and Cladodes

Prickly pear fruit and cladodes are valued because of their high nutrient content, vitamins, and other health components (Yahia 2009a; Hegwood 1990).

Cladodes (called nopal in Mexico) are low in calories, high in fiber, and traditionally consumed as vegetable in Mexico and the southern region of the US. Nopal contains about 920 g/kg water, 40–60 g/kg fiber, 10–20 g/kg proteins, and about 10 g/kg minerals, primarily calcium. Vitamin C content is about 100–150 mg/kg, and β-carotenes (provitamin A) are about 300 μg/kg. Cladodes are used in various pharmaceutical applications for their therapeutic, dermatological, and medical properties. Ethanol extract of *Opuntia ficus-indica* shows potential analgesic and anti-inflammatory effects (Park and others 1998). Galati and others (2001) reported preventive and curative effects of *O. ficus-indica* cladodes preparations on rats affected by ethanol-induced ulcers. The cactus consumption gives rise to cytoprotection phenomena by breaking up the epithelial cells and stimulating an increase in mucus production. When *O. ficus-indica* cladodes are administered as a preventive therapy, they keep the gastric mucosa in a normal condition by preventing mucus dissolution caused by ethanol and favoring mucus production. An increase of mucus production is also observed during the course of the curative treatment. The treatment with *O. ficus-indica* cladodes provokes an increase in the number of secretory cells. Probably, the gastric fibroblasts are involved in the antiulcer activity.

The consumption of prickly pear cactus fruit is recommended for their beneficial and therapeutic properties (Yahia 2009a). Aqueous extracts of cactus pear fruit (*O. ficus-indica* L. Mill) was reported by Butera and others (2002) to possess a high total antioxidant capacity, expressed as Trolox equivalents, and exhibit a marked antioxidant capacity in several *in vitro* assays, including the oxidation of red blood cell membrane lipids and the oxidation of human LDLs induced by copper and 2,2′-azobis(2-amidinopropane hydrochloride). Antioxidant components reported by these authors included vitamin C, negligible amounts of carotenoids, and vitamin E. However, Corral-Aguayo and others (2008) reported that the antioxidant capacity of extracts of prickly pear fruit ranked the lowest compared to seven other fruits. Some prickly pear fruit contains two betalain pigments, the purple-red betanin and the yellow indicaxanthin, both with radical-scavenging and reducing properties (Forni and others 1992; Fernandez-Lopez and Almela 2001; Stintzing and others 2002; Castellanos-Santiago and Yahia 2008). Qualitative and quantitative analysis of betalain pigments in ten cultivars/lines of prickly pear (*Opuntia* spp.) fruit were conducted with reverse phase high-performance liquid chromatography-diode array detection (HPLC-DAD) coupled with electrospray mass spectrometry (ESI-MS) (Castellanos-Santiago and Yahia 2008). Betacyanins and betaxanthins were identified by comparison with the UV/Vis and mass spectrometric characteristics as well as the retention times of semisynthesized reference betaxanthins. Carotenoids and chlorophylls were also identified and quantified based on their molecular mass determined by applying HPLC-DAD coupled with positive atmospheric pressure chemical ionization mass spectrometry (APCI-MS). A total of 24 known/unknown betalains were present in the studied prickly pear fruit, including 18 betaxanthins and 6 betacyanins. The ratio and concentration of betalain pigments are responsible for the color in the different cultivars, showing the highest betalain content in fruit of purple color, comparable to that found in red beet (*Beta vulgaris* L. cv. Pablo). All cultivars/lines had a similar carotenoid profile, in which lutein was the most abundant compound in "Camuesa," while neoxanthin was the most abundant compound in the line "21441." Chlorophyll a was the most abundant in all cultivars/lines with the highest quantity in "21441." Daily supplementation with 500 g cactus fruit (*O. ficus-indica*) pulp for 2 weeks greatly improved the oxidation stress status of healthy humans (Tesoriere and others 2004). The effects included remarkable reduction in plasma markers of oxidative damage to lipids, such as isoprostanes and malondialdehyde (MDA), an improvement in the oxidative status of LDL, considerably higher concentrations of major plasma antioxidants, and improvement in the redox status of erythrocytes. The fruit also enhances renal function (Cacioppo 1991).

Reports indicate that other parts of prickly pear cactus are also used in folk medicine as emollient, moisturizing, wound-healing, hypocholesterolemic, and hypoglycemic agents and in gastric mucosal diseases (Hegwood 1990; Frati and others 1990a, 1990b; Cruse 1973; Meyer and McLaughlin 1981; Harvala and others 1982; Camacho-Ibanez and others 1983; Brutsch 1990; Fernandez and others 1992, 1994; Yahia 2009a). In Sicilian folk medicine, a flower infusion has an effect generally defined as depurative, and in particular it is used because of its diuretic and relaxant action on the renal excretory tract (Arcoleo and others 1961, 1966; Sisini 1969). Therefore, it is stipulated that a flower infusion may help the expulsion of renal calculus. Galati and others (2002) reported that flower infusion shows a modest increase in diuresis and natriuresis. Treatment with cladode infusions increases diuresis but does not significantly influence the uric acid pattern. The fruit infusion instead had diuretic and antiuric activity. The diuretic action observed may depend on stimulation of the urinary tract and is linked to the activation of neurohumoral mechanisms, mediators of stimuli acting on glomerules, or the pyelo-ureteral peristalsis. These effects might be due to the influence that the electrolytes, present in considerable quantities in the plant, exert on renal epithelium. In particular, *O. ficus-indica is* rich in K^+ ions, which are present in concentrations of 548 mg kg^{-1} in the cladodes, 21.7 mg kg^{-1} in the flowers, and 18 mg kg^{-1} in the fruit (d'Aquino 1998).

Mushrooms

Some types of mushrooms contain moderate quantities of good-quality protein and are good sources of dietary fiber, vitamins C and B, and minerals (Breene 1990). Extensive clinical studies have demonstrated that some species have medicinal and therapeutic value, by injection or oral administration, in the prevention/treatment of cancer, viral diseases (influenza, polio), hypercholesterolemia, blood platelet aggregation, and hypertension (Breene 1990). It has been reported that the inclusion of cultivated mushrooms, particularly shiitake and enokitake, in the diet is likely to provide some protection against some manifestation of cancer (Mori and others 1987).

Pickled Vegetables

Roussin's red (dimethylthiotetranitrosodiiron) has been identified as a nonalkylating *N*-nitroso compound present in pickled vegetables from Linxian, a high-risk area for esophageal cancer in China (Cheng and others 1981). *In vitro* experiments showed that Roussin's red had no significant mutagenic and transforming activities at the doses used, but it did enhance transformation in C3H/10TI/2 cells initiated with 3-methylcholanthrene (0.1 μg/mL), and it decreased the number of sebaceous glands and increased the epidermal thickness in short-term skin tests. This indicated that Roussin's red resembled 12-*O*-tetradecanoylphorbol-13-acetate and may be considered a new naturally occurring tumor promoter. Examination of dietary data in relation to the place-of-birth-specific incidence rates showed positive associations of stomach cancer with consumption of rice, pickled vegetables, and dried/salted fish, and a negative association with vitamin C intake (Kolonel and others 1981). These results are consistent with the particular hypothesis that stomach cancer is caused by endogenous nitrosamine formation from dietary precursors, and that vitamin C may protect against the disease. Extracts of pickled vegetables commonly consumed in Linhsien County, a high-incidence area for esophageal cancer in northern China, were studied for mutagenicity (Lu and others 1981). The liquid residue from ether extracts produced a dose-dependent increase of mutants in *Salmonella typhimurium* TA98 and TA100 strains; mutagenicity required the presence of a fortified liver microsomal activation system induced by Aroclor 1254 in adult male BD VI inbred rats. An amount of extract equivalent to 2.8 g fresh pickled vegetables produced sixfold (75 revertants/g) and twofold (45 revertants/g) increases in revertant frequencies in strains TA98 and TA100, respectively. Roussin's red methyl ester, a tetranitroso compound, $[(NO)_2Fe(CH_3S)]_2$, was isolated and identified from the ether extracts and was shown to be mutagenic in strain TA100 in the presence of a liver activation system, producing 25 revertants/mmol. A type of Japanese mixed vegetable pickle was found to be mutagenic to *Salmonella typhimurium* strains TA100 and TA98 with S9 mix (Takahashi and others 1979). Its activity was one-sixth that of a similar Chinese pickle from Linhsien County, China. Mutagens in the Japanese pickle were isolated, purified, and identified; the main component was kaempferol, and a minor component was isorhamnetin. As flavonoids are ubiquitous in vegetables, kaempferol and isorhamnetin do not seem to be specific to the pickle. Whole pickles of each kind of vegetable used in the Japanese pickle were extracted with methanol and subjected to mutagenicity tests; extracts of all vegetables except carrots were slightly mutagenic, with green leaves of vegetables and yellow chrysanthemum flowers showing the highest specific activity.

Enhancement of Phytochemicals in Fruits and Vegetables

In addition to increasing the consumption of fruits and vegetables, the enhancement of the nutritional and health quality of fruits and vegetables is an important goal in the effort to improve global health and nutrition. Erbersdobler (2003) and Boeing and others (2004) advocated the intake of functional foods that are enriched with phytochemicals.

Several crop production techniques have been reported to enhance phytochemical content in several fruits and vegetables (Schreiner 2005). Several postharvest practices and techniques can also affect the content of phytochemicals (Beuscher and others 1999; Goldmann and others 1999; Huyskens-Keil and Schreiner 2004).

Genetic engineering can make a substantial contribution to improved nutrition in crops. The most famous example of using biotechnology to improve nutrition is the development of vitamin A enriched "golden" rice (Ye and others 2000). Biotechnology has also been applied to improvements in the nutritional quality of a range of fruit and vegetables (Dalal and others 2006), targeting protein quality and quantity, desirable fatty acids, vitamins and minerals, and antioxidants. Many vegetables and fruits contain significant amounts of protein but are deficient in particular amino acids, such as the case of potato, which is deficient in lysine, tyrosine, methionine, and cysteine, and therefore the introduction of a transgene encoding a seed albumin protein into potato resulted in tubers with sufficient quantities of all essential amino acids (Chakraborty and others 2000). Cassava has also been improved by introducing an artificial protein encoding a maximal content of all essential amino acids (Sautter and others 2006). Conventional breeding has resulted in similar successes, such as maize with higher quality protein that is enriched for lysine and tryptophan (Hoisington 2002). Polyunsaturated fatty acids have been implicated as being beneficial to human health. Enhancement of the γ-linoleic acid content of tomato was achieved through introduction of a gene encoding a $\Delta 6$ desaturase enzyme (Cook and others 2002). Plants, including fruits and vegetables, naturally contain a large array of antioxidants, including carotenoids, vitamins C and E, and flavonoids, and efforts to increase the antioxidant content of fruits and vegetables have taken both breeding-based and transgenic approaches. Natural high-pigment mutants are available in tomato that can be used in breeding strategies (Long and others 2006), and wild accessions of tomato exhibit substantial metabolic diversity that may similarly be exploited (Fernie and others 2006). A series of introgression lines have been generated in cultivated tomato containing genomic segments of a wild tomato relative, and the population exhibited a broad range of phenotypes, including characteristics that have potentially important attributes related to health and nutrition. Analysis of these lines identified more than 800 quantitative trait loci (QTL) underlying levels of various metabolites including ascorbate and α-tocopherol (Schauer and others 2006). Using structural flavonoid genes (encoding stilbene synthase, chalcone synthase, chalcone reductase, chalcone isomerase, and flavone synthase) from different plant sources, Schijlen and others (2006) were able to produce transgenic tomatoes accumulating new phytochemicals. Biochemical analysis showed that the fruit peel contained high levels of stilbenes (resveratrol and piceid), deoxychalcones (butein and isoliquiritigenin), flavones (luteolin-7-glucoside and luteolin aglycon), and flavonols (quercetin glycosides and kaempferol glycosides). These researchers have demonstrated that, because of the presence of the novel flavonoids, the transgenic tomato fruits displayed altered antioxidant profiles. In addition, total antioxidant capacity of tomato fruit peel with high levels of flavones and flavonols increased more than threefold. These results on genetic engineering of flavonoids in tomato fruit demonstrate the possibilities for changing the levels and composition of health-related polyphenols in a crop plant (Fraser and others 2007). Introduction of transgenes has been shown to increase antioxidant content of tomatoes (Bovy and others 2002; Fraser and others 2002). QTL for lycopene content and β-carotene content have been identified in carrot (Santos and Simon 2002), and varietal differences in antioxidant content of onion suggests the presence of genetic diversity that could be exploited by selective breeding (Yang and others 2004). Diaz de la Garza and others (2004) used genetic engineering to generate a tenfold increase in the folic acid content of ripe tomatoes. Substantial improvements can be made using marker-assisted and traditional breeding. The use of nontransgenic approaches such as TILLING (targeted induced local lesions in genomes) can allow for relatively easy identification of novel alleles in either mutagenized or natural populations. This technology was initially used in reverse genetic studies by plant biologists, and it is now being applied to crop improvement (Slade and Knauf 2005). Naturally occurring germplasm can provide a rich source of genetic variation, and it is likely that new strategies for improving the nutritional value of fruits and vegetables will result from screening for variation in nutritional and health composition in wild crops and the broadest possible spectrum of existing cultivars.

Conclusions

Accumulating evidence demonstrates that fruit and vegetable consumption has health-promoting properties. Fruits and vegetables are rich sources of diverse phytochemicals. The majority of evidence linking fruit and vegetable intake to health continues to be observational, and data in some areas are still contradictory. Therefore, although it is evident that fruit and vegetable consumption is important for health, there are still needs for more controlled, clinical intervention trials in order to investigate and to confirm the effect of the consumption of fruits and vegetables on various diseases, as well as studies to reveal the mechanisms behind the effect of the different phytochemical components of fruits and vegetables.

References

Agrawal P and Agrawal GP. 1982. Alterations in the phenols of papaya fruits infected by *Colletotrichum spp*. Proc Indian Natl Sci Acad, Part B 48:422–426.

Agarwal S and Rao AV. 1998. Tomato lycopene and low density lipoprotein oxidation: a human dietary intervention study. Lipids 33:981–984.

Agudo A, Slimani N, Ocke MC, Naska A, Miller AB, Kroke A, Bamia C, Karalis D, Vineis P, Palli D, Bueno-de-Mesquita HB, Peeters PHM, Engeset D, Hjartaker A, Navarro C, Garcia CM, Wallstrom P, Zhang JX, Welch AA, Spencer E, Stripp C, Overvad K, Clavel-Chapelon F, Casagrande C and Riboli E. 2002. Consumption of vegetables, fruit and other plant foods in the European Perspective Investigation in Cancer and Nutrition (EPIC) cohorts from 10 European countries. Publi Health Nutr 5(6B):1179–1196.

Aldoori W and Ryan-Harschman M. 2002. Preventing diverticular disease. Review of recent evidence on high-fiber diets. Can Fam Phys 48:1632–1637.

Alonso AM, Guillen DA, Barroso CG, Puertas B and Gracia A. 2002. Determination of antioxidant activity of wine byproducts and its correlation with polyphenolic content. J Agric Food Chem 50:5832–5836.

Al-Saikhan MS, Howard LR and Miller JC Jr. 1995. Antioxidant activity and total phenolics in different genotypes of potato (*Solanum tuberosum* L). J Food Sci 60:341–343, 7.

Amagase J and Milner JA. 1993. Impact of various sources of garlic and their constituents on 7,12-dimethylbenz[α]anthracene binding to mammary cell DNA. Carcinogenesis 14:1627–1631.

American Dietetic Association. 2002. Health implications of dietary fiber. JADA 102:994–1000.

Ames BN, Gold LW and Willett WC. 1995. The causes and prevention of cancer. Proc Natl Acad Sci USA 92(12):5258–5265.

Antova T, Pattenden S, Nikiforov B, Leonardi G, Boeva B, Fletcher T, Rudnai P, Slachtova H, Tabak C, Zlotkowska R, Houthuijs D, Brunekreef B and Holikova J. 2003. Nutrition and respiratory health in six central and Eastern European countries. Thorax 58:231–236.

Arcoleo A, Ruccia M and Cusmano S. 1961. Sui pigmenti flavonici delle *Opuntiae*. Nota I. Isoranmetina dai fiori di *O. ficus indica* Mill. Ann Chim 51:751–758.

Arcoleo A, Ruccia M and Natoli MC. 1966. *Beta-sitosterolo* dai fiori di *Opuntia ficus-indica* Mill. (Cactaceae). Atti Accad Sci Lett Art (Palermo) 25:323–232.

Ashton OB, Wong M, McGhie TK, Vather R, Wang Y, Requejo-Jackman C, Ramankutty P and Woolf AB. 2006. Pigments in avocado tissue and oil. J Agric Food Chem 54(26):10151–10158.

Baker RA. 1994. Potential dietary benefits of citrus pectin and fiber. Food Technol 1994(November):133–139.

Barcelo S, Gardiner JM, Gesher A and Chipman JK. 1996. CYP2E1-mediated mechanism of anti-genotoxicity of the broccoli constituent of sulforaphane. Carcinogenesis 17:277–282.

Barden CL and WJ Bramlage. 1994. Accumulation of antioxidants in apple peels as related to postharvest factors and superficial scald susceptibility of the fruit. J Am Soc Hort Sci 119:264–269.

Bazzano LA, He J, Ogden GL, Loria C, Vupputuri S, Meyers L, Whelton PK. 2001. Legume consumption and risk of coronary heart disease in US men and women. Arch Intern Med 26:2573–2578.

Bazzano LA, He J, Ogden G, Vupputuri S, Loria C, Meyers L, Meyers L and Whelton PK. 2002. Fruit and vegetable intake and risk of cardiovascular disease in US adults: the first national health and nutrition examination survey epidemiologic follow-up study. Am J Clin Nutr 76:93–99.

Bazzano LA, Li TY, Joshipura KJ and Hu FB. 2008. Intake of fruit, vegetables, and fruit juices and risk of diabetes in women. Diabetes Care 31(7):1311–1317.

Benavente-Garcia O, Castillo J, Marin FR, Ortuno A and Del Rio JA. 1997. Uses and properties of citrus flavonoids. J Agric Food Chem 45:4505–4515.

Beom JL, Yong SL and Myung HC. 1998. Antioxidant activity of vegetables and their blends in iron-catalyzed model system. J Food Sci Nutr 3:309–314.

Bes-Rastrollo M, Martínez-González MA, Sánchez-Villegas A, de la Fuente Arrillaga C and Martínez JA. 2006. Association of fiber intake and fruit/vegetable consumption with weight gain in a Mediterranean population. Nutrition 22(5):504–511.

Beuscher R, Howard L and Dexter P. 1999. Postharvest enhancement of fruits and vegetables for improved human health. Hort Sci 34(7):1167–1170.

Blanck HM, Gillespie C, Kimmons JE, Seymour JD and Serdula MK. 2008. Trends in fruit and vegetable consumption among US men and women, 1994–2005. Prev Chronic Dis 5(2):A35.

Block G, Patterson B and Subar A. 1992. Fruit, vegetables, and cancer prevention. A review of epidemiological evidence. Nutr Cancer 18:1–29.

Blomhoff R, Carlsen MH, Andersen LF and Jacobs DR Jr. 2006. Health benefits of nuts: potential role of antioxidants. Br J Nutr 96 Suppl 2:S52–S60.

Bocco A, Guvelier ME, Richard H and Berset C. 1998. Antioxidant activity and phenolic composition of citrus peel and seed extracts. J Agric Food Chem 46:2123–2129.

Boeing H, Barth C, Kluge S and Walter D. 2004. Tumorentstehung-hemmende und fordende Ernahrungsfaktoren. In Deutsche Gesellschaft fur Ernahrung, editor. Ernahrungsbericht 2004. Bonn: DGE Medien-Service, pp. 235–282.

Boone CW, Kelloff GJ and Maolone WE. 1990. Identification of candidate cancer chemopreventive agents and their evaluation in animal models and human clinical trials: a review. Cancer Res 50:2–9.

Botterweck AA, van den Brandt PA and Goldbohm RA. 1998. A prospective cohort study on vegetable and fruit consumption and stomach cancer risk in The Netherlands. Am J Epidemiol 148(9):842–853.

Bovy A, de Vos R, Kemper M, Schijlen E, Almenar PM, Muir S, Collins G, Robinson S, Verhoeyen M, Hughes S, Santos-Buelga C and van Tunen A. 2002. High-flavonol tomatoes resulting from the heterologous expression of the maize transcription factor genes LC and C1. Plant Cell 14:2509–2526.

Boyer J and Liu RH. 2004. Apple phytochemicals and their health benefits. Nutr J 12(3):5.

Breene WM. 1990. Nutritional and medicinal value of specialt muchrooms. J Food Prot 53(10):883–894.

Breinholt V, Hendricks J, Pereira C, Arbogast D and Bailey G. 1995. Dietary chlorophyllin is a potent inhibitor of aflatoxin B_1 hepatocarcinogenesis in rainbow trout. Cancer Res 55(1):57–62.

Breithaupt D and Bamedi A. 2002. Carotenoids and carotenoid esters in potato (*Solanum tuberosum* L.): new insight into an ancient vegetable. J Agric Food Chem 50:7175–7181.

Brevick A, Vollset S, Tell G, Refsum H, Ueland P, Locken E, Drevon C and Andersen L. 2005. Plasma concentration of folate as a biomarker for the intake of fruit and vegetables: the Hordaland Homocysteine Study. Am J Clin Nutr 81:434–439.

Brown CR, Edwards CG, Yang CP and Dean BB. 1993. Orange flesh trait in potato: inheritance and carotenoid content. J Am Soc Hort Sci 118:145–150.

Brutsch MO. 1990. Some lesser-known uses of the prickly pear *(Opuntia* spp.). Proceedings of Transkei and Ciskei Research Society Conference.

Bueline T, Cosma M and Appenzeller M. 2001. Diet acids and alkali influence calcium retention in bone. Osteoporosis Int 12:493–499.

Buiatti E, Palli D, Decarli A, Amadori D, Avellini C, Biachi S, Bonaguri C, Cipriani F, Cocco P and Giacosa A. 1990. A case-control study of gastric cancer and diet in Italy: II. Association with nutrients. Int J Cancer 45:896–901.

Burda S, Oleszek W and Lee CY. 1990. Phenolic compounds and their changes in apples during maturation and color change. J Agric Food Chem 38:945–948.

Burke J, Curran-Celentano J and Wenzel A. 2005. Diet and serum carotenoid concentrations affect macular pigment optical density in adults 45 yaers and older. J Nutr 135:1208–1214.

Burns J, Fraser PD and Bramley PM. 2003. Identification and quantification of carotenoids, tocopherols and chlorophylls in commonly consumed fruits and vegetables. Phytochemistry 62:939–947.

Butera D, Tesoriere L, Di Gaudio F, Bongiorno A, Allegra M, Pintaudi A, Kohen R and Livrea M. 2002. Antioxidant activities of Sicilian prickly pear (*Opuntia ficus-indica*) fruit extracts and reducing properties of its betalains: betanin and indicaxanthin. J Agric Food Chem 50:6895–6901.

Cacioppo O. 1991. Fico d'india e pitaya. Verona: L'informatore Agrario.

Camacho-Ibanez R, Meckes-Lozoya M and Mellado-Campos V. 1983. The hypoglycemic effect of *Opuntia streptacantha* studied in different animal experimental models. J Ethnopharmacol 7:175–181.

Campbell JK, Canene-Adams K, Lindshield BL, Boileau TW, Clinton SK and Erdman JW Jr. 2004. Tomato phytochemicals and prostate cancer risk. J Nutr 134(12 Suppl):3486S–3492S.

Canini A, Alesiani D, D'Arcangelo G and Tagliatesta P. 2007. Gas chromatography-mass spectrometry analysis of phenolic compounds from *Carica papaya* L. leaf. J Food Comp Anal 20:584–590.

Cao G, Sofic E and Prior RL. 1996. Antioxidant capacity of tea and common vegetables. J Agric Food Chem 44:3426–3431.

Carlton PS, Kresty LA, Siglin JC, Morse MA, Lu J, Morgan C and Stoner GD. 2001. Inhibition of *N*-nitrosomethylbenzylamine-induced tumorigensis in the rat esophagus by dietary freeze-dried strawberries. Carcinogen 3:441–446.

Carranza MJ, Herrera AJ, Alvizouri MM, Alvarado JM and Chávez CF. 1997. Effects of a vegetarian diet vs a vegetarian diet enriched with avocado in hypercholesterolemic patients. Arch Med Res 28(4):537–541.

Casadesus G, Shukitt-Hale B, Stellwagen H, Shu X, Lee H-G, Smith M and Joseph J. 2004. Modulation of hippocampal plasticity and cognitive behavior by short-term blueberry supplementation in aged rats. Nutr Neurosci 7:309–316.

Castellanos-Santiago E and Yahia EM. 2008. Identification and quantification of betalaines from the fruits of 10 Mexican prickly pear cultivars by high-performance liquid chromatography and electron spray ionization mass spectrometry. J Agric Food Chem 56:5758–5764.

Celik F and Topcu F. 2006. Nutritional risk factors for the development of chronic obstructive pulmonary disease (COPD) in male smokers. Clin Nutr 25(6):955–961.

Cerhan J, Saag K, Merlino L, Mikuls T and Criswell L. 2003. Antioxidant micronutrients and risk of rheumatoid arthritis in a cohort of older women Am J Epidemiol 157:345–354.

Chakraborty S, Chakraborty N and Datta A. 2000. Increased nutritive value of transgenic potato by expressing a nonallergenic seed albumin gene from *Amaranthus hypochondriacus*. Proc Natl Acad Sci USA 97:3724–3729.

Chaovanalikit A and Wrolstad RE. 2004. Total anthocyanins and total phenolics of fresh and processed cherries and their antioxidant properties. J Food Sci 69:FCT67–FCT72.

Chen MF, Chen LiT and Boyce HW, Jr. 1995. Cruciferous vegetables and glutathione: their effects on colon mucosal glutathione level and colon tumor development in rats induced by DMH. Nutr Cancer 23(1):77–83.

Cheng SJ, Sala M, Li MH, Courtois I and Chouroulinkov I. 1981. Promoting effect of Roussin's red identified in pickled vegetables from Linxian China. Carcinogenesis 2:313–319.

Christen W, Liu S, Schaumberg D and Buring J. 2005. Fruit and vegetable intake and risk of cataract in women. Am J Clin Nutr 81:1417–1422.

Chu Y-F, Sun J, Wu X and Lui RH. 2002. Antioxidant and antiproliferative activities of common vegetables. J Agric Food Chem 50:6910–6916.

Colditz GA, Branch LG and Lipnic RJ. 1985. Increased green and yellow vegetables intake and lowered cancer death in an elderly population. Am J Clin Nutr 41:32–36.

Cook D, Grierson D, Jones C, Wallace A, West G and Tucker G. 2002. Modification of fatty acid composition in tomato (*Lycopersicon esculentum*) by expression of a borage delta6-desaturase. Mol Biotechnol 21:123–128.

Corral-Aguayo R, Yahia EM, Carrillo-Lopez A and Gonzalez-Aguilar G. 2008. Correlation between some nutritional components and the total antioxidant capacity measured with six different assays in eight horticultural crops. J Agric food Chem 56:10498–10504.

Crowell PL and Gould MN. 1994. Chemoprevention and therapy of cancer by D-limonine. Crit Rev Oncog 5:1–22.

Cruse R. 1973. Desert plant chemurgy: a current review. Econ Bot 27:210–230.

Curry EA. 1997. Effect of postharvest handling and storage on apple nutritional status using antioxidants as a model. HortTech 7(3):240–243.

Dalal M, Dani RG and Kumar PA. 2006. Current trends in the genetic engineering of vegetable crops. Scientia Hort 107:215–225.

Daood HG, Vinkler M, Markus F, Hebshi EA and Biacs PA. 1996. Antioxidant vitamin content of spice red pepper (paprika) as affected by technological and varietal factors. Food Chem 55:365–372.

d'Aquino A. 1998. Tesi per il conseguimento del Dottorato di Ricerca in Farmacognosia (IX Ciclo). *Opuntia ficus indica* Mill. Ricerche Farmacognostiche. Facolta di Farmacia- Universita di Messina.

Dashwood RH, Arbogst DN, Fong AT, Pereira C, Hendricks JD and Baily GS. 1989. Quantitative interrelationships between aflatoxin B1 carcinogen dose, indole-3-carbinol anto-carcinogen dose, target organ DNA adduction and final tumor response. Carcinogenesis 10:175–181.

Davies BH, Matthews S and Kirk JTO. 1970. The nature and biosynthesis of the carotenoids of different colour varieties of *Capsicum annuum*. Phytochemistry 9:797–800.

Davis P and Iwahashi CK. 2001. Whole almonds and almond fractions reduce aberrant crypt foci in a rat model of colon carcinogenesis. Cancer Letters 165:27–33.

Davis P, Iwahashi CK and Yokoyama W. 2003. Whole almonds activate gastrointestinal (GI) tract anti-proliferative signaling in APCmin (multiple intestinal neoplasia) mice. FACEB J 17(5):AQ1153.

de Carvalho EB, Vitolo MR, Gama CM, Lopez FA, Taddei JA and de Morais MB. 2006. Fiber intake, constipation, and overweight among adolescents living in Sao Paulo City. Nutrition 22(7–8):744–749.

Devereux G and Seaton A. 2005. Diet as a risk factor for atopy and asthma. J Allergie Clin Inmunol 115:1109–1117.

Di Mascio P, Kaiser S and Sies H. 1989. Lycopene as the most efficient biological carotenoid singlet oxygen quencher. Arch Biochem Biophys 274:532–538.

Di Mascio P, Murphy MC and Sies H. 1991. Antioxidant defense systems, the role of carotenoid, tocopherol and thiols. Am J Clin Nutr 53 (Suppl):194–200.

Diaz de la Garza R, Quinlivan EP, Klaus SM, Basset GJ, Gregory JFR and Hanson AD. 2004. Folate biofortification in tomatoes by engineering the pteridine branch of folate synthesis. Proc Natl Acad Sci USA 101:13720–13725.

Donovan JL, Meyer AS and Waterhouse AL. 1998. Phenolic composition and antioxidant activity of prune juice. J Agric Food Chem 46:1247–1252.

Dorais M, Papadopoulos AP and Gosselin A. 2001a. Greenhouse tomato fruit quality. Hort Rev 26:239–319.

Dorais M, Papadopoulos AP and Gosselin A. 2001b. Influence of electric conductivity management on greenhouse tomato yield and fruit quality. Agronomie 21:367–383.

Dosil-Díaz O, Ruano-Ravina A, Gestal-Otero JJ and Barros-Dios JM. 2008. Consumption of fruit and vegetables and risk of lung cancer: a case-control study in Galicia, Spain. Nutrition 24(5):407–413.

Downey MO, Dokoozlian NK and Krstic MP. 2006. Cultural practice and environmental impacts on the flavonoid composition of grapes and wine: a review of recent research. Am J Enol Vitic 57:257–268.

Duthie GG, Duthie SJ and Kyle JAM. 2000. Plant polyphenols in cancer and heart disease: implications as nutritional antioxidants. Nutr Res Rev 13(1):79–106.

Eberhardt MV, Lee C and Liu RH. 2000. Antioxidant activity of fresh apples. Nature 405:903–904.

Eichholzer M, Tonz O and Zimmerman R. 2006. Folic acid: a public health challenge. Lancet 367:1352–1361.

Engelhart M, Geerlings M, Ruitemberg D, van Swieten J, Hoffman A, Wittman J and Breteler M. 2002. Dietary antioxidants and risk of Alzheimer's disease. JAMA 287; 3223–3229.

Erbersdobler H. 2003. Workstoffe. In: Erberdobler H and Meyer A, editors. Praxishandbuch Functional Food. Hamburg: B. Behr's Verlag, pp. 1–14.

Escarpa A and Gonzalez MC. 2001. An overview of analytical chemistry of phenolic compounds in foods. Crit Rev Anal Chem 31:57–139.

Farchi S, Forastieri F, Agabiti N, Corbo G, Pistelli R, Fortes C, Dell'Orco V and Perucci C. 2003. Dietary factors associated with wheezing and allergic rhinitis in children. Eur Respir J 22:772–780.

Fernandez LM, Lin ECK, Trejo A and McNamara DJ. 1992. Prickly pear *(Opuntia* sp.) pectin reverses low density lipoprotein receptor suppression induced by a hypercholesterolemic diet in Guinea Pigs. J Nutr 122:2330–2340.

Fernandez LM, Lin ECK, Trejo A and McNamara DJ. 1994. Prickly pear (*Opuntia* sp.) pectin alters hepatic cholesterol metabolism without affecting cholesterol absorption in guinea pigs fed a hypercholesterolemic diet. J Nutr 124:817–824.

Fernandez-Lopez JA and Almela L. 2001. Application of high performance liquid chromatography to the characterization of the betalain pigments in prickly pear fruits. J Chromatogr 913:415–420.

Fernie AR, Tadmor Y and Zamir D. 2006. Natural genetic variation for improving crop quality. Curr. Opin. Plant Biol 9:196–202.

Feskanich D, Ziegler RG, Michaud DS, Giovanucci EL, Speizer FE, Willett WC and Colditz GA. 2000. Prospective study of fruit and vegetable consumption and risk of lung cancer among men and women. Int J Natl Cancer Inst 92(22):1812–1823.

Forni E, Polesello A, Montefiori D and Maestrelli A. 1992. A high performance liquid chromatographic analysis of the pigments of blood-red prickly pear (*Opuntia ficus-indica*). J Chromatogr 593:177–183.

Franceschi S, Bidoli E, La Vecchis C, Talamini R, D'Avanzo B and Negri E. 1994. Tomatoes and risk of digestive-tract cancers. Int J Cancer 59:181–184.

Frankel EN, Bosanek CA, Mayer AS, Silliman K and Kirk LL. 1998. Commercial grape juices inhibit the *in vitro* oxidation of human low-density lipoproteins. J Agric Food Chem 46:834–838.

Frankel EN and Meyer AS. 1998. Antioxidants in grapes and grape juices and their potential health effects. Pharmaceutical Biol 36:1–7.

Frankel EN, Waterhouse AL and Kinsella JE. 1993. Inhibition of human LDL oxidation by resveratrol. Lancet 341:1103–1104.

Fraser PD, Enfissi EMA, Halket JM, Truesdale MR, Yu D, Gerrish C, Bramley PM. 2007. Manipulation of phytoene levels in tomato fruit: effects on isoprenoids, plastids, and intermediary metabolism. Plant Cell 19:3194–3211.

Fraser PD, Romer S, Shipton CA, Mills PB, Kiano JW, Misawa N, Drake RG, Schuch W and Bramley PM. 2002. Evaluation of transgenic tomato plants expressing an additional phytoene synthase in a fruit-specific manner. Proc Natl Acad Sci USA 99:1092–1097.

Frati, A. 1992. Medical implication of prickly pear cactus. In: Felker P and Moss JR, editors. Proceedings of the Third Annual Texas Prickly Pear Council, pp. 29–34. Kingsville, TX.

Frati AC, Gordillo BE, Altamirano P, Ariza CR, Cortes-Franco R and Chavez-Negrete A. 1990a. Acute hypoglycemic effect of *Opuntia streptacantha* Lemaire in NIDDM. Diabetes Care 13; 455–456.

Frati AC, Hernandez de la Riva H, Ariza CR and Torres MD. 1983. Effects of nopal *(Opuntia* sp.) on serum lipids, glycemia and body weight. Arch Invest Med (Mexico) 14:117–125.

Frati AC, Jimenez E and Ariza CR. 1990b. Hypoglycemic effect of *Opuntia ficus-indica* in non-insulin dependent diabetes mellitus patients. Phytother Res 4:195–197.

Frison S and Sporns P. 2002. Variation of the flavanol glycoside composition of almond seedcoats as determined by MALDI-TOF mass spectrometry. J Agric Food Chem 50(23):6818–6822.

Galati EM, Pergolozzi S, Miceli N, Monforte MT and Tripodo MM. 2001. Study on the increment of production of gastric mucus in rats treated with *Opuntia ficus-indica* (L) Mill. cladodes. J Ethnopharmacol 83(3):229–233.

Galati EM, Tripodo, MM, Trovato A, Miceli N and Monforte MT. 2002. Biological effect of *Opuntia ficus-indica* (L) Mill (Cactacea) waste matter. Note I: diuretic activity. J Ethnopharmacol 79(1):17–21.

Galvis-Sanchez AC, Gil-Izquierdo A and Gil MI. 2003. Comparative study of six pear cultivars in terms of their phenolic and vitamin C contents and antioxidant capacity. J Sci Food Agric 83:995–1003.

Gandini S, Merzeninch H, Robertson C and Boyle P. 2000. Meta analysis of studies on breast cancer risk and diet: the role of fruit and vegetable consumption and the intake of associated micronutrients. Eur J Cancer 36:636–646.

Gao L and Mazza G. 1995. Characterization, quantification, and distribution of anthocyanins and colorless phenolic in sweet cherries. J Agric Food Chem 43:343–346.

Garcia-Solis P, Yahia EM and Aceves C. 2008. Study of the effect of 'Ataulfo' mango (*Mangifera indica* L) intake on mammary carcinogenesis and antioxidant capacity in plasma of *N*-methyl-nitrosourea (MNU)-treated rats. Food Chem 111:309–315.

Garcia-Solis P, Yahia EM, Morales-Tlalpan V and Diaz-Munoz M. 2009. Screening of antiproliferative effect of aqueous extracts of plant foods in Mexico on the breast cancer cell line MCF-7. Int J Food Sci Nutr. In press.

Gazzani G, Pappeti A, Massolini G and Daglia M. 1998. Anti- and prooxidant activity of water soluble components of some common diet vegetables and the effect of thermal treatment. J Agric Food chem. 46:4118–4122.

Geeson JD, Browne KM, Maddison K, Sheferd J and Guaraldi F. 1985. Modified atmosphere packaging to extend the shelf life of tomatoes. J Food Technol 20:339–349.

Gerster H. 1997. The potential role of lycopene for human health: review article. J Am Coll Nutr 16:109–126.

Giovannucci E. 1999. Tomatoes, tomato-based products, lycopene, and cancer: review of the epidemiologic literature. J Natl Cancer Inst 91(4):317–331.

Giovannucci E, Ascherio A, Rimm EB, Stampfer MJ, Colditz GA and Willett WC. 1995. Intake of carotenoids and retinol in relation to risk of prostate cancer. J Natl Cancer Inst 87:767–776.

Giovannucci E and Clinton SK. 1998. Tomatoes, lycopene, and prostate cancer. Proc Soc Exp Biol Med 218(2):129–139.

Giovannucci E, Rimm EB, Liu Y, Stampfer MJ and Willett WC. 2003. A prospective study of cruciferous vegetables and prostate cancer. Cancer Epidimiol Biomarkers Prev 12(12):1403–1409.

Goldberg I. 1994. Introduction. In: Goldberg I, editor. Functional Foods: Designer Foods, Pharmafoods, Nutraceuticals. New York: Chapman and Hall, pp. 1–16.

Goldmann IL, Kader AA and Heintz C. 1999. Influence of production, handling, and storage on phytonutrient content of foods. Nutr Rev 57:46–52.

Goodman MT, Kolonel TJ, Wilkens, LR, Yoshizawa CN, Le Marchand L and Hankin JH. 1992. Dietary factors in lung cancer prognosis. Eur J Cancer 28:495–501.

Graham S, Dayal H, Swanson M, Mittelman A and Wilkinson G. 1978. Diet in the epidemiology of cancer of the colon and rectum. J Natl Cancer Inst 61:709–714.

Graham S and Mettlin C. 1981. Fiber and other constituents of vegetables in cancer epidemiology. Progr Cancer Res Ther 17:189–215.

Gross J. 1987. Pigments in Fruits. London: Academic Press.

Grube BJ, Eng ET, Yeh CK, Kwon A and Shiuan C. 2001. White button mushroom phytochemicals inhibit aromatase activity and breast cancer cell proliferation. J Nutr 13:3288–3293.

Guester H. 1997. The potential role of lycopene for human health. J Am Clin Nutr 16:109–126.

Gutfinger T and Letan A. 1974. Studies of unsaponificables in several vegetable oils. Lipids 9:658.

Hageman G, Krul C, van Herwinjnen M, Schilderman P and Kleinjans J. 1997. Assessment of the anticarcinogenic potential of raw garlic in humans. Cancer Lett 114:161.

Haila KM, Lievonen SM and Heinon IM. 1996. Effects of lutein, lycopene, annatto, and γ tocopherol on oxidation of triglycerides. J Agric Food Chem 44:2096–2100.

Hall AP, More JF and Morgan AF. 1955. A vitamin content of California grown avocados. J Agric Food Chem 3:250–252.

Hamer M and Chida Y. 2007. Intake of fruit, vegetables, and antioxidants and risk of type 2 diabetes: systematic review and meta-analysis. J Hypertens 25(12):2361–2369.

Harwood M, Danielewska-Nikiel B, Borselleca JF, Flamm GW, Williams GM and Lines TC. 2007. A critical review of the data related to the safety of quercetin and lack of evidence of in vivo toxicity, including lack of genotoxic/carcinogenic properties. Food Chem Toxicol 45:2179–2205.

Harvala C, Alkofahi A and Philianos S. 1982. Sur l'action enxymatique d'un produit extrait des graines d'*Opuntia ficus-indica* Miller. Plant Meedic Phitotherap 4:298–302.

Hatono S, Jimenez A and Wargovich MJ. 1996. Chemopreventive effect of *S*-allylcysteine and its relationship to the detoxification enzyme glutathione *S*-transferase. Carcinogenesis 17:1041–1044.

He K, Hu FB, Colditz GA, Manson JE, Willett WC and Liu S. 2004. Changes in intake of fruits and vegetables in relation to risk of obesity and weight gain among middle-aged women. Int J Obes Relat Metab Disord 28(12):1569–1574.

Hegwood DA. 1990. Human health discoveries with *Opuntia sp.* (prickly pear). Hort Sci 25(12):1515–1516.

Heinonen IM, Meyer AS and Frankel EN. 1998. Antioxidant activity of berry phenolics on human low-density lipoprotein and liposome oxidation. J Agric Food Chem 46:4107–4112.

Helzlsouer KJ, Comstock GW and Morris JS. 1989. Selenium, lycopene, alpha tocopherol, beta carotene, retinol, and subsequent bladder cancer. Cancer Res 49:6144–6148.

Hennekens CH, Burig JE, Manson JE, Stampfer M, Rosner B, Cook NR, Belanger C, LaMotte F, Gaziano JM, Ridker PM, Willett W and Peto R. 1996. Lack of effect of long-term supplementation with beta carotene on the incidence of malignant neoplasms and cardiovascular disease. N Engl J Med 334:1145–1149.

Hirayama T. 1990. Lifestyle and Mortality. Contributions to epidemiology and Biostatistics, Volume 6. Basel, Switzerland: Karger.

Hoisington D. 2002. Opportunities for nutritionally enhanced maize and wheat varieties to combat protein and micronutrient malnutrition. Food Nutr Bull 23:376–377.

Howard LR, Smith RT, Wagner AB, Villalon B and Burns EE. 1994. Provitamin A and ascorbic acid content of fresh pepper cultivars (*Capsicum annuum*) and processed jalapenos. J Food Sci 59:362–365.

Howard LR, Talcott ST, Brenes CH and Villalon B. 2000. Changes in phytochemical and antioxidant activity of selected pepper cultivars (*Capsicum* species) as influenced by maturity. J Agric Food Chem 48:1713–1720.

Howe G, Hirohata T, Hislop T, Iscovich J, Yuan J, Katsouyanni K, Lubin F, Marubini E, Modan B, Rohan T, Toniolo P and Shunzhang Y. 1992. Dietary factors and risk of breast cancer: combined analysis of 12 case-control studies. J Nat Cancer Inst 82:561–569.

Huet R. 1982. Constituents of citrus fruits with pharmacodynamic effects: citroflavonoids. Fruits 37:267–71.

Hufnagel JC and Hofmann T. 2008. Orosensory-directed identification of astringent mouth feel and bitter-testing compounds in red wine. J Agric Food Chem 56:1376–1386.

Hung HC, Merchant A, Willett W, Ascherio A, Rosner BA, Rimm E and Joshipura KJ. 2003. The association between fruit and vegetable consumption and peripheral arterial disease. Epidemiology 14(6):659–665.

Huyskens-Keil S and Schreiner M. 2004. Quality dynamics and quality assurance of fresh fruit and vegetables in pre- and postharvest. In: Dris R and Jain S, editors. Production Practices and Quality Assessment of Food Crops. Dordrecht: Kluwer Academic, pp. 401–449.

Hyson DA, Scheeman BO and Davis PA. 2002. Almonds and almond oil have similar effects on plasma lipid and LDL oxidation in healthy men and women. J Nutr 132(4):703–707.

Ip C, Lisk DJ. 1994. Enrichment of selenium in allium vegetables for cancer prevention. Carcinogenesis 15:1881–1885.

Ip C, Lisk DJ and Stoewasand GS. 1992. Mammary cancer prevention by regular garlic and selenium enriched garlic. Nutr Cancer 17:279–286.

Itoh T, Tamura T, Matsumato T and Dupaigne P. 1975. Studies on avocado oil in particular on the un-saponifiable sterol fraction. Fruits 30:687–695.

Iwanzik W, Tevini M, Stute R and Hilbert R. 1983. Carotinoidgehalt und Zusammensetzung verschiedener deutscher karto isorten und deren Bedeutung fur die Fleischfarbe der Knolle. Pot Res 26:149–162.

Jajoo R, Song L, Rasmussen H, Harris S and Dawson-Hughes B. 2006. Dietary acid-base balance, bone resorption and calcium excretion. J Nutrition 25:223–230.

Jenkins DJ, Hu FB, Tapsell LC, Josse AR and Kendall CW. 2008. Possible benefit of nuts in type 2 diabetes. J Nutr 138(9):1752S–1756S.

Johnston CS, Taylor CA and Hampl JS. 2000. More Americans are eating "5 a day" but intakes of dark green and cruciferous vegetables remain low. J Nutr 130(12):3063–3067.

Jones DP, Coates RJ, Flagg EW, Eley JW, Block G, Greenberg RS, Gunter EW and Jackson B. 1992. Glutathione in foods listed in the National Cancer Institute's health habits and history food frequency questionnaire. Nutr Cancer 17(1):57–75.

Joseph J, Arendash G, Gordon M, Diamond D, Shukitt-Hale B and Morgan D. 2003. Blueberry supplementation enhances signaling and prevents behavioral deficits in an Alzheimer disease model. Nutr Neurosci 6:153–163.

Joseph J, Shukitt-Hale B, Casadesus G and Fisher D. 2005. Oxidative stress and inflammation in brain aging: nutritional considerations. Neurochem Res 30:927–935.

Joshipura K, Hu F, Manson J, Stamfer M, Rimm E, Speizer F, Coltiz G, Asherio A, Rosner B, Spiegelman D and Willett W. 2001. The effect of fruit and vegetable intake on risk on coronary heart disease. Ann Int Med 134:1106–1114.

Ju Z and WJ Bramlage. 1999. Phenolics and lipid-soluble antioxidants in fruit cuticle of apples and their antioxidant activities in model systems. Postharv Biol Technol 16:107–118.

Jung KJ, Wallig MA and Singletary KW. 2006. Purple grape juice inhibits 7,12-dimethylbenz[*a*]anthracene (DMBA)-induced rat mammary tumorigenesis and *in vivo* DMBA-DNA adduct formation. Cancer Lett 233:279–288.

Kahkonen MP, Hopia AI, Vurela HJ, Rauha J-P, Pihlaja K, Kujala TS and Heinon M. 1999. Antioxidant activity of plant extracts containing phenolic compounds. J Agric Food Chem 47:3954–3962.

Kalt W, Forney CF, Martin A and Prior RL. 1999. Antioxidant capacity, vitamin C, and anthocyanins after fresh storage of small fruits. J Agric Food Chem 47:4638–4644.

Kanazawa K and Sakakibara H. 2000. High content of dopamine, a strong antioxidant, in Cavendish banana. J Agric food Chem 48:844–848.

Kang J, Asherio A and Grodstein F. 2005. Fruit and vegetable consumption and cognitive decline in aging women. Am Neurol 2005:713–720.

Kelloff GJ, Boone CW, Crowell JA, Steele VE, Lubet RA, Doody LA, Maolone WF, Hawk ET and Sigman CC. 1996. New agents for cancer chemoprevention. J Cell Biochem 26s:1–28.

Kelly Y, Sacker A and Marmot A. 2003. Nutrition and respiratory health in adults: findings from the Health Survey for Scotland. Eur Respir J 21:664–671.

Kirsh VA, Peters U, Mayne ST, Subar AF, Chatterjee N, Johnson CC and Hayes RB. 2007. Prospective study of fruit and vegetable intake and risk of prostate cancer. J Natl Cancer Inst 99(15):1200–1209.

Kolonel LN, Nomura MY, Hirohata T, Hankin JH and Hinds MW. 1981. Association of diet and place of birth with stomach cancer incidence in Hawaii Japanese and Caucasians. Am J Clin Nutr 34(11):2478–2485.

Kotake-Nara E, Kushiro M, Hong Z, Sugawara T, Miyashita K and Nagao A. 2001. Carotenoids affect proliferation of human prostate cancer cells. J Nutrition 131(12):3303–3306.

Kris-Etherton P, Hecker K, Bonanome A, Coval S, Binkoski A, Hilpert K, Griel A and Etherton T. 2002. Bioactive compounds in foods: their role in the prevention of cardiovascular disease and cancer. Am J Med 113:71–88.

Kris-Etherton PM, Hu FB, Ros E and Sabaté J. 2008. The role of tree nuts and peanuts in the prevention of coronary heart disease: multiple potential mechanisms. J Nutr 138(9):1746S–1751S.

Lam LKT, Zhang J and Hasegawa S. 1994. Citrus limonoid reduction of chemically induced tumorigenesis. Food Technol November 1994:104–108.

Lanham-New S. 2006. Fruits and vegetables: the unexpected natural answer to the question of osteoporosis prevention? Am J Clin Nutr 83:1254–1255.

Larrauri JA, Ruperez P and Galixto FS. 1997. Pineapple shell as a source of dietary fiber with associated polyphenols. J Agric Food chem. 45:4028–4031.

Larsson SC, Bergkvist L and Wolk A. 2006. Fruit and vegetable consumption and incidence of gastric cancer: a prospective study. Cancer Epidimiol Biomarkers Prev 15(10):1998–2001.

Lau F, Shukitt-Hale V and Joseph J. 2005. The beneficial effects of fruit polyphenols on brain aging. Neurobiol Aging 265:S128–S132.

Laval-Martin D, Quennement J and Moneger R. 1975. Pigment evolution in *Lycopersicon esculentum* fruits during growth and ripening. Phytochemistry 14:2357–2362.

Law MR and Morris JK. 1998. By how much does fruit and vegetable consumption reduce the risk of ischaemic heart disease? Eur J Clin Nutr 52:549–556.

Le Marchand L, Yoshizawa CN, Kolonel LN, Hankin LH, Goodman MT. 1989. Intake of flavonoids and lung cancer. J Natl Cancer Inst 81:1158–1164.

Lea AGH. 1978. The phenolics of ciders: oligomeric and polymeric procyanidins. J Sci Food Agric 29:471–477.

Lea AGH. 1992. Flavor, color and stability in fruit products: the effect of polyphenols. In: Hemingway RW and Leaks PE, editors. Plant Polyphenols. New York: Plenum Press, pp. 827–847.

Lea AGH, Timberlake CF. 1974. The phenolics of cider. J Sci Food Agric 25:1537–1545.

Lee ES, Steiner M, Lin R. 1994. Thioallyl compounds: potent inhibitors of cell proliferation. Biochim Biophys Acta 1221:73–77.

Lee LM, Cook NR, Manson JE, Buring JE, Hennekens CH. 1999. β-Carotene supplementation and incidence of cancer and cardiovascular disease: the Women's Health Study. J Natl Cancer Inst 91:2102–2106.

Lee KW, Kim WJ, Kim DO, Lee HJ, Lee CY. 2003. Major phenolics in apples and their contribution to the total antioxidant capacity. J Agric Food Chem 51:6516–6520.

Lee Y, Howard LR and Villalon B. 1995. Flavonoids and antioxidant activity of fresh pepper (*Capsicum annuum*) cultivars. J Food Sci 60:473–476.

Lentheric I, Pinto E, Vendrell M and Larrigaudiere C. 1999. Harvest date affects the antioxidative system in pear fruits. J Hort Sci Technol 74(6):791–795.

Leontowicz M, Gorinstein S, Leontowicz H, Krzeminski R, Lojek A, Katrich E, Ciz M, Martin-Bellozo O, Soliva-Fortuny R, Haruenkit R and Trakhtenberg S. 2003. Apple and pear peel and pulp and their influence on plasma lipids and antioxidant potentials in rats fed cholesterol-containing diets. J Agric Food Chem. 51:5780–5785.

Lewis D, Antoniak M, Venn A, Davies L, Goodwin A, Salfield N, Britton J and Fogarty A. 2005. Secondhand smoke, dietary food intake, road traffic exposures, and the prevalence of asthma: a cross-sectional study in young children. Am J Epidemiol 161:406–411.

Li S, Pan M-H, Lo C-Y, Tan D, Wang Y, Shahidi F and Ho C-T. 2009. Chemistry and health effects of polymethoxyflavones and hydroxylated polymethoxyflavones. J Funct Foods 1:2–12.

Lin P-H, Ginty F, Appel L, Aickin M, Bohannon A, Garnero P, Barcaly D and Svetkey L. 2003. The DASH diet and sodium reduction improve markers of bone turnover and calcium metabolism in adults. J Nutrition 133:3130–3136.

Lin XY, Liu JZ and Milner JA. 1994. Dietary garlic suppresses DNA adducts caused by *N*-nitroso compounds. Carcinogenesis 15:349–352.

Linseisen J, Rohrmann S, Miller AB, Bueno-de-Mesquita HB, Büchner FL, Vineis P, Agudo A, Gram IT, Janson L, Krogh V, Overvad K, Rasmuson T, Schulz M, Pischon T, Kaaks R, Nieters A, Allen NE, Key TJ, Bingham S, Khaw KT, Amiano P, Barricarte A, Martinez C, Navarro C, Quirós R, Clavel-Chapelon F, Boutron-Ruault MC, Touvier M, Peeters PH, Berglund G, Hallmans G, Lund E, Palli D, Panico S, Tumino R, Tjønneland A, Olsen A, Trichopoulou A, Trichopoulos D, Autier P, Boffetta P, Slimani N and Riboli E. 2007. Fruit and vegetable consumption and lung cancer risk: updated information from the European Prospective Investigation into Cancer and Nutrition (EPIC). Intl J Cancer 121(5):1103–1114.

Liu RH. 2004. Potential synergy of phytochemicals in cancer prevention: mechanism of action. J Nutr 134:3479S-3485S.

Liu RH, Liu J and Chen B. 2005. Apples prevent mammary tumors in rats. J Agric Food Chem 53:2341–2343.

Long M, Millar DJ, Kimura Y, Donovan G, Rees J, Fraser PD, Bramley PM and Bolwell GP. 2006. Metabolite profiling of carotenoid and phenolic pathways in mutant and transgenic lines of tomato: identification of a high antioxidant fruit line. Phytochemistry 67:1750–1757.

Lu SH, Camus AM, Tomatis L and Bartsch H. 1981. Mutagenicity of extracts of pickled vegetables collected in Linhsien County, a high-incidence area for esophageal cancer in Northern China. J Natl Cancer Inst 66(1):33–36.

AmMa J, Yang H, Basile MJ and Kennelly EJ. 2004. Analysis of polyphenolic antioxidants from the fruits of three *Pouteria* species by selected ion monitoring liquid chromatography/mass spectrometry. J Agric Food Chem 52:5873–5878.

Macdonald H, New S, Fraser W, Campbell M and Reid D. 2005. Low dietary potassium intakes and high dietary estimates of net endogenous acid production are associated with low bone mineral density in premenopausal women and increased markers of bone resorption in postmenopausal women. Am J Clin Nutr 81:923–933.

Macheix JJ, Fleuriet A and Billot J. 1990. Fruit Phenolics. Boca Raton, FL: CRC Press.

Mahattanatawe K, Manthey JA, Luzio G, Talcott ST, Goodner K and Baldwin EA. 2006. Total antioxidant activity and fiber content of select Florida-grown tropical fruits. J Agric Food Chem 54:7355–7363.

Markus F, Daood HG, Kapitany J and Biacs PA. 1999. Changes in the carotenoid and antioxidant content of spice red pepper (paprika) as a function of ripening and some technological factors. J Agric Food Chem 47:100–107.

Mazliak P. 1971. Avocado lipid constituents. Fruits 26:615–623.

Mazza G. 1998. Functional Foods. Pennsylvania: Technomic.

Mayr U, Treutter D, Santos-Buelga C, Bauer H and Fuecht W. 1995. Developmental changes in the phenol concentrations of 'Golden Delicious' Apple fruits and leaves. Phytochemistry 38:115.

McDannell R and McLean AEM. 1988. Chemical and biological properties of indole glucosinolates (glucobrassicins): a review. Food Chem Toxicol 26:59–70.

Medellin MLC, Salvidar SOS and De la Garza JV. 1998. Effect of raw and cooked nopal (*Opuntia ficus-indica*) ingestion on growth and total cholesterol, lipoproteins and blood glucose in rats. Arch Latinoamer Nutr 48:316–323.

Mejia LA, Hudson E, Gonzalez E and Vazquez F. 1988. Carotenoid content and vitamin A activity of some common cultivars of Mexican peppers (*Capsicum annuum*) as determined by HPLC. J Food Sci 53:1448–1451.

Mei X, Lin X, Liu JZ, Lin XY, Song PJ, Hu JF and Liang XJ. 1989. The blocking effect of garlic on the formation of *N*-nitrosoproline in humans. Acta Nutrimenta Sinica 11:141–146.

Mei X, Wang ML, Xu HX, Pan XY, Gao CY, Han N and Fu MY. 1982. Garlic and gastric cancer. I. The influence of garlic on the level of nitrate and nitrite in gastric juice. Acta Nutrimenta Sinica 4:53–56.

Mercadante AZ, Rodriguez-Amaya DB and Britton G. 1997. HPLC and mass spectrometric analysis of carotenoids from mango. J Agric Food Chem 45:120–123.

Mertens-Talcott SU, Bomser JA, Romero C, Talcott ST and Percival SS. 2005. Ellagic acid potentiates the effect of quercetin on p21waf1/cip1, p53, and MAP-kinases without affecting intracellular generation of reactive oxygen species *in vitro*. J Nutr 135:609–614.

Meyer BN and McLaughlin JL. 1981. Economic uses of *Opuntia*. Cactus Succulent J 53:107–112.

Meyers KJ, Watkins CB, Pritts MP and Liu RH. 2003. Antioxidant and antiproliferative activities of strawberries. J Agric Food Chem 51:6887–6892.

Michels KB, Giovannucci E, Joshipura KJ, Rosner BA, Stampfer MJ, Fuchs CS, Codlitz GA, Speizer FE and Willett W. 2000. Prospective study of and incidence of colon and rectal cancers. J Natl Cancer Inst 92:1749–1752.

Middleton E Jr. and Kandaswami C. 1994. Potential health-promoting properties of citrus flavonoids. Food Technol November 1994:115–119.

Miean KH and Mohamed S. 2001. Flavonoid (myricetin, quercetin, kaempferol, luteolin, and apigenin) content of edible tropical plants. J Agric Food Chem 49:3106–3112.

Millburry P, Chen C-Y, Kwak H-K and Blumberg J. 2002. Almond skins polyphenolics act synergistically with alpha-tocopherol to increase the resistance of low-density lipoproteins to oxidation. Free Rad Res 36(1 suppl):78–80.

Miller NJ and Rice-Evans CA. 1997. The relative contributions of ascorbic acid and phenolic antioxidants to the total antioxidant activity of orange and apple fruit juices and blackcurrant drinks. Food Chem 60:331–337.

Minguez-Mosquera MI and Hornero-Mendez D. 1994. Comparative study of the effect of paprika processing on the carotenoids in peppers (*Capsicum annuum*) of the Bola and Agridulce varieties. J Agric Food Chem 42:1555–1560.

Moeller S, Taylor A, Tucker K, McCullogh M, Chylack L, Hankinson S, Willet W and Jacques P. 2004. Overall adherence to the Dietary Guidelines for Americans is associated with reduced prevalence of early age-related nuclear lens opacities in women. J Nutr 134:1812–1819.

Mori K, Toyomasu T, Nanba H and Kuroda H. 1987. Antitumor action and fruit bodies of edible mushrooms orally administered to mice. Mush J Tropics 7:121–126.

Muller H. 1997. Determination of the carotenoid content in selected vegetables and fruits by HPLC and photodiode array detection. Z Lebensm Unters Forsch 204:88–94.

Nagura J, Iso H, Watanabe Y, Maruyama K, Date C, Toyoshima H, Yamamoto A, Kikuchi S, Koizumi A, Kondo T, Wada Y, Inaba Y and Tamakoshi A. 2009. Fruit, vegetable and bean intake and mortality from cardiovascular disease among Japanese men and women: the JACC Study. Br J Nutr 13:1–8.

Nakamura K, Nagata C, Oba S, Takatsuka N and Shimizu H. 2008. Fruit and vegetable intake and mortality from cardiovascular disease are inversely associated in Japanese women but not in men. J Nutr 138(6):1129–1134.

Naska A, Vasdekis V, Trichopoulou A, Friel S, Leonhauser I and Moreira G. 2000. Fruits and vegetable availability among ten European countries: how does it compare with the "five-a-day" recommendation? Brit J Nutr 84:549–556.

Ness A, Egger M and Powles J. 1999. Fruit and vegetable and ischaemic heart disease: systematic review or misleading meta-analysis? Eur J Clin Nutr 53:900–902.

Ness A and Powles JW. 1997. Fruit and vegetables, and cardiovascular disease: a review. Int J Epidemiol 26:1-13.

Nettleton JA, Steffen LM, Ni H, Liu K and Jacobs DR Jr. 2008. Dietary patterns and risk of incident type 2 diabetes in the Multi-Ethnic Study of Atherosclerosis (MESA). Diabetes Care 31(9):1777–1782.

Okabo H, Sasaki S, Horiguchi H, Oguma E, Miyamoto K, Hosoi Y, Kim M- and Kayama F. 2006. Dietary patterns associated with bone mineral density in premenopausal Japanese farmwomen. Am J Nutr 83(5):1185–1192.

Oleszek W, Lee CY, Jaworzki AW and Price KR. Identification of some phenolic compounds in apples. J Agric Food Chem 36:430–436.

Oloyede OI. 2005. Chemical profile of unripe pulp of *Carica papaya.* Pakistan J Nutr 4:379–381.

Olmedilla B, Granado F, Southon S, Wright AJA, Blanco I, Gil-Martinez E, Berg H, Corridan B, Roussel AM, Chopra M and Thurnham DI. 2001. Serum concentrations of carotenoids and vitamins A, E, and C in control subjects from five European countries. British J Nutr 85(2):227–238.

Omenn GS, Goodman GE, Thornkist MD, Balmes J, Cullen MR, Glass A, Keogh JP, Meyskens FL, Valanis B, Williams JH, Barnhart S, Cherniack MG, Brodkin CA and Hammar S. 1996. Risk factors for lung cancer and for intervention effects in CARET, the Beta-Carotene and Retinol Efficacy Trial. J Nat Cancer Inst 88:1550–1559.

Ornelas-Paz JJ, Yahia EM and Gardea A. 2007. Identification and quantification of xanthophyll esters, carotenes and tocopherols in the fruit of seven Mexican mango cultivars by liquid chromatography-$APcI^+$-time of flight mass spectrometry. J Agric Food Chem 55:6628–6635.

Ornelas-Paz JJ, Yahia EM and Gardea-Bejar A. 2008. Relationship between fruit external and internal color and carotonoids content in Manila and Ataulfo mangoes determined by liquid chromatography-$APcI^+$-time of flight mass spectrometry. Postharv Biol Technol 50:145–152.

Osuna-Garcia JA, Wall MM and Waddell CA. 1998. Endogenous levels of tocopherols and ascorbic acid during fruit ripening of New Mexican-type chile (*Capsicum annuum*). J Agric Food Chem 46:5093–5096.

Park EH, Kahng JH and Paek EA. 1998. Studies on the pharmacological actions of cactus: identification of its anti-inflammatory effect. Arch Pharmaceutical Res 21(1):30–34, 63.

Patel B, Welch A, Bingham S, Luben R, Day N, Khaw K, Lomas D and Wareham N. 2006. Dietary antioxidants and asthma in adults. Thorax 61:388–393.

Pattison D, Symmons D, Lunt M, Welch A, Birngham S, Day N and Silman A. 2005. Dietary β-cryptoxanthin and inflammatory polyarthritis: results from population-based prospective study. Am J Clin Nutr 82:451–455.

Pattison D, Symmons D and Young A. 2004b. Does diet have a role in the aetiology of rheumatoid arthritis? Proc Nutr Soc 63:137–143.

Pendlington S, Dupont MS and Trussel FJ. 1965. The carotenoid composition of *Solanum tuberosum.* Biochem J 94:25–26.

Percival SS, Talcott ST, Chin ST, Mallak AC, Lound-Singleton A and Pettit-Moore J. 2006. Neoplastic transformation of BALB/3T3 cells and cell cycle of HL-60 cells are inhibited by mango (*Mangifera indica L.*) juice and mango juice extract. J Nutr 136:1300–1304.

Pinent M, Blay M, Bladé MC, Salvadó MJ, Arola L and Ardévol A. 2004. Grape seed-derived procyanidins have an antihyperglycemic effect in streptozotocin-induced diabetic rats and insulinomimetic activity in insulin-sensitive cell lines Endocrinology 145(11):4985–4990.

Plumb GW, Lamer N, Chambers SJ, Wanigatunga S, Heany RK, Plumb JA, Aruoma OI, Halliwell B, Miller NJ and Williamson G. 1996a. Are whole extracts and purified glucosinolates from cruciferous vegetables antioxidants? Free Rad Res 25:75–86.

Plumb GW, Chambers SJ, Lambert N, Bartolome B, Heaney RK, Wangatunga S, Aruoma OI, Halliwell B and Williamson G. 1996b. Antioxidant actions of fruit, herb and spice extracts. J Food Lipids 3:171–178.

Plumb GW, Chambers SJ, Lambert N, Wanigatunga S and Williamson G. 1997a. Influence of fruit and vegetable extracts on lipid peroxidation in microsomes containing specific cytochrome P450s. Food Chem 60:161–164.

Plumb GW, Price KR, Rhodes MJC and Williamson G. 1997b. Antioxidant properties of the major polyphenolic compounds in broccoli. Free Rad Res 27:429–435.

Prior RL, Cao G. 2000. Antioxidant phytochemicals in fruits and vegetables: diet and health implications. Hort Sci 35:588–592.

Prior RL, Cao G, Martin A, Sofic E, McEwen J, OBrien C, Lischner N, Ehlenfeldt M, Kalt W, Krewer G and Mainland CM. 1998. Antioxidant capacity as influenced by total phenolic and anthocyanidin content, maturity, and variety of Vaccinium species. J Agric Food chem. 46:2686–2693.

Pryne C, Mishra G, O'Connell MA, Muniz G, Laskey MA, Yan L, Prentice A and Ginty F. 2006. Fruit and vegetable intakes and bone mineral status: a cross-sectional study in 5 age and sex cohorts. Am J Clin Nutr 83:1420–1428.

Radhika G, Sudha V, Mohan Sathya R, Ganesan A and Mohan V. 2008. Association of fruit and vegetable intake with cardiovascular risk factors in urban south Indians. Br J Nutr 99(2):398–405.

Raicht RF, Cohen BI, Fazzini EP, Sarwal AN and Takahashi M. 1980. Protective effect of plant sterols against chemically induced colon tumors in rats. Cancer Res 40:403–405.

Ramarathnam N, Ochi H, Takeuchi M. 1997. Antioxidative defense system in vegetable extracts. In Shahidi F, editor. Natural Antioxidants. Chemistry, Health Effects, and Applications. Champaign, IL: AOCS Press, pp. 76–87.

Rechkemmer G. 2001. Funktionelle Lebensmittel-Zukunft de Ernahrung oder Marketing-Strategie. Forschungereport Sonderheft 1:12–15.

Reddy BS, Mori H and Nicolais M. 1981. Effect of dietary wheat bran and dehydrated citrus fiber on azoxymethane-induced intestinal carcinogenesis in Fischer 344 rats. J Natl Cancer Inst 66:553–557.

Ren Y, Waldron KW, Pacy JE and Ellis PR. 2001. Chemical and histochemical characterization of cell wall polysaccharides in almond seeds in relation to lipid bioavailability. In: Pfannhauser W, Fenwick GR and Khokhar S, editors. Biologically Active Phytochemicals in Foods. Cambridge, UK: Royal Soc Chem, pp. 448–452.

Retelny VS, Neuendorf A and Roth JL. 2008. Nutrition protocols for the prevention of cardiovascular disease. Nutr Clin Pract 23(5):468–476.

Ribaya-Mercado JD, Garmyn M, Gilchrest BA and Russel RM. 1995. Skin lycopene is destroyed preferentially over β-carotene during ultraviolet irradiation in humans. J Nutr 125:1854–1859.

Ricardo-da-Silva JM, Rosec JP, Bourzeix M and Heredia N. 1990. Separation and quantitative determination of grape and wine procyanidins by HPLC. J Sci Food Agric 53:85–92.

Rissanen TH, Voutilainen S, Virtanen JK, Venho B, Vanharanta M, Mursu J and Salonen JT. 2003. Low intake of fruit, berries and vegetables is associated with excess mortality in men: the Kuopio Ischaemic Heart Disease Risk Factor (KIHD) Study. J Nutr 133:199–204.

Rivera-Pastarna D, Yahia EM and Gonzalez-Aguilar G. 2009. Identification and quantification of carotenoids and phenolic compounds in papaya using mass spectroscopy. In preparation.

Robards K, Prenzler PD, Tucker G, Swatsitang P and Glover W. 1999. Phenolic compounds and their role in oxidative processes in fruits. Food Chem 66:401–436.

Rodriguez de Sotillo D, Hadley M, Holm ET. 1994. Phenolic in aqueous potato peel extract: extraction, identification and degradation. J Food Sci 59:649–651.

Rodriguez-Saona LE, Giusti MW and Wrolstad RE. 1998. Anthocyanin pigment composition of red-fleshed potatoes. J Food Sci 63:458–465.

Romieu I, Varraso R, Avenel V, Leynaert B, Kauffman F and Clavel-Chapelon F. 2006. Fruit and vegetable intakes and asthma in the E3N study. Thorax 61:209–215.

Rouseff RL and Nagy S. 1994. Health and nutritional benefits of citrus fruit components. Food Technol November 1994:125–130.

Rungapamestry V, Duncan AJ, Fuller Z and Ratcliffe B. 2007. Effect of cooking *Brassica* vegetables on the subsequent hydrolysis and metabolic fate of glucosinolates. Proc Nutr Soc 66(1):69–81.

Sang S, Kikuzaki H, Lapsley K, Rosen RT, Nakatani N and Ho CT. 2002. Sphingolipid and other constituents from almond nuts (*Prunus amygdalus* Batsch). J Agric Food chem. 50:4709–4712.

Santos CA and Simon PW. 2002. QTL analyses reveal clustered loci for accumulation of major provitamin A carotenes and lycopene in carrot roots. Mol Genet Genomics 268:122–129.

Sarkar D, Sharma A and Talukder G. 1994. Chlorophyll and chlorophyllin as modifiers of genotoxic effects. Mutat Res 318(3):239–247.

Sautter C, Poletti S, Zhang P and Gruissem W. 2006. Biofortification of essential nutritional compounds and trace elements in rice and cassava. Proc Nutr Soc 65:153–159.

Schauer N, Semel Y, Roessner U, Gur A, Balbo I, Carrari F, Pleban T, Perez-Melis A, Bruedigam C, Kopka J, Willmitzer L, Zamir D and Fernie AR. 2006. Comprehensive metabolic profiling and phenotyping of interspecific introgression lines for tomato improvement. Nat Biotechnol 24:447–454.

Schijlen E, Ric de Vos CH, Jonker H, van den Broeck H, Molthoff J, van Tunen A, Martens S and Bovy A. 2006. Pathway engineering for healthy phytochemicals leading to the production of novel flavonoids in tomato fruit. Plant Biotechnol J 4(4):433–444.

Schreiner M. 2005. Vegetable crop management strategies to increase the quantity of phytochemicals. Eur J Nutr 44:85–94.

Serrano M, Guillen F, Martinez-Romero D, Castillo S and Valero D. 2005. Chemical constituents and antioxidant activity of sweet cherry at different ripening stages. J Agric Food Chem 53:2741–2745.

Shannon J, Cook LS and Stanford JL. 2003. Dietary intake and risk of postmenopausal breast cancer (United States). Cancer Causes Control 14:19–27.

Shi J and Le Maguer M. 2000. Lycopene in tomatoes: chemical and physical properties affected by food processing. Crit Rev Food Sci Nutr 40(1):1–42.

Shukitt-Hale B, Carey A, Simon L, Mark D and Joseph J. 2006. Effect of Concord grape juice on cognitive motor deficits in aging. Nutrition 22:295–302.

Siddapa GS and Bhatia BS. 1954. The identification of sugars in fruits by paper chromatography. Indian J Hort 11:104.

Silaste M, Rantala M, Alftham G, Aro A and Kessaniemi Y. 2003. Plasma homocysteine concentration is decreased by dietary intervention. Br J Nutr 89:295–301.

Simmone AH, Simmone EH, Eitenmiller RR, Mills HA and Green NR. 1997. Ascorbic acid and provitamin A contents in some unusually colored bell peppers. J Food Compos Anal 10:299–311.

Sisini A. 1969. Sulla glucose-6-fosfato isomerasi in *Opuntia ficus-indica*. Boll Soc Ital Biol Sper 45:794–796.

Slade AJ and Knauf VC. 2005. TILLING moves beyond functional genomics into crop improvement. Transgenic Res 14:109–115.

Smith-Warner SA, Spiegelman D, Shiaw-Shyuan Y, Adami HO, Beeson WL, Brandt PA, Folsom AR, Fraser GE, Freudenheim JL, Goldbohm RA, Graham S, Miller AB, Potter JD, Rohan TE, Speizer FE, Toniolo P, Willet WC, Wolk A and Zeleniuch-Jacquotte A, Hunter DJ. 2001. Intake of fruits and vegetables and risk of breast cancer. A pooled analysis of cohort studies. J Am Med Assoc 285(6):769–776.

Somers TC and Ziemelis G. 1985. Spectral evaluation of total phenolic components in *Vitis vinifera*: grapes and wines. J Sci Food Agric 36:1275–1284.

Sondheimer E and Karash CB. 1956. The major anthocyanin pigments of the wild strawberry (*Fragaria vesca*). Nature 178:648–649.

Spanos GA, Worlstad RE and Heatherbell DA. 1990. Influence of processing and storage on the phenolic composition of apple juice. J Agric Food Chem 38:1572–1579.

Stahl W and Sies H. 1996. Lycopene: a biologically important carotenoid for humans? Arch Biochem Biophys 336:1–9.

Stark MJ, Burke YD, McKinzie JH, Ayoubi AS and Crowell PL. 1995. Chemotherapy of pancreatic cancer with the monoterpene perillyl alcohol. Cancer Lett 96:15–21.

Stavric B. 1994. Role of chemopreventers in human diet. Clin Biochem 27(5):319–332.

Steinmetz KA, Kushi LH, Bostick RM, Folsom AR and Potter JD. 1994. Vegetables, fruit and colon cancer in the Iowa Women's Health Study. Am J Epidemiol 139:1–15.

Steinmetz K and Potter J. 1991. Vegetables, fruit, and cancer. Cancer Causes Control 2:325–442.

Steinmetz KA and Potter JD. 1996. Vegetables, fruit and cancer prevention: a review. J Am Diet Assoc 96:1027–1039.

Steinmetz KA, Potter JD and Folsom AR. 1993. Vegetables, fruit, and lung cancer in the Iowa Women's Health Study. Cancer Res 53:536–543.

Stintzing FC, Schieber A and Carle R. 2002. Identification of betalains from yellowbeet (*beta vulgaris* L.) and cactus pear (*Opuntia ficus-indica* L. Mill.) by high performance liquid chromatography–electrospray ionization mass spectroscopy. J Agric Food Chem 50:2302–2307.

Stoewsand GS. 1995. Bioactive organosulfur phytochemicals in *Brassica oleracea* vegetables. A review. Food Chem Toxicol 33(6):537–543.

Stram DO, Hankin JH, Wilkens LR, Park S, Henderson BE, Nomura AM, Pike MC and Kolonel LN. 2006. Prostate cancer incidence and intake of fruits, vegetables and related micronutrients: the multiethnic cohort study (United States). Cancer Causes Control 17(9):1193–1207.

Sugie S, Okumura A, Tanaka T and Mori H. 1993. Inhibitory effects of benzylisothiocyanate and benzylthiocyanate on diethylnitrosamine-induced hepatocarcinogenesis in rats. Jpn J Cancer Res 84:865–870.

Sun J, Chu Y-F, Wu X and Lui RH. 2002. Antioxidant and antiproliferative activities of common fruits. J Agric Food Chem 50:6910–6916.

Suzuki S, Takada T, Sugawara Y, Muto T and Kominami R. 1991. Quercetin induces recombinational mutations in cultured cells as detected by DNA fingerprinting. Jpn J Cancer Res 82(10):1061–1064.

Svendsen M, Blomhoff R, Holme I, Tonstad S. 2007. The effect of an increased intake of vegetables and fruit on weight loss, blood pressure and antioxidant defense in subjects with sleep related breathing disorders. Eur J Clin Nutr 61(11):1301–1311.

Syngletary KW, Jackson SJ, Milner JA. 2005. Non-nutritive components in foods as modifiers of the cancer process. In: Bendich A and Deckelbaum RJ, editors. Preventive Nutrition: the Comprehensive Guide for Health Professionals, 3rd ed. Totowa, NJ: Humana Press.

Tabak C, Wijga A, de Meer G, Janssen N, Brunekreef B and Smith H. 2006. Diet and asthma in Dutch school children (ISAAC-2). Thorax 61(12):1048–1053.

Takahashi Y, Nagao M, Fujino T, Yamaizumi Z and Sugimura T. 1979. Mutagens in Japanese pickle identified as flavonoids. Mut Res 68:117–123.

Tanaka T, Mori Y, Morishita Y, Hara A, Ohno T, Kojinna T and Mori H. 1990. Inhibitory effect of sinigrin and indole-3-carbinol on diethylenitrosamine-induced hepatocarcinogenesis in male AC/N rats. Carcinogenesis 11:1403–1406.

Tango JS, Dacosta ST, Antunes AJ and Figueriedo IB. 1972. Composition of fruit oil of different varieties of avocados grown in Sao Paulo. Fruits 27:143–146.

Terry P, Giovannucci E, Michels KB, Bergkvist L, Hansen H, Holmberg L and Wolk A. 2001. Fruit, vegetables, dietary fiber, and risk of colorectal cancer. J Natl Cancer Inst 93(7):525–533.

Tesoriere L, Butera D, Pintaudi AM, Allegra M and Livrea MA. 2004. Supplementation with cactus pear (*Opuntia ficus-indica*) fruit decreases oxidative stress in healthy humans: a comparative study with vitamin C. Am J Clin Nutr 80:391–395.

Thomas P. 1975. Effect of postharvest temperatures on quality, carotenoids and ascorbic acid contents in Alphonso mangos on ripening. J Food Sci 40:704–706.

Thompson RS, Jaques D, Haslam E and Tanner RJN. 1972. Plant proanthocyanidins. Part 1. J Chem Soc Perkin Trans 1:1837–1841.

Tsugani S, Tsuda M, Gey F and Watanabe S. 1992. Cross-sectional study with multiple measurements of biological markers for assessing stomach cancer risks at the population level. Environ Health Perspect 98:207–210.

Turnball WH, Walton J and Leeds AR. 1993. Acute effects of mycoprotein on subsequent energy intake and appetite variables. Am J Clin Nutr 58(4):507–512.

Tussedre PL, Frankel EN, Waterhouse AL, Peleg H and German JB. 1996. Inhibition of *in vitro* human LDL oxidation by phenolic antioxidants from grapes and wines. J Sci Food Agric 70:55–61.

Tylavsky F, Holliday K, Danish R, Womack C, Norwood J and Carbone L. 2004. Fruit and vegetable intakes are an independent predictor of bone size in early pubertal children. Am J Clin Nutr 79:311–317.

USDA. 1991. Composition of Foods: Fruits and Fruit Products—Raw, Processed, Prepared. US Dept Agr Handb 8–9, 283 pp.

USDA Economic Research Service. 2002. Food consumption per capita. www.ers.usda.gov/data/foodconsumption/

VanEenwyk J, Davis FG and Bowen PE. 1991. Dietary and serum carotenoids and cervical intraepithelial neoplasia. Int J Cancer 48:34–38.

Van Gils C, Peeters PH, Bueno-de-Mesquita HB, Boshuizen HC, Lahmann PH, Clavel-Chapelon F, Thiebaut A, Kesse E, Sieri S, Palli D, Tumino R, Panico S, Vineis P, Gonzalez CA, Ardanaz E, Sanchez MJ, Amiano P, Navarro C, Quiros JR, Key TJ, Allen N, Khaw KT, Bingham SA, Psaltopoulou T, Koliva M, Trichopoulou A, Nagel G, Linseisen J, Boeing H, Berglund G, Wirfalt E, Hallmans G, Lenner P, Overvad

K, Tjonneland A, Olsen A, Lund E, Engeset D, Alsaker E, Norat TA, Kaaks R, Slimani N and Riboli E. 2005. Consumption of vegetables and fruit and risk of breast cancer. JAMA 293:183–193.
Vatanparast H, Baxter-Jones A, Faulkner R, Baile D and Whiting S. 2005. Positive effects of vegetable and fruits consumption and calcium intake on bone mineral accrual in boys during growth from childhood to adolescence: the University of Saskatchewan Pediatric Bone Mineral Accrual Study. Am J Clin Nutr 82:700–706.
Velioglu YS, Mazza G, Gao L and Oomah DB. 1998. Antioxidant activity and total phenolics in selected fruits, vegetables, and grain products. J Agric Food Chem 46:4113–4117.
Verhagen H, Poulsen HE, Loft S, Van-Poppel G, Willems MI and Van-Bladeren PJ. 1995. Reduction of oxidative DNA-damage in humans by Brussels sprouts. Carcinogenesis 16(4):969–970.
Villegas R, Shu XO, Gao YT, Yang G, Elasy T, Li H and Zheng W. 2008. Vegetable but not fruit consumption reduces the risk of type 2 diabetes in Chinese women. J Nutr 138(3):574–580.
Vinson JA, Hao Y, Su X and Zubik L. 1998. Phenol antioxidant quantity and quality in foods: vegetables. J Agric Food Chem 46:3630–3634.
Vioque J, Weinbrenner T, Castelló A, Asensio L and Garcia de la Hera M. 2008. Intake of fruits and vegetables in relation to 10-year weight gain among Spanish adults. Obesity (Silver Spring) 16(3):664–670.
Voorips LE, Goldbohm RA, van Poppel G, Sturmans F, Hermus RJ and van den Brandt PA. 2000. Vegetable and fruit consumption and risk of colon and rectal cancer in a prospective cohort study. Am J Epidemiol 152:1081–1092.
Walda I, Tabak C, Smith H, Rasanen L, Fidanza F, Menotti A, Nissinen A, Feskens E and Kromhout D. 2002. Diet and 20-year chronic obstructive pulmonary disease mortality in middle-aged men from three European countries. Eur J Clin Nutr 2002:638–643.
Wang H, Cao G and Prior RL. 1996. Total antioxidant capacity of fruits. J Agric Food Chem 44:701–705.
Wang SY, Feng R, Lu Y, Bowman L and Ding M. 2005. Inhibitory effect on activator protein-1, nuclear factor-kappab, and cell transformation by extracts of strawberries (fragaria × *ananassa* duch.). J Agric Food Chem 53:4187–4193.
Wang SY and Jiao HJ. 2000. Scavenging capacity of berry crops on superoxide radicals, hydrogen peroxide, hydroxyl radicals, and singlet oxygen. J Agric Food Chem 48:5677–5684.
Wang SY and Lin HS. 2000. Antioxidant activity in fruit and leaves of blackberry, raspberry, and strawberry is affected by cultivar and maturity. J Agric Food Chem 48:140–146.
Wargovich MJ. 2000. Anticancer properties of fruits and vegetables. HortSci 35(4):573–575.
Watson L, Margetts B, Howarth P, Dorwarth M, Thompson R and Little P. 2002. The association between diet and chronic obstructive pulmonary disease in subjects selected from general practice. Eur Respir J 20:313–318.
Wattenberg LW. 1975. Effects of dietary constituents on the metabolism of chemical carcinogens. Cancer Res 35(11):3326–3331.
Wattenberg LW and Coccia JB. 1991. Inhibition of 4-(methylnitrosamino)-1-(3-pyridyl)butanone carcinogenesis in mice by D-limonene and citrus fruit oil. Carcinogenesis 12:115–117.
Wattenberg LW and Loub WD. 1978. Inhibition of polycyclic aromatic hydrocarbon–induced neoplasia by naturally occurring indoles. Cancer Res 38:1410–1413.
Whanger PD, Ip C, Polan CE, Uden PC and Welbaum G. 2000. Tumorigenesis, metabolism, speciation, bioavailability, and tissue deposition of selenium in selenium-enriched ramps (*Allium tricoccum*). J Agric Food Chem 48(11):5723–5730.
Whiting GC and Coggins RA. 1975a. 4-*p*-Coumaroyl quinic acid in apple fruits. Phytochemistry 14:593–597.
Whiting GC and Coggins RA. 1975b. Estimation of the monomeric phenolics of ciders. J Sci Food Agric 26:1833–1839.
William R, Spencer J and Rice-Evans C. 2004. Flavonoids: antioxidants or signaling molecule? Free Radic Biol Med 36:838–849.
Wolfe K, Wu X and Liu RH. 2003. Antioxidant activity of apple peels. J Agric Food Chem 51:609–614.
World Cancer Research Fund Food. 1997. Nutrition and the Prevention of Cancer: A Global Perspective. Washington DC: American Institute for Cancer Research.
Wright ME, Park Y, Subar AF, Freedman ND, Albanes D, Hollenbeck A, Leitzmann MF and Schatzkin A. 2008. Intakes of fruit, vegetables, and specific botanical groups in relation to lung cancer risk in the NIH-AARP Diet and Health Study. Am J Epidemiol 168(9):1024–1034.

Yahia EM. 2006. Handling tropical fruits. In Scientists Speak. Alexandria, VA: World Foods Logistics Organization, pp. 5–9.
Yahia EM. 2009a. Prickly pear. Chapter 13. In: Rees D, Farrell G, Orchard JE, editors. Crop Postharvest: Science and Technology, Volume 3. Oxford: Wiley-Blackwell. In press.
Yahia EM. 2009b. Avocado. Chapter 8. In: Rees D, Farrell G, Orchard JE, editors. Crop Postharvest: Science and Technology, Volume 3. Oxford: Wiley-Blackwell. In press.
Yahia EM and Brecht JK. 2009. Tomato. Chapter 2. In: Rees D, Farrell G, Orchard JE, editors. Crop Postharvest: Science and Technology, Volume 3. Oxford: Wiley-Blackwell. In press.
Yahia EM, Contreras M and Gonzalez G. 2001b. Ascorbic acid content in relation to ascorbic acid oxidase activity and polyamine content in tomato and bell pepper fruits during development, maturation and senescence. Lebensm Wiss u-Technol 34:452–457.
Yahia EM, Ornelas-Paz JJ and Gardea A. 2006. Extraction, separation and partial identification of Ataulfo mango fruit carotenoids. Acta Hortic 712:333–338.
Yahia EM, Soto G, Puga V and Steta M. 2001a. Hot air treatment effect on the postharvest quality and ascorbic acid content in tomato fruit. In: Artes F, Gil MI and Conesa MA, editors. Improving Postharvest Technologies of Fruits, Vegetables and Ornamentals, Volume 2. Paris: International Institute of Refrigeration, pp. 550–556.
Yang J, Meyers KJ, van der Heide J and Liu RH. 2004. Varietal differences in phenolic content and antioxidant and antiproliferative activities of onions. J Agric Food Chem 52:6787–6793.
Ye H, Lozada M and West B. 2005. Diverticulosis coli: Update on a "Western" disease. Adv Anat Pathol 12:74–80.
Ye X, Al-Babili S, Kloti A, Zhang J, Lucca P, Beyer P, Potrykus I. 2000. Engineering the provitamin A (beta-carotene) biosynthetic pathway into (carotenoid-free) rice endosperm. Science 287:303–305.
You WC, Blot WJ, Chang YS, Ershow AG, Yang ZT, An Q, Henderson B, Xu WG, Fraumeni JF and Wang TG. 1989. *Allium* vegetables and reduced risk of stomach cancer. J Natl Cancer Inst 81:162–164.
Zhang Y and Talalay P. 1994. Anticarinogenic activities of organic isothiocyanate: chemistry and mechanisms. Cancer Res 54:1976s–1981s.
Zhang LX, Cooney RV and Bertram JS. 1992. Carotenoids up-regulate connexin-43 gene expression independent of their provitamin-A or antioxidant properties. Cancer Res 52:5707–5712.
Ziegler RG, Blot WJ, Hoover R, Blattner WA and Fraumeni JF Jr. 1981. Protocol for a study of nutritional factors and the low risk of colon cancer in Southern retirement areas. Cancer Res 41:3724–3726.

2 Phenolic Compounds: Chemistry and Occurrence in Fruits and Vegetables

Cristina Andrés-Lacueva*, Alex Medina-Remon, Rafael Llorach, Mireia Urpi-Sarda, Nasiruddin Khan, Gemma Chiva-Blanch, Raul Zamora-Ros, Maria Rotches-Ribalta, and Rosa M. Lamuela-Raventós

Chemistry and Classification of Polyphenols

Polyphenols are the most abundant antioxidants in human diets. They are secondary metabolites of plants. These compounds are designed with an aromatic ring carrying one or more hydroxyl moieties. Several classes can be considered according to the number of phenol rings and to the structural elements that bind these rings.

In this context, two main groups of polyphenols, termed flavonoids and nonflavonoids, have been traditionally adopted. As seen in Figure 2.1, the flavonoids group comprises the compounds with a C6-C3-C6 structure: flavanones, flavones, dihydroflavonols, flavonols, flavan-3-ols, anthocyanidins, isoflavones, and proanthocyanidins. The nonflavonoids group is classified according to the number of carbons that they have (Fig. 2.2) and comprises the following subgroups: simple phenols, benzoic acids, hydrolyzable tannins, acetophenones and phenylacetic acids, cinnamic acids, coumarins, benzophenones, xanthones, stilbenes, chalcones, lignans, and secoiridoids.

Flavonoids

Flavonoids have a skeleton of diphenylpropanes, two benzene rings (A and B) connected by a three-carbon chain forming a closed pyran ring with the benzene A ring (see Fig. 2.1).

Flavonoids in plants usually occur glycosylated mainly with glucose or rhamnose, but they can also be linked with galactose, arabinose, xylose, glucuronic acid, or other sugars. The number of glycosyl moieties usually varies from one to three; nevertheless, flavonoids have been identified with four and also five moieties (Vallejo and others 2004).

Flavonols and *flavones* have a double bond between C2 and C3 in the flavonoid structure and an oxygen atom at the C4 position. Furthermore, flavonols also have a hydroxyl group at the C3 position. *Dihydroflavonols* have the same structure as flavonols without the double bond between C2 and C3.

Flavanones are represented by the saturated three-carbon chain and an oxygen atom in the C4 position.

* Corresponding author: Cristina Andrés-Lacueva. Associate Professor, Nutrition and Food Science Department, Pharmacy School, University of Barcelona. Av Joan XXIII s/n, Barcelona 08028 (Spain).

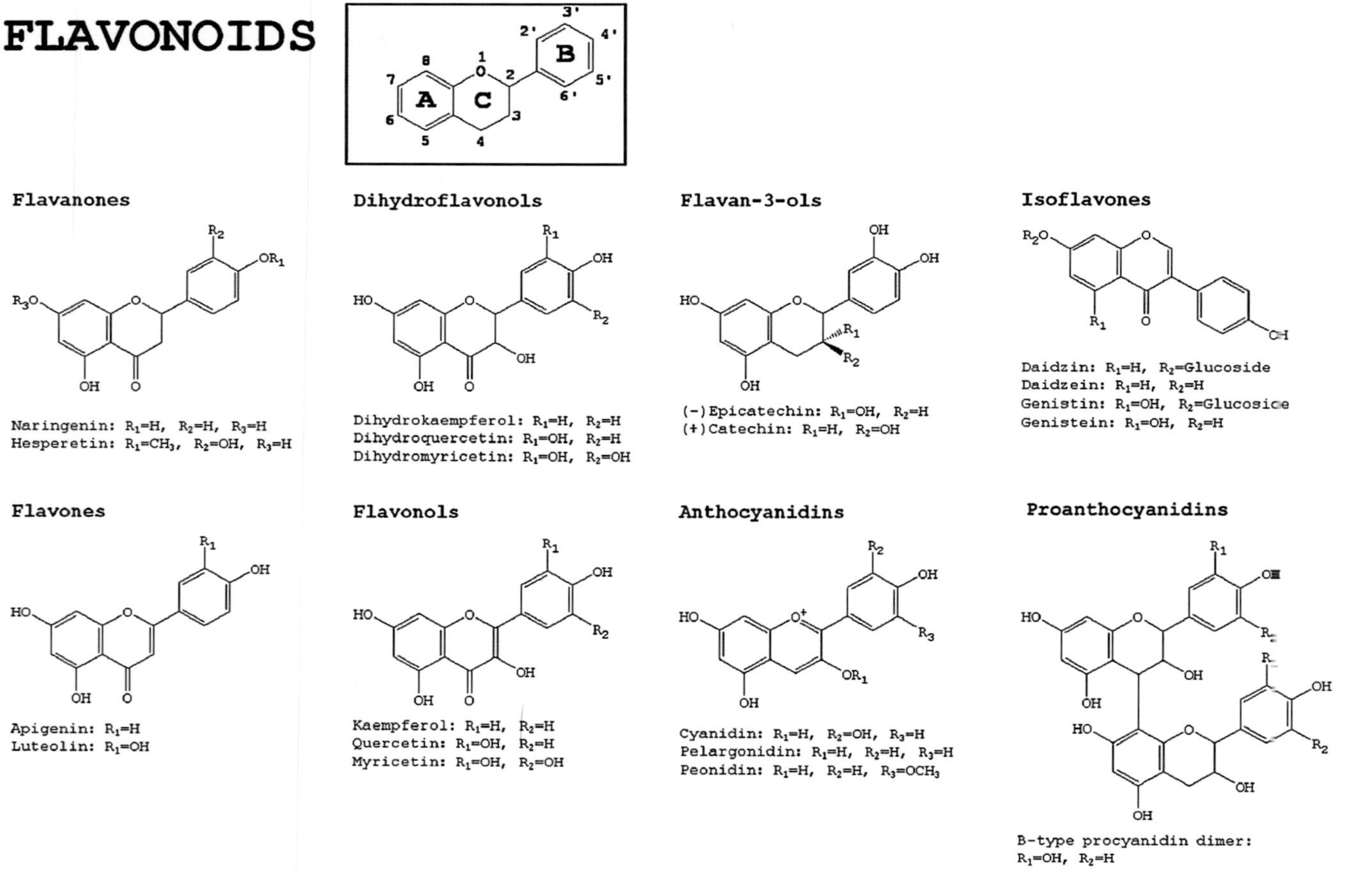

FLAVONOIDS

Flavanones

Naringenin: R_1=H, R_2=H, R_3=H
Hesperetin: R_1=CH_3, R_2=OH, R_3=H

Dihydroflavonols

Dihydrokaempferol: R_1=H, R_2=H
Dihydroquercetin: R_1=OH, R_2=H
Dihydromyricetin: R_1=OH, R_2=OH

Flavan-3-ols

(−)Epicatechin: R_1=OH, R_2=H
(+)Catechin: R_1=H, R_2=OH

Isoflavones

Daidzin: R_1=H, R_2=Glucoside
Daidzein: R_1=H, R_2=H
Genistin: R_1=OH, R_2=Glucoside
Genistein: R_1=OH, R_2=H

Flavones

Apigenin: R_1=H
Luteolin: R_1=OH

Flavonols

Kaempferol: R_1=H, R_2=H
Quercetin: R_1=OH, R_2=H
Myricetin: R_1=OH, R_2=OH

Anthocyanidins

Cyanidin: R_1=H, R_2=OH, R_3=H
Pelargonidin: R_1=H, R_2=H, R_3=H
Peonidin: R_1=H, R_2=H, R_3=OCH_3

Proanthocyanidins

B-type procyanidin dimer:
R_1=OH, R_2=H

Figure 2.1. Chemical structures of flavonoids.

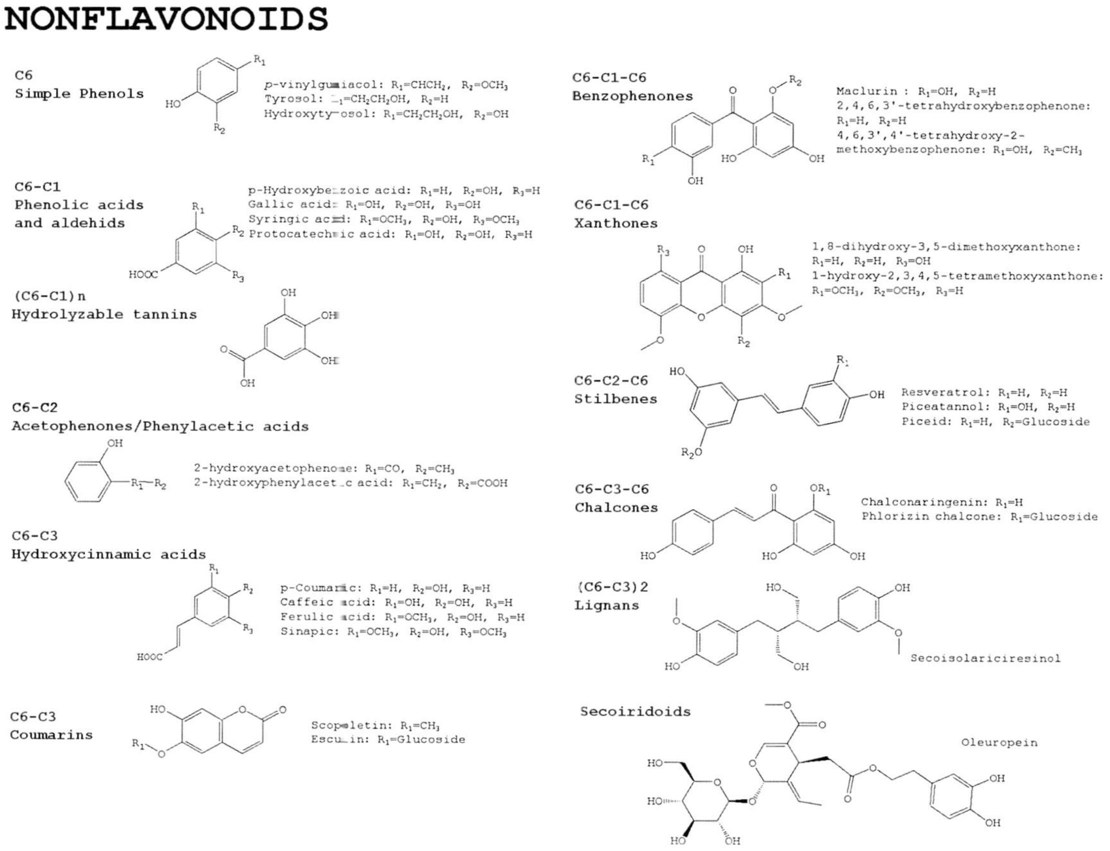

NONFLAVONOIDS

C6
Simple Phenols

p-vinylguaiacol: R_1=CHCH$_2$, R_2=OCH$_3$
Tyrosol: R_1=CH$_2$CH$_2$OH, R_2=H
Hydroxytyrosol: R_1=CH$_2$CH$_2$OH, R_2=OH

C6-C1
Phenolic acids and aldehids

p-Hydroxybenzoic acid: R_1=H, R_2=OH, R_3=H
Gallic acid: R_1=OH, R_2=OH, R_3=OH
Syringic acid: R_1=OCH$_3$, R_2=OH, R_3=OCH$_3$
Protocatechuic acid: R_1=OH, R_2=OH, R_3=H

(C6-C1)n
Hydrolyzable tannins

C6-C2
Acetophenones/Phenylacetic acids

2-hydroxyacetophenone: R_1=CO, R_2=CH$_3$
2-hydroxyphenylacetic acid: R_1=CH$_2$, R_2=COOH

C6-C3
Hydroxycinnamic acids

p-Coumaric: R_1=H, R_2=OH, R_3=H
Caffeic acid: R_1=OH, R_2=OH, R_3=H
Ferulic acid: R_1=OCH$_3$, R_2=OH, R_3=H
Sinapic: R_1=OCH$_3$, R_2=OH, R_3=OCH$_3$

C6-C3
Coumarins

Scopoletin: R_1=CH$_3$
Esculin: R_1=Glucoside

C6-C1-C6
Benzophenones

Maclurin : R_1=OH, R_2=H
2,4,6,3'-tetrahydroxybenzophenone: R_1=H, R_2=H
4,6,3',4'-tetrahydroxy-2-methoxybenzophenone: R_1=OH, R_2=CH$_3$

C6-C1-C6
Xanthones

1,8-dihydroxy-3,5-dimethoxyxanthone: R_1=H, R_2=H, R_3=OH
1-hydroxy-2,3,4,5-tetramethoxyxanthone: R_1=OCH$_3$, R_2=OCH$_3$, R_3=H

C6-C2-C6
Stilbenes

Resveratrol: R_1=H, R_2=H
Piceatannol: R_1=OH, R_2=H
Piceid: R_1=H, R_2=Glucoside

C6-C3-C6
Chalcones

Chalconaringenin: R_1=H
Phlorizin chalcone: R_1=Glucoside

(C6-C3)2
Lignans

Secoisolariciresinol

Secoiridoids

Oleuropein

Figure 2.2. Chemical structures of nonflavonoids.

Isoflavones also have a diphenylpropane structure in which the B ring is located in the C3 position. They have structural analogies to estrogens, such as estradiol, with hydroxyl groups at the C7 and C4 positions (Shier and others 2001).

Anthocyanins are based on the flavylium salt structure and are water soluble pigments in plants. They are found in the form of glycosides in plants and foods of their respective aglycones, called *anthocyanidins*. The most common sugars encountered are glucose, galactose, rhamnose, xylose, arabinose, and fructose, which are linked mainly in the C3 position as glycosides and in C3, C5 as diglycosides. Glycosylation at the C7, C3′, and C5′ positions has also been observed (Clifford 2000a).

Flavan-3-ols or *flavanols* have a saturated three-carbon chain with a hydroxyl group in the C3 position. In foods they are present as monomers or as *proanthocyanidins*, which are polymeric flavanols (4 to 11 units) known also as condensed tannins. In foods they are never glycosylated.

Nonflavonoids

Simple phenols (C6), the simplest group, are formed with an aromatic ring substituted by an alcohol in one or more positions as they may have some substituent groups, such as alcoholic chains, in their structure. *Phenolic acids* (C6-C1) with the same structure as simple phenols have a carboxyl group linked to benzene. *Hydrolyzable tannins* are mainly glucose esters of gallic acid. Two types are known: the gallotannins, which yield only gallic acid upon hydrolysis, and the ellagitannins, which produce ellagic acid as the common degradation product.

Acetophenones are aromatic ketones, and *phenylacetic acids* have a chain of acetic acid linked to benzene. Both have a C6-C2 structure.

Hydroxycinnamic acids are included in the phenylpropanoid group (C6-C3). They are formed with an aromatic ring and a three-carbon chain. There are four basic structures: the coumaric acids, caffeic acids, ferulic acids, and sinapic acids. In nature, they are usually associated with other compounds such as chlorogenic acid, which is the link between caffeic acid and quinic acid.

Coumarins belong to a group of compounds known as the benzopyrones, all of which consist of a benzene ring joined to a pyrone. They may also be found in nature, in combination with sugars, as glycosides. They can be categorized as simple furanocoumarins, pyranocoumarins, and coumarins substituted in the pyrone ring (Murray and others 1982).

Benzophenones and *xanthones* have the C6-C1-C6 structure. The basic structure of benzophenone is a diphenyl ketone, and that of xanthone is a 10-oxy-10*H*-9-oxaanthracene. More than 500 xanthones are currently known to exist in nature, and approximately 50 of them are found in the mangosteen with prenyl substituents.

Stilbenes have a 1,2-diphenylethylene as their basic structure (C6-C2-C6). Resveratrol, the most widely known compound, contains three hydroxyl groups in the basic structure and is called 3,4′,5-trihydroxystilbene. In plants, piceid, the glucoside of resveratrol, is the major derivative of resveratrol. Stilbenes are present in plants as *cis* or *trans* isomers. *Trans* forms can be isomerized to *cis* forms by UV radiation (Lamuela-Raventós and others 1995).

Chalcones with a C6-C3-C6 structure are flavonoids lacking a heterocyclic C-ring. Generally, plants do not accumulate chalcones. After its formation, naringenin chalcone is rapidly isomerized by the enzyme chalcone isomerase to form the flavanone, naringenin. The most common chalcones found in foods are phloretin and its 2′-*O*-glucoside, chalconaringenin and arbutin.

Lignans are compounds derived from two β-β′-linked phenylpropanoid (C6-C3) units and are widely distributed in the plant kingdom. They are classified into eight subgroups: furofuran, furan, dibenzylbutane, dibenzylbutyrolactone, aryltetralin, arylnaphthalene, dibenzocyclooctadiene, and dibenzylbutyrolactol. These subgroups are based upon the way in which oxygen is incorporated into the skeleton and the cyclization pattern. Furthermore, they vary largely in the oxidation levels of both the aromatic rings and the propyl side chains.

Secoiridoids are complex phenols produced from the secondary metabolism of terpenes as precursors of several indole alkaloids (Soler-Rivas and others 2000). They are characterized by the presence of elenolic acid, in its glucosidic or aglyconic form, in their molecular structure. Oleuropein, the best-known secoiridoid, is a heterosidic ester of elenolic acid and 3,4- dihydroxyphenylethanol containing a molecule of glucose, the hydrolysis of which yields elenolic acid and hydroxytyrosol (Soler-Rivas and others 2000).

Methods of Identification and Quantification

Sample Handling

In order to obtain a correct phenolic fingerprint of fruit and vegetables, it is necessary to take into account the complexity and variability of their matrix. This can be grouped into two kinds of factors: the physical and structural aspects and the biological aspects. The first factor is evident when comparing soft fruits, such as berries or grapes, with other fruits such as oranges or apples, and also when comparing soft vegetables, such as lettuce or watercress, with hard vegetables such as carrot or pumpkin. The structural fragility is apparent in their susceptibility to mechanical damage during handling or manipulation, which is frequently necessary before extraction. Mechanical damage could trigger enzymatic reactions related to the browning that is the consequence of the transformation of phenolic compounds to melanins. Two enzymes, such as polyphenol oxidases (PPO) and peroxidases (POD), are considered important factors in the process of phenolic oxidation (Tomás-Barberán and Espín 2001). Processes such as peeling, cutting, or crushing, which are fundamental to facilitating the correct extraction of phenolics, might cause alterations that later lead to incorrect identification and quantification of phenolic compounds. Enzymatic inactivation could be achieved using heated solvents, lowering the pH, adding the chemicals, and using a high level of organic solvents. In this context, Arts and Hollman (1998) observed a browning in the extract obtained from apple and grape with concentrations of methanol below 40%. Additionally, this low methanol concentration showed a decrease of ~70% of catechin yield that the authors attributed to the effect of PPO on the phenolic content.

In order to preserve, as much as possible, the phenolic content in fruit and vegetable samples, the literature proposed the application of cold temperatures, even reaching to freezing, when lyophilization is the objective. These procedures also could inactivate the enzymes. The freeze-drying is largely the main preservation technique used in the studies related to the identification and quantification of the phenolic compounds of fruit and vegetables. Asami and others (2003) found that the total phenolic content of freeze-dried samples of marionberries and strawberries was higher than that of the air-dried samples. In this context, frozen samples should be thawed before phenolic extraction. This procedure could provoke a loss of phenolic compounds because some of them show important thermolability, and also the thawing could provoke the activation of the enzyme that alters the phenolic content. Among the different techniques proposed for thawing, the microwave seems to be the most effective. A recent work (Oszmianski and others 2009) looking into the effect of freeze-thaw treatment on the polyphenol content of frozen strawberries showed that using a microwave thawing for 5 minutes instead of 20 hr at 20°C had some protective effect on many polyphenolic compounds, such as anthocyanins or ellagic acid.

Extraction

The methods of extraction need to be able to produce a correct *photograph* or *fingerprint* of the real phenolic content of samples. In fact, the extraction conditions must be as mild as possible to avoid modifications. In this context several factors, such as the complexity of the matrix, the variation of the solubility of phenolic compounds, the presence of interfering substances, and the time of extraction, as well as the temperature, could provoke modifications. Additionally, the fruit or vegetable phenolic profile is a mixture that, from the qualitative point of view, varies from lower mass phenolics, such as gallic acid, to highly polymerized phenolics, such as procyanidins and tannins. Also, from the quantitative point of view, the profile varies from traces to several hundreds of milligrams. Some such phenolics have even been detected uniquely in one fruit or vegetable, which makes them exclusive. Examples of these exclusive phenolics are dihydrochalcones (e.g., phloridzin, a characteristic phenolic compound from apple and its derivatives; Tomás-Barberán and Clifford 2000), or isoflavones, such as genistein and daidzein, which are restricted to the Fabaceae (e.g., soya; Cassidy and others 2000). All of these factors show that a unique solvent and/or method for total phenolic extraction does not exist. Usually, the methods of extraction involve a number of extraction steps with two or more different solvents, or a single extraction step with a mix of organic solvent with water. After this, further steps are required to evaporate, concentrate, and possibly purify.

Several extraction methods or techniques have been proposed for phenolic extraction. Often, freeze-dried and frozen samples are subjected to milling, grinding, and homogenization, which facilitate the solvent–compounds connection. Recently, extraction protocols have been extensively reviewed (Tura and Robards 2002; Stalikas 2007). For solid samples (e.g., unspoiled fruit and vegetables) the most frequent methods are based on the solid-liquid extraction process and include Soxhlet, sonication, solid-phase extraction (Hernández-Montes and others 2006), supercritical fluid extraction, and microwave. Devanand (2006) found significant differences in chlorogenic acid extraction from freeze-dried samples of eggplant using sonication, stirring, a shaker and rotator shaker, reflux, and a pressurized liquid extractor. Herrera and Luque de Castro (2005) applied ultrasound-assisted extraction, subcritical water, and microwave-assisted extraction to extract the phenolic compounds from strawberries. The ultrasound-assisted extraction was much the fastest and produced less loss of analytes than the other methods. Several studies have shown the effective use of solid-phase extraction as a method to extract the phenolic compounds from raw plant extracts or even biological samples (Michalkiewicz and others 2008; Mezadri and others 2008; Xia and others 2007; Chen and others 2001; Suárez and others 1996).

According to the literature, the most common solvents are water (mainly hot water), methanol, ethanol, acetone, and ethyl acetate. It is also common to use a mixture of water and organic solvents. Zhao and Hall (2008) studied the influence of the different solvents (water, ethanol, and acetone), and mixtures of them, on the extraction of phenolic compounds from raisins. The authors showed that the highest total phenolic content was achieved with the extracts obtained from solvent-to-water ratios of 60:40 (v/v), this being the extract obtained from ethanol:water (60:40, v/v,) which yielded the highest total phenolic content. On the other hand, Mané and others (2007) found that an acidified mixture of acetone–water–methanol was the best solvent for the simultaneous extraction of major polyphenol groups from different grape parts, including grape skin, pulp, and seeds.

Other important factors that could lead to error during the identification or quantification of phenolic compounds are the possible artifacts related to the isomerization, hydrolysis, and oxidation produced during the extraction. With regard to these factors, light could cause isomerization, which is termed photoisomerization. An example of this is the *trans-cis* photoisomerization of resveratrol. Resveratrol and piceid (resveratrol glucoside) are stilbenes present in grape products, where mainly *trans* isomers are detected. The effect of the UV causes a conversion to the cis isomers. In fact, within 10 minutes of exposure to sunlight of standard solutions, ~90% of isomerization is achieved (Romero-Pérez and others 1999). The use of a laboratory with UV-filtered light for sample preparation and extraction has been proposed to avoid light degradation (Teow and others 2007). Likewise, some caffeoylquinic derivatives could undergo isomerization in warm aqueous media. The extraction of artichoke by-products with boiling water has been linked to the appearance of different isomers such as 1,3-*O*-dicaffeoylquinic acid (Llorach and others 2003). In this context, the temperature of extraction and the time of extraction have been linked to both a positive effect, due to an increase in the solubility that could facilitate the extraction from the matrix, and a negative effect, due to the higher temperature that could provoke either a loss or a transformation of phenolic compounds. Regarding the oxidation, some chemicals, such as *tert*-butylhydroquinone, 2,6-di-*tert*-butyl-4-methyphenol (BHT), ascorbic acid, or sulfites, have been proposed as preventers of oxidation during phenolic extraction (Escribano-Bailón and Santos-Buelga 2003). However, Bradshaw and others (2001) found that ascorbic acid plays a role in inducing browning in catechin.

Separation

Over the past two decades, capillary electrophoresis (CE) and related techniques have rapidly developed for the separation of a wide range of analytes, ranging from large protein molecules to small inorganic ions. Gas chromatography has been considered as a powerful tool due to its sensitivity and selectivity, especially when coupled with mass spectrometry. Nevertheless, liquid chromatography is the most used method to separate and analyze phenolic compounds in plant and tissue samples.

Liquid chromatography is carried out in columns. The common columns are packed using reversed-phase C_{18}-bonded silica gel as stationary phase. Elution systems are usually binary, with one of the solvents being an acidified aqueous solvent. The second is an organic solvent (e.g., methanol), acidified with the same acid used in the aqueous solvent (Ibern-Gómez and others 2002). Usually, it includes gradient elution and, occasionally, isocratic elution. Other aspects, such as the pH of chromatography, or the buffer, if used, could drastically affect the separation of phenolic compounds and also further ionization.

Mass Spectrometry Methods for Identification and Quantification

Mass spectrometry has become a very important technique in the identification and quantification of phenolics in fruit and vegetables. Different factors, such as sensitivity and specificity, have been cited to explain the acceptance of this method by the scientific community. Additionally, this technique might easily combine with different separation techniques such as CE, gas chromatography (GC), and liquid chromatography (LC), including HPLC and UPLC (ultra performance liquid chromatography).

An important part of the mass spectrometry methods is the ionization sources and the analyzers. Several ionization sources are available. Among these, electrospray ionization (ESI) is used the most because of the wide range of molecules that it covers. Other important sources are those related to atmospheric pressure ionization, including atmospheric-pressure photoionization (APPI) and atmospheric-pressure chemical ionization (APCI), that have shown interesting results analyzing flavonoids (de Rijke and others 2003). Additionally, the molecules could be ionized by a loss of protons (negative ionization) or by a gain of protons (positive ionization). Both methods have been applied in studies related to phenolic analysis. Other sources of ionization exist, such as fast atom bombardment (FAB), which has been applied successfully in the study of flavonoid glycosides (Stobiecki 2000). In addition, matrix-assisted laser desorption ionization (MALDI) has also been used for the study of polyphenol composition (Prasain and others 2004).

Various analyzers have been used to analyze phenolic compounds. The choice of the MS analyzer is influenced by the main objective of the study. The triple quadrupole (QqQ) has been used to quantify, applying multiple reaction monitoring experiments, whereas the ion trap has been used for both identification and structure elucidation of phenolic compounds. Moreover, time-of-flight (TOF) and Fourier-transform ion cyclotron resonance (FT-ICR) are mainly recommended for studies focused on obtaining accurate mass measurements with errors below 5 ppm and sub-ppm errors, respectively (Werner and others 2008). Nowadays, hybrid equipment also exists, including different ionization sources with different analyzers, for instance electrospray or atmospheric pressure chemical ionization with triple quadrupole and time-of-flight (Waridel and others 2001).

Mass spectrometry applied to characterization of phenolic compounds has been widely reviewed (Fulcrand and others 2008; Harnly and others 2007; de Rijke and others 2006; Prasain and others 2004). Therefore, here we describe the most common mass spectroscopic methods used for the analysis of phenolic compounds.

Capillary Electrophoresis–Mass Spectrometry

Separation by capillary electrophoresis is based on the differences in electrophoretic motilities in a solution of charged species in an electric field of small capillaries. Its application in the analysis of phenolic compounds in fruit and vegetables is relatively recent compared to gas chromatography or liquid chromatography. Taking into account the format of the buffers used in the capillary, it is possible with this technique to distinguish different CE techniques called capillary zone electrophoresis (CZE), capillary gel electrophoresis, and micellar electrokinetic chromatography (MEKC)(Prasain and others 2004). CE as a separation technique has been successfully applied to flavonoid studies (de Rijke and others 2006; Prasain and others 2004). Huck and others (2005) have greatly reviewed the main challenge concerning CE-MS. CE-ESI-MS has been successfully applied to separate and identify the phenolic compounds in olives (Lafont and others 1999). The authors reported that using SIM CE/ESI-MS, a limit of detection in the picogram range may be achieved for some of the detected phenolic compounds.

Gas Chromatography–Mass Spectrometry (GC-MS)

Gas chromatography has been applied over the past 20 years as a separation technique to study phytochemicals. This technique has been considered as a powerful tool due to its sensitivity and selectivity, especially when coupled with mass spectrometry. However, a particular disadvantage of this technique is that the majority of polyphenolic compounds are nonvolatile. Samples used in GC are heavily processed before being ready for analysis. This process includes cleanup, solid-phase extraction, and often a derivatization process. Derivatization is carried out to generate a volatile phenolic compound. A variety of reagents are used for derivatization, including diazomethane and methyl chloroformate. However, the most frequent derivative is the trialkylsilyl group, of which the most common alkyl substituent is the methyl group (trimethylsilyl derivative) (Robbins 2003). According to the literature, the *N,O*-bis(trimethylsilyl)acetamide (BSA), *N*-methyl-*N*-(trimethylsilyl)trifluoroacetamide (MSTFA), and *N,O*-bis(trimethylsilyl)-trifluoroacematide (BSTFA) are the main reagents used in the derivatization process (Robbins 2003). Gas chromatography is often coupled to mass spectrometers. Recently Stalikas (2007) has reviewed the use of this technique in the field of phenolic acids and flavonoids analysis. Traditionally, GC-MS studies have been carried out using a flame ionization detector (FID); however, in recent years there have also been studies that used electron impact ionization. Additionally, an important variability has been shown concerning the chromatographic conditions, including different kinds of columns and temperature range. The author reported 12 representative studies, including seven related to fruit and vegetables, as examples of sample preparation and gas chromatography methods. These studies include the analysis of phenolic acids, such as gallic acid, vanillic acid, or coumaric acid from plant extracts, and flavonoids such as kaempferol or quercetin from *Vitis vinifera*, or pelargonidin and cyanidin from grapefruit. In this context, the GC-FID has been applied to identify and quantify the phenolic content in mangosteen fruits (Zadernowski and others 2009).

Liquid Chromatography–Mass Spectrometry

Liquid chromatography, coupled to the different ionization sources, is generally the technique most used to characterize the phenolic profile in fruit and vegetable products. With regard to the source ionization, it seems that ESI is used more frequently than other sources, such as APCI or APPI. Another important aspect of this technique is the ionization of phenolic compounds. Negative ionization seems to be more suitable than positive; however, positive mode could provide additional information, especially in studies dealing with the identification of unknowns. The LC is frequently coupled to tandem mass (MS-MS), producing a fragmentation of targeted or untargeted compounds and resulting in a different daughter fragment that could be used to correctly identify and also quantify phenolic compounds. A review of the fragments observed in both positive and negative ionization modes for some selected flavonoid classes has been carried out by de Rijke and others (2006).

The main MS/MS techniques are precursor ion, product ion, and neutral loss. In addition, it is possible to carry out MS^n experiments using an ion trap (Kang and others 2007). In this context, de Rijke and others (2003) carried out a study with 15 flavonoids, comparing different ionization sources and different analyzers. Among the results, the authors showed that the main fragmentations observed in the MS spectra on the ion trap, or the tandem MS spectra on the triple-quadrupole, were generally the same.

The LC-MS/MS technique has been used to quantify and identify phenolic compounds. In order to quantify, multiple reaction monitoring (MRM), in which there is a combination of the precursor ion and one of its daughter fragments, is used to characterize a particular compound. This behavior should be as specific as possible in samples with a complex mixture of phenolic compounds. This technique has been largely used to quantify phenolic compound metabolites in urine and plasma (Urpí-Sardà and others 2005, 2007). In this context, LC-ESI-MS/MS with negative mode has been applied for the identification of a variety of phenolic compounds in a cocoa sample (Sánchez-Rabaneda and others 2003; Andrés-Lacueva and others 2000).

During the past few years an important challenge in the LC-MS of flavonoids has been to optimize the analytical procedure in order to achieve structure elucidation (Cuyckens and Claeys 2002). Specifically, an important effort has been made to study the glycosylation pattern of flavonoids. Stobiecki reported the application of different MS techniques to flavonoid analysis (Stobiecki 2000). Likewise, Claeys and co-workers carried out an exhaustive study, under different conditions, of the interglycosidic linkage in *O*-diglycosides by tandem MS techniques (MS/MS) (Cuyckens and Claeys 2004, 2005; Cuyckens and others 2003). In this context, Ferreres and others (2004) have shown that it is possible to differentiate the (1→2) and (1→6) interglucosidic linkages and to distinguish among the flavonoid isomers with two glucoses, three glucoses, and four glucoses. These authors have proposed the $LC\text{-}MS^n$ as a powerful tool to characterize the O-glycosylated *C*-glycosyl flavones. The study of the relative abundance of the main ions from the MS2 and/or MS3 fragmentation events allows for the differentiation of the position of the O-glycosylation, either on phenolic hydroxyl or on the sugar moiety of C-glycosylation (Ferreres and others 2007).

Several studies reflect the widespread use of the $LC\text{-}MS^n$ for the characterization of phenolic acids, predominantly chlorogenic acids (Clifford and others 2003, 2005, 2006a,b,c). For example, using $LC\text{-}MS^3$ it is possible to discriminate between different isomers of coumaroylquinic acid, caffeoylquinic acid, and feruloylquinic acid. In addition, a hierarchical key was proposed to facilitate the process of identification when standards are not available (Clifford and others 2003).

In some studies the LC has been coupled to triple quadrupole and time-of-flight detectors. Moco and others (2006) used this technique to study phytochemicals including flavonoids from tomato samples. In fact, the results have been compiled in a database called "MoTo."

Direct Infusion Mass Spectrometry (DIMS)

In some cases liquid chromatography fails to separate some polyphenol compounds (i.e., proanthocyanidins), hampering their correct analysis. Some studies have suggested direct mass spectrometry as a possible solution. A recent study of the evaluation of the main phenolic components of grape juices was carried out using direct infusion mass spectrometry as well as ESI-MS in negative mode (Gollücke and others 2008). McDougall and others (2008) reported that DIMS in both negative and positive modes could be applied to rapidly assess differences in the polyphenol content of berries.

Matrix-Assisted Laser Desorption Ionization Time-of-Flight Mass Spectrometry (MALDI-TOF)

The MALDI-TOF technique was first developed for the analysis of large biomolecules (Karas and others 1987). This technique presents some interesting characteristics. Of these, the high speed of analysis and the sensitivity of the technique have been pointed out as important advantages compared with other methods. In MALDI the samples are cocrystallized with a matrix that is usually composed of organic compounds, such as 3,5-dimethoxy-4-hydroxycinnamic acid (sinapic acid), 2′,4′,6′-trihydroxyacetophenone, α-cyano-4-hydroxycinnamic acid (alpha-cyano or alpha-matrix), and 2,5-dihydroxybenzoic acid (DHB). After the cocrystallization, the laser is fired and the matrix absorbs energy and allows a soft ionization of the samples. Afterward the ions are analyzed by a TOF mass spectrometer.

This technique has been successfully applied for the analysis of different kinds of polyphenols from different sources. Reed and others (2005) studied the oligomeric polyphenols in foods. Recently MALDI-TOF MS was applied to characterize almond skin proanthocyanidins, revealing the existence of a series of A- and B-type procyanidins and propelargonidins up to heptamers (Monagas and others 2007). The application of MALDI TOF to soya products provides an isoflavone profile in a few minutes and serves as a powerful tool to identify and study the processing changes of isoflavones in these products (Wang and Sporns 2000b). Likewise, it has been used to characterize flavonol glycosides (Wang and Sporns 2000a).

Other Methods for Identification and Quantification

Nuclear Magnetic Resonance Spectroscopy (NMR)

LC-NMR plays a central role in the on-line identification of the constituents of crude plant extracts (Wolfender and others 2003). This technique alone, however, will not provide sufficient spectroscopic information for a complete identification of natural products, and other hyphenated methods, such as LC-UV-DAD and LC-MS/MS, are needed for providing complementary information. Added to this, LC NMR experiments are time-consuming and have to be performed on the LC peak of interest, identified by prescreening with LC-UV-MS. NMR applied to phenolic compounds includes ^{1}H NMR, ^{13}C NMR, correlation spectroscopy (COSY), heteronuclear chemical shift correlation NMR (C-H HECTOR), nuclear Overhauser effect in the laboratory frame (NOESY), rotating frame of reference (ROESY), total correlation spectroscopy (TOCSY) (Escribano-Bailón and Santos-Buelga 2003), heteronuclear multiple-quantum coherence (HMQC), and heteronuclear multiple-bond correlation (HMBC) (Es-Safi and others 2008).

Electrochemical Methods

Electrochemical detection is sensitive and selective, and it gives useful information about polyphenolic compounds in addition to spectra obtained by photodiode array detectors. Differences in electrochemically active substituents on analogous structures can lead to characteristic differences in their voltammetric behavior. Because the response profile across several cell potentials is representative of the voltammetric properties of a compound, useful qualitative information can be obtained using electrochemical detection (Aaby and others 2004).

These methods can be used to determine redox potentials of phenolics, identify the mechanism of oxidation, identify a flavonoid based on comparison with a standard, and determine redox potentials for unknown phenolics (Escribano-Bailón and Santos-Buelga 2003).

Coulometric Array Detection

The multichannel coulometric detection system serves as a highly sensitive tool for the characterization of antioxidant phenolic compounds because they are electroactive substances that usually oxidize at low potential. The coulometric efficiency of each element of the array allows a complete voltammetric resolution of analytes as a function of their oxidation potential. Some of the peaks may be resolved by the detector even if they coelute (Floridi and others 2003).

Photodiode Array Detectors (DAD)

Food and plant phenolics are commonly detected using DAD detectors (Tan and others 2008). Photodiode array detection allows collection of the entire UV spectrum during the elution of a chromatographic peak, which makes it possible to identify a phenolic compound by its spectra. Simple phenols, phenolic acids, flavanones, benzophenones, isoflavones, and flavan-3-ols have maximum absorbance at 280 nm, hydroxycinnamic acids at 320 nm, flavonols, flavones, and dihydroflavonols at 365 nm, and anthocyanins at 520 nm (Ibern-Gómez and others 2002; Merken Hand Beecher 2000). Hydrolyzable tannins show a characteristic shoulder at 300 nm, suitable for identifying them (Arapitsas and others 2007). For stilbenes, maximum absorbance of *trans*-forms are at 306 nm and at 285 nm for *cis*-forms (Lamuela-Raventós and others 1995).

Spectrophotometric Assays for the Determination of Specific Phenolic Groups

A number of spectrophotometric methods for the quantification of phenolic compounds in plant materials have been developed. Based on different principles, these assays are used to determine various structural groups present in phenolic compounds. Spectrophotometric methods may quantify all extractable phenolics as a group (Marshall and others 2008), or they may determine a specific phenolic substance such as sinapine (Ismail and Eskin 1979) or a given class of phenolics such as phenolic acids (Brune and others 1989).

Determination of Total Phenolics

Folin-Denis Assay

The Folin-Denis assay is used as a procedure for the quantification of total phenolics in plant materials, food, and beverages. Reduction of phosphomolybdic-phosphotungstic acid (Folin-Denis reagent) to a blue-colored complex in an alkaline solution occurs in the presence of phenolic compounds (Folin and Denis 1912).

Folin-Ciocalteu Assay

The Folin-Ciocalteu assay is the most widely used method to determine the total content of food phenolics (Heck and others 2008). Folin-Ciocalteu reagent is not specific and detects all phenolic groups found in extracts, including those found in extractable proteins. A disadvantage of this assay is the interference of reducing substances, such as ascorbic acid (Singleton and others 1999). The content of phenolics is expressed as gallic acid or catechin equivalents.

Determination of Proanthocyanidins

Vanillin Assay

The vanillin method is based on the condensation of the vanillin reagent with proanthocyanidins in acidic solutions. Protonated vanillin, a weak electrophilic radical, reacts with the flavonoid ring at the 6- or 8-position. The vanillin reaction is affected by the acidic nature and concentrations of substrate, the reaction time, the temperature, the vanillin concentration, and water content (Sun and others 1998).

Proanthocyanidin Assay

The proanthocyanidin assay is carried out in a solution of butanol-concentrated hydrochloric acid, where proanthocyanidins (condensed tannins) are converted to anthocyanidins (products of autoxidation of carbocations formed by cleavage of interflavanoid bonds) (Matus-Cádiz and others 2008).

Determination of Hydrolyzable Tannins

The most widely used method is based on the reaction between potassium iodate and hydrolyzable tannins (Hartzfeld and others 2002). This method provides a good estimate for gallotannins but underestimates the content of ellagitannins.

Other analytical assays proposed for the quantification of hydrolyzable tannins in plant materials include the rhodanine assay for the estimation of gallotannins (Berardini and others 2004) and sodium nitrate for the quantitative determination of ellagic acid (Wilson and Hagerman 1990).

Determination of Anthocyanins

Quantification of anthocyanins takes advantage of their characteristic behavior in acidic media; anthocyanins exist in these media as an equilibrium between the colored oxonium ion and the colorless pseudobase form. Using an average extinction coefficient, the total content of anthocyanins may be estimated from the absorption of the total extracts at 520 nm (Moskowitz and Hrazdina 1981).

Determination of Flavan-3-ols

DMACA (4-(Dimethylamino)-cinnamaldehyde) Assay

DMCA assay is used to quantify catechins, and it is based on the formation of a green chromophore between catechin and 4-(dimethylamino)-cinnamaldehyde (DMACA) (Polster and others 2003).

New Methods in Development

New detection methods of phenolic compounds are being developed. Based on the principle of the enzyme-linked immunosorbent assay (ELISA), a method has been developed to quantify phenolic compounds such as isoflavones (Vergne and others 2007).

Occurrence

Polyphenols represent a wide variety of diverse structures from different subclasses, which is why it is difficult to estimate the total polyphenol content in fruit and vegetables. Many phenolic compounds escape HPLC/UV quantification because of the presence of unidentified compounds leading to underestimation of total polyphenol content. The fruits with the highest polyphenol concentrations are strawberries, lychees, and grapes (>180 mg of gallic acid equivalent (GAE)/100 g fresh weight (FW)); the vegetables with the highest concentration are artichokes, parsley, and Brussels sprouts (>250 mg of GAE/100 g FW); melons and avocados have the lowest polyphenol concentration (Brat and others 2006).

Flavonoids

Flavonols

Flavonols are the most frequent flavonoids in foods (Manach and others 2004). Capers are the main source of flavonols (containing up to 490 mg/100 g FW) (US Department of Agriculture 2007a). Other abundant sources (ranging between 10 and 100 mg/100 g FW) are onions, kales, berries, and some herbs and spices (US Department of Agriculture 2007a). Cocoa, brewed tea, and red wine are also good dietary sources of flavonols, at 30 (Lamuela-Raventós and others 2001), 4.5, and 3.1 mg/100 mL FW, respectively (US Department of Agriculture 2007a). Flavonols are mainly accumulated in the outer tissues of fruits and vegetables because their synthesis is stimulated by sunlight (Manach and others 2004). Flavonols are found in glycosylated forms, and their bioavailability depends on their sugar moiety (Hollman and others 1997). Quercetin and kaempferol are the main sources of flavonols (Manach and others 2004). The highest dietary sources of quercetin are capers, followed by onions, asparagus, berries, and lettuce (Table 2.1). In many other vegetables and fruits, quercetin is frequently present in low concentrations around 0.1 and 5 mg/100 g FW (US Department of Agriculture 2007a). Vegetables (0.1–26.7 mg/100 g FW) and some spices, such as chives, tarragon, and fennel (6.5–19 mg/100 g FW), are characteristic sources of kaempferol, whereas fruits are a poor source (down to 0.1 mg/100 g) (US Department of Agriculture 2007a). Myrcetin, which is the third most abundant flavonol, is found in some spices, such as parsley, fennel, and oregano

Table 2.1. Flavonoids in fruits and vegetables

Subclass	Polyphenol	Food	Content mg/100 g FW	Reference
Flavonols	Quercetin	Onions	21.4	US Department of Agriculture 2007a
		Asparagus	12.4	
		Berries	10.7	
		Lettuce	7.1	
	Kaempferol	Kale	26.7	US Department of Agriculture 2007a
		Endive	10.1	
		Spinach	7.6	
		Berries	0.4	
	Myricetin	Berries	5.7	US Department of Agriculture 2007a
		Grape	0.4	
		Red cabbage	0.2	
	Isorhamnetin	Onions	5.0	US Department of Agriculture 2007a
		Pears	0.3	
Flavones	Apigenin	Artichokes	4.7	US Department of Agriculture 2007a
		Celery	2.3	
		Red onion	0.4	
		Lettuce	0.16	
	Luteolin	Pepper	5.0	US Department of Agriculture 2007a
		Artichokes	2.3	
		Red grape	1.3	
		Oranges	1.1	
Flavonones	Eriodictyol	Lemon, raw	21.4	US Department of Agriculture 2007a
		Lemon, juice	4.9	
		Orange, juice	0.2	
	Hesperetin	Lime, raw	43.0	US Department of Agriculture 2007a
		Lemon, raw	27.9	
		Orange, raw	27.3	
		Citric fruit, juice	10.5	
	Naringenin	Grapefruit, raw	21.3	US Department of Agriculture 2007a
		Orange, raw	15.3	
		Artichokes, raw	12.5	
		Ripe tomatoes, raw	0.7	
Isoflavones	Genistein	Soya bean	64.8	US Department of Agriculture 2007b
		Soya milk	6.1	
		Tofu	13.3	
	Daidzein	Soya bean	34.5	US Department of Agriculture 2007b
		Soya milk	4.5	
		Tofu	8.5	
	Glycitein	Soya bean	13.8	US Department of Agriculture 2007b
		Soya milk	0.6	
		Tofu	2.3	

Table 2.1. (*Continued*)

Subclass	Polyphenol	Food	Content mg/100 g FW	Reference
Flavan-3-ols	Catechin	Peaches	12.2	US Department of Agriculture 2007a
		Berries	11.2	
		Red grape	10.1	
		Bananas	6.1	
	Epicatechin	Red grape	8.7	US Department of Agriculture 2007a
		Apricot	5.5	
		Apples	5.5	
	Epigallocatechin	Peaches	1.1	US Department of Agriculture 2007a
		Apples	1.0	
		Berries	0.6	
	Epicatechin 3 gallate	Red grape	2.8	US Department of Agriculture 2007a
		Plums	0.8	
		Apples/pears	0.01	
	Epigallocatechin 3-gallate	Apples	0.5	US Department of Agriculture 2007a
		Berries	0.6	
		Plums	0.4	
	Gallocatechin	Berries	0.5	US Department of Agriculture 2007a
		Plums	0.1	
		Pomegranate/ persimmons	0.2	
Proantho-cyanidins	Dimers	Berries	9.5	Gu and others 2004; US Department of Agriculture 2004
		Plums	30.1	
		Apples	11.3	
		Peaches/apricot/ nectarines	6.2	
		Kiwi	1.1	
	Trimers	Berries	7	Gu and others 2004; US Department of Agriculture 2004
		Plums	20.9	
		Apples	7.1	
		Peaches/apricot/ nectarines	2.3	
		Kiwi	0.9	
	4–6mers	Berries	23.5	Gu and others 2004; US Department of Agriculture 2004
		Plums	57.8	
		Apples	24.5	
		Peaches/apricot/ nectarines	9.5	
		Kiwi	3.2	

Table 2.1. *(Continued)*

Subclass	Polyphenol	Food	Content mg/100 g FW	Reference
	7–10mers	Berries	21.9	Gu and others 2004; US Department of Agriculture 2004
		Plums	33.8	
		Apples	20.8	
		Peaches/apricot/ nectarines	5.6	
		Kiwi	2.6	
	Polymers	Berries	151.7	Gu and others 2004; US Department of Agriculture 2004
		Plums	57.3	
		Apples	29.7	
		Peaches/apricot/ nectarines	10.1	
		Kiwi	0	
Antho-cyanidins	Cyanidin	Berries	189.9	US Department of Agriculture 2007a
		Plums	12.0	
		Red cabbage	72.9	
	Delphinidin	Berries	97.9	US Department of Agriculture 2007a
		Eggplant	13.8	
		Red grape	3.7	
	Malvidin	Blueberries	61.4	US Department of Agriculture 2007a
		Red grape	34.7	
	Pelargonidin	Strawberries	31.3	US Department of Agriculture 2007a
		Radish	25.7	
	Petunidin	Berries	33.9	US Department of Agriculture 2007a
		Red grapes	2.1	
	Peonidin	Berries	21.5	US Department of Agriculture 2007a
		Cherries	4.5	
		Red grape	2.9	

(2–19.8 mg/100 g FW), and it is also present in brewed tea (0.5–1.6 mg/100 mL FW) and red wine (0–9.7 mg/100 mL FW) (US Department of Agriculture 2007a). In fruits it is only present in high concentrations in berries, whereas in most fruits and vegetables it is found in a content of less than 0.2 mg/100 g FW (see Table 2.1). Isorhamnetin is the least abundant flavonol; it has been detected in a few foods, such as some spices: fennel 9.3 mg/100 g FW, chives 5.0–8.5 mg/100 g FW, tarragon 5 mg/100 g FW; in almonds it ranged between 1.2 and 10.3 mg/100 g FW (US Department of Agriculture 2007a). In vegetables and fruits it is only present in onions and pears (see Table 2.1).

Flavones

Flavones are widely present, in a small quantity, in foods of plant origin. Some spices, such as parsley, thyme, and oregano, have a range of 25 to 630 mg/100 g FW and are the most important sources of flavones (US Department of Agriculture 2007a). Flavones consist basically of apigenin and luteolin glycosides (Manach and others 2004). Apigenin is mainly present in vegetables, such as artichokes, celery, red onion, and lettuce (see Table 2.1). However, luteolin is present in both vegetables and fruits (see Table 2.1).

Flavanones

Citric fruits, both raw and as derived products, such as juices and jams, are the main sources of flavanones (Manach and others 2004). In lesser concentrations we also found eriodictyol in almonds (0.03–0.6 mg/100 g FW), hesperitin in mint (0–21.9 mg/100 g FW), and naringenin in artichokes and ripe tomatoes (0–22.9 and 0–1.5 mg/100 g FW, respectively) (US Department of Agriculture 2007a). Naringenin is the most abundant flavanone present in grapefruit, and hesperitin in lime, whereas orange contains notable amounts of both hesperetin and naringenin. Eriodicyol is the characteristic flavonone of lemon and lemon juice (see Table 2.1). Flavanones are usually glycosylated at position 7 by a disaccharide (neohesperidose, rutinose) or, in a minor percentage, by a monosaccharide (glucose) (Tomás-Barberán and Clifford 2000).

Isoflavones

Isoflavones occur almost exclusively in leguminous plants (Manach and others 2004). Soya bean and its processed products, such as soya milk, tofu, tempeh, and miso, are the main source of genistein, daidzein, and glycetin (US Department of Agriculture 2007b). We can also find isoflavones in lower concentrations in beans and broadbeans (0.01 to 0.04 mg/100 g FW) (US Department of Agriculture, 2007b). Isoflavones have not been detected to date in fruits and vegetables (US Department of Agriculture, 2007b). In foods, isoflavones occur in four forms: aglycone, 7-*O*-glucoside, 6″-*O*-acetyl-7-*O*-glucoside, and 6″-*O*-malonyl-7-*O*-glucoside (Coward and others 1998).

Anthocyanidins

Anthocyanidins provide the characteristic red-blue colors of most fruits and vegetables. Berries are the main dietary source of anthocyanidins (66.8–947.5 mg/100 g FW) (US Department of Agriculture 2007a). Other fruits, such as red grapes, cherries, and plums, and some vegetables, such as red cabbage, red onions, radish, and eggplant, are also sources of anthocyanidins, with contents ranging between 2 and 150 mg/100 g FW (US Department of Agriculture 2007a). Anthocyanidins are poorly distributed (<10 mg/100 g FW) in other fruits, such as peaches, nectarines, and some kinds of pears and apples (US Department of Agriculture 2007a). The anthocyanidin content increases as the fruit ripens. These polyphenols are found mainly in the skin, except in berries, where they are present in both skin and flesh (Manach and others 2004).

Berries, such as blueberries, bilberries, and black currants, are the main sources of cyanidin, delphinidin, malvidin, peonidin, and petunidin. Malvidin is the characteristic anthocyanidin of red grape and red wine. Plums, cherries, and red cabbage are rich in cyanidin, whereas eggplant is a good source of delphinidin. Pelargonidin is the most abundant anthocyanidin occurring in strawberries and radishes (see Table 2.1).

Flavan-3-ols

Flavan-3-ols are found in many types of fruit, red wine, beer, and nuts, but tea and chocolate are by far the richest sources (Manach and others 2004). A few vegetables present flavan-3-ols contained at very low concentrations (down to 1.5 mg/100 g FW) (US Department of Agriculture 2007a). In contrast to the main classes of flavonoids, flavan-3-ols are found as aglycones in foods (Manach and others 2004). Catechin and epicatechin are the most frequently occurring flavan-3-ols in foods, such as tea (0–70 mg/100 mL FW), cocoa powder (19.7–127.7 mg/100 g FW) (Andrés-Lacueva and others 2008), red wine (0.2–55.6 mg/100 mL FW), nuts (0–4 mg/100 g FW), beer (0–10 mg/100 mL FW), and fruits (US Department of Agriculture 2007a). Catechin and epicatechin are present in many fruits at concentrations of 0.5–3 and 0.5–6 mg/100 g FW, respectively (see Table 2.1). Epigallocatechin, epicatechin 3-gallate, epigallocatechin 3-gallate, and gallocatechin are found in several fruits, such as berries, red grapes, plums, apples, and peaches, normally at very low concentrations (less than 1 mg/100 g FW) (see Table 2.1); however, they are present in high amounts in tea (6.8–395 mg/100 g FW) (US Department of Agriculture 2007a). Chocolate is also a good source of epigallocatechin. Theaflavin (0–5.3 mg/100 mL FW), thearubigins (7.8–139 mg/100 mL FW), theaflavin 3-3′-digallate (0–4.9 mg/100 mL FW), theaflavin 3′-gallate (0–4.1 mg/100 mL FW), and theaflavin 3-gallate (0–3.2 mg/100 mL FW) are flavan-3-ols that are detected only in tea (US Department of Agriculture 2007a). These concentrations occur in brewed tea; in tea leaves the content is between 50- and 100-fold more than brewed tea (US Department of Agriculture 2007a).

Proanthocyanidins

Proanthocyanidins (PAs), also known as condensed tannins, are oligomeric and polymeric flavan-3-ols. Procyanidins are the main PAs in foods; however, prodelphinidins and propelargonidins have also been identified (Gu and others 2004). The main food sources of total PAs are cinnamon, 8084 mg/100 g FW, and sorghum, 3937 mg/100 g FW. Other important sources of PAs are beans, red wine, nuts, and chocolate, their content ranging between 180 and 300 mg/100 g FW. In fruits, berries and plums are the major sources, with 213.6 and 199.9 mg/100 g FW, respectively. Apples and grapes are intermediate sources of PAs (60 to 90 mg/100 g FW), and the content of PAs in other fruits is less than 40 mg/100 g FW. In the majority of vegetables PAs are not detected, but they can be found in small concentrations in Indian squash (14.8 mg/100 g FW) (Gu and others, 2004; US Department of Agriculture, 2004).

Table 2.1 shows the PA content in groups of fruits classified in dimers, trimers, 4-6mers (tetramers, pentamers, and hexamers), 7-10mers (heptamers, octamers, nonamers, and decamers), and finally polymers (more than 10 monomers). Polymers and 4-6mers are the most common PAs in fruits.

Nonflavonoids

Simple Phenols

Before the ageing process wine contains small quantities of volatile phenols, which increase significantly during the time of contact with the wood, especially over the first 12 months. The use of oak wood during the ageing of wines has a great influence on wine composition, especially on volatile substances that are extracted from the wood, affecting its organoleptic properties. Some of these volatile compounds susceptible to migration from oak wood to wine are eugenol, guaiacol, and 4-ethylguaiacol (Fernandez de Simon and others 2003). The formation of 4-vinylguaiacol in coffee beans starts immediately at the beginning of the roasting process. The evolution of this compound during roasting is highly dependent on temperature (Baggenstoss and others 2008).

Ferulic acid has been extensively studied as a precursor of *p*-vinylguaiacol, the most detrimental off-flavor that forms in orange juice during storage (Rapisarda and others 1998).

Virgin olive oil contains considerable amounts of simple phenols that have a great effect on the stability/sensory and nutritional characteristics of the product. Some of the most representative are hydroxytyrosol (3,4-dihydroxyphenylethanol) and tyrosol (4-hydroxyphenylethanol); however, phenolic compounds are removed when the oil is refined (Tovar and others 2001). The phenolic content of virgin olive oil is influenced by the variety, location, degree of ripeness, and type of oil extraction procedure used, and that is why hydroxytyrosol can be considered as an indicator of maturation for olives (Esti and others 1998). Hydroxytyrosol concentrations are correlated with the stability of the oil, whereas those of tyrosol are not (Visioli and Galli 1998).

Phenolic acids

The total phenolic acid content in rowanberry, as determined by HPLC, is 103 mg/100 g FW. Besides rowanberry, the best phenolic acid sources among berries are chokeberry (96 mg/100 g FW), blueberry (85 mg/100 g FW), sweet rowanberry (75 mg/100 g FW), and saskatoon berry (59 mg/100 g FW). Among fruits, the highest contents (28 mg/100 g FW) are determined in dark plum, cherry, and one apple variety (Valkea kuulas). Coffee (97 mg/100 mL) and green and black teas (30–36 mg/100 mL) are the best sources among beverages (Mattila and others 2006). Sinapic acid was notable (4.25 mg/100 g FW) in Chinese cabbage, and protocatechuic acid had the highest concentration of all the phenolic acids in white wine (Li and others 1993).

Benzoic Aldehydes

Benzoic aldehydes mainly cover syringaldehyde and vanillin. Natural vanilla is prepared from the seeds (beans) of *Vanilla planifolia*, which may contain about 21 mg/100 g FW total phenols, including the major components vanillin (19.4 mg/100 g FW), 4-hydroxybenzaldehyde (1 mg/100 g FW), and vanillic acid (0.4 mg/100 g FW) (Clifford 2000b). In mango, vanillin has been found as "free" as well as vanillyl glucoside (Sakho and others 1997). It has also been found in lychees (Ong and Acree 1998) and wines (Moreno and others 2007). For analysis of both brandy and wine aged in oak barrels, the limits of detection were found to be 27.5, 14.25, 14.75, and

19.75 μg/100 mL for syringaldehyde, coniferaldehyde, sinapaldehyde, and vanillin, respectively (Panossian and others 2001).

Hydrolyzable Tannins

The potent antioxidant properties of pomegranate juice have been attributed to its high content of punicalagin isomers, unique ellagitannins that can reach levels exceeding 200 mg/100 mL juice (Cerdá and others 2003a,b). Total hydrolyzable tannins in barley flour, oak wood, and green tea were determined spectrophotometrically as 870, 1120, and 590 mg/100 g, respectively (Taubert and others 2005). Grape seed extract has a high tannin content of 535.6 mg/100 g (Ahn and others 2002). Persimmon is the edible fruit of a number of species of trees of the genus *Diospyros* in the ebony wood family (Ebenacea). Persimmon seed extract has a tannin content of 577.37 ± 0.66 mg/100 g (Ahn and others 2002). Although citrus fruits do not themselves contain tannins, orange-colored juices often contain food dyes with tannins. Apple juice, grape juice, and berry juices are all high in tannins. Most legumes contain tannins. Red-colored beans contain the most tannins, and white-colored beans have the least. Chickpeas, have a smaller amount of tannins (Reed 1995).

Acetophenones and Phenylacetic Acids

Phenylacetic acid has been detected in fermented soya bean made with the strain *Bacillus licheniformis* as a starter, but has not been present in extracts of nonfermented soya bean. The phenylacetic acid produced by *Bacillus licheniformis* during the fermentation of soya bean is one of the main compounds of antimicrobial activity of Chungkook-Jang, a traditional Korean fermented-soya bean food with antimicrobial properties (Kim and others 2004).

Hydroxycinnamic Acids

The most common hydroxycinnamic acid derivatives are *p*-coumaric (4-hydroxycinnamic), caffeic (3,4-dyhydroxycinnamic), ferulic (4-hydroxy-3-methoxycinnamic), and sinapic (4-hydroxy-3,5-dimethoxycinnamic) acids, which frequently occur in foods as simple esters with quinic acid or glucose (Mattila and Kumpulainen 2002).

Hydroxycinnamic acids are found in all parts of fruit and vegetables, even though the highest concentrations are observed in the outer part of mature fruits, although concentration decreases during ripening (Manach and others 2004). An overlong storage period of blood orange fruits induces extensive hydrolysis of hydroxycinnamic derivatives to free acids, and these, in turn, could develop the malodorous vinylphenols, which are an indication of too-advanced senescence in blood orange fruits (Rapisarda and others 2001).

Caffeic acid is generally the most abundant hydroxycinnamic acid in fruit and vegetables. The richest sources are coffee (drink), lettuce, carrots, blueberries, blackberries, cranberries, sweet potatoes (whole, cooked, and raw) and potatoes (Table 2.2). Prunes, peaches, orange juice, apples, tomatoes, grapes, and grape products (Betés-Saura and others 1996) also contain small quantities of caffeic acid.

Ferulic acid is the most abundant hydroxycinnamic acid in cereal grains. The content in wheat grain is approximately 800–2000 mg/100 g DW. It is found chiefly in the outer part of the grain, in the *trans* form, which is transformed into arabinoxylans and

Table 2.2. Nonflavonoids in foods

Subclass	Polyphenol	Foods	Content (mg/100g FW)	Reference
Phenolic alcohols	Tyrosol	Olive oils	0.29–2.44[£]	De la Torre-Carbot and others 2005
	Hydroxytyrosol	Olive oils	0.70–6.35[£]	De la Torre-Carbot and others 2005
Volatile phenols	Eugenol	Passion fruit	92	Chassagne and others 1997
		Passion fruit juice	172	Chassagne and others 1997
		Passion fruit peel	0.0035	Chassagne and others 1997
		Wines	0.0089[£]	Fernandez de Simon and others 2003
	Guaiacol	Wines	0.002–0.0043[£]	Fernandez de Simon and others 2003
	4-Vinylguaiacol	Raw coffee	0.0117	Czerny and Grosch 2000
		Roasted coffee	3.9	Czerny and Grosch 2000
	4-Ethylguaiacol	Raw coffee	0.0021	Czerny and Grosch 2000
		Roasted coffee	0.406	Czerny and Grosch 2000
		Wines	0.0018–0.0029[£]	Fernandez de Simon and others 2003
	Methyl salicylate	Passion fruit juice	0.076	Chassagne and others 1997
Phenolic (benzoic) acids	Ellagic acid	Strawberries	0.4–2.9	Da Silva-Pinto and others 2007
		Muscadine grapes	66.7	Pastrana-Bonilla and others 2003
	Gallic acid	Muscadine grapes	8.6	Pastrana-Bonilla and others 2003
	4-Hydroxybenzoic acid	White currant	1.8	Maatta and others 2001
		Red currant	0.3	Maatta and others 2001
	Protocatechuic acid	Raspberry	6–10	Macheix and Fleuriet 1990
Benzoic aldehydes	Syringaldehyde	Ripe walnut fruit	33.83	Colaric and others 2005
	Vanillin	Orange	0.02	Goodner and others 2000
		Tangerine	0.35	
		Lemon	0.041	
		Lime	0.035	
		Grapefruit juice	0.06	

Hydrolyzable tannins	Ellagitannins	Pomegranate juice	156.1[£]	Seeram and others 2006
		Strawberry	19.8 ± 0.2	Mertens-Talcott and others 2003
		Blueberry	0.9 ± 0.1	
		Raspberry	17.9 ± 0.3	
		Blackberry	42.4 ± 0.4	
		Longan seed	23.3	Soong and Barlow 2006
	Gallotannins	Longan seed	156	Soong and Barlow 2006
		Mango kernel	1550	Berardini and others 2004
Hydroxycinnamic acids	*p*-Coumaric acid	Blueberries	2.4–15.8	Sellappan and others 2002
		Cranberries	2.2–25.4	Zuo and others 2002
		Sweet cherries	1.0–6.8	Kim and others 2005
		Sour cherries	0.9–4.1	Kim and others 2005
		Orange juices	7.9–4.46	Rapisarda and others 1998
		Citrus[§]	1.8–19.3[¥]	Bocco and others 1998
		Broccoli	130.6 ± 42	Yeh and Yen 2005
		Eggplant	173.3 ± 40	Yeh and Yen 2005
		Asparagus	18.3 ± 16	Yeh and Yen 2005
	Caffeic acid	Blueberries	0–6.32	Sellappan and others 2002
		Blackberries	1.38–3.64	Sellappan and others 2002
		Cranberries	0.38–15.6	Zuo and others 2002
		Potatoes	3.6	Im and others 2008
		Sweet potato	0.3–2.2	Truong and others 2007
		Carrot	14	Mattila and Kumpulainen 2002
		Lettuce	4–55	Llorach and others 2004
		Coffee drink	96	Mattila and Kumpulainen 2002
	Ferulic acid	Blueberries	3.02–16.97	Sellappan and others 2002
		Blackberries	2.99–3.51	Sellappan and others 2002
		Cranberries	0.8–8.8	Zuo and others 2002

Table 2.2. (*Continued*)

Subclass	Polyphenol	Foods	Content (mg/100g FW)	Reference
		Orange juices	3.07–6.37[£]	Rapisarda and others 1998
		Citrus[§]	3.6–158[¥]	Bocco and others 1998
		Carrot	1.57	Kang and others 2008
		Beet root	1.3–14.3	Ng and others 1998
		Coffee drink	9	Mattila and Kumpulainen 2002
		Broccoli	105.9	Yeh and Yen 2005
		Eggplant	93.6	Yeh and Yen 2005
		Asparagus	46.1	Yeh and Yen 2005
	Sinapic acid	Cranberries	0–21.18	Zuo and others 2002
		Orange juices	0.78–3.59[£]	Rapisarda and others 1998
		Citrus[§]	3.0–95.4[¥]	Bocco and others 1998
Hydroxycinnamates (esters of hydroxy-cinnamic acids)	Chlorogenic acids	Sweet cherries	3.2–12.0	Kim and others 2005
		Sour cherries	0.6–5.8	Kim and others 2005
		Apricot	3.0–16.5	Ruiz and others 2005
		Nectarines	2.3–27.7	Tomás-Barberán and others 2001
		Peaches	2.4–24.2	Tomás-Barberán and others 2001
		Prunes	41.1–43.6	Donovan and others 1998
		Apples	1.93–119.5	Wojdylo and others 2008
		Pear	6–59	Cui and others 2005
		Quince pulp	0.56–18.56	Silva and others 2002
		Potato	0.35–18.71	Im and others 2008
		Sweet potato	4.6–13.6	Truong and others 2007
		Coffee drink	96	Mattila and Kumpulainen 2002
		Green coffee beans	1158–2741[¥]	Guerrero and others 2001
		Tomato	8.5	Buta and Spaulding 1997
Coumarins	Scopoletin	Noni fruit juice	6100[£]	Surono and others 2008
Stilbenes	*trans*-Resveratrol	Red grapes	0.25	Zamora-Ros and others 2007

		White grapes	0.07	
	trans-Piceid	Red grapes	0.06	Zamora-Ros and others 2007
		White grapes	0.025	
Chalcones	Chalconaringenin	Cherry tomatoes	15.3	Slimestad and Verheul 2005
Dihydro-chalcones	Phloretin xylogalactoside	Apples	2.1	Burda and others 1990
	Phloridzin	Apples	1	Burda and others 1990
Lignans	Lariciresinol	*Brassica*	0.599	Milder and others 2005
		Kale	0.972	
		Apricot	0.105	
		Strawberry	0.117	
		Peach	0.080	
		Pear	0.155	
	Pinoresinol	*Brassica*	0.1691	Milder and others 2005
		Kale	0.315	
		Apricot	0.314	
		Strawberry	0.212	
		Peach	0.186	
		Pear	0.034	
	Secoisolarici-resinol	*Brassica*	0.019	Milder and others 2005
		Pear	0.004	
		Kale	0.038	
		Strawberry	0.005	
		Peach	0.027	
		Apricot	0.031	
	Matairesinol	*Brassica*	0.012	Milder and others 2005
Secoiridoids	Oleuropein aglycon	Olive fruit	1.45	Gómez-Caravaca and others 2005

FW: fresh weight; [£]mg/100ml, [¥] mg/100g dry material, [§]Ci rus peel and seeds hemicelluloses in the aleurone and pericarp (Manach and others 2004). Ferulic acid is much less common than caffeic acid in fruit and vegetables. Broccoli, eggplant, and asparagus are major sources of ferulic acid. Low concentrations of ferulic acid have been reported in blueberries, blackberries, cranberries, orange juice, carrots, potatoes, beetroot, apples, coffee, and others (see Table 2.2). *p*-Coumaric acid has been reported in high quantities in broccoli, eggplant, and asparagus (Yeh and Yen 2005). Other sources of *p*-coumaric acid are blueberries, cranberries, prunes, sweet cherries, and orange juice. The highest contents of sinapic acid are found in citrus peel and seeds and in some cranberry varieties. However, the contents in these fruits are clearly higher than in orange juice (see Table 2.2).

The most familiar hydroxycinnamate or ester of hydroxycinnamic acid is chlorogenic acid (5-*O*-caffeoylquinic acid). Other forms of caffeoylquinic acids are also found, namely, 3-*O*-caffeoylquinic acid (*neo*-chlorogenic acid) and/or 4-*O*-caffeoylquinic acid (*crypto*-chlorogenic acid) (Mattila and Kumpulainen 2002). Chlorogenic acids are widely distributed in plants. They are a family of esters formed between quinic acid and certain *trans*-cinnamic acids, most commonly caffeic, *p*-coumaric, and ferulic. The coffee bean is remarkably rich, containing at least 30 chlorogenic acids that are not acylated at C1 of the quinic acid moiety (Clifford and others 2006a,b,c). Coffee drink, apples, prunes, pears, and sweet cherries are the highest sources of chlorogenic acids. Moderately good values are found in sour cherries, apricots, nectarines, peaches, quince, and potatoes (see Table 2.2). The concentration of chlorogenic acid in the core of pears is greater than that in the peel. However, the mean concentration of chlorogenic acid in the Oriental pear is 16.3 mg/100 g FW, less than that found in the Occidental pear (30.9 mg/100 g FW) (Cui and others 2005). Chlorogenic acid is the predominant phenolic in potato tubers, constituting up to 90% of the total phenolic content. Chlorogenic acid in fresh-cut potatoes after 6 days in cold storage varies between 7 and 30 mg/100 g FW (Truong and others 2007). For dry lyophilized potato powder, chlorogenic acid levels range from 3.28 to 637 mg/100 g FW (Im and others 2008). Steam-cooked potato strips retain 42% of the initial chlorogenic acid, whereas frying preserves only 24%. Boiled and microwaved potato retains 35% and 55% of the original amount, respectively (Tudela and others 2002).

Coumarins

Coumarin content in tonka seeds shows values of up to 300 mg/100 g FW (Clifford 2000b). A new coumarin, isoschininallylol, was isolated from the fruits of *Poncirus trifoliata* Raf. (Xu and others 2008). Bismurrangatin and murramarin A, two new coumarins, were isolated from the vegetative branches of *Murraya exotica* (Negi and others 2005). The hydroxycoumarin scopoletin was isolated from seed kernels of *Melia azedarach* L., from which three other compounds, vanillin, 4-hydroxy-3-methoxycinnamaldehyde, and (+/–)-pinoresinol, have also been isolated (Carpinella and others 2005). Coumarins are found at high levels in some essential oils, particularly cinnamon bark oil (700 mg/100 g), cassia leaf oil (up to 8730 mg/100 g), and lavender oil. Coumarin is also found in fruits (e.g., bilberries and cloudberries), green tea, and other foods, such as chicory (Lake 1999). Simple coumarins (esculetin and scopoletin) may also be present in carrots, but in some cases the roots seem to be free of the furanocoumarins (Mercier and others 1993). Parsnip root may contain about 4 mg/

100 g total furanocoumarins (Wawrzynowicz and Waksmundzka Hajnos 1990). Celery may contain a significant amount (up to 8.5 mg/100 g) but has a very variable quantity of furanocoumarins, with simple coumarins (esculetin and scopoletin) contributing a modest 0.1 mg/100 g (Diawara and others 1995). The total coumarin content (mainly meranzin derivatives) of the flavedo of bitter orange, grapefruit, and pomelo is 710, 250, and 510 mg/100 g, respectively (McHale and others 1987). Scopoletin content is particularly high in noni fruit juice (see Table 2.2).

Benzophenones

Benzophenones are distributed scarcely in foods. They are mainly present in the *Garcinia* genus, such as in fruits of *Garcinia indica* (Yamaguchi and others 2000) and *Garcinia cambogia* (Masullo and others 2008), characteristic products of India. Recently, benzophenone derivatives have also been identified in several by-products of mango at low concentrations (0–15 mg/100 g FW) (Barreto and others 2008).

Xanthones

A total of 515 xanthones have been identified in 20 families of higher plants, mainly in the Bonnetiaceae and Clusiaceae families (Vieira and Kijjoa 2005). Mangosteen fruit, a typical southeast Asian fruit, is the characteristic dietary source of xanthones (less than 7–8 mg/100 g FW) (Walker 2007). Mangostin is a C-glucosylxanthone that is also found in by-products of mango at high concentration levels (Barreto and others 2008).

Stilbenes

Stilbenes are present at low concentration levels in a few human foods. The most representative stilbenes are resveratrol and its glycoside piceid; both stilbenes can be found in *cis* and *trans* forms (Zamora-Ros and others 2007). Resveratrol and piceid are characteristic polyphenols of grape (see Table 2.2) and grape products (Zamora-Ros and others 2007; Manach and others 2004; Waterhouse and Lamuela-Raventós 1994), and their composition is affected by grape variety, degree of maturity at harvest, fungal pressure, climate, and wine-making technology (González-Barrio and others 2006; Romero-Pérez and others 2001; Mattivi and others 1995). They are also found in peanuts (0.006 mg/100 g FW), pistachios (0.007 mg/100 g FW), and berries (0.008 mg/100 g FW), but red wines (0.558 mg/100 mL FW) (Lamuela-Raventós and others 1995) are by far the richest dietary sources (Zamora-Ros and others 2007).

Chalcones

Chalcones and dihydrochalcones have been reported in a restricted number of foods (Robards and others 1999; Tomás-Barberán and Clifford 2000). Chalconaringenin occurs in tomato skin, but the acid extraction conditions of the usual polyphenol analyses convert the chalcone to the corresponding flavanone (naringenin) in the tomato. The most common dihydrochalcones found in foods are phloretin glucoside (phloridzin) and phloretin xylogalactoside, which are characteristic of apples (see Table 2.2) and derived products such as apple juice, cider, and pomace (Robards and others 1999; Tomás-Barberán and Clifford 2000).

Lignans

Lignans are a diverse group of plant-derived compounds that form the building blocks for plant cell walls. The richest source of lignans is flaxseed. Flaxseed contains mainly secoisolariciresinol (0.29 0.21 mg/100 g), but pinoresinol, lariciresinol, and matairesinol are also present in substantial amounts (0.55–3.32 mg/100 g). Lignan content in beverages (wine, beer, tea, coffee) has been found as follows: South African red wine (91.3 μg/100 mL), French red wine (78.9 μg/100 mL), Ceylonese black tea (77.1 μg/100 mL), English-blend black tea (71.2 μg/100 mL), coffee (18–31 μg/100 mL), soya milk (37.7 μg/100 mL), chocolate milk (2.2 μg/100 mL) (Milder and others 2005). Lignans are also found in *Brassica* vegetables, kale, and fruits such as apricot, strawberry, peach, and pear (see Table 2.2).

Secoiridoids

Oleuropein (the most abundant bitter principle) and its analog ligstroside, both secoiridoid biophenols, were extracted from Hojiblanca black olives (Piperno and others 2004). Secoiridoids are present exclusively in plants of the Oleraceae family (Tripoli and others 2005). Oleuropein is present in high amounts (6000 ± 90 mg/100 g DW) in the leaves of the olive tree (Le Tutour and others 1992), but it is also present in all constituent parts of the fruit peel, pulp, and seed (Servili and others 1999).

References

Aaby K, Hvattum E and Skrede G. 2004. Analysis of flavonoids and other phenolic compounds using high-performance liquid chromatography with coulometric array detection: relationship to antioxidant activity. J Agric Food Chem 52(15):4595–4603.

Ahn HS, Jeon TI, Lee JY, Hwang SG, Lim Y and Park DK. 2002. Antioxidative activity of persimmon and grape seed extract: *in vitro* and *in vivo*. Nutr Res 22: 1265–1273.

Andrés-Lacueva C, Lamuela-Raventós RM, Jauregui O, Casals I, Izquierdo-Pulido M and Permanyer J. 2000. An LC method for the analysis of cocoa phenolics. LC GC Eur 13: 902–904.

Andrés-Lacueva C, Monagas M, Khan N, Izquierdo-Pulido M, Urpi-Sarda M, Permanyer J and Lamuela-Raventós RM. 2008. Flavanol and flavonol contents of cocoa powder products: influence of the manufacturing process. J Agric Food Chem 56(9):3111–3117.

Arapitsas P, Menichetti S, Vincieri FF and Romani A. 2007. Hydrolyzable tannins with the hexahydroxydiphenoyl unit and the m-depsidic link: HPLC-DAD-MS identification and model synthesis. J Agric Food Chem 55(1):48–55.

Arts ICW and Hollman PCH. 1998. Optimization of a quantitative method for the determination of catechins in fruits and legumes. J Agric Food Chem 46(12):5156–5162.

Asami DK, Hong YJ, Barrett DM and Mitchell AE. 2003. Comparison of the total phenolic and ascorbic acid content of freeze-dried and air-dried marionberry, strawberry, and corn grown using conventional, organic, and sustainable agricultural practices. J Agric Food Chem 51(5):1237–1241.

Baggenstoss J, Poisson L, Kaegi R, Perren R and Escher F. 2008. Coffee roasting and aroma formation: application of different time-temperature conditions. J Agric Food Chem 56(14):5836–5846.

Barreto JC, Trevisan MT, Hull WE, Erben G, de Brito ES, Pfundstein B, Wurtele G, Spiegelhalder B and Owen RW. 2008. Characterization and quantitation of polyphenolic compounds in bark, kernel, leaves, and peel of mango (*Mangifera indica* L.). J Agric Food Chem 56(14):5599–5610.

Berardini N, Carle R and Schieber A. 2004. Characterization of gallotannins and benzophenone derivatives from mango (Mangifera indica L. cv. 'Tommy Atkins') peels, pulp and kernels by high-performance liquid chromatography/electrospray ionization mass spectrometry. Rapid Commun Mass Spectrom 18(19):2208–2216.

Betés-Saura C, Andrés-Lacueva C and Lamuela-Raventós RM. 1996. Phenolics in white free-run juice and wines from Penedès by high-performance liquid chromatography: changes during vinification. J Agric Food Chem 44(10):3040–3046.

Bocco A, Cuvelier ME, Richard H and Berset C. 1998. Antioxidant activity and phenolic composition of citrus peel and seed extracts. J Agric Food Chem 46(6):2123–2129.

Bradshaw MP, Prenzler PD and Scollary GR. 2001. Ascorbic acid-induced browning of (+)-catechin in a model wine system. J Agric Food Chem 49(2):934–939.

Brat P, George S, Bellamy A, Du CL, Scalbert A, Mennen L, Arnault N and Amiot M J. 2006. Daily polyphenol intake in France from fruit and vegetables. J Nutr 136(9):2368–2373.

Brune M, Rossander L and Hallberg L. 1989. Iron absorption and phenolic compounds: importance of different phenolic structures. Eur J Clin Nutr 43(8):547–557.

Burda S, Oleszek W and Lee CY. 1990. Phenolic compounds and their changes in apples during maturation and cold storage. J Agric Food Chem 38(4):945–948.

Buta JG and Spaulding DW. 1997. Endogenous levels of phenolics in tomato fruit during growth and maturation. J Plant Growth Regul 16(1):43–46.

Carpinella MC, Ferrayoli CG and Palacios SM. 2005. Antifungal synergistic effect of scopoletin, a hydroxycoumarin isolated from *Melia azedarach* L. fruits. J Agric Food Chem 53(8):2922–2927.

Cassidy A, Handley B and Lamuela-Raventós RM. 2000. Isoflavones, lignans and stilbenes – origins, metabolism and potential importance to human health. J Sci Food Agric 80(7):1044–1062.

Cerdá B, Cerón JJ, Tomás-Barberán FA and Espín JC. 2003a. Repeated oral administration of high doses of pomegranate ellagitannin punicalagin to rats for 37 days is not toxic. J Agric Food Chem 51(11):3493–3501.

Cerdá B, Llorach R, Cerón JJ, Espín JC and Tomás-Barberán FA. 2003b. Evaluation of the bioavailability and metabolism in the rat of punicalagin, an antioxidant polyphenol from pomegranate juice. Eur J Nutr 42(1):18–28.

Chassagne D, Crouzet J, Bayonove CL and Baumes RL. 1997. Glycosidically bound eugenol and methyl salicylate in the fruit of edible *Passiflora* species. J Agric Food Chem 45(7):2685–2689.

Chen H, Zuo Y and Deng Y. 2001. Separation and determination of flavonoids and other phenolic compounds in cranberry juice by high-performance liquid chromatography. J Chromatogr A 913(1–2):387–395.

Clifford MN. 2000a. Anthocyanins—nature, occurrence and dietary burden. J Sci Food Agric 80(7):1063–1072.

Clifford MN. 2000b. Miscellaneous phenols in foods and beverages—nature, occurrence and dietary burden. J Sci Food Agric 80(7):1126–1137.

Clifford MN, Johnston KL, Knight S and Kuhnert N. 2003. Hierarchical scheme for LC-MSn identification of chlorogenic acids. J Agric Food Chem 51(10):2900–2911.

Clifford MN, Knight S, Kuhnert N. 2005. Discriminating between the six isomers of dicaffeoylquinic acid by LC-MS(n). J Agric Food Chem 53(10):3821–3832.

Clifford MN, Knight S, Surucu B and Kuhnert N. 2006a. Characterization by LC-MS(n) of four new classes of chlorogenic acids in green coffee beans: dimethoxycinnamoylquinic acids, diferuloylquinic acids, caffeoyl-dimethoxycinnamoylquinic acids, and feruloyl-dimethoxycinnamoylquinic acids. J Agric Food Chem 54(6):1957–1969.

Clifford MN, Marks S, Knight S and Kuhnert N. 2006b. Characterization by LC-MSn of four new classes of *p*-coumaric acid-containing diacyl chlorogenic acids in green coffee beans. J Agric Food Chem 54(12):4095–4101.

Clifford MN, Zheng W and Kuhnert N. 2006c. Profiling the chlorogenic acids of aster by HPLC-MS(n). Phytochem Anal 17(6):384–393.

Colaric M, Veberic R, Solar A, Hudina M and Stampar F. 2005. Phenolic acids, syringaldehyde, and juglone in fruits of different cultivars of *Juglans regia* L. J Agric Food Chem 53(16):6390–6396.

Coward L, Smith M, Kirk M and Barnes S. 1998. Chemical modification of isoflavones in soyfoods during cooking and processing. Am J Clin Nutr 68(6):1486S–1491S.

Cui T, Nakamura K, Ma L, Li JZ and Kayahara H. 2005. Analyses of arbutin and chlorogenic acid, the major phenolic constituents in oriental pear. J Agric Food Chem 53(10):3882–3887.

Cuyckens F and Claeys M. 2002. Optimization of a liquid chromatography method based on simultaneous electrospray ionization mass spectrometric and ultraviolet photodiode array detection for analysis of flavonoid glycosides. Rapid Commun Mass Spectrom 16(24):2341–2348.

Cuyckens F and Claeys M. 2004. Mass spectrometry in the structural analysis of flavonoids. J Mass Spectrom 39(1):1–15.

Cuyckens F and Claeys M. 2005. Determination of the glycosylation site in flavonoid mono-*O*-glycosides by collision-induced dissociation of electrospray-generated deprotonated and sodiated molecules. J Mass Spectrom 40(3):364–372.

Cuyckens F, Shahat AA, Van den Heuvel H, Abdel-Shafeek KA, El-Messiry MM, Seif-El Nasr MM, Pieters L, Vlietinck AJ and Claeys M. 2003. The application of liquid chromatography-electrospray ionization mass spectrometry and collision-induced dissociation in the structural characterization of acylated flavonol *O*-glycosides from the seeds of *Carrichtera annua*. Eur J Mass Spectrom 9(4):409–420.

Czerny M and Grosch W. 2000. Potent odorants of raw arabica coffee. Their changes during roasting. J Agric Food Chem 48(3):868–872.

Da Silva Pinto M, Lajolo FM and Genovese MI. 2007. Bioactive compounds and antioxidant capacity of strawberry jams. Plant Foods Hum Nutr 62(3):127–131.

De la Torre-Carbot K, Jauregui O, Gimeno E, Castellote AI, Lamuela-Raventós RM and Lopez-Sabater M. 2005. Characterization and quantification of phenolic compounds in olive oils by solid-phase extraction, HPLC-DAD, and HPLC-MS/MS. J Agric Food Chem 53(11):4331–4340.

de Rijke E, Out P, Niessen WM, Ariese F, Gooijer C and Brinkman UA. 2006. Analytical separation and detection methods for flavonoids. J Chromatogr A 1112(1–2):31–63.

de Rijke E, Zappey H, Ariese F, Gooijer C and Brinkman UA. 2003. Liquid chromatography with atmospheric pressure chemical ionization and electrospray ionization mass spectrometry of flavonoids with triple-quadrupole and ion-trap instruments. J Chromatogr A 984(1):45–58.

Devanand LL. 2006. Significance of sample preparation in developing analytical methodologies for accurate estimation of bioactive compounds in functional foods. J Sci Food Agric 86(14):2266–2272.

Diawara MM, Trumble JT, Quirós CF and Hansen R. 1995. Implications of distribution of linear furanocoumarins within celery. J Agric Food Chem 43(3):723–727.

Donovan JL, Meyer AS and Waterhouse AL. 1998. Phenolic composition and antioxidant activity of prunes and prune juice (*Prunus domestica*). J Agric Food Chem 46(4):1247–1252.

Escribano-Bailón M and Santos-Buelga C. 2003. Polyphenol extraction from foods. In: Santos-Buelga C, Williamson G, editors. Methods in Polyphenol Analysis. Cambridge: The Royal Society of Chemistry, pp. 1–12.

Es-Safi NE, Meudec E, Bouchut C, Fulcrand H, Ducrot PH, Herbette G and Cheynier V. 2008. New compounds obtained by evolution and oxidation of malvidin 3-*O*-glucoside in ethanolic medium. J Agric Food Chem 56(12):4584–4591.

Esti M, Cinquanta L and La Notte E. 1998. Phenolic compounds in different olive varieties. J Agric Food Chem 46(1):32–35.

Fernandez de Simon B, Cadahia E and Jalocha J. 2003. Volatile compounds in a Spanish red wine aged in barrels made of Spanish, French, and American oak wood. J Agric Food Chem 51(26):7671–7678.

Ferreres F, Gil-Izquierdo A, Andrade PB, Valentão P and Tomás Barberán FA. 2007. Characterization of *C*-glycosyl flavones O-glycosylated by liquid chromatography–tandem mass spectrometry. J Chromatogr A 1161 (1–2):214–223.

Ferreres F, Llorach R and Gil-Izquierdo A. 2004. Characterization of the interglycosidic linkage in di-, tri-, tetra- and pentaglycosylated flavonoids and differentiation of positional isomers by liquid chromatography/electrospray ionization tandem mass spectrometry. J Mass Spectrom 39(3):312–321.

Floridi S, Montanari L, Marconi O and Fantozzi P. 2003. Determination of free phenolic acids in wort and beer by coulometric array detection. J Agric Food Chem 51(6):1548–1554.

Folin O, Denis W. 1912. On phosphotungstic-phosphomolybdic compounds as color reagents. J Biol Chem 12: 239–243.

Fulcrand H, Mane C, Preys S, Mazerolles G, Bouchut C, Mazauric JP, Souquet JM, Meudec E, Li Y, Cole RB and Cheynier V. 2008. Direct mass spectrometry approaches to characterize polyphenol composition of complex samples. Phytochemistry. 69(18):3131–3138.

Gollücke APB, Catharino RR, de Souza JC, Eberlin MN and de Queiroz Tavares D. 2008. Evolution of major phenolic components and radical scavenging activity of grape juices through concentration process and storage. Food Chem 112(4):868–873.

Gómez-Caravaca AM, Carrasco-Pancorbo A, Cañabate-Díaz B, Segura-Carretero A and Fernández-Gutiérrez A. 2005. Electrophoretic identification and quantitation of compounds in the polyphenolic fraction of extra-virgin olive. Electrophoresis 26(18):3538–3551.

González-Barrio R, Beltran D, Cantos E, Gil MI, Espín JC and Tomás-Barberán F A. 2006. Comparison of ozone and UV-C treatments on the postharvest stilbenoid monomer, dimer, and trimer induction in var. 'Superior' white table grapes. J Agric Food Chem 54(12):4222–4228.

Goodner KL, Jella P and Rouseff RL. 2000. Determination of vanillin in orange, grapefruit, tangerine, lemon, and lime juices using GC-olfactometry and GC-MS/MS. J Agric Food Chem 48(7):2882–2886.

Gu L, Kelm MA, Hammerstone JF, Beecher G, Holden J, Haytowitz D, Gebhardt S and Prior RL. 2004. Concentrations of proanthocyanidins in common foods and estimations of normal consumption. J Nutr 134(3):613–617.

Guerrero G, Suárez M and Moreno G. 2001. Chlorogenic acids as a potential criterion in coffee genotype selections. J Agric Food Chem 49(5):2454–2458.

Harnly JM, Bhagwat S and Lin LZ. 2007. Profiling methods for the determination of phenolic compounds in foods and dietary supplements. Anal Bioanal Chem 389(1):47–61.

Hartzfeld PW, Forkner R, Hunter MD and Hagerman AE. 2002. Determination of hydrolyzable tannins (gallotannins and ellagitannins) after reaction with potassium iodate. J Agric Food Chem 50(7):1785–1790.

Heck CI, Schmalko M and Gonzalez de Mejia E. 2008. Effect of growing and drying conditions on the phenolic composition of mate teas (*Ilex paraguariensis*). J Agric Food Chem 56(18):8394–8403.

Hernández-Montes E, Pollard SE, Vauzour D, Jofre-Montseny L, Rota C, Rimbach G, Weinberg PD and Spencer JPE. 2006. Activation of glutathione peroxidase via Nrf1 mediates genistein's protection against oxidative endothelial cell injury. Biochem Biophys Res Commun 346(3):851–859.

Herrera MC and Luque de Castro MD. 2005. Ultrasound-assisted extraction of phenolic compounds from strawberries prior to liquid chromatographic separation and photodiode array ultraviolet detection. J Chromatogr A 1100(1):1–7.

Hollman PC, van Trijp JM, Mengelers MJ, de Vries JH and Katan M B. 1997. Bioavailability of the dietary antioxidant flavonol quercetin in man. Cancer Lett 114(1–2):139–140.

Huck WC, Stecher G, Scherz H and Bonn G. 2005. Analysis of drugs, natural and bioactive compounds containing phenolic groups by capillary electrophoresis coupled to mass spectrometry. Electrophoresis 26(7–8):1319–1333.

Ibern-Gómez M, Andrés-Lacueva C, Lamuela-Raventós RM and Waterhouse AL. 2002. Rapid HPLC analysis of phenolic compounds in red wines. Am J Enol Vitic 53(3):218–221.

Im HW, Suh BS, Lee SU, Kozukue N, Ohnisi-Kameyama M, Levin CE and Friedman M. 2008. Analysis of phenolic compounds by high-performance liquid chromatography and liquid chromatography/mass spectrometry in potato plant flowers, leaves, stems, and tubers and in home-processed potatoes. J Agric Food Chem 56(9):3341–3349.

Ismail F and Eskin NAM. 1979. A new quantitative procedure for determination of Sinapine. J Agric Food Chem 27(4):917–918.

Kang J, Hick LA and Price WE. 2007. A fragmentation study of isoflavones in negative electrospray ionization by MS^n ion trap mass spectrometry and triple quadrupole mass spectrometry. Rapid Commun Mass Spectrom 21(6):857–868.

Kang YH, Parker CC, Smith AC and Waldron KW. 2008. Characterization and distribution of phenolics in carrot cell walls. J Agric Food Chem 56(18):8558–8564

Karas M, Bachmann D, Bahr U, Hillenkamp F. 1987. Matrix-assisted ultraviolet laser desorption of non-volatile compounds. Int J Mass Spectrom Ion Process 78: 53–68.

Kim DO, Heo HJ, Kim YJ, Yang HS and Lee CY. 2005. Sweet and sour cherry phenolics and their protective effects on neuronal cells. J Agric Food Chem 53(26):9921–9927.

Kim Y, Cho JY, Kuk JH, Moon JH, Cho JI, Kim YC and Park KH. 2004. Identification and antimicrobial activity of phenylacetic acid produced by *Bacillus licheniformis* isolated from fermented soybean, Chungkook-Jang. Curr Microbiol 48(4):312–317.

Lafont F, Aramendia M, García I, Borau V, Jiménez C, Marinas JM and Urbano F. 1999. Analyses of phenolic compounds by capillary electrophoresis electrospray mass spectrometry. Rapid Commun Mass Spectrom 13(7):562–567.

Lake B. 1999. Coumarin metabolism, toxicity and carcinogenicity. relevance for human risk assessment. Food Chem Toxicol 37(4):423–453.

Lamuela-Raventós RM, Andrés-Lacueva C, Permanyer J and Izquierdo-Pulido M. 2001. More antioxidants in cocoa. J Nutr 130(8S Suppl):2109S–2114S.

Lamuela-Raventós RM, Romero-Pérez A, Waterhouse A and de la Torre-Boronat M. 1995. Direct HPLC analysis of *cis*- and *trans*-resveratrol and piceic isomers in Spanish red *Vitis vinifera* wines. J Agric Food Chem 43(2):281–283

Le Tutour B and Guedon D. 1992. Antioxidative activities of *Olea europaea* leaves and related phenolic compounds. Phytochemistry 31: 1173–1178.

Li P, Wang XQ, Wang HZ and Wu YN. 1993. High performance liquid chromatographic determination of phenolic acids in fruits and vegetables. Biomed Environ Sci 6(4):389–398.

Llorach R, Espín JC, Tomás-Barberán FA and Ferreres F. 2003. Valorization of cauliflower (*Brassica oleracea* L. var. botrytis) by-products as a source of antioxidant phenolics. J Agric Food Chem 51(8):2181–2187.

Llorach R, Tomás-Barberán FA and Ferreres F. 2004. Lettuce and chicory byproducts as a source of antioxidant phenolic extracts. J Agric Food Chem 52(16):5109–5116.

Maatta K, Kamal-Eldin A and Törrönen R. 2001. Phenolic compounds in berries of black, red, green, and white currants (*Ribes* sp.). Antioxid Redox Signal 3(6):981–993.

Macheix JJ, Fleuriet A and Billot J. 1990. Fruit Phenolics. Boca Raton, FL: CRC Press.

Manach C, Scalbert A, Morand C, Rémésy C and Jiménez L. 2004. Polyphenols: food sources and bioavailability. Am J Clin Nutr 79(5):727–724.

Mané C, Souquet JM, Olle D, Verries C, Veran F, Mazerolles G, Cheynier V and Fulcrand H. 2007. Optimization of simultaneous flavanol, phenolic acid, and anthocyanin extraction from grapes using an experimental design: application to the characterization of champagne grape varieties. J Agric Food Chem 55(18):7224–7233.

Marshall A, Bryant D, Latypova G, Hauck B, Olyott P, Morris P and Robbins M. 2008. A high-throughput method for the quantification of proanthocyanidins in forage crops and its application in assessing variation in condensed tannin content in breeding programmes for *Lotus corniculatus* and *Lotus uliginosus*. J Agric Food Chem 56(3):974–981.

Masullo M, Bassarello C, Suzuki H, Pizza C and Piacente S. 2008. Polyisoprenylated benzophenones and an unusual polyisoprenylated tetracyclic xanthone from the fruits of *Garcinia cambogia*. J Agric Food Chem 56(13):5205–5210.

Mattila P, Hellström J and Törrönen R. 2006. Phenolic acids in berries, fruits, and beverages. J Agric Food Chem 54(19):7193–7199.

Mattila P and Kumpulainen J. 2002. Determination of free and total phenolic acids in plant-derived foods by HPLC with diode-array detection. J Agric Food Chem 50(13):3660–3667.

Mattivi F, Reniero F and Korhammer S. 1995. Isolation, characterization, and evolution in red wine vinification of resveratrol monomers. J Agric Food Chem 43(7):1820–1823.

Matus-Cádiz MA, Daskalchuk TE, Verma B, Puttick D, Chibbar RN, Gray GR, Perron CE, Tyler RT and Hucl P. 2008. Phenolic compounds contribute to dark bran pigmentation in hard white wheat. J Agric Food Chem 56(5):1644–1653.

McDougall G, Martinussen I and Stewart D. 2008. Towards fruitful metabolomics: high throughput analyses of polyphenol composition in berries using direct infusion mass spectrometry. J Chromatogr B 871(2):362–369.

McHale D, Khopkar PP and Sheridan JB. 1987. Coumarin glycosides from citrus flavedo. Phytochemistry 26(9):2547–2549.

Mercier J, Ponnampalam R, Berard LS and Arul J. 1993. Polyacetylene content and UV-induced 6-methoxymellein accumulation in carrot cultivars. J Sci Food Agric 63(3):313–316.

Merken HM and Beecher GR. 2000. Measurement of food flavonoids by high-performance liquid chromatography: a review. J Agric Food Chem 48(3):577–599.

Mertens-Talcott SU, Talcott S T and Percival S S. 2003. Low concentrations of quercetin and ellagic acid synergistically influence proliferation, cytotoxicity and apoptosis in MOLT-4 human leukemia cells. J Nutr 133(8):2669–2674.

Mezadri T, Villaño D, Fernández-Pachón MS, García-Parrilla MC and Troncoso AM. 2008. Antioxidant compounds and antioxidant activity in acerola (*Malpighia emarginata* DC.) fruits and derivatives. J Food Comp Anal 21(4):282–290.

Michalkiewicz A, Biesaga M and Pyrzynska K. 2008. Solid-phase extraction procedure for determination of phenolic acids and some flavonols in honey. J Chromatogr A 1187(1–2): 18–24.

Milder IE, Arts IC, van de Putte B, Venema DP and Hollman PC. 2005. Lignan contents of Dutch plant foods: a database including lariciresinol, pinoresinol, secoisolariciresinol and matairesinol. Br J Nutr 93(3):393–402.

Moco S, Bino RJ, Vorst O, Verhoeven HA, de Groot J, van Beek TA, Vervoort J and de Vos CH. 2006. A liquid chromatography-mass spectrometry-based metabolome database for tomato. Plant Physiol 141(4):1205–1218.

Monagas M, Garrido I, Lebrón-Aguilar R, Bartolome B and Gómez-Cordobés C. 2007. Almond (*Prunus dulcis* (Mill.) D.A. Webb) Skins as a potential source of bioactive polyphenols. J Agric Food Chem 55(21):8498–8507.

Moreno NJ, Marco AG and Azpilicueta CA. 2007. Influence of wine turbidity on the accumulation of volatile compounds from the oak barrel. J Agric Food Chem 55(15):6244–6251.

Moskowitz AH and Hrazdina G. 1981. Vacuolar contents of fruit subepidermal cells from vitis species. Plant Physiol 68(3):686–692.

Murray R, Mendez J and Brown S. 1982. The Natural Coumarins: Occurrence, Chemistry and Biochemistry. Chichester, UK: John Wiley & Sons.

Negi N, Ochi A, Kurosawa M, Ushijima K, Kitaguchi Y, Kusakabe E, Okasho F, Kimachi T, Teshima N, Ju-Ichi M, Abou-Douh AM, Ito C and Furukawa H. 2005. Two new dimeric coumarins isolated from *Murraya exótica*. Chem Pharm Bull (Tokyo) 53(9):1180–1182.

Ng A, Harvey AJ, Parker ML, Smith AC and Waldron KW. 1998. Effect of oxidative coupling on the thermal stability of texture and cell wall chemistry of beet root (*Beta vulgaris*). J Agric Food Chem 46(8):3365–3370.

Ong PKC and Acree TE. 1998. Gas chromatography/olfactory analysis of lychee (*Litchi chinensis* Sonn.). J Agric Food Chem 46(6):2282–2286.

Oszmianski J, Wojdylo A and Kolniak J. 2009. Effect of l-ascorbic acid, sugar, pectin and freeze-thaw treatment on polyphenol content of frozen strawberries. LWT-Food Sci Technol.42(2):581–586.

Panossian A, Mamikonyan G, Torosyan M, Gabrielyan E and Mkhitaryan S. 2001. Analysis of aromatic aldehydes in brandy and wine by high-performance capillary electrophoresis. Anal Chem 73(17):4379–4383.

Pastrana-Bonilla E, Akoh CC, Sellappan S and Krewer G. 2003. Phenolic content and antioxidant capacity of muscadine grapes. J Agric Food Chem 51(18):5497–5503.

Piperno A, Toscano M and Uccella NA. 2004. The Cannizzaro-like metabolites of secoiridoid glucosides in some olive cultivars. J Sci Food Agric 84(4):341–349.

Polster J, Dithmar H and Walter F. 2003. Are histones the targets for flavan-3-ols (catechins) in nuclei? Biol Chem 384(7):997–1006.

Prasain JK, Wang CC and Barnes S. 2004. Mass spectrometric methods for the determination of flavonoids in biological samples. Free Radic Biol Med 37(9):1324–1350.

Rapisarda P, Bellomo SE and Intelisano S. 2001. Storage temperature effects on blood orange fruit quality. J Agric Food Chem 49(7):3230–3235.

Rapisarda P, Carollo G, Fallico B, Tomaselli F and Maccarone E. 1998. Hydroxycinnamic acids as markers of Italian blood orange juices. J Agric Food Chem 46(2):464–470.

Reed JD. 1995. Nutritional toxicology of tannins and related polyphenols in forage legumes. J Anim Sci 73(5):1516–1528.

Reed JD, Krueger CG and Vestling MM. 2005. MALDI-TOF mass spectrometry of oligomeric food polyphenols. Phytochemistry 66(18):2248–2263.

Robards K, Prenzler PD, Tucker G, Swatsitang P and Glover W. 1999. Phenolic compounds and their role in oxidative processes in fruits. Food Chem 66(4):401–436.

Robbins RJ. 2003. Phenolic acids in foods: An overview of analytical methodology. J Agric Food Chem 51(10):2866–2887.

Romero-Pérez AI, Ibern-Gómez M, Lamuela-Raventós RM and de la Torre-Boronat MC. 1999. Piceid, the major resveratrol derivative in grape juices. J Agric Food Chem 47(4):1533–1536.

Romero-Pérez AI, Lamuela-Raventós RM, Andrés-Lacueva C and Torre-Boronat MC. 2001. Method for the quantitative extraction of resveratrol and piceid isomers in grape berry skins. Effect of powdery mildew on the stilbene content. J Agric Food Chem 49(1):210–215.

Ruiz D, Egea J, Gil MI and Tomás-Barberán FA. 2005. Characterization and quantitation of phenolic compounds in new apricot (*Prunus armeniaca* L.) varieties. J Agric Food Chem 53(24):9544–9552.

Sakho M, Chassagne D, Jaus A, Chiarazzo E and Crouzet J. 1997. Enzymatic maceration: effects on volatile components of mango pulp. J Food Sci 63(6):975–978.

Sánchez-Rabaneda F, Jáuregui O, Casals I, Andrés-Lacueva C, Izquierdo-Pulido M and Lamuela-Raventós RM. 2003. Liquid chromatographic/electrospray ionization tandem mass spectrometric study of the phenolic composition of cocoa (*Theobroma cacao*). J Mass Spectrom. 38(1):35–42.

Seeram NP, Henning SM, Zhang Y, Suchard M, Li Z and Heber D. 2006. Pomegranate juice ellagitannin metabolites are present in human plasma and some persist in urine for up to 48 hours. J Nutr 136(10):2481–2485.

Sellappan S, Akoh CC and Krewer G. 2002. Phenolic compounds and antioxidant capacity of Georgia-grown blueberries and blackberries. J Agric Food Chem 50(8):2432–2438.

Servili M, Baldioli M, Selvaggini R, Macchioni A and Montedoro GF. 1999. Phenolic compounds of olive fruit: one and two-dimensional nuclear magnetic resonance characterization of nüzhenide and its distribution in the constitutive parts of fruit. J Agric Food Chem 47(1):12–18.

Shier WT, Shier AC, Xie W and Mirocha CJ. 2001. Structure-activity relationships for human estrogenic activity in zearalenone mycotoxins. Toxicon 39 (9):1435–1438.

Silva BM, Andrade PB, Ferreres F, Domingues AL, Seabra RM and Ferreira MA. 2002. Phenolic profile of quince fruit (*Cydonia oblonga* Miller) (pulp and peel). J Agric Food Chem 50(16):4615–4618.

Singleton VL, Orthofer R and Lamuela-Raventós RM. 1999. Analysis of total phenols and other oxidation substrates and antioxidants by means of Folin-Ciocalteu reagent. In: Abelson JN, Simon MI, editors. Methods in Enzymology, Volume 299. San Diego, CA: Academic Press, pp. 152–178.

Slimestad R and Verheul MJ. 2005. Content of chalconaringenin and chlorogenic acid in cherry tomatoes is strongly reduced during postharvest ripening. J Agric Food Chem 53(18):7251–7256.

Soler-Rivas C, Espín JC and Wichers HJ. 2000. Oleuropein and related compounds. J Sci Food Agric 80(7):1013–1023.

Soong YY and Barlow PJ. 2006. Quantification of gallic acid and ellagic acid from longan (*Dimocarpus longan* Lour.) seed and mango (*Mangifera indica* L.) kernel and their effects on antioxidant activity. Food Chem 87(3):524–530.

Stalikas C. 2007. Extraction, separation, and detection methods for phenolic acids and flavonoids. J Sep Sci 30(18):3268–3295.

Stobiecki M. 2000. Application of mass spectrometry for identification and structural studies of flavonoid glycosides. Phytochemistry 54(3):237–256.

Suárez B, Picinelli A and Mangas JJ. 1996. Solid-phase extraction and high-performance liquid chromatographic determination of polyphenols in apple musts and ciders. J Chromatogr A 727(2):203–209.

Sun BS, Ricardo-da-Silva JM and Spranger MI. 1998. Critical factors of vanillin assay for catechins and proanthocyanidins. J Agric Food Chem 46: 4267–4274.

Surono IS, Nishigaki T, Endaryanto A and Waspodo P. 2008. Indonesian biodiversities, from microbes to herbal plants as potential functional foods. J Fac Agric, Shun Shu Univ. 44(1–2):23–27.

Tan XJ, Li Q, Chen XH, Wang ZW, Shi ZY, Bi KS and Jia Y. 2008. Simultaneous determination of 13 bioactive compounds in Herba Artemisiae Scopariae (Yin Chen) from different harvest seasons by HPLC-DAD. J Pharm Biomed Anal 47(4–5):847–853.

Taubert D, Grimberg G and Schomig E. 2005. Tannic acid in plant dust causes airway obstruction. Thorax 60(9):789–791.

Teow CC, Truong V-D, McFeeters RF, Thompson RL, Pecota KV and Yencho GC. 2007. Antioxidant activities, phenolic and [beta]-carotene contents of sweet potato genotypes with varying flesh colours. Food Chem 103(3):829–838.

Tomás-Barberán FA and Clifford MN. 2000. Flavanones, chalcones and dihydrochalcones—nature, occurrence and dietary burden. J Sci Food Agri 80(7):1073–1080.

Tomás-Barberán FA and Espín JC. 2001. Phenolic compounds and related enzymes as determinants of quality in fruits and vegetables. J Sci Food Agric 81(9):853–876.

Tomás-Barberán FA, Gil MI, Cremin P, Waterhouse AL, Hess-Pierce B and Kader AA. 2001. HPLC-DAD-ESIMS analysis of phenolic compounds in nectarines, peaches, and plums. J Agric Food Chem. 49(10):4748–4760.

Tovar MJ, Motilva MJ and Romero MP. 2001. Changes in the phenolic composition of virgin olive oil from young trees (*Olea europaea* L. cv. Arbequina) grown under linear irrigation strategies. J Agric Food Chem 49(11):5502–5508.

Tripoli E, Giammanco M, Tabacchi G, Di Majo D, Giammanco S and La Guardia M. 2005. The phenolic compounds of olive oil: structure, biological activity and beneficial effects on human health. Nutr Res Rev 18(1):98–112.

Truong VD, McFeeters RF, Thompson RT, Dean LL and Shofran B. 2007. Phenolic acid content and composition in leaves and roots of common commercial sweetpotato (*Ipomea batatas* L.) cultivars in the United States. J Food Sci 72(6):C343–C349.

Tudela JA, Cantos E, Espín JC, Tomás-Barberán FA and Gil MI. 2002. Induction of antioxidant flavonol biosynthesis in fresh-cut potatoes. Effect of domestic cooking. J Agric Food Chem 50(21):5925–5931.

Tura D and Robards K. 2002. Sample handling strategies for the determination of biophenols in food and plants. J Chromatogr A 975(1):71–93.

Urpí-Sardà M, Jáuregui O, Lamuela-Raventós RM, Jaeger W, Miksits M, Covas MI and Andrés-Lacueva C. 2005. Uptake of diet resveratrol into the human low-density lipoprotein. Identification and quantification of resveratrol metabolites by liquid chromatography coupled with tandem mass spectrometry. Anal Chem 77(10):3149–3155.

Urpi-Sarda M, Zamora-Ros R, Lamuela-Raventós R, Cherubini A, Jauregui O, de la Torre R, Covas MI, Estruch R, Jaeger W and Andrés-Lacueva C. 2007. HPLC–tandem mass spectrometric method to characterize resveratrol metabolism in humans. Clin Chem 53(2):292–299.

US Department of Agriculture. 2004. USDA database for the proanthocyanidin content of selected foods. Beltsville, MD: USDA.

US Department of Agriculture. 2007a. USDA database for the flavonoid content of selected foods. Beltsville, MD: USDA.

US Department of Agriculture. 2007b. USDA–Iowa state university database on the isoflavone content of foods. Beltsville, MD: USDA.

Vallejo F, Tomás-Barberán FA and Ferreres F. 2004. Characterisation of flavonols in broccoli (*Brassica oleracea* L. var. italica) by liquid chromatography-UV diode-array detection-electrospray ionisation mass spectrometry. J Chromatogr A 1054(1–2):181–193.

Vergne S, Titier K, Bernard V, Asselineau J, Durand M, Lamothe V, Potier M, Perez P, Demotes-Mainard J, Chantre P, Moore N, Bennetau-Pelissero C and Sauvant P. 2007. Bioavailability and urinary excretion of isoflavones in humans: effects of soy-based supplements formulation and equol production. J Pharm Biomed Anal 43(4):1488–1494.

Vieira LMM and Kijjoa A. 2005. Naturally-occurring xanthones: recent developments. Curr Med Chem 12(21):2413–2446.

Visioli F and Galli C. 1998. Olive oil phenols and their potential effects on human health. J Agric Food Chem 46(10):4292–4296.

Walker EB. 2007. HPLC analysis of selected xanthones in mangosteen fruit. J Sep Sci 30(9):1229–1234.

Wang J and Sporns P. 2000a. MALDI-TOF MS analysis of food flavonol glycosides. J Agric Food Chem 48(5):1657–1662.

Wang J and Sporns P. 2000b. MALDI-TOF MS analysis of isoflavones in soy products. J Agric Food Chem 48(12):5887–5892.

Waridel P, Wolfender J-L, Ndjoko K, Hobby KR, Major HJ and Hostettmann K. 2001. Evaluation of quadrupole time-of-flight tandem mass spectrometry and ion-trap multiple-stage mass spectrometry for the differentiation of *C*-glycosidic flavonoid isomers. J Chromatogr A 926(1):29–41.

Waterhouse AL and Lamuela-Raventós RM. 1994. The occurrence of piceid, a stilbene glucoside, in grape berries. Phytochemistry 37: 571–573.

Wawrzynowicz T and Waksmundzka Hajnos M. 1990. The application of systems with different selectivity for the separation and isolation of some furocoumarins. J Liq Chromatogr 13(20):3925–3940.

Werner E, Heilier JF, Ducruix C, Ezan E, Junot C and Tabet JC. 2008. Mass spectrometry for the identification of the discriminating signals from metabolomics: Current status and future trends. J Chromatogr B Analyt Technol Biomed Life Sci 871(2):143–163.

Wilson TC and Hagerman AE. 1990. Quantitative determination of ellagic acid. J Agric Food Chem 38: 1678–1683.

Wojdylo A, Oszmianski J and Laskowski P. 2008. Polyphenolic compounds and antioxidant activity of new and old apple varieties. J Agric Food Chem 56(15):6520–6530.

Wolfender JL, Ndjoko K and Hostettmann K. 2003. Application of LC-NMR in the structure elucidation of polyphenols. In: Santos-Buelga C, Williamson G, editors. Methods in Polyphenol Analysis. Cambridge, UK: Royal Society of Chemistry, pp. 128–156.

Xia YQ, Guo TY, Zhao HL, Song MD, Zhang BH and Zhang BL. 2007. A novel solid phase for selective separation of flavonoid compounds. J Sep Sci 30(9):1300–1306.

Xu GH, Kim JA, Kim SY, Ryu JC, Kim YS, Jung SH, Kim MK and Lee SH. 2008. Terpenoids and coumarins isolated from the fruits of *Poncirus trifoliata*. Chem Pharm Bull (Tokyo) 56(6):839–842.

Yamaguchi F, Ariga T, Yoshimura Y and Nakazawa H. 2000. Antioxidative and anti-glycation activity of garcinol from *Garcinia indica* fruit rind. J Agric Food Chem 48(2):180–185.

Yeh CT and Yen GC. 2005. Effect of vegetables on human phenolsulfotransferases in relation to their antioxidant activity and total phenolics. Free Radic Res 39(8):893–904.

Zadernowski R, Czaplicki S and Naczk M. 2009. Phenolic acid profiles of mangosteen fruits (*Garcinia mangostana*). Food Chem 112(3):685–689.

Zamora-Ros R, Andrés-Lacueva C, Lamuela-Raventós RM, Berenguer T, Jakszyn P, Martínez C, Sánchez MJ, Navarro C, Chirlaque MD, Tormo MJ, Quirós JR, Amiano P, Dorronsoro M, Larrañaga N, Barricarte A, Ardanaz E and González CA. 2007. Concentrations of resveratrol and derivatives in foods and estimation of dietary intake in a Spanish population: European Prospective Investigation into Cancer and Nutrition (EPIC)-Spain cohort. Br J Nutr 100(1):188–196.

Zhao B and Hall CA III. 2008. Composition and antioxidant activity of raisin extracts obtained from various solvents. Food Chem 108(2):511–518.

Zuo Y, Wang C and Zhan J. 2002. Separation, characterization, and quantitation of benzoic and phenolic antioxidants in American cranberry fruit by GC-MS. J Agric Food Chem 50(13):3789–3794.

3 Synthesis and Metabolism of Phenolic Compounds

Mikal E. Saltveit

Introduction

Plant metabolism can be separated into primary pathways that are found in all cells and deal with manipulating a uniform group of basic compounds, and secondary pathways that occur in specialized cells and produce a wide variety of unique compounds. The primary pathways deal with the metabolism of carbohydrates, lipids, proteins, and nucleic acids and act through the many-step reactions of glycolysis, the tricarboxylic acid cycle, the pentose phosphate shunt, and lipid, protein, and nucleic acid biosynthesis. In contrast, the secondary metabolites (e.g., terpenes, alkaloids, phenylpropanoids, lignin, flavonoids, coumarins, and related compounds) are produced by the shikimic, malonic, and mevalonic acid pathways, and the methylerythritol phosphate pathway (Fig. 3.1). This chapter concentrates on the synthesis and metabolism of phenolic compounds and on how the activities of these pathways and the compounds produced affect product quality.

Phenolic compounds encompass a diverse group of molecules that include flavonoids (flavones, anthocyanidins), stilbenes, tannins, lignans, and lignin. Of the nearly 10,000 phenolics found in plants, some are soluble in organic solvents, some are water-soluble, and others are large insoluble polymers. The synthesis of plant phenols can proceed by several different pathways, and therefore they comprise a diverse metabolic group. Their chemical diversity is matched by their varied role in the plant (Table 3.1). Some function in mechanical support, whereas others protect the plant from harmful ultraviolet solar radiation and excessive water loss. Some attract pollinators and seed dispersers, whereas others serve as signals that induce defensive reactions to biotic or abiotic stresses. Some of these compounds can suppress the growth of nearby competing plants (i.e., allelopathy); others provide protection against herbivores and pathogens.

Induced products of secondary metabolism often are present in large amounts. Phenolic compounds are second only to cellulose in making up the bulk of organic matter, with phenolics (mainly lignin) accounting for about 40% of the organic carbon in the biosphere. The evolution and domination of land by vascular plants would have been impossible without phenolic compounds to protect against ultraviolet-B (UV-B) damage, lignin to provide mechanical support, and the phenolic-containing compounds suberin and cutin to produce epidermal barriers that minimize water loss.

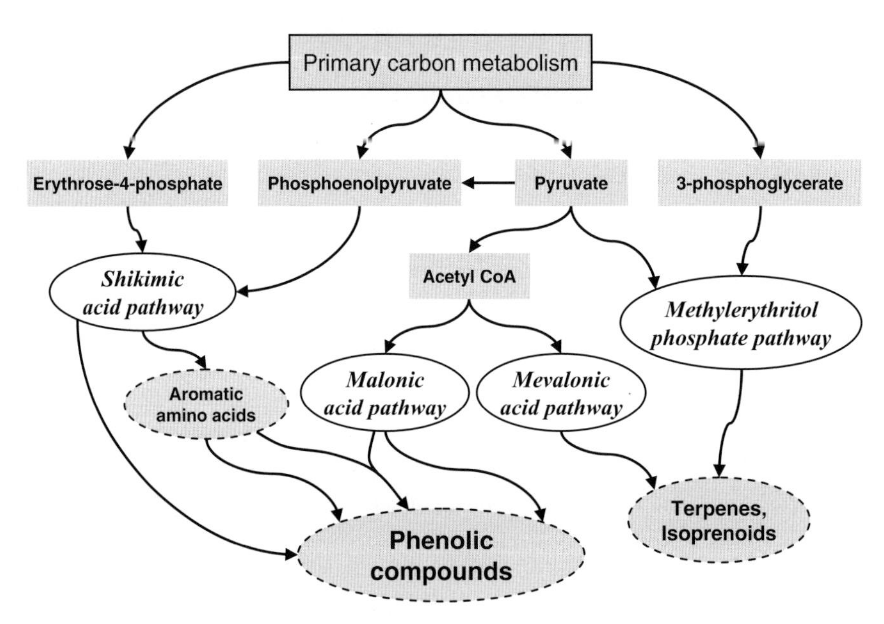

Figure 3.1. Overview of the major biosynthetic pathways giving rise to secondary metabolism.

Structure of some simple phenolic compounds

Addition of a hydroxyl group to a six-carbon benzene ring yields 1-hydroxybenzene or phenol, which forms the basis of all subsequent phenolic compounds (Fig. 3.2). Resorcinol or 1,4-dihydroxybenzene is formed when a second hydroxyl group is added at the 4-position on the benzene ring.

Replacement of the hydroxyl group on the phenyl ring with a carboxyl group forms a molecule of benzoic acid. Addition of a hydroxyl at the 2-position on a benzoic acid molecule forms 2-hydroxybenzoic acid or salicylic acid. The slightly more complex phenylpropanoid skeleton contains a linear three-carbon chain (the propanoic group) added to the benzene ring (the phenyl group). Addition of ammonia to carbon 2 of this three-carbon side chain yields the amino acid phenylalanine (Fig. 3.3). Phenylalanine

Table 3.1. Important classes of phenylpropanoids

Compound	Property
Coumarins	Antibiotics, discourages herbivores
Cutin	External barrier to water and gas diffusion in aerial parts
Flavonoids	Antimicrobial, signals, pigments, UV protection
Lignan	Antibiotics, discourages herbivores [$(C_6\text{-}C_3)_2$]
Lignin	Strengthens cell wall [$(C_6\text{-}C_3)_n$]
Suberin	External and internal barrier to water and gas diffusion (e.g., in roots)
Stilbenes	Antibiotics, fungicides
Tannin	Fungicides, discourages herbivores

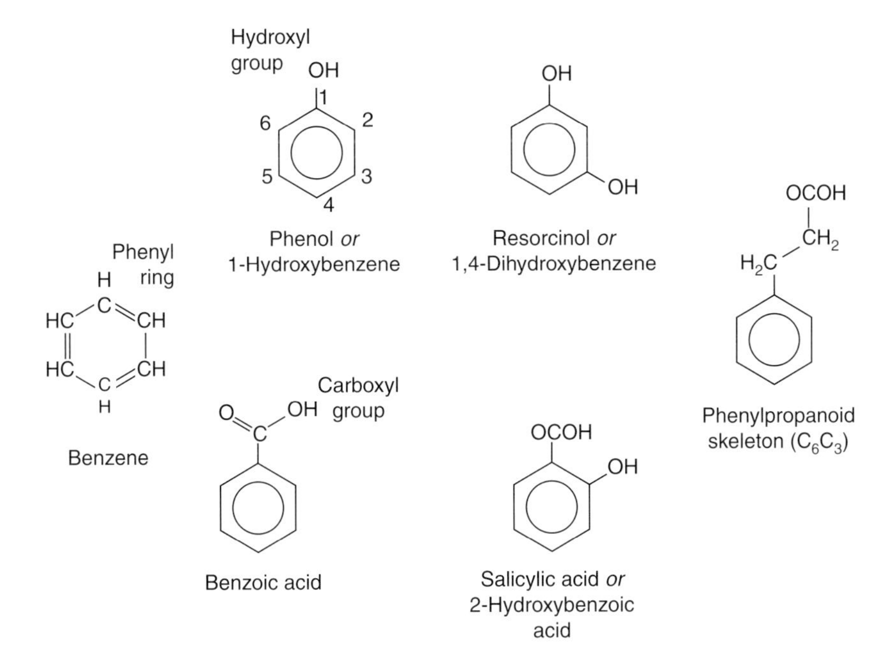

Figure 3.2. Structure of simple phenolic compounds.

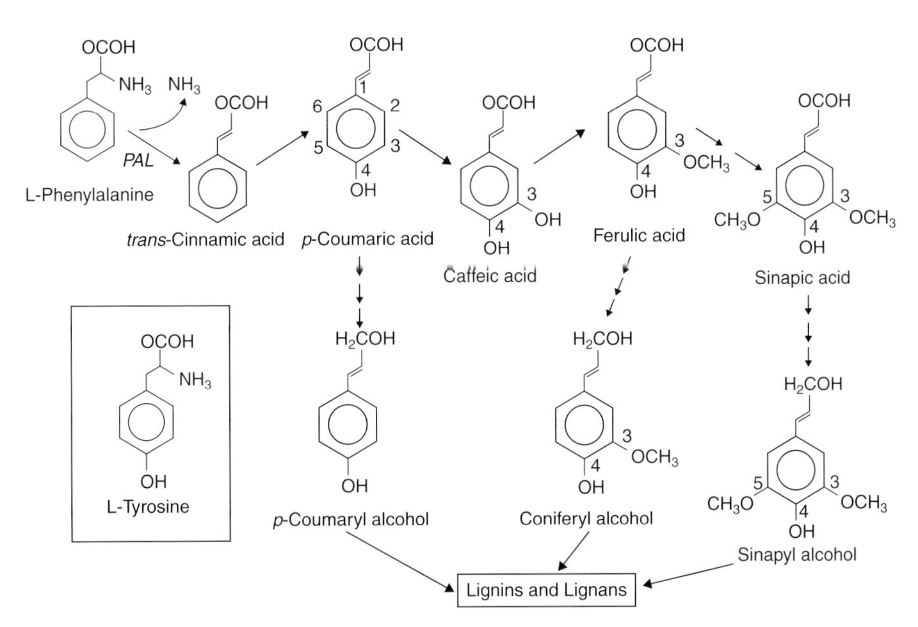

Figure 3.3. General phenylpropanoid pathway. Each arrow represents one enzymatic reaction.

can be thought of as a molecule of alanine (a three-carbon amino acid in place of the propanoic group) attached to a phenyl ring; hence the name *phenyl + alanine*. Addition of a hydroxyl group at the 4-position on the phenyl ring of phenylalanine produces another aromatic amino acid called tyrosine. Molecules containing the six-carbon ring structure of benzene are called aromatic compounds.

Synthesis of phenylpropanoids

Precursors of phenylpropanoids are synthesized from two basic pathways: the shikimic acid pathway and the malonic pathway (see Fig. 3.1). The shikimic acid pathway produces most plant phenolics, whereas the malonic pathway, which is an important source of phenolics in fungi and bacteria, is less significant in higher plants. The shikimate pathway converts simple carbohydrate precursors into the amino acids phenylalanine and tyrosine. The synthesis of an intermediate in this pathway, shikimic acid, is blocked by the broad-spectrum herbicide glyphosate (i.e., Roundup). Because animals do not possess this synthetic pathway, they have no way to synthesize the three aromatic amino acids (i.e., phenylalanine, tyrosine, and tryptophan), which are therefore essential nutrients in animal diets.

Many secondary phenolic compounds are derived from the amino acids phenylalanine and tyrosine and therefore contain an aromatic ring and a three-carbon side chain (see Fig. 3.3). Phenylalanine is the primary substrate for phenylpropanoid synthesis in most higher vascular plants, with tyrosine being used to a lesser extent in some plants. Because of their common structure, compounds derived from these amino acids are collectively called phenylpropanoids.

Simple phenolic compounds include: (1) the phenylpropanoids, *trans*-cinnamic acid, *p*-coumaric acid and their derivatives; (2) the phenylpropanoid lactones called coumarins (Fig. 3.4); and (3) benzoic acid derivatives in which two carbons have been cleaved from the three carbon side chain (Fig. 3.2). More complex molecules are elaborated by additions to these basic carbon skeletons. For example, the addition of quinic acid to caffeic acid produces chlorogenic acid, which accumulates in cut lettuce and contributes to tissue browning (Fig. 3.5).

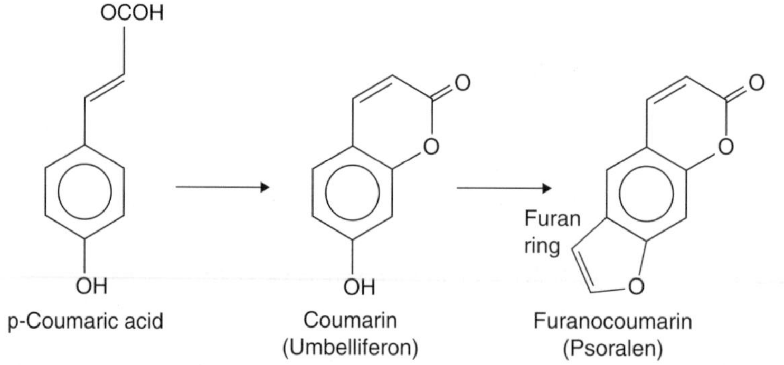

Figure 3.4. Synthesis of coumarin and furanocoumarin from *p*-coumaric acid.

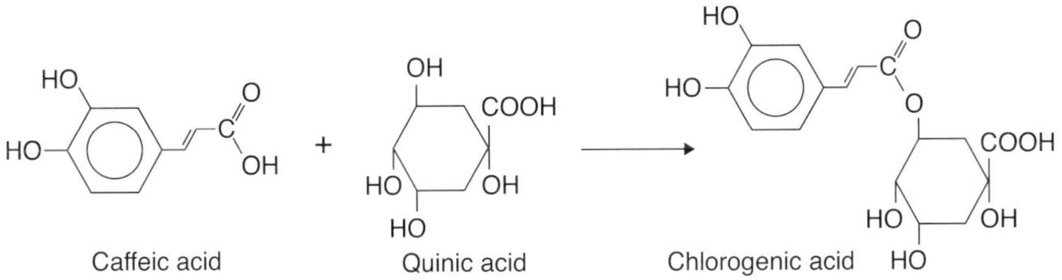

Figure 3.5. Addition of quinic acid to caffeic acid produces chlorogenic acid.

The key reaction that links primary and secondary metabolism is provided by the enzyme phenylalanine ammonia lyase (PAL) which catalyzes the deamination of L-phenylalanine to form *trans*-cinnamic acid with the release of NH_3 (see Fig. 3.3). Tyrosine is similarly deaminated by tyrosine ammonia lyase (TAL) to produce 4-hydroxycinnamic acid and NH_3. The released NH_3 is probably fixed by the glutamine synthetase reaction. These deaminations initiate the main phenylpropanoid pathway.

PAL activity is under the control of both internal developmental and external environmental regulation. Production of lignin for strengthening the secondary wall of xylem cells and production of pigments to attract pollinators to flowers necessitate an increase in PAL activity. Biotic (e.g., pathogen attack) and abiotic (e.g., mechanical injury) stresses also increase PAL activity. Both activators appear to act through increased transcription of PAL mRNA. Multiple genes encode for PAL, some of which are only activated in specific tissue or under certain environmental signals. Translation and subsequent protein modification yields the active PAL enzyme, which then stimulates the phenylpropanoid pathway.

A series of subsequent reactions after PAL first introduces a hydroxyl at the 4-position of the ring of cinnamic acid to form *p*- or 4-coumaric acid (i.e., 4-hydroxycinnamic acid). Addition of a second hydroxyl at the 3-position yields caffeic acid, whereas O-methylation of this hydroxyl group produces ferulic acid (see Fig. 3.3). Two additional enzymatic reactions are necessary to produce sinapic acid. These hydrocinnamic acids are not found in significant amounts in plant tissue because they are rapidly converted to coenzyme A esters, or glucose esters. These activated intermediates form an important branch point because they can participate in a wide range of subsequent reactions.

Coumarins

Coumarins are widespread in plants where they serve a defensive role. Severe internal bleeding in mammals can follow the consumption of clover that contains high levels of coumarins. Its effect in impeding blood clotting was exploited in the development of the rodenticide warfarin and in development of human blood thinners to treat and prevent stroke. Attachment of a furan ring to the basic coumarin structure produces furanocoumarins (see Fig. 3.4), which are notable for their phytotoxicity. Sunlight in the ultraviolet region (320–400 nm) transforms nontoxic furanocoumarins into activated compounds that can disrupt DNA transcription and repair. These compounds are abundant in the Umbelliferae family, which includes celery, parsnips, and parsley. Stress and disease can induce a 100-fold increase in the level of these compounds in celery. Celery harvesters can become sensitized to these compounds and develop skin rashes. Herbivorous insects have developed strategies to avoid these compounds by producing webs or rolling up leaves so they are not exposed to the activating wavelengths of sunlight.

Formation of lignans and lignin

A series of four reactions reduces the carboxylic group of 4-coumaric acid, ferulic acid, and sinapic acid into their corresponding alcohols, that is, 4-coumaric alcohol, coniferyl alcohol, and sinapyl alcohol, respectively. These three alcohols are called monolignols. Dimerization of monolignols produces lignans, which are defense compounds against bacteria and fungi. Some lignans, such as podophyllotoxin, have been used as treatments in humans to combat cancer and acquired immunodeficiency syndrome (AIDS).

Lignin is an integral component of the cell walls of all vascular plants; the name derives from the Latin word for wood. Oxidation of monolignols by a free radical mediated reaction produces random intermolecular bonds. This polymerization reaction occurs within the cell wall and produces a strong, hydrophobic three-dimensional matrix that covalently links the cellulose and proteins within the cell wall. Because of the random nature of the chemical bonds in lignin, it is difficult to attack enzymatically, and few organisms can degrade lignin. The composition of lignin varies among species. Conifers have lignin with high coniferyl content, whereas cereal lignans have higher coumaryl content.

Lignified cell walls are analogous to reinforced concrete or fiberglass, wherein the cellulose fibers are analogous to the metal reinforcing rods or the glass fibers, while lignin is analogous to the concrete or resin that fills in the gaps and binds it all together. Lignin is a major component of secondary cell walls and provides the mechanical strength that allows land plants to attain significant height and conduct water under tension through reinforced xylem cells. Lignification of adjoining cells is a common response to infections and wounding of herbaceous tissues. A physical and physiological barrier is thereby erected to thwart microbial assault and strengthen the damaged tissue.

Lignin must be removed during the production of cellulose and paper from wood pulp. This is an expensive operation and produces pollution. Genetic engineering is being used to produce trees with reduced amounts of and more easily extracted lignin.

Synthesis of suberin and cutin

The conquest of the land by plants necessitated the development of a coating, the cuticle, that would reduce water loss. Suberin and cutin vary in their proportion of fatty acids, fatty alcohols, hydroxyfatty acids, and dicarboxylic acids. The cuticle is synthesized and excreted by the epidermis of aerial portions of the plant, such as the primary stems, leaves, flower organs, and fruits. The two major hydrophobic layers that contribute to the cuticle are composed of phenolic molecules combined with lipid polymers. Cutin is a polymer found in the outer cell wall of the epidermis, which is overlaid by an excreted layer of wax platelets. Cutin monomers include C16 and C18 fatty acids, whereas waxes are biosynthetically distinct long chains (C24 to C34) of saturated hydrocarbons.

In contrast to the exterior localization of cutin, suberin can be deposited in both external and internal tissues. External deposition occurs in the periderm of secondary roots and stems and on cotton fibers, whereas internal deposition occurs in the root endodermis and the bundle sheath of monocots. The Casparian strip of the root endodermis contains suberin, which produces a barrier isolating the apoplast of the root cortex from the central vascular cylinder. Suberin also produces a gas-impermeable barrier between the bundle sheath and mesophyll cells in C_4 plants. The bark of trees contains periderm-derived cork cells that have a high suberin content.

These gas- and water-impermeable cell layers protect the plant from desiccation, but they also hamper the uptake of carbon dioxide necessary for photosynthesis and oxygen necessary for respiration. Specialized tissues have evolved to allow passive (lenticels) and active (guard cells) modification of the permeability of the external cuticle to gas exchange.

Flavonoids

Flavonoids are the largest class of phenylpropanoids in plants. The basic flavonoid structure is two aromatic rings (one from phenylalanine and the other from the condensation of three malonic acids) linked by three carbons (Fig. 3.6). Chalcone is converted to naringenin by thc enzyme chalcone isomerase, which is a key enzyme in flavonoid synthesis. This enzyme, like PAL and chalcone synthase (CHS), is under precise control and is inducible by both internal and external signals. Naringenin is the

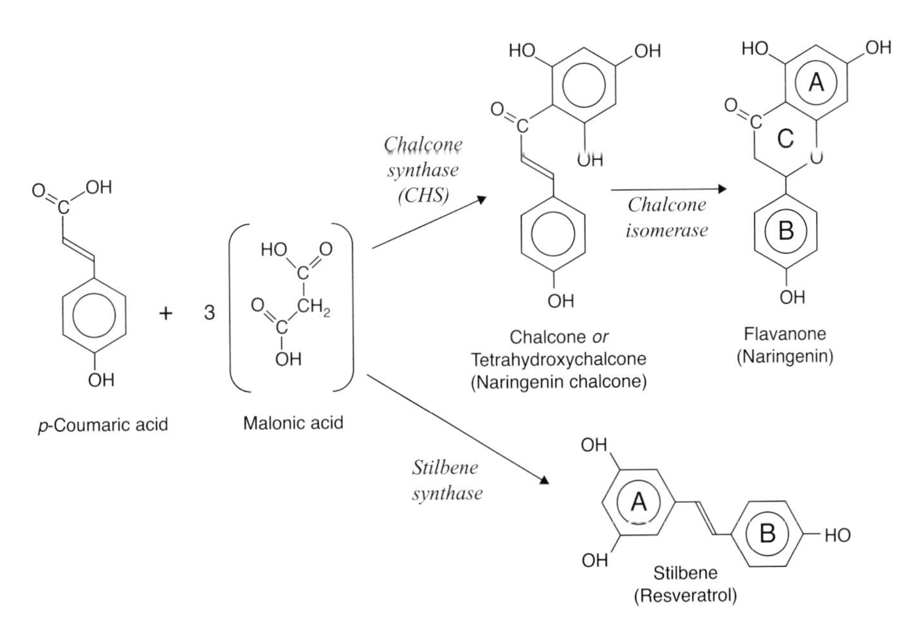

Figure 3.6. Synthesis of chalcone, flavanone, and stilbene from *p*-coumaric and malonic acids.

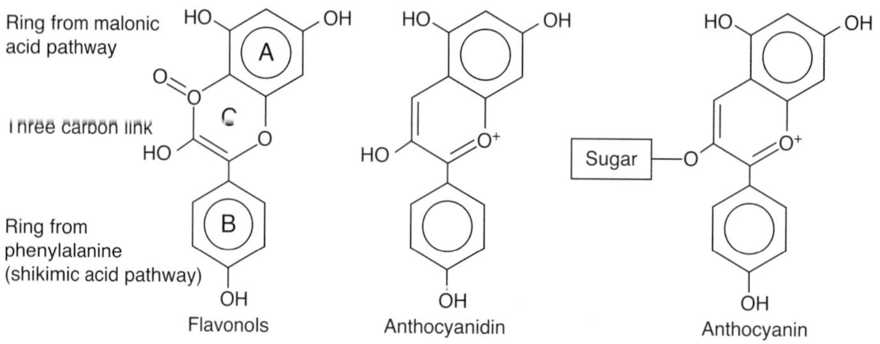

Figure 3.7. Structural relationships among flavonols, anthocyanidin, and anthocyanin.

progenitor of basically all subsequent flavonoid structures. Four of the major groups of the flavonoids are the anthocyanins, the flavones, the flavonols, and the isoflavones (Fig. 3.7). They mainly differ in the degree of oxidation of the three-carbon link.

Anthocyanins are colored flavonoids that attract animals when a flower is ready for pollination or a fruit is ready to eat. They are glycosides (i.e., the molecule contains a sugar) that range in color from red, pink, and purple to blue depending on the number and placement of substitutes on the B ring (see Fig. 3.7), the presence of acid residues, and the pH of the cell vacuole where they are stored. Without the sugar these molecules are called anthocyanidins. The color of some pigments results from a complex of different anthocyanin and flavone molecules with metal ions.

Flavones and *flavonols* are two major groups of flavonoids that absorb light of shorter wavelengths than those visible to humans. Insects such as bees can see into the ultraviolet spectrum and are attracted by the specific pattern these pigments make in some flowers. The presence of these two groups of pigments in the leaves of all green plants provides protection against excessive UV-B (280–320 nm) radiation. They absorb strongly in the UV-B region, while letting photosynthetically active wavelengths pass through undiminished. Exposure to UV-B radiation stimulates the synthesis and accumulation of flavones and flavonols. Flavonoids also play a role in development. Secretion of flavonoids into soil by roots of legumes facilitates symbiosis with nitrogen-fixing bacteria.

Isoflavones are a group of flavonoids in which the position of the B ring is shifted (Fig. 3.8). Unlike most other flavonoids, isoflavonoids have a limited taxonomic distribution, being mainly found in legumes. However, their chemical structures exhibit an unusually wide range of modifications. The poisonous isoflavone rotenone, which inhibits electron transport in the mitochondria and thereby respiration in animals, is found in the leaves of tropical legumes. Natives of South America immobilized fish by throwing crushed leaves of certain legumes into the water. Other legumes contain isoflavones that mimic the effect of estrogens. Genistein is an example of an isoflavone with strong estrogen activity. Sheep have become infertile when grazing on clover rich in isoflavonoids. The anticancer benefits of food made from soybeans (e.g., tofu) may result from the isoflavonoids they contain. These phytoestrogens are of considerable

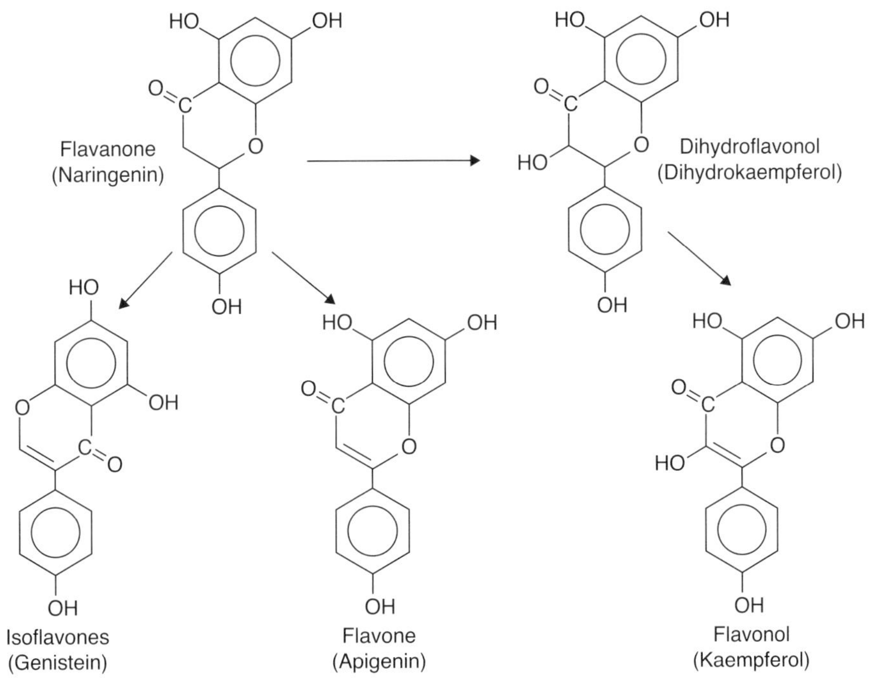

Figure 3.8. Structural relationships among the flavonoids.

medical interest. Other isoflavonoids have antimicrobial activity (i.e., phytoalexins) and are synthesized in response to bacteria of fungal attack.

Many flavonoids have antioxidant properties and as such are thought to protect humans against cardiovascular disease and cancer. Foods containing flavonoids (e.g., green tea, soy, and red wine) are touted as beneficial for health.

Stilbenes

Condensation of coumaric acid with malonic acid yields the basic chalcone and stilbane skeletons (see Fig. 3.6). Stilbenes are found in most vascular plants, where they exhibit fungicidal and to a lesser extent antibiotic properties. They function as both constitutive and inducible defense substances. Some stilbenes inhibit fungal spore germination and hyphal growth, whereas others are toxic to insects and parasitic nematodes (roundworms). They also possess antifeeding and nematicide properties in mammals. For example, resveratrol (a stilbene in red wine) suppresses tumor formation in mammals.

Tannin

Tannins, like lignins, are polymerized phenolics with defensive properties. Their name comes from their use in tanning rawhides to produce leather. In tanning, collagen

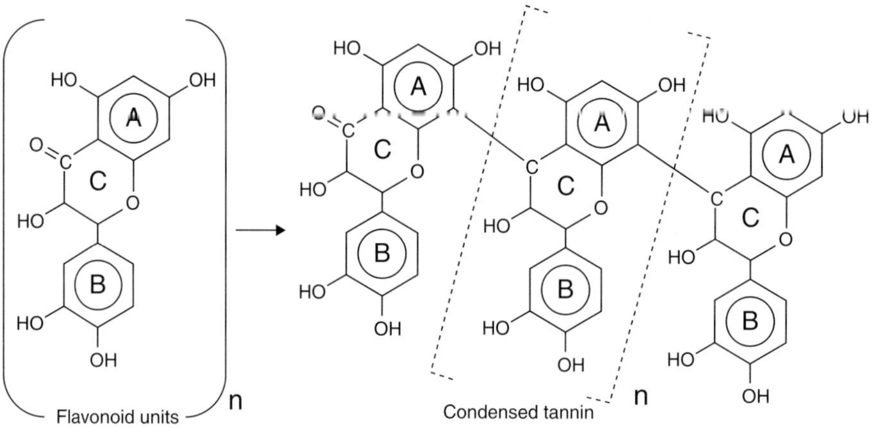

Figure 3.9. Structure of condensed tannins.

proteins are bound together with phenolic groups to increase the hide's resistance to water, microbes, and heat.

There are two categories of tannins: condensed and hydrolyzable tannins. The polymerization of flavonoid molecules produces condensed tannins, which are commonly found in woody plants (Fig. 3.9). Hydrolyzable tannins are also polymers, but they are a more heterogeneous mixture of phenolic acids (especially gallic acid) and simple sugars. Though widely distributed, their highest concentration is in the bark and galls of oaks.

Tannins have an unpleasant astringent taste that results from their nonspecific binding to proteins in tissues lining the mouth. Eating plants high in tannins results in their binding proteins in the food, thereby rendering these plants less digestible. Tannins may also bind to digestive enzymes in the gut, thereby reducing their activity. Injury or infection of plant tissue can stimulate the synthesis and accumulation of these phenolic molecules to levels that can render the plant toxic to herbivores.

Herbivores that commonly feed on tannin-rich plants have evolved interesting methods to lessen the effect of ingested tannins on their digestive systems. For example, the salivary proteins of rabbits and other rodents are high in the amino acid proline, which has a very high affinity for tannins. Eating food high in tannins stimulates the secretion of these proteins and diminishes the toxic effect of the tannins.

Secondary metabolism and product quality

Plants have the innate capacity, gained over eons of evolution, to actively respond to biotic and abiotic stresses in their environment. Many of the stresses that occur naturally are frequently mimicked by the operations that constitute the cultivation, harvesting, marketing, and consumption of fruits and vegetables. Metabolic responses induced by stress and that are adaptive in nature can result in reduced product quality.

This chapter has focused on the synthesis and metabolism of phenolic compounds that constitute the most important induced secondary products affecting product quality.

The perception of an environmental stress and its transduction into a physiological response is an area of active research. The importance of appropriately responding to a multitude of stresses has resulted in the evolution of complex signaling pathways that include the plant hormones abscisic acid and ethylene and products of lipid and phenolic metabolism such as jasmonic and salicylic acid. The interrelationships among the signals and their induced responses make it difficult to genetically engineer plants to respond appropriately to stresses for maximal yield and quality while limiting those portions of the response that detrimentally affect product quality.

Constitutive reactions and preharvest stresses can result in the accumulation of phenolic compounds to levels that can rapidly cause discoloration of the commodity on injury. The high level of phenolic compounds found in some cultivars of fruits and vegetables (e.g., apples and potatoes) causes them to discolor when subjected to stresses that result in cellular decompartmentation. The mixing of phenolic compounds with polyphenol oxidase and peroxidase enzymes in the presence of oxygen produces colored pigments. Discoloration can be minimized by selecting cultivars with constitutively low levels of phenolic compounds, by enhancing membrane stability, by using low temperatures to slow the enzymatic and nonenzymatic reactions, and by excluding oxygen from the post-trauma environment. In these instances, phenolic accumulation resulted from stresses experienced prior to the stress that actually caused the discoloration.

In contrast, other fruits and vegetables have low levels of phenolic compounds at harvest and do not immediately discolor when stressed. In these commodities (e.g., lettuce), a stress induces the synthesis and accumulation of phenolic compounds that form colored pigments in the cells near the injury that induced the response. Limiting the generation and transmission of stress signals will ameliorate the loss of product quality in the latter instance, whereas it will have no effect in the former. In a similar manner, the accumulation of stress-induced monolignols (i.e., 4-coumaric, coniferyl, and sinapyl alcohols) often precedes their polymerization to form lignin, which can toughen affected tissues.

Exposure to volatile stress signals such as ethylene and jasmonic acid can induce the synthesis and accumulation of high levels of phenolic compounds. The effect is not readily apparent because the accumulated phenolic compounds remain sequestered in organelles within the induced cells. However, increased membrane permeability resulting from even a minor injury or stress can result in levels of discoloration that are abnormally high for that cultivar.

Although the constitutive and induced synthesis of a wide variety of phenolic compounds is critical for optimal growth and development of fruits and vegetables, overproduction and excessive accumulation can result in reduced product quality. The many complex and interconnected pathways make their manipulation through traditional breeding and genetic engineering a daunting task. Management of cultural procedures and techniques of harvest and storage to reduce phenolic metabolism may encounter consumer resistance when chemicals or practices are employed that are considered unhealthy to the individual or deleterious to the environment. Understanding the synthesis and metabolism of phenolic compounds should provide the insights needed to formulate techniques that will improve product quality.

General References

Bell EA. 1981. The physiological role(s) of secondary (natural) products. In: Conn EE, editor. The Biochemistry of Plants. Volume 7, Secondary Plant Products. New York: Academic Press, pp. 1–19.

Croteau R, Kutchan TM and Lewis NG. 2000. Natural products (Secondary metabolites). In: Buchanan BB, Gruissem W, Jones RL, editors. Biochemistry and Molecular Biology of Plants. Somerset, NJ: John Wiley & Sons, pp. 1250–1318.

Heldt HW and Heldt F. 2005. A large diversity of isoprenoids has multiple functions in plant metabolism. In: Plant Biochemistry, 3rd ed. San Diego, CA: Elsevier Academic Press, pp. 413–434.

Heldt HW and Heldt F. 2005. Phenylpropanoids comprise a multitude of plant secondary metabolites and cell wall components. In: Plant Biochemistry, 3rd ed. San Diego, CA: Elsevier Academic Press, pp. 435–454.

Heldt HW and Heldt F. 2005. Secondary metabolites fulfill specific ecological functions in plants. In: Plant Biochemistry, 3rd ed. San Diego, CA: Elsevier Academic Press, pp. 402–412.

Taiz L and Zeiger E. 2006. Secondary metabolites and plant defense. In: Plant Physiology, 4th ed. Sunderland, MA: Sinauer Associates, pp. 316–344.

4 Enzymatic and Nonenzymatic Degradation of Polyphenols

José Manuel López-Nicolás and Francisco García-Carmona

Introduction

In recent years, numerous papers have been published about one of the most important groups of phytochemicals, the polyphenols (Manach and others 2004). These compounds, which possess an array of healthy properties, but also some disadvantages that will be discussed in this chapter, are present in a variety of plants used in both human and animal diets. However, the structure of this type of compound means that they can be oxidized by several pro-oxidant agents. The objective of this chapter is to describe the main enzymatic agents responsible for the degradation of polyphenols. In order to understand the mechanisms of degradation that will be described in the following sections, a brief summary of the main properties of the polyphenols is required.

General Features

One of the largest groups of compounds are the polyphenols, with nearly 10,000 phenolic structures currently known (Urquiaga and Leighton 2000). This type of molecule possesses an aromatic ring bearing one or more hydroxyl substituents and is a product of the secondary metabolism of plants. Table 4.1 shows the major classes of plant polyphenols, according to the number of carbon atoms present in the molecule (Manach and others 2004). Indeed, polyphenols may be simple molecules (e.g., phenolic acids) or complex structures (e.g., highly polymerized compounds such as condensed tannins).

Occurrence of Polyphenols in Foods

The presence of polyphenols in foods has been widely reported, fruits and several beverages being the main sources of these compounds in the diet (Urquiaga and Leighton 2000). Although several classes of phenolic molecules such as quercetin can be found in most plant foods (wine, tea, cereals, legumes, fruit juices, etc.), others are found only in a specific type of food (flavanones in citrus fruit, isoflavones in soya, phloridzin in apples, etc.). However, in nature, it is common for several types of polyphenols to be found in the same food product. Such is the case with apples, which contain flavanols, chlorogenic acid, hydroxycinnamic acids, phloretin glycosides, quercetin glycosides and anthocyanins. Moreover, several parameters, such as ripeness at the time of harvest, environmental factors, processing, storage and the plant variety can influence polyphenol composition. However, the polyphenol composition is still not known in the case of most fruits and some cereal varieties, and further investigations are required (Urquiaga and Leighton 2000).

Table 4.1. Types of phenolic compounds in plants

Number of Carbon Atoms	Basic Skeleton	Class
6	C_6	Simple phenols Benzoquinones
7	C_6-C_1	Phenolic acids
8	C_6-C_2	Acetophenones Tyrosine derivatives Phenylacetic acids
9	C_6-C_3	Hydroxycinnamic acids Phenylpropenes Coumarins Isocoumarins Chromones
10	C_6-C_4	Naphthoquinones
13	C_6-C_1-C_6	Xanthones
14	C_6-C_2-C_6	Stilbenes Anthraquinones
15	C_6-C_3-C_6	Flavonoids Isoflavonoids
18	$(C_6$-$C_3)_2$	Lignans Neolignans
30	$(C_6$-C_3-$C_6)_2$	Biflavonoids
N	$(C_6$-$C_3)_n$ $(C_6)_n$ $(C_6$-C_3-$C_6)_n$	Lignins Catechol melanins Flavolans (condensed tannins)

As regards their physiological roles, many polyphenols (because of their antimicrobial and antioxidant properties) are directly involved in the response of plants to various types of stress (López-Nicolás and others 2006; López-Nicolás and García-Carmona 2008).

Healthy Properties of Polyphenols

The widely studied antioxidant properties of phenolic compounds mean that these molecules have several biological effects, including the inhibition of LDL oxidation *in vitro* and *in vivo* and the protection of DNA from oxidative damage; they also have antithrombotic, anti-inflammatory, antimicrobial, anticarcinogenic, and antimutagenic properties (Manach and others 2004). Moreover, several studies have reported that polyphenols can inactivate carcinogens or inhibit the expression of mutant genes and the activity of enzymes involved in the activation of procarcinogens. Other biological actions attributed to polyphenols include their capacity to induce mutagenesis in microbial assays and to act as cocarcinogens or promoters in inducing skin carcinogenesis (Urquiaga and Leighton 2000).

Degradation of Polyphenols

Because of the easily oxidizable structure of the polyphenols previously described, many studies have been published about the enzymatic degradation of these antioxidant compounds. This chapter exhaustively reviews the main publications concerning the degradation of this type of antioxidant compound by several enzymes.

The effectiveness of enzymes in the degradation of polyphenols can be attributed partially to their high redox potentials, which enable them to facilitate reaction with chemically resistant (stable) starting compounds. The reaction products obtained after the oxidation of polyphenols by the enzymes discussed in this chapter may be important in the chemical, pharmaceutical, agrochemical, and food industries. Moreover, these oxidative enzymes can influence primary or secondary metabolism. Briefly, the key advantages in using oxidases for chemical applications are (Burton 2003): (1) oxidation reactions that are readily achieved or controlled, (2) stereoselectivity, (3) regiospecificity, (4) addition of functionality, (5) introduction of chirality, (6) increased degree of hydroxylation, and (7) increased molecule hydrophilicity and polarity. For these reasons, a review of the main oxidant enzymes that degrade polyphenols to their correspondent products is necessary.

Main Groups of Degradative Enzymes of Polyphenols

Many papers have been published about the enzymatic degradation of polyphenols through the action of oxidizing enzymes. Thus, various classifications have been provided for these types of biocatalytic molecules, according to their coenzyme requirements or according to the nature of the oxidizing substrate (the electron acceptor) and the reaction products (Fig. 4.1).

An excellent classification of this type of enzymes was that of Burton (2003). Because of their importance in polyphenol degradation, we have studied four enzymes closely related with the oxidation of phenolic compounds: polyphenoloxidase, peroxidase, laccase, and lipoxygenase.

Oxidases and peroxidases react with oxygen (as molecular oxygen or peroxide, respectively) leading to reactive oxygen intermediates that then react further with reducing substrates. As a result, the reactions may be nonspecific or may produce several different products. Oxygenases introduce one or two oxygen atoms into their substrates and are often more selective than oxidases and peroxidases, particularly in terms of regiospecificity.

Monooxygenases

This class of enzymes catalyzes the introduction of one atom of oxygen into a substrate molecule, generally using NADH or NADPH to provide a reducing potential to supply electrons to the substrate:

$$\text{Substrate} + \text{donor} - \text{H} + O_2 + H^+ \rightarrow \text{Substrate} - \text{O} + \text{donor} + H_2O$$

The systems where this type of reaction is produced may be metal-, heme- or flavin-dependent. In flavin-dependent monooxygenases, a flavin–oxygen intermediate reacts with the substrate, producing water in a second step and requiring cofactors for regeneration of the flavin moiety. The non-heme-dependent oxygenases include the

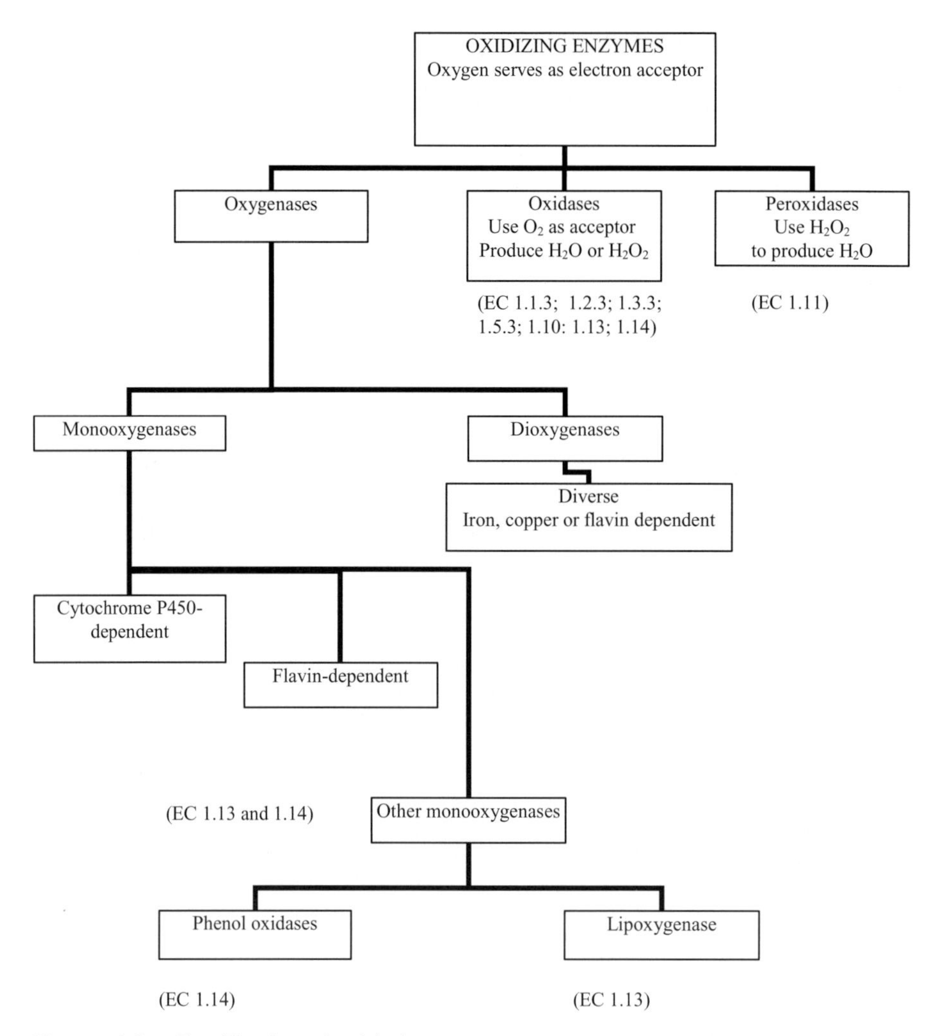

Figure 4.1. Classification of oxidative enzymes.

non-cofactor-dependent copper oxidases, such as polyphenoloxidase and laccase and the flavin–molybdenum–cobalt-dependent xanthine oxidase and aldehyde oxidase.

Dioxygenases

Heme-dependent iron–sulfur plant dioxygenases and Rieske iron–sulfur non-heme dioxygenases are the two main classes of dioxygenases incorporating dioxygen substrates. The reactions catalyzed by these enzymes include aromatic ring cleavage and hydroperoxidation or *cis*-dihydroxylation. As regards their substrate specificity, some compounds such as arenes (toluene, naphthalene) or carboxylates (benzoate, phthalate) are substrates of this type of enzyme. This group of enzymes also includes the ring-cleaving dioxygenases, such as catechol dioxygenase, which incorporate both atoms as hydroxyl groups on adjacent carbons of aromatic rings, generating *cis*-dihydrodiols that can be further oxidized to ring-open products.

Other Oxidases

This group includes three types of enzymes that are able to degrade polyphenols and that are studied in detail in this chapter: polyphenoloxidases, laccases, and lipoxygenases. Polyphenol oxidase (EC 1.14.18.1) is one of the first copper oxidases to be reported. This enzyme is a monooxygenase that uses molecular oxygen as the electron acceptor in the site-specific hydroxylation and oxidation of phenols and catechols (Sánchez-Ferrer and others 1995). Laccases are blue copper oxidases in which four copper atoms are bound in three different redox sites. The catalyzed reaction by laccase is through a free-radical mechanism, using molecular oxygen, in relatively nonspecific reactions. These two classes of enzyme are studied in depth in subsequent sections. Lipoxygenases (linoleate:oxygen reductase, EC 1.13.11.12) are able to oxidize different polyphenol substrates such as stilbenes. LOX possesses two enzymatic activities (dioxygenase and hydroperoxidase) that play a fundamental role in the metabolism of several biological molecules in plants and animals and are reviewed in this chapter.

Peroxidases

One of the most important enzymes responsible for polyphenol degradation is peroxidase (E.C. 1.11.1.7). This oxidative enzyme, which is found in cells of plants, animals, and microbes, is a heme protein with the ferric protoheme group associated to the prosthetic group, whereas other peroxidases have magnesium, vanadium, selenium, or the flavin group at their active sites. As for the action mechanism of these enzymes, peroxidases react nonselectively via free-radical mechanisms, using hydrogen peroxide as the electron acceptor. In the reaction catalyzed by peroxidase, the ferric iron becomes oxidized, forming a reactive Fe(IV)–O species and a radical intermediate of the heme group; the reducing substrate then reacts with the heme radical, producing the oxidized product and regenerating the Fe(III) ion.

Polyphenol Oxidase

As previously indicated, four enzymes related to the enzymatic degradation of polyphenols, polyphenoloxidase, peroxidase, laccase, and lipoxygenase, are extensively reviewed in this chapter. The first enzyme described is polyphenoloxidase.

General Features

Polyphenoloxidase (PPO, EC 1.14.18.1) is one of the most studied oxidative enzymes because it is involved in the biosynthesis of melanins in animals and in the browning of plants. The enzyme seems to be almost universally distributed in animals, plants, fungi, and bacteria (Sánchez-Ferrer and others 1995) and catalyzes two different reactions in which molecular oxygen is involved: the *o*-hydroxylation of monophenols to *o*-diphenols (monophenolase activity) and the subsequent oxidation of *o*-diphenols to *o-quinones* (diphenolase activity). Several studies have reported that this enzyme is involved in the degradation of natural phenols with complex structures, such as anthocyanins in strawberries and flavanols present in tea leaves. Several polyphenols such as catechin (Jiménez-Atiénzar and others 2004), quercetin (Jiménez and García-Carmona 1999), kojic acid (Cabanes and others 1994), fisetin (Jiménez and others 1998), eriodictyol (Jiménez-Atiénzar and others 2005), and 3,4-dihydroxymandelate (Cabanes and others 1988) have been reported to be substrates of PPO.

Structure of Polyphenoloxidase

The crystal structure of one PPO in its active form, from *Ipomoea batatas*, has been solved (Klabunde and others 1998). Monomeric 39,000 Mr catechol oxidase from sweet potato is ellipsoid in shape and measures 55 × 45 × 45 Å. The secondary structure is primarily α-helical with the core of the enzyme formed by a four-helix bundle composed of α-helices α-2, α-3, α-6, and α-7. The helical bundle accommodates the catalytic dinuclear copper center and is surrounded by helices α-1 and α-4 and several short β-strands. Two disulfide bridges (Cys11–Cys28 and Cys27–Cys89) help to anchor the loop-rich N-terminal region of the protein (residues 1–50) to helix α-2. Each of the two active-site coppers is coordinated by three histidine residues from the four helices of the α-bundle. CuA is coordinated by His88, His109, and His118. His88 is located in the middle of helix α-2; His109 and His118 are at the beginning and in the middle of helix α-3. The second catalytic copper, the CuB site, is coordinated by His240, His244, and His274. These residues are found at the middle of helices α6 and α7.

No fungal PPO has yet been crystallized, in either its active or its latent form. From structural studies, it is also apparent that PPOs have distinct features and that not only the amino acid sequences of PPOs differ; differences also exist at the highly conserved active site. The amino acid sequence of a considerable number of PPOs from plants, fungi, and other organisms derived from cloning of the enzyme has now been published, and reports on the molecular weight of plant PPO are very diverse and variable. It must be assumed that part of this variability is due to partial proteolysis of the enzyme during its isolation. Furthermore, because there is obviously a family of genes coding for plant PPO, some multiplicity must be a result of genetic variability.

Sources of Polyphenoloxidase

Among the plant kingdom, PPO is present in many organisms such as grape (Sánchez-Ferrer and others 1989), lettuce (Chazarra and others 2001), peach (Cabanes and others 2007), banana (Sojo and others 2000), quince (Orenes-Piñero and others 2005), cucumber (Gandía-Herrero and others 2003), and eggplant (Pérez-Gilabert and García-Carmona, 2000). In most cases, these enzymes show a wide variation in molecular weights, and multiple isoforms have often been found.

With regard to fungal PPO, this enzyme has been recently reviewed in great detail (Pérez-Gilabert and others 2001; Mayer 2006). Perhaps the isolation of the latent form of PPO in the ascocarp of *Terfezia claveryi* (a truffle), should be recalled, the general behavior of this PPO falling in line with that of other fungal PPOs. Finally, both cresolase and diphenolase activities of PPO have been reported in bacteria such as streptomycetes (Clausa and Decker 2006).

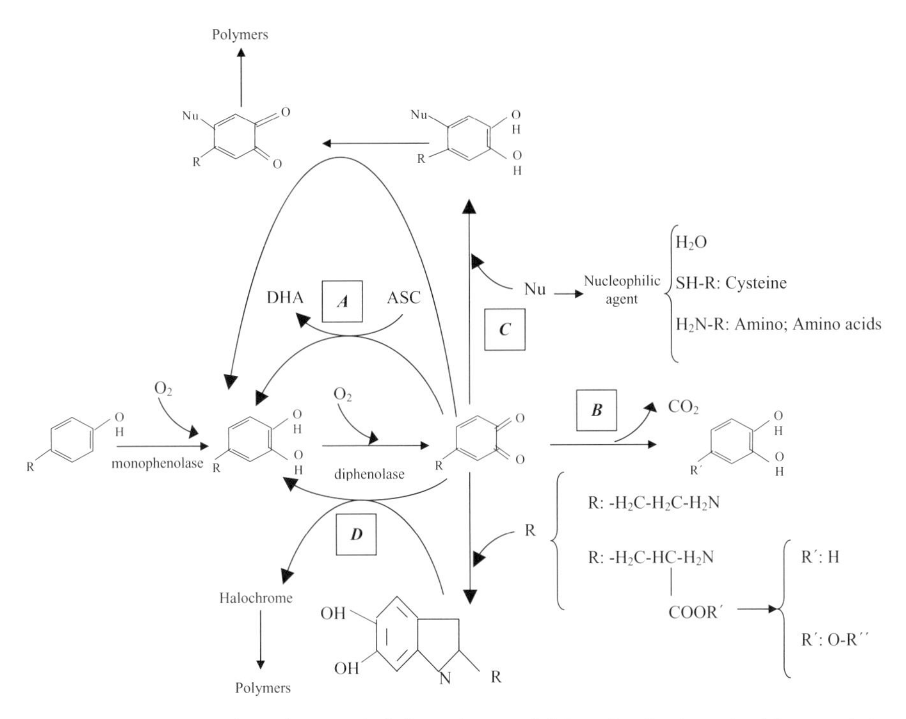

Figure 4.2. Monophenolase and diphenolase activities of polyphenoloxidase and stabilization of the formed quinones. (A) Quinone reduction. (B) Descarboxylation. (C) Modification by an exomolecular nucleophile. (D) Modification by an endomolecular nucleophile.

Action Mechanism of Polyphenoloxidase

As mentioned previously, PPO shows two catalytic activities: the conversion of monophenols into *o*-diphenols (monophenolase activity) and the oxidation to the corresponding *o-quinones* (diphenolase activity) (Fig. 4.2) (Sánchez-Ferrer and others, 1995).

Monophenolase Activity

The monophenolase activity of PPO is generally defined as the first step in the melanization pathway and consists of the *o*-hydroxylation of the monophenol to *o*-diphenol. This activity distinguishes PPO from other phenol-oxidizing enzymes, such as laccase and peroxidase, and is characterized by the following facts:

- The close association of the two activities (monophenolase and diphenolase) is borne out by the reciprocal competitive inhibition pattern of monophenolic and diphenolic substrates and by coincidence of the various steps in the catalytic cycle.
- Monophenolase activity shows a characteristic lag period before the maximum velocity of the hydroxylation step is reached. The time required to reach the steady-state rate depends on several factors: enzyme source; concentration of monophenol;

enzyme concentration; and finally, the presence of catalytic amounts of *o*-diphenol or transition metal ions.

- The inactivation reaction is similar to that presented by diphenolase activity, only slower.

Diphenolase Activity

The diphenolase activity involves the oxidation of two *o*-diphenols to two *o*-quinones with a concomitant $4e^-$ reduction of O_2, which yields two molecules of water. This activity is characterized by the following experimental facts:

- When a noncyclizable diphenol (e.g., 4-methylcatechol) is used, both the accumulation of the quinone product and the decrease in the oxygen concentration are linear with time, and there is no lag period.
- When a cyclizable diphenol (e.g., L-DOPA) is used, the decrease in oxygen is linear at any pH, although the appearance of the cyclizable quinone (DOPAchrome) is pH-dependent. The appearance of this lag period at low pH indicates the presence of a chemical intermediate (*o*-DOPAquinone-H^+). Thus, when reactions occur at neutral or slightly acid pH, the cycling branch (deprotonation of o-DOPAquinone-H^+ to render DOPAchrome by a redox recycling reaction) is the principal pathway. However, at strongly acid pH values lower than 5, the alternative hydroxylation branch that involves the formation of *p*-topaquinone through the addition of water to the quinone ring takes on more importance.
- There is an irreversible enzymatic inactivation reaction, which occurs during the oxidation of the cyclizable and noncyclizable diphenols to *o*-quinones. This inactivation process has been interpreted as being the result of a direct attack of an *o*-quinone on a nucleophilic residue (His) near the active enzyme center or of an attack of a copper-bound hydroxyl radical generated by the Cu(I)–peroxide complex. However, the latter hypothesis seems to be more probable, because inactivation also occurs in the presence of reducing agents that remove the *o*-quinones generated.
- The oxidation of *o*-diphenols occurs via a $2e^-$ process because no semiquinone intermediate appears during the enzyme turnover.
- Only the *oxy*- and *met*-forms of the enzyme are capable of binding the diphenols.

The *o*-quinones formed by PPO are unstable molecules that can be stabilized by different mechanisms (see Fig. 4.2):

(A) In the presence of a reducing agent, such as ascorbic acid or NADH, *o*-quinones are reduced to *o*-diphenols that are stable compounds, and the reducing agent is oxidized to dehydroascorbic acid or NAD^+.

(B) Several PPO substrates such as 3,4-dihydroxymandelate can be decarboxylated by PPO. The product of the enzyme action is an *o*-quinone, which, owing to its instability, evolves to the final product, 3,4-dihydroxybenzaldehyde, by a chemical reaction of oxidative decarboxylation.

(C) In the presence of a nucleophilic exomolecular agent such as H_2O, cysteine, amino acids, or amino groups, the nucleophilic agent is added to the *o*-quinone structure that evolves to a modified diphenol in position 5, which produces several polymeric compounds.

(D) In the presence of an intramolecular nucleophile, the *o*-quinones can evolve to cyclic indolic structures that evolve to halochrome and, finally, some polymeric structures can be formed.

Inhibitors of Polyphenoloxidase

In recent decades, several methods have been developed to prevent PPO activity in foods. This enzyme is the main compound responsible for food browning, a phenomenon that must usually be avoided because it can alter the nutritional and sensory properties of foods. Traditionally, sulfites have been used for this purpose, but they have been associated with severe allergy-like reactions in certain populations, and the Food and Drug Administration has restricted their use to only a few applications. Moreover, heat treatment is not suitable for inhibiting this sort of reaction, and therefore several other methods to reduce PPO activity have been studied, including the addition of ascorbic acid or chemical agents, the exclusion of oxygen, refrigeration, and various other nonthermal treatments (Sánchez-Ferrer and others 1995). As regards the inhibition of PPO by phenolic compounds, the most interesting ones were chalcones and related compounds such as glabridin and isoliquiritigenin (Nerya and others 2003). However, the precise action mechanism of these chalcone derivatives is not entirely understood. Moreover, other phenolics with a structure similar to PPO substrates are also able to inhibit the enzymatic activity of this enzyme. Among this group of inhibitors, there are several flavanols that are able to act as copper chelators, and flavanones or procyanidins that inhibit the PPO from apples (Mayer 2006).

A novel approach to analyzing inhibitory and noninhibitory compounds was adopted in an attempt to predict which structures might inhibit PPO. As a result, stilbenes with hydroxyl groups in their structure were found to inhibit PPO activity (López-Nicolás and Garcia Carmona 2008).

Several studies on the use of cyclodextrins (CDs) as PPO inhibitors have been published (López-Nicolas and others 2007). CDs are naturally occurring cyclic oligosaccharides derived from starch with six, seven, or eight glucose residues linked by $\alpha(1\rightarrow4)$ glycosidic bonds in a cylindrical cavity with a hydrophobic internal surface and a hydrophilic outer surface, designated α-, β- and γ-CDs, respectively (López-Nicolas and Garcia-Carmona 1995). The hydrophobic cavity is able to form inclusion complexes with a wide range of organic guest molecules, including PPO substrates. Thus, the free concentration of PPO substrates is lower than in the absence of CDs, and the PPO activity is inhibited.

Roles of Polyphenoloxidase

In this section the main physiological roles and potential technological applications of PPO are reviewed.

Use of Polyphenoloxidase for the Synthesis of Organic Compounds

One of the most important applications of PPO, although rarely reported, is its role in synthetic processes, such as the biosynthesis of betalains. Several researchers reported the hydroxylation of tyrosine to dopa, which can then be oxidized to dopaquinone, through a PPO from *Portulaca grandiflora* and from *Beta vulgaris*. Thus, a dioxygenase activity complements the constitutive PPO activity and the initiation of this dioxygenase activity leads to betalain formation. A further indication of the involvement of PPO in betalain biosynthesis comes from the works of Gandía-Herrero and others (2005, 2007), who describe how PPO is able to hydrolyze tyramine to produce dopamine. This dopamine, in the presence of betalamic acid, produces dopamine betaxanthin, which, upon subsequent oxidation, yields 2-descarboxybetanidin.

Polyphenoloxidase in Browning Reactions

Browning results from both the enzymatic and nonenzymatic oxidation of phenolic compounds (Martinez and Whitaker 1995). Browning usually impairs the sensory properties of products because of the associated changes in color, flavor, and softening (probably due to the action of pectic enzymes). Moreover, once cell walls and cellular membranes lose their integrity, enzymatic oxidation proceeds much more rapidly. Browning is sometimes desirable, as it can improve the sensory properties of some products such as dark raisins and fermented tea leaves. Browning is undesirable in other fruits and vegetables such as lettuce and potato. The formation of shrimp black spot is another example of undesirable browning due to PPO activity. The initial products of oxidation are quinones, which rapidly condense to produce relatively insoluble brown polymers (melanins). Some nonenzymatic causes of browning in foods include the Maillard reaction, autooxidation reactions involving phenolic compounds, and the formation of iron–phenol complexes. The most important factors that determine the rate of enzymatic browning of fruits and vegetables are the concentrations of both active PPO and phenolic compounds, pH, temperature, and oxygen availability of the tissue. Understanding the details of the enzymatic browning process is necessary if it is to be controlled and in order to obtain a final product that is acceptable to consumers.

Polyphenoloxidase as a Defense against Stress and Pathogens

As indicated previously, many papers have been published about the different properties of PPO. For example, its role in the defense mechanism of plants has been the subject of recent investigations, several authors reporting the changes in expression of specific genes coding for PPO during injury, during herbivore or pathogen attack, or during exposure to external stresses.

For example, the resistance of plants to the pathogen *Pseudomonas syringae* was studied by Thipyapong and others (2004) in tomato plants into which antisense PPO cDNA was inserted. Their results showed a strong reduction of PPO activity and a dramatic increase in the susceptibility of plants, although the overall growth and development of the tomato plants was not affected by the downregulation of PPO.

Polyphenoloxidase in Defense against Herbivores

Several authors have reported that PPO plays a role in defense against herbivores, factors such as gene expression, enzyme formation, and activation as well as substrate formation being related to the defense mechanism. For example, the response of potato and bean plants to the Colorado potato beetle has brought to light the fact that both wounding and regurgitation from the larvae can increase PPO activity in the leaves of both *Solanum tuberosum* and *Phaseolus vulgaris* (Mayer 2006). Moreover, these authors showed that bean leaves respond much more strongly than potato leaves to the formation of peroxidase in this system. In another study, a surprising correlation between drought resistance and PPO expression was reported (Mayer 2006). Thus, tomato plants in which PPO expression was reduced showed better water stress tolerance than either the nontreated plants or those in which PPO was overexpressed. However, further investigations are necessary in this field because no satisfactory explanation has been provided for the underlying mechanism.

The possible function of a tobacco flower–specific gene, which, when cloned and characterized, coded for a PPO, is discussed by Goldman and others (1998). Again, the usual suggestions were that defense functions or control of the formation of phenolic compounds acting as signaling molecules are involved. However, once again, no direct evidence was produced. It is nevertheless curious that there are a number of reports locating specific PPOs in flower parts, a finding that deserves more attention.

Polyphenoloxidase in Fungi and Defense against Fungal Attack

Although relatively few works study the induction of PPO in fungi compared with those published on the induction of plant PPO, we review the most interesting works in this field. Soler-Rivas and others (2000) showed that the infection of the mushroom *Agaricus bisporus* with *Pseudomonas tolaasii* produced a discoloration of the cap that was accompanied by the induction of fungal PPO. Moreover, similar results were found after the treatment of mushrooms with extracts containing the bacterial toxin tolaasin or with purified tolaasin. In the opinion of these authors, the induction of PPO activity could be attributed to the conversion of a latent PPO to an active form, whereas the addition of pure tolaasin induces transcription of a gene coding for PPO.

Several investigations have studied whether PPO plays a role in the formation of melanin. Thus, a review about the role of fungal melanin in pathogenesis was published by Jacobson (2000). Although the review focused on the importance of melanin in human fungal infections and fungal pathogenicity, the implications for fungus–plant interactions are obvious. An interesting function for fungal PPO comes from an examination of fungal interactions. The activity of several enzymes such as peroxidase, laccase, and PPO was followed in a medium where different species of fungi were allowed to grow together in either pure or mixed cultures on Petri dishes. The results showed that some of the fungi studied released PPO when confronted with another fungal species, although PPO activity was not detected specifically in the interaction zone (Mayer 2006).

Peroxidases

General Features

Peroxidases (PXs) (EC 1.11.1.7) are heme proteins that contain iron(III) protoporphyrin IX (ferriprotoporphyrin IX) as the prosthetic group, with a molecular weight ranging from 30,000 to 150,000 Da. As indicated in a previous section, PXs belong to the oxidoreductases group and catalyze the reduction of peroxides, such as hydrogen peroxide, and the oxidation of a variety of organic and inorganic compounds, including polyphenols. The term PX represents a group of specific enzymes, NADH peroxidase (EC 1.11.1.1), glutathione PX (EC 1.11.1.9), and iodide PX (EC 1.11.1.8), as well as a group of nonspecific enzymes that are simply known as PXs. In nature, PX activity has been identified in many plants, microorganisms, and animals. In plants, PXs participate in the lignification process and in various defense mechanisms.

Structure of Peroxidise

As stated earlier, several PXs have been described. However, in this section we describe the main features of horseradish PX isoenzyme C (HRP C), representative of these enzymes. This isoenzyme contains a single polypeptide of 308 amino acid residues, where the N-terminal residue is blocked by pyroglutamate and the C terminus is heterogenous, with some molecules lacking the terminal residue, Ser308. In the structure there are four disulfide bridges between cysteine residues and a buried salt bridge between Asp99 and Arg123. Moreover, nine potential N-glycosylation sites can be recognized in the primary sequence from the motif Asn–amino acid residue–Ser/Thr. A branched heptasaccharide accounts for 75% to 80% of the glycans, but the carbohydrate profile of HRP C is heterogeneous, and many minor glycans have also been characterized (Veitch 2004). These invariably contain two GlcNAc terminals and several mannose residues. However, the total carbohydrate content of this isoenzyme depends to a certain extent on the source of the enzyme.

HRP C contains two different types of metal center (i.e., iron(III) protoporphyrin IX-heme group and two calcium atoms) that are fundamental for the integrity of the enzyme. The heme group is attached to the enzyme at His170 by a coordinate bond between the histidine side-chain NE2 atom and the heme iron atom. The second axial coordination site is unoccupied in the resting state of the enzyme but available to hydrogen peroxide during enzyme turnover. Small molecules such as carbon monoxide, cyanide, fluoride, and azide bind to the heme iron atom at this distal site, giving six-coordinated PX complexes.

The three-dimensional structure of HRP C is largely α-helical, although there is also a small region of β-sheet. There are two domains, the distal and proximal, between which the heme group is located. These domains probably originated as a result of gene duplication, a proposal supported by their common calcium binding sites and other structural elements (Veitch 2004).

Action Mechanism of Peroxidise

The catalysis mechanism of horseradish PX and, in particular, the C isoenzyme, has been investigated extensively. However, although there is a wide range of PX isoenzymes, the following equation describes most of the reactions catalyzed by PX:

$$\begin{array}{c}\text{HRP C}\\ H_2O_2 + 2AH_2 \rightarrow 2H_2O + 2AH^{\cdot}\end{array}$$

in which AH_2 and $AH^{\cdot}$ represent a reducing substrate and its radical product, respectively. As for the substrates of PX, several molecules, including polyphenols, indoles, amines, and sulfonates, have been reported.

The generation of radical species in the two one-electron reduction steps can result in a complex profile of reaction products, including trimers and higher oligomers that may themselves act as reducing substrates in subsequent turnovers. Several of the functions of PX are related to the formation of these radical products. Among these functions,

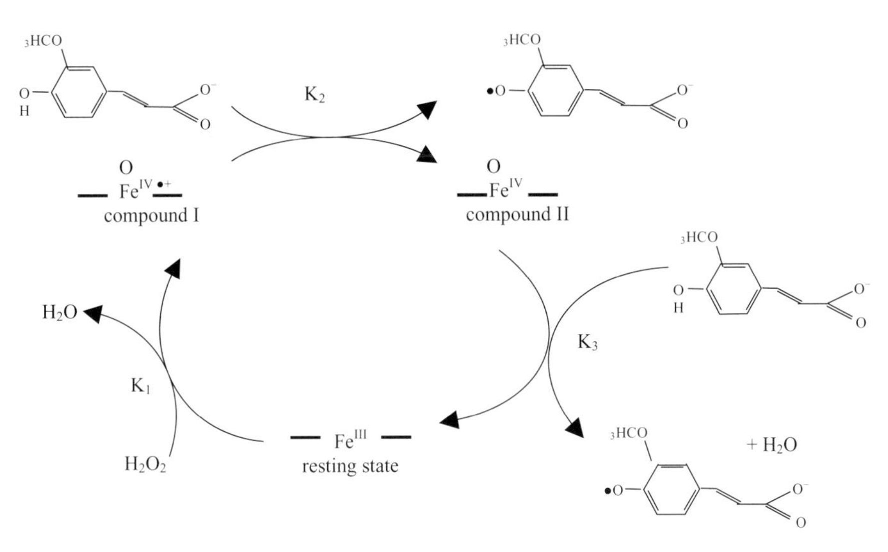

Figure 4.3. The catalytic cycle of horseradish peroxidase with ferulate as reducing substrate. The rate constants K_1, K_2, and K_3 represent the rate of compound I formation, rate of compound I reduction, and rate of compound II reduction, respectively.

we can cite various cross-linking reactions (expressed in response to external factors such as the wounding of plant tissue) including the formation of diferulate linkages from polymer-attached ferulate groups of polysaccharides or pectins, the formation of dityrosine linkages, the cross-linking of phenolic monomers in the formation of suberin and the oxidative coupling of phenolic compounds as part of the biosynthesis of lignin.

The first step in the catalytic cycle is the reaction between H_2O_2 and the Fe(III) resting state of the enzyme to generate compound I, a high-oxidation-state intermediate comprising an Fe(IV) oxoferryl center and a porphyrin-based cation radical (Fig. 4.3).

A transient intermediate (compound 0) formed prior to compound I has been detected in reactions between HRP C and H_2O_2 at low temperatures and described as an Fe(III)–hydroperoxy complex. In formal terms, compound I is two oxidizing equivalents above the resting state. The first one-electron reduction step requires the participation of a reducing substrate and leads to the generation of compound II, an Fe(IV) oxoferryl species that is one oxidizing equivalent above the resting state. Both compound I and compound II are powerful oxidants, with redox potentials estimated to be close to +1 V. The second one-electron reduction step returns compound II to the resting state of the enzyme. The reaction of excess hydrogen peroxide with the resting-state enzyme gives compound III, which can also be prepared by several other routes. This intermediate is best described as a resonance hybrid of iron(III)–superoxide and iron(II)–dioxygen complexes. A high-resolution crystal structure of 95% pure compound III published recently shows dioxygen bound to heme iron in a bent conformation (Veitch 2004).

Applications of Peroxidise

Application in the Paper Pulp Industry

Biopulping is a process in which extracellular enzymes produced by a white-rot fungus degrade lignin (Hamid and Khalil-ur-Rehman 2009), leading to a modified lignin, which is responsible for the characteristic brown color. Among the major lignin-degrading enzymes we can cite manganese PX (MnPX), laccase, and, to a lesser extent, lignin PX (LiPX). Finally, the modified lignin can be enzymatically degraded, using a biobleaching process (Hamid and Khalil-ur-Rehman 2009). However, more studies are needed to determine whether selective lignin degradation and efficient biopulping require a proper balance between lignin and cellulose degradation.

Synthesis of Polymers

The production of conducting polymers is of remarkable interest because of their wide range of applications, including anticorrosive protection, optical displays, and light-emitting diodes (Hamid and Khalil-ur-Rehman 2009). HRP C and soybean peroxidise have been used to polymerize phenolic compounds such as cardanol or aromatic amine compounds, whereas new types of aromatic polymers have been synthesized in water and in water-miscible organic solvents (Hamid and Khalil-ur-Rehman 2009). Moreover, HRP C and an anionic PX have been used in the synthesis of polyelectrolyte complex polyaniline, one of the most extensively investigated conducting polymers because of its high environmental stability and promising electronic properties.

Lignin is the second most abundant biopolymer on earth, and it has the potential for application in the production of polymeric dispersants, soil-conditioning agents, phenolic resins or adhesives, and laminates. In the presence of H_2O_2, PX catalyzes the oxidation of phenols that eventually give rise to higher molecular weight polymers, and this characteristic can be used as an attractive alternative to the conventional formaldehyde method for the production of lignin-containing phenolic resins. White-rot basidiomycetes such as *Phanerochaete chrysosporium, Phleba ratia,* and *Pleurotus* spp., among others, secrete the lignin-degrading enzyme, MnP, which oxidizes phenolic compounds directly, via phenoxy radicals, which represents the first step in the degradation of lignin. Furthermore, this MnP produced by the basidiomycete *Bjerkandera adusta* has been used by several authors (Hamid and Khalil-ur-Rehman 2009) for acrylamide polymerization.

Production of Biofuel

One of the most innovative research projects for the production of fuels is closely related to the action of PX. Ethanol and other biofuels produced from lignocellulosic biomass represent a renewable, more carbon-balanced alternative to both fossil fuels and corn-derived or sugarcane-derived ethanol. However, the production of fuels is limited by the lignin present in plant cell walls. This lignin impedes the breakdown of cell-wall polysaccharides to simple sugars and subsequent conversion of these sugars to useable fuel. So, two forms of PX, LiPX and MnPX, can metabolize lignin to carbon dioxide and thus facilitate the process involved in the production of fuels. In the oxidative depolymerization process, phenolic compounds are formed and are prone to polymerize again. MnPX converts the phenolic LiPX breakdown products to form quinones, which are further metabolized in a process that probably involves reduction to the corresponding hydroquinones (Hamid and Khalil-ur-Rehman 2009).

Elimination of Phenolic Contaminants

There are several chemical compounds found in the waste waters of a wide variety of industries that must be removed because of the danger they represent to human health. Among the major classes of contaminants, several aromatic molecules, including phenols and aromatic amines, have been reported. Enzymatic treatment has been proposed by many researchers as an alternative to conventional methods. In this respect, PX has the ability to coprecipitate certain difficult-to-remove contaminants by inducing the formation of mixed polymers that behave similarly to the polymeric products of easily removable contaminants. Thus, several types of PX, including HRP C, LiP, and a number of other PXs from different sources, have been used for treatment of aqueous aromatic contaminants and decolorization of dyes. Thus, LiP was shown to mineralize a variety of recalcitrant aromatic compounds and to oxidize a number of polycyclic aromatic and phenolic compounds. Furthermore, MnP and a microbial PX from *Coprinus macrorhizus* have also been observed to catalyze the oxidation of several monoaromatic phenols and aromatic dyes (Hamid and Khalil-ur-Rehman 2009).

Several authors have attempted to optimize the HRP C conditions to remove phenols from aqueous solutions. Among the main parameters evaluated to improve the elimination of contaminants are the reactor configuration, enzyme immobilization, and the use of additives to protect the enzyme from entrapment in the precipitating polymers (Hamid and Khalil-ur-Rehman 2009).

However, an important problem arises during the peroxidative removal of phenols from aqueous solutions: PX is inactivated by free radicals, as well as by oligomeric and polymeric products formed in the reaction, which attach themselves to the enzyme (Nazari and others 2007). This suicide peroxide inactivation has been shown to reduce the sensitivity and efficiency of PX. Several techniques have been introduced to reduce the extent of suicide inactivation and to improve the lifetime of the active enzyme, such as immobilization. Moreover, Nazari and others (2007) reported a mechanism to prevent and control the suicide peroxide inactivation of horseradish PX by means of the activation and stabilization effects of Ni^{2+} ion, which was found to be useful in processes such as phenol removal and peroxidative conversion of reducing substrates, in which a high concentration of hydrogen peroxide may lead to irreversible enzyme inactivation.

Decolorization and Deodorization of Various Compounds

Various industrial sectors related to textiles, paper, and photography use dyes of synthetic origin with a complex aromatic molecular structure, which are frequently discharged in industrial effluents. One solution to these environmental problems is to use oxidative enzymes that destroy colored compounds and that may be of practical interest for the decolorization of synthetic dyes. Enzymes such as LiPX and MnPX are involved in the decolorization of synthetic azo dyes, such as Acid Orange II.

Odoriferous polyphenol compounds may be present initially in manure or result from the anaerobic transformation of animal waste. Recently, HRP C and minced horseradish roots have been shown to be an effective alternative for the deodorization of manures (Hamid and Khalil-ur-Rehman 2009).

Laccases

General Features

Laccase is one of the main oxidizing enzymes responsible for polyphenol degradation. It is a copper-containing polyphenoloxidase (*p*-diphenoloxidase, EC 1.10.3.2) that catalyzes the oxidation of several compounds such as polyphenols, methoxy-substituted phenols, diamines, and other compounds, but that does not oxidize tyrosine (Thurston, 1994). In a classical laccase reaction, a phenol undergoes a one-electron oxidation to form a free radical. In this typical reaction the active oxygen species can be transformed in a second oxidation step into a quinone that, as the free radical product, can undergo polymerization.

Structure of Laccase

The laccase molecule is a dimeric or tetrameric glycoprotein, which contains four copper atoms per monomer, distributed in three redox sites. More than 100 types of laccase have been characterized. These enzymes are glycoproteins with molecular weights of 50–130 kDa. Approximately 45% of the molecular weight of this enzyme in plants are carbohydrate portions, whereas fungal laccases contain less of a carbohydrate portion (10–30%). Some studies have suggested that the carbohydrate portion of the molecule ensures the conformational stability of the globule and protects it from proteolysis and inactivation by radicals (Morozova and others 2007).

Sources of Laccases

Several investigations have demonstrated that laccases can be classified into two clearly distinct major groups: those from higher plants and those from fungi (Mayer and Staples 2002). Moreover, other authors have reported the presence of laccase-like enzymes in bacteria or in insects (Mayer 2006). As to the presence of this enzyme in higher plants, laccase appears to be far more limited than in fungi (Mayer and Staples 2002). *Pinus taeda* tissue contains eight laccases, all expressed predominantly in xylem tissue. In another paper, several researchers characterized in great detail the presence of laccase in *Rhus vernicifera* (Mayer and Staples 2002). Moreover, the family Anacardiaceae appears to contain laccase in the resin, whereas some organisms such as *Acer pseudoplatanus* have been shown to produce and secrete laccase (Mayer and Staples 2002). Many years ago, the presence of a laccase in the leaves of *Aesculus parviflora* and in green shoots of tea was reported. More recently, five distinct laccases have been shown to be present in the xylem tissue of *Populus euramericana* (Mayer and Staples 2002). On the other hand, the presence of laccase has been documented in virtually every fungus examined, and the presence of both constitutive and inducible laccases has been reported (Mayer and Staples 2002).

Action Mechanism of Laccase

Although several works suggested that these enzymes cannot oxidize nonphenolic compounds because the redox potentials of laccases are lower than those of the

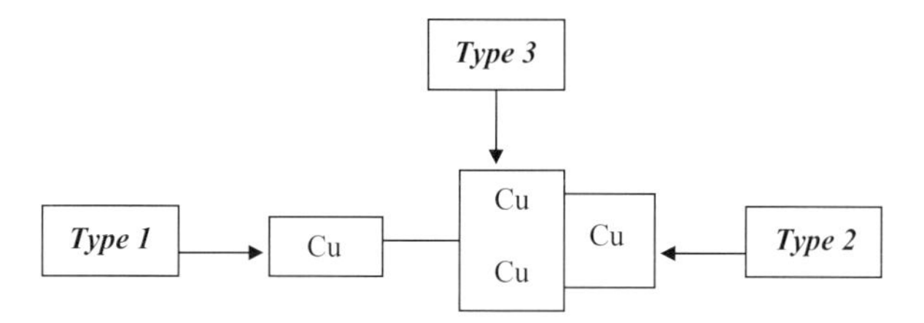

Figure 4.4. Schematic representation of laccase trinuclear copper active center.

substrates, the same papers have reported that, in the presence of small molecules capable of acting as electron transfer mediators, laccases can indeed oxidize nonphenolic structures (Baiocco and others 2003). This broadened the range of compounds that can be oxidized by these enzymes, and it is now clear that laccases are able to catalyze the oxidation of *o*- and *p*-diphenols, aminophenols, polyphenols, polyamines, lignins, and aryl diamines as well as some inorganic ions coupled to the reduction of molecular dioxygen to water.

Laccase Active Site

Laccase contains four copper atoms that have been classified as type 1 or blue (T1), type 2 or normal (T2), and type 3 (T3) or coupled binuclear copper sites, where the coppers are antiferromagnetically coupled through a bridging ligand (Fig. 4.4).

Several papers have shown that the T2 and T3 centers combine to function as a trinuclear copper cluster with respect to exogenous ligand interaction, including reacting with dioxygen (Durán and others 2002). Moreover, these studies have demonstrated that whereas the T2 is 3-coordinate with two histidines and water as ligands, the T3 coppers are each 4-coordinate, having three histidine ligands and a bridging hydroxide. Finally, the structural model of bridging between the T2 and T3 has provided insight into the catalytic reduction of oxygen to water. Concerning the role of these Cu types in the reduction of oxygen, it has been elucidated that the T2 copper is required for the reduction of oxygen because bridging to this center is involved in the stabilization of the peroxide intermediate. Moreover, many papers have demonstrated that the T2 Cu is required for dioxygen reactivity in laccase and that dioxygen is reduced in the absence of the T1 Cu, demonstrating that the T2/T3 trinuclear Cu site represents the active site for the binding and multielectron reduction of dioxygen. On the other hand, the T1 Cu is clearly not necessary for dioxygen reactivity (Durán and others 2002).

Key indexes of laccases include standard redox potentials of the three copper centers of the enzyme: T1, T2, and T3 (Tatiana and others 2003). The potential of T1 has been determined for many laccases, varying within the range 430–780 mV. On the grounds of T1 potential, all copper-containing oxidases are divided into three groups: high-, medium-, and low-potential enzymes. Low-potential laccases include enzymes with T1 potentials below 470 mV; medium potential, 470–710 mV; and high potential, above 710 mV. The potentials of T2 are only known for the low-potential plant laccase from *R. vernisifera* (390 mV) and high-potential fungal laccase from *T. hirsute* (400 mV). The potentials of T3 of the laccases from *R. vernisifera* and *T. versicolor* are 460 and 785 mV, respectively. Laccases directly oxidize compounds whose ionization potentials

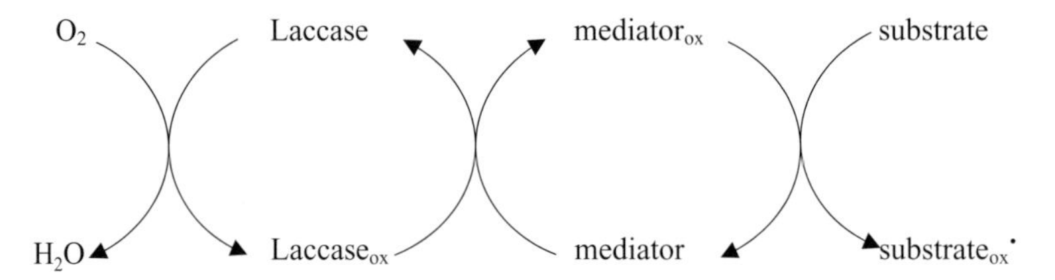

Figure 4.5. Role of mediators in laccase-catalyzed reaction.

do not exceed (or exceed by a minimal amount) the redox potential of the T1 copper ion.

Laccase Substrates

As indicated previously, laccases are nonspecific with regard to reducing substrates. This type of enzyme is able to catalyze the oxidation of various organic substances, including *o*- and *p*-diphenols, aminophenols, polyphenols, polyamines, methoxyphenols, lignins, aryl diamines, and some inorganic ions, with the simultaneous direct reduction of dioxygen to water without intermediate production of hydrogen peroxide. To classify the organic substrates of laccases, several papers have organized these compounds into three groups: *o*-, *m*-, or *p*-substituted compounds. Much research has focused on investigating the optimum substrates for laccase, and some papers have concluded that *o*-substituted compounds (e.g., guaiacol, *o*-phenylenediamine, pyrocatechol, dihydroxyphenylalanine, pyrogallol, caffeic acid, gallic acid, and protocatechuic acid) are the best substrates for this enzyme, unlike *p*- (*p*-phenylenediamine, *p*-cresol, and hydroquinone) or *m*-substituted compounds (*m*-phenylenediamine, orcinol, resorcinol, and phloroglucin).

Laccase Redox Mediators

The discovery of ABTS as a laccase substrate mediating or enhancing the enzyme action was essential to increase the range of molecules that can be converted by laccases (Fig. 4.5). Such a mediator requires several conditions: (1) it must be a good laccase substrate; (2) its oxidized and reduced forms must be stable; (3) it must not inhibit the enzymatic reaction; and (4) its redox conversion must be cyclic.

The oxidized mediator form, produced in the course of the enzymatic reaction, can nonenzymatically oxidize compounds (including nonphenolic lignin structures) with ionization potentials exceeding the potentials of laccases (Morozova and others 2007).

Applications of Laccase

Several applications of laccase have been reported in recent years (Rodríguez-Couto and Toca-Herrera 2006). As shown in this section, many of these applications are similar to those published for peroxidase enzyme.

Role of Laccase in the Food Industry

As described previously for peroxidase, laccase can be applied to several processes related to the food industry because of its ability to eliminate undesirable phenolics.

This capacity is applicable in various sectors of the food industry such as bioremediation, beverage processing, ascorbic acid determination, sugar-beet pectin gelation, and baking; it also acts as a biosensor.

Bioremediation of food industry wastewater: Bioremediation is a general concept that includes all those processes and actions that take place as an attempt to biotransform an environment, already altered by contaminants, to its original status. Laccase is a well-known enzyme in bioremediation because of its ability to degrade phenolic compounds (Morozova and others 2007). As mentioned for peroxidase, aromatic compounds, including phenols and aromatic amines, constitute one of the major classes of pollutants and are heavily regulated in many countries. This ability of laccases has been applied in different areas of both the food and textile industries, such as breweries and olive oil factories.

Wine and beer industry: Polyphenols can alter color and flavor of products such as wines. There are many aggressive ways of removing polyphenolic compounds, such as using polyvinylpolypyrrolidone (PVPP) or sulfur dioxide. However, polyphenol removal should be selective to avoid the undesirable alteration of the wine's organoleptic characteristics. For this reason, one option is to use laccases that polymerize the polyphenolic compounds during the wine-making process and then to remove these polymers by clarification (Morozova and others 2007). Several papers have reported that laccase is able to remove undesirable polyphenols and produce stable wines with a good flavor.

A similar problem occurs with beer stabilization. A serious problem in the brewing industry is the tendency of some beers to develop hazes during long-term storage due to protein precipitation that is usually stimulated by small quantities of naturally occurring proanthocyanidin polyphenols. In the same way as observed for wine, the excess polyphenols are traditionally removed by treatment with insoluble PVPP, with the same resulting problems. To resolve the problems, several authors have proposed the use of laccase, which forms polyphenol complexes that may be removed by filtration or other separation means.

Sugar-beet pectin gelation: There is a continuous search in the food industry to find new functional ingredients. Sugar-beet pectin is a food ingredient with specific functional properties. It may form gels through the oxidative cross-linking of ferulic acid. For example, Micard and Thibault (1999) showed that it is possible to cross-link the beet pectin through the oxidative coupling of the feruloyl groups using laccase. Interesting work on a biomimetic approach to improve emulsion stability has also been reported (Littoz and Mcclements, 2008). The objective of the study was to prepare and characterize stable oil-in-water emulsions containing oil droplets coated by multilayered biopolymer interfaces that were cross-linked by the enzyme laccase.

Baking: Several authors have reported that the use of laccase results in increased volume, improved crumb structure, and greater softness of the baked product, as well as increased strength and stability and reduced stickiness, thereby improving the machinability of the dough (Minussi and others 2002). Laccase accelerates dough formation and dough breakdown, and the effects on mixing properties are enhanced when ferulic acid is added. Moreover, other researchers reported the effect of different enzymes, such as laccase, xylanase, and their combination, on various rheological properties of wheat doughs, including extensibility and resistance to stretching. Laccase treatment resulted in dough hardening, whereas xylanases softened flour and gluten doughs.

Recently, Caballero and others (2007) studied the improvement of dough rheology, bread quality, and bread shelf life by using enzyme combinations. They systematically analyzed the individual and synergistic effects of some gluten-cross-linking enzymes (transglutaminase, glucose oxidase, and laccase), along with polysaccharide and gluten-degrading enzymes (alpha-amylase, xylanase, and protease), in bread-making systems.

Role of Laccase in the Pulp and Paper Industry

In a previous section we noted that the industrial preparation of paper requires separation and degradation of lignin in wood pulp. Thus, the capability of laccases to form reactive radicals in lignin can also be used in targeted modification of wood fibers. For example, laccases can be used in the enzymatic adhesion of fibers in the manufacturing of lignocellulose-based composite materials such as fiberboards (Felby and others 2002). Laccases have been proposed to activate the fiber-bound lignin during manufacturing of the composites, thus resulting in boards with good mechanical properties without toxic synthetic adhesives. Another possibility is to functionalize lignocellulosic fibers by laccases in order to improve the chemical or physical properties of the fiber products.

Role of Laccase in the Textile Industry

In the textile industry, laccases are used to bleach textiles, modify the surface of fabrics, and synthesize dyes (Cristóvão and others 2009). Processes based on the use of this oxidative enzyme might replace traditionally high chemical-, energy-, and water-consuming textile operations. Recently, Rodríguez-Couto and Toca-Herrera (2006) reviewed the applications of laccases in the textile industry. For these authors, the main applications of this enzyme in the textile industry are wastewater treatment, denim finishing, cotton bleaching, rove scouring, wool dyeing, antishrink treatment for wool, and dye synthesis. However, one of the problems in the use of laccase is the lack of sufficient enzyme stocks. Another additional problem is the cost and toxicity of redox mediators. For these reasons much research is necessary to optimize the use of this enzyme in the textile industry.

Role of Laccase in Soil Bioremediation

Polycyclic aromatic hydrocarbons together with other xenobiotics are a major source of contamination in soil, and their correct degradation is of great environment importance.

Synthetic Chemistry

The production and synthesis of polymers and medical agents is one of the most interesting potential uses for laccases in synthetic chemistry, where their application has been proposed for oxidative deprotection and the production of complex polymers and pharmaceutical agents.

Role of Laccase in the Cosmetics Industry

The cosmetic world has not been indifferent to the application of laccase: for example, laccase-based hair dyes are less irritating and easier to handle than current hair dyes, because laccases replace H_2O_2 as an oxidizing agent in the dye formulation (Rodríguez-Couto and Toca-Herrera 2006).

Lipoxygenases

General Features

Lipoxygenases (LOXs) (linoleate oxygen oxidoreductase, EC 1.13.11.12) are a group of nonheme iron-containing enzymes that are widely distributed in plants and animals (Siedow 1991; Baysal and Demirdoven 2007). These enzymes have a dual enzymatic function associated with a single protein: *dioxygenase* and *hydroperoxidase*. As dioxygenase, LOX catalyzes the stereospecific dioxygenation of polyunsaturated fatty acids (PUFAs) containing a 1,4-*cis*,*cis* pentadiene system to a pentadienyl radical intermediate that reacts with molecular oxygen to yield *cis,trans*-conjugated diene hydroperoxides. This process involves redox cycling of the iron present in the enzyme molecule. The second activity, hydroperoxidase, was first described in 1943 and later confirmed in LOXs obtained from different sources and tissues. As hydroperoxidase, LOX is capable of oxidizing a large number of reducing substrates, which are transformed into free radicals (Pérez-Gilabert and others 1994; Núñez-Delicado and others 1999). Among these compounds, polyphenols such as resveratrol have been reported to act as substrates of the hydroperoxidase activity of LOX. This second reaction is slow in the presence of fatty acid hydroperoxide alone, but increases if a suitable electron donor, a cosubstrate, is included in the reaction medium, or if the fatty acid hydroperoxide is replaced by H_2O_2. The low specificity of this activity for the peroxide and cosubstrate suggests that this enzyme plays an important role in the oxidative metabolism of xenobiotics. The fact that the substrates of dioxygenase activity (fatty acids) and many substrates of hydroperoxidase activity (different phenols such as stilbenoids) are compounds of low aqueous solubility makes the kinetic characterization of this enzyme a difficult task, which has been resolved by the use of surfactants or substrate-complexing agents such as cyclodextrins (López-Nicolás and others 1994, 1997).

Sources of Lipoxygenases

Although LOX from soybean seed is the best characterized of plant LOXs, this enzyme is present in a wide variety of plant and animal tissues (Liavonchanka and Feussner, 2006). The enzyme occurs in a variety of isoenzymes, which often vary in their optimum pH and in product and substrate specificity. Given the occurrence of multiple LOX isoenzymes in soybean leaves and the proposed roles of these enzymes in the plant metabolism, it is possible that individual isoenzymes play specific functions (Feussner and Wasternack 2002). The molecular structure of soybean LOX is the most reported, and four isoenzymes have been isolated (Baysal and Demirdoven 2007). Soy isoenzyme 1 produces 9- and 13-hydroperoxides (1:9) when the enzyme acts on free PUFA at pH 9.0, its optimum pH (López-Nicolás and others 1999). Soy isoenzyme 2 acts on triglycerides as well as free PUFA leading to 9- and 13-hydroperoxide

(1:1) at an optimum pH of 6.8. The activity of soy isoenzyme 3 is very similar to that of isoenzyme 2, but, contrary to that observed for isoenzyme 2, its activity is inhibited by calcium ions, whereas LOX-2 is stimulated by the metal. Finally, LOX isoenzyme 4 is very similar to isozyme 3 but can be separated by gel chromatography or electrophoresis (Baysal and Demirdoven 2007).

Potato LOX has the potential to be used as an alternative model to the mammalian enzyme because of its great availability(López-Nicolás and others 2000). To date, three isoenzymes of potato LOX have been isolated. Several works have reported linoleic acid as the optimum substrate for potato LOX-1, 9-hydroperoxide being the main product of the reaction. Another LOX substrate, linolenic acid, has been reported as the preferred substrate for both potato LOX-2 and -3, which produce 13-hydroperoxide as the main product.

Other sources of LOX are tomato, which possesses a 9-LOX activity; cucumber, which has a LOX enzyme similar to potato and tomato enzymes in both pH optimum and substrate specificity; apples, which present high activity in the core and peel; strawberry; pea; and others (Baysal and Demirdoven 2007). Finally, LOX has been found in eggplant, where linoleic acid was the best substrate, the 9-hydroperoxy isomer was the main product, and the 13-hydroperoxy isomer was only a minor product at pH 7 (López-Nicolás and others 2001).

In plant tissues, various enzymes convert the hydroperoxides produced by LOX to other products, some of which are important as flavor compounds. These enzymes include hydroperoxide lyase, which catalyzes the formation of aldehydes and oxo acids; hydroperoxide-dependent peroxygenase and epoxygenase, which catalyze the formation of epoxy and hydroxy fatty acids, and hydroperoxide isomerase, which catalyzes the formation of epoxyhydroxy fatty acids and trihydroxy fatty acids. LOX produces flavor volatiles similar to those produced during autoxidation, although the relative proportions of the products may vary widely, depending on the specificity of the enzyme and the reaction conditions.

Finally, Pérez-Gilabert and others (2005) purified, partially characterized, and studied the kinetic properties of LOX from desert truffle (*Terfezia claveryi* Chatin) ascocarps.

Action Mechanism of Lipoxygenase

Iron Content

LOX molecules contain one atom of iron, which is in the high-spin Fe(II) state in the native resting form of LOX and must be oxidized to Fe(III) by the reaction product (Liavonchanka and Feussner, 2006). One of the main characteristics of this LOX reaction is the presence of a lag period as a consequence of this requirement for oxidation of the iron in the enzyme. The mechanism of LOX as it acts on PUFA substrates can be observed in Figure 4.6.

The active enzyme abstracts a hydrogen atom stereospecifically from the intervening methylene group of a PUFA in a rate-limiting step, with the iron being reduced to Fe(II). The enzyme–alkyl radical complex is then oxidized by molecular oxygen to an enzyme–peroxy radical complex under aerobic conditions, before the electron is transferred from the ferrous atom to the peroxy group. Protonation and dissociation from

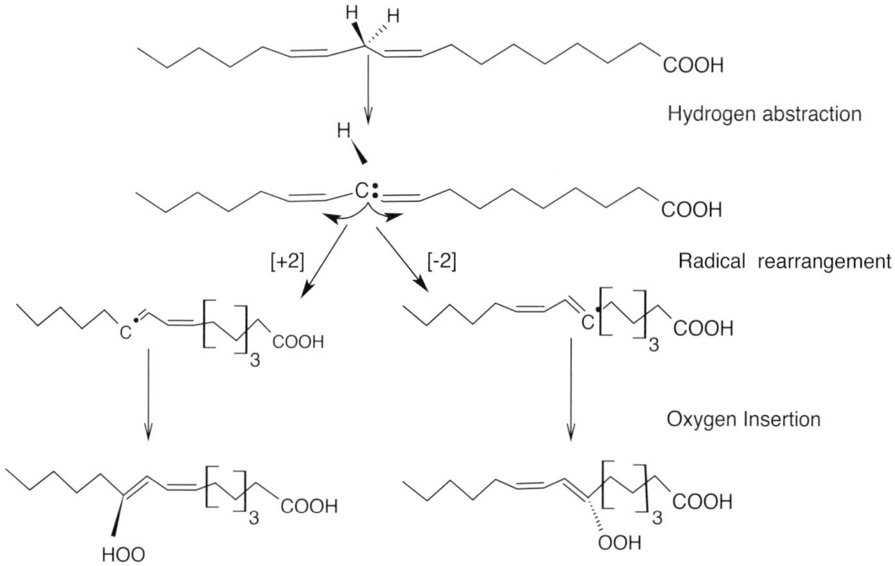

Figure 4.6. The lipoxygenase reaction.

the enzyme allow the formation of the hydroperoxide. Under anaerobic conditions, the alkyl radical dissociates from the enzyme–alkyl radical complex, and a mixture of products including dimers, ketones, and epoxides is produced by radical reactions.

Positional Specificity of Lipoxygenases

The mechanism of positional specificity of LOXs has been explained by two models. In the first, a space-related hypothesis was established based on data for mammalian LOX, showing that the substrate generally penetrates the active site with its methyl end first. Then, the depth of the substrate-binding pocket determines the site of hydrogen abstraction, and the positional specificity of molecular oxygen insertion depends on this position. However, in plant LOX reactions, only one double allylic methylene group in the natural substrates, linoleic and linolenic acids, seems to be accessible, rendering the space-related hypothesis unlikely. According to a second hypothesis, then, the substrate orientation is regarded as the key step in the determination of the position of dioxygen insertion, which then leads to varying regiospecificities of different isozymes.

In the case of 13-LOXs, the active site is again penetrated by the substrate using its methyl end first, whereas with 9-LOXs, the substrate is forced into an inverse orientation, favoring penetration with its carboxy group first. Consequently, a radical rearrangement at either [+2] or [−2], respectively, may be facilitated in both cases by the same mechanism within the active site.

Recent investigations have reported that, at least for some plant LOXs, a combined version of both models may explain the underlying reactions because the inverse orientation of the substrate is determined by the space available in the substrate-binding pocket.

In short, the space within the active site and the orientation of the substrate are both important determinants for the positional specificity of plant LOXs and are modified by additional factors such as substrate concentration, the physicochemical state of the substrate, pH, or temperature (López-Nicolás and others 2000). However, it should be stressed that for other LOXs regiospecificity may be determined in a more complex manner.

Roles of Lipoxygenase

Several works have demonstrated that LOX from different sources is associated with several of the physiological processes that have been reviewed in this section (Porta and Rocha-Sosa, 2002).

Maturation and Germination of Seeds

LOXs are normally present in the seeds of plants, although LOXs do not have a clear physiological role in seed development, because no negative effects on crop performance were observed in LOX-deficient seeds as compared with a normal line. For this reason, the possible function of seed LOXs as storage proteins is considered.

Several papers have reported that, during germination, new LOXs are synthesized in the seedling and the cotyledons; the LOX mRNAs synthesized during germination can also be found in the mature plant. Their levels were increased by the application of abscisic acid and jasmonic acid, or by stress such as wounding, pathogen infection, or water deficit (Porta and Rocha-Sosa 2002).

Researches into oilseed plants germinated in the dark showed that storage lipids are mobilized from lipid bodies in the cotyledons, and the free fatty acids that are released are further metabolized via β-oxidation. Moreover, in germinating cucumber seeds, a specific LOX associated with lipid bodies is capable of adding oxygen to the sterified fatty acids, thus generating triacylglycerol, containing one, two, or three 13-HPOD acid residues (Porta and Rocha-Sosa 2002).

Three LOXs are present in the mature seeds of soybean. Although these isoforms disappear during the first days of germination, three new isoenzymes are synthesized in the cotyledons. In contrast with the observations made in cucumber, seed or seedling soybean LOXs are not associated with lipid bodies. Moreover, in germinating soybean seedlings, there is no substantial oxygenation of polyunsaturated fatty acids, which suggests that in soybean at least, LOX is not used for lipid mobilization during germination (Porta and Rocha-Sosa 2002).

Lipoxygenases Are Involved in Vegetative Growth

Several researchers have reported that LOXs are involved in the control of tuber growth and development, probably by initiating the synthesis of oxylipins that regulate cell growth during tuber formation. Kolomiets and others (2001) showed that the production of transgenic plants expressing an antisense, tuber-specific LOX (POTLX-1) gene provides some clues about the function of LOX in potato tubers. *In situ* localization showed that Lox1 class mRNA is found in the distal, most actively growing portion of the developing tuber. Antisense POTLX-1 plants displayed reduced LOX activity and a severalfold reduction in tuber yield. Tubers that formed were misshapen and small.

Production of Oxylipins

Wounding and herbivore attack: Different situations such as mechanical wounding or insect feeding produced the induction of LOX transcripts in wounded and systemic leaves in the same plant. The function of LOX in wounding seems to be related to the synthesis of a number of different compounds with signaling activity.

Pathogen attack: The induction of LOX genes during plant–pathogen interactions has been reported in several species, suggesting that the function of LOX in the defense against pests seems to be related to the synthesis of a number of different compounds with signaling functions or antimicrobial activity.

Inhibition of LOX Oxidation

Various disadvantages related to nutritional and sensory food properties are associated with LOX activity. Indeed, LOX is involved in off-flavor and -odor production, the loss of pigments such as carotenes and chlorophylls, and the destruction of essential fatty acids. For these reasons, several authors have studied the effect on the LOX activity of adding different inhibitors. Although LOXs can be thermally inactivated above 60°C with a resulting improvement in the shelf life of foods, this increase in the system's heat also increases nonenzymatic oxidation, which may exceed the oxidation due to LOX. For these reasons, alternatives have been developed such as new strategies for food packaging (use of controlled modified atmospheres or new packaging materials and equipment); the use of complexing agents such as cyclodextrins; or minimizing the exposure of polyphenol- and lipid-containing food products to air, light, and high temperatures during processing and storage (López-Nicolás and others 1995).

Nonenzymatic Degradation of Polyphenols

As indicated in previous sections, degradative reactions involving polyphenols in food processing and storage include biochemical and chemical processes. Although the most important biochemical mechanism for the degradation of polyphenols is enzymatic oxidation, other factors such as chemical reactions, fermentations, cooking, manufacture, or storage may also catalyze transformations and degradation of polyphenolic compounds.

As reported previously, browning reactions, which are some of the most important phenomena occurring in food during processing and storage, represent an interesting research area with implications for food stability and technology, as well as for nutrition and health. Browning reactions can involve different compounds and proceed through different chemical pathways. The major groups of reactions leading to browning are enzymatic phenol oxidation and the so-called nonenzymatic browning. The latter is favored by heat treatments and includes a wide number of reactions such as the Maillard reaction, caramelization, chemical oxidation of phenols, and maderization.

The Maillard reaction, first described during the early 20th century, originally referred to the browning reaction that occurs between amino acids and sugars during the cooking and processing of foods. However, in recent years different compounds such as polyphenols have been related to the Maillard reaction (Manzocco and others 2001). This reaction contributes to the sensory, nutritional, health, and toxicological properties of different foods. Although in its early stages the Maillard reaction leads to the formation of the well-known Amadori and Heyn's products, little information is available on the chemical structure of the hundreds of brown products that are formed by a series of consecutive and parallel reactions, including oxidations, reductions, and aldol condensations.

The Maillard reaction is also known to contribute to the natural and normal aging of tissue proteins and other biomolecules such as polyphenols. Recent publications have reported that polyphenols can react with amino groups complicating the pathway of the browning reactions and modifying the healthy properties of polyphenols (Manzocco and others 2001). Concerning the effect of chemical degradation of polyphenols by the Maillard reaction on their antioxidant activity, the chemical oxidation of these compounds is generally responsible for a loss in antioxidant capacity. However, recent observations suggest that partially oxidized polyphenols can exhibit higher antioxidant activity than that of nonoxidized phenols. For this reason, further investigations are required in this field.

Other factors can influence in the degradation of polyphenols. A group of polyphenols easily altered by several nonenzymatic factors are isoflavones. These compounds are found in soybeans as 6'-*O*-malonyl-β-glucoside conjugates. An interesting study published by Coward and others (1998) showed that the isoflavone glucoside conjugates were easily altered during extraction, processing, and cooking by different chemical reactions. Another example of the interaction of polyphenols with other compounds that can influence their stability has been reported recently studying the interactions between polyphenols and proteins in an apple polyphenol–milk model system, which may mask or enhance antioxidant capacity (Wegrzyn and others, 2008). Individual polyphenols and dairy components (including caseins and whey proteins) form polyphenol–protein complexes with high stability and antioxidant capacity (Wegrzyn and others, 2008).

Another factor that can modify the presence of polyphenols in foods is the fermentation of diverse products. Several works have reported that fermentation can reduce the level of bitterness and astringency of the cocoa bean, which could be attributed to the loss of polyphenols during fermentation (Maleyki and others 2008). Moreover, the oxidation of polyphenols to insoluble tannins during fermentation was responsible for the formation of flavor precursors for chocolate processing. Nonenzymatic oxidation of polyphenols could also occur at this stage. Finally, the use of heat during manufacturing of various products such as chocolates and cocoa-based foods can change the enantiomeric composition of polyphenols (Maleyki and others 2008). In addition, the epimerization of polyphenols could be caused by the conditions applied during the extraction procedures. It has been shown that 2 days of sun-drying of fresh unfermented cocoa beans (without fermentation) causes a 50% decrease in polyphenol content.

Acknowledgments

This work was supported by AGL2007-65907 (MEC, FEDER, Spain) and by Programa de ayudas a Grupos de Excelencia de Región de Murcia, de la Fundación Séneca, Agencia de Ciencia y Tecnología de la Región de Murcia (Plan Regional de Ciencia y Tecnología 2007/2010).

References

Baiocco P, Barreca AM, Fabbrini M, Galli C and Gentili P. 2003. Promoting laccase activity towards non-phenolic substrates: a mechanistic investigation with some laccase–mediator systems. Org Biomol Chem 1(1):191–197.

Baysal T and Demirdoven A. 2007. Lipoxygenase in fruits and vegetables: a review. Enzyme Microb Technol 40:491–496.

Burton SG. 2003. Oxidizing enzymes as biocatalysts. Trends Biotechnol 21(12):543–549.

Caballero PA, Gómez M and Rosell CM. 2007. Improvement of dough rheology, bread quality and bread shelf-life by enzymes combination. J Food Eng 81:42–53.

Cabanes J, Chazarra S and Garcia-Carmona F. 1994. Kojic acid, a cosmetic skin whitening agent, is a slow-binding inhibitor of catecholase activity of tyrosinase. J Pharm Pharmacol 46(12):982–985.

Cabanes J, Escribano J, Gandía-Herrero F, García-Carmona F nd Jiménez-Atiénzar M. 2007. Partial purification of latent polyphenol oxidase from peach (*Prunus persica* L. Cv. Catherina). Molecular properties and kinetic characterization of soluble and membrane-bound forms. J Agric Food Chem 55(25):10446–10451.

Cabanes J, Sánchez-Ferrer A, Bru R and García-Carmona F. 1988. Chemical and enzymic oxidation by tyrosinase of 3,4-dihydroxymandelate. Biochem J 256:681–684.

Chazarra S, García-Carmona F and Cabanes, J. 2001. Evidence for a tetrameric form of iceberg lettuce (*Lactuca sativa* L.) polyphenol oxidase: purification and characterization. J Agric Food Chem 49(10):4870–4875.

Clausa H and Decker H. 2006. Bacterial tyrosinases. Syst Appl Microbiol 29:3–14.

Coward L, Smith M, Kirk M and Barnes, S. 1998. Chemical modification of isoflavones in soyfoods during cooking and processing. Am J Clin Nutr 68:1486–1491.

Cristóvão RO, Tavares AP, Ferreira LA, Loureiro JM, Boaventura R and Macedo EA. 2009. Bioresource modeling the discoloration of a mixture of reactive textile dyes by commercial laccase. Technology 100:1094–1099.

Durán N, Rosa MA, D'Annibale A and Gianfreda L. 2002. Applications of laccases and tyrosinases (phenoloxidases) immobilized on different supports: a review. Enzyme Microb Technol 31:907–931.

Felby C, Hassingboe J and Lund M. 2002. Pilot-scale production of fiberboards made by laccase oxidized wood fibers: board properties and evidence for cross-linking of lignin. Enzyme Microb Technol 31(6):736–741.

Feussner I and Wasternack C. 2002. The lipoxygenase pathway. Annu Rev Plant Physiol Plant Mol Biol 53:275–297.

Gandía-Herrero F, Escribano J and García-Carmona F. 2005. Characterization of the monophenolase activity of tyrosinase on betaxanthins: the tyramine-betaxanthin/dopamine-betaxanthin pair. Planta 222(2):307–318.

Gandía-Herrero F, Escribano J and García-Carmona F. 2007. Characterization of the activity of tyrosinase on betanidin. J Agric Food Chem 55(4):1546–1551.

Gandía-Herrero, F, Jiménez, M, Cabanes, J, García-Carmona, F, Escribano, J. 2003. Tyrosinase inhibitory activity of cucumber compounds: enzymes responsible for browning in cucumber. J Agric Food Chem 51(26):7764–7769.

Goldman MHS, Seurinck J, Marins M, Goldman GH and Marian C. 1998. A tobacco flower-specific gene encodes a polyphenol oxidase. Plant Mol Biol 36:479–485.

Hamid M and Khalil-ur-Rehman A. 2009. Potential applications of peroxidases. Food Chem doi: 10.1016/j.foodchem.2009.02.035

Jacobson ES. 2000. Pathogenic roles for fungal melanins. Clin Microbiol Rev 13:708–717.

Jiménez M, Escribano-Cebrián J and García-Carmona, F. 1998. Oxidation of the flavonol fisetin by polyphenol oxidase. Biochim Biophys Acta 27: 1425(3):534–542.

Jiménez M and García-Carmona F. 1999. Oxidation of the flavonol quercetin by polyphenol oxidase. J Agric Food Chem 47(1):56–60.

Jiménez-Atiénzar M, Cabanes J, Gandía-Herrero F and Garcia-Carmona F. 2004. Kinetic analysis of catechin oxidation by polyphenol oxidase at neutral pH. Biochem Biophys Res Commun 319:902–910.

Jiménez-Atiénzar M, Escribano J, Cabanes J, Gandía-Herrero F and García-Carmona F. 2005. Oxidation of the flavonoid eriodictyol by tyrosinase. Plant Physiol Biochem 43(9):866–873.

Klabunde T, Eicken C, Sacchettini JC and Krebs B. 1998. Crystal structure of a plant catechol oxidase containing a dicopper center. Nat Struct Biol 5:1084–1090.

Kolomiets MV, Hannapel DJ, Chen H, Tymeson M and Gladon RJ. 2001. Lipoxygenase is involved in the control of potato tuber development. Plant Cell 13:613–626.

Liavonchanka A and Feussner, I. 2006. Lipoxygenases: occurrence, functions and catalysis. J Plant Physiol 163:348–357.

Littoz F and Mcclements DJ. 2008. Bio-mimetic approach to improving emulsion stability: Cross-linking adsorbed beet pectin layers using laccase. Food Hydrocolloids 22(7):1203–1211.

López-Nicolás JM, Bru R and García-Carmona F. 1997. Enzymatic oxidation of linoleic acid by lipoxygenase forming inclusion complexes with cyclodextrins as starch model molecules. J Agric Food Chem 45:1144–1148.

López-Nicolás JM, Bru R, Sánchez-Ferrer A and García-Carmona F. 1995. Use of "soluble lipids" for biochemical processes: linoleic acid: cyclodextrin inclusion complexes in aqueous solutions. Biochem J 308:151–154.

López-Nicolás JM, Bru R, Sánchez-Ferrer A and García-Carmona F. 1994. An octaethylene glycol monododecyl ether-based mixed micellar assay for lipoxygenase acting at neutral pH. Anal Biochem 221(2):410–415.

López-Nicolás JM, Bru-Martinez R and García-Carmona F. 2000. Effect of calcium on the oxidation of linoleic acid by potato (*Solanum tuberosum* var. Desiree) tuber 5-lipoxygenase. J Agric Food Chem 48(2):292–296.

López-Nicolás JM and García-Carmona F. 2008. Aggregation state and pK_a values of (E)-resveratrol as determined by fluorescence spectroscopy and UV-visible absorption. J Agric Food Chem 56(17):7600–7605.

López-Nicolás JM, Núñez-Delicado E, Pérez-López AJ, Carbonell A and Cuadra-Crespo P. 2006. Determination of stoichiometric coefficients and apparent formation constants for β-cyclodextrin complexes of trans-resveratrol using reversed-phase liquid chromatography. J Chromatogr A 1135:158–165.

López-Nicolás JM, Pérez-Gilabert M and García-Carmona F. 2001. Eggplant lipoxygenase (*Solanum melongena*): product characterization and effect of physicochemical properties of linoleic acid on the enzymatic activity. J Agric Food Chem 49(1):433–438.

López-Nicolás JM, Pérez-López AJ, Carbonell-Barrachina A and García-Carmona F. 2007. Kinetic study of the activation of banana juice enzymatic browning by the addition of maltosyl-beta-cyclodextrin. J Agric Food Chem 55(23):9655–9662.

Maleyki A, Jalil M and Ismail A. 2008. Polyphenols in cocoa and cocoa products: is there a link between antioxidant properties and health? Molecules 13:2190–2219.

Manach C, Scalbert A, Morand C, Rémésy C and Jiménez, L. 2004. Polyphenols: food sources and bioavailability. Am J Clin Nutr 79(5):727–747.

Manzocco L, Calligaris S, Mastrocola D, Nicoli MC and Lerici CR. 2001. Review of nonenzymatic browning and antioxidant capacity in processed foods. Trends Food Sci Technol 11:340–346.

Martinez MV and Whitaker JR. 1995. The biochemistry and control of enzymatic browning. Trends Food Sci Technol 6(6):195–200.

Mayer AM. 2006. Polyphenol oxidases in plants and fungi: going places? A review. Phytochemistry 67:2318–2331.

Mayer AM and Staples RC. 2002. Laccase: new functions for an old enzyme. Phytochemistry 60:551–565.

Micard V and Thibault JF. 1999. Oxidative gelation of sugar-beet pectins: use of laccases and hydration properties of the cross-linked pectins. Carbohyd Polym 39:265–273.

Minussi RC, Pastore GM and Durany N. 2002. Potential applications of laccase in the food industry. Trends Food Sci Technol 13:205–216.

Morozova OV, Shumakovich GP, Shleev SV and Yaropolov YI. 2007. Laccase–mediator systems and their applications: a review. Appl Biochem Microbiol 43(5):523–535.

Nazari K, Esmaeili N, Mahmoudi A, Rahimi H and Moosavi-Movahedi AA. 2007. Peroxidative phenol removal from aqueous solutions using activated peroxidase biocatalyst. Enz Microb Technol 41:226–233.

Nerya O, Vaya J, Musa R, Izrael S, Ben-Arie R and Tamir S. 2003. Glabrene and isoliquiritigenin as tyrosinase inhibitors from licorice roots. J Agric Food Chem 51:1201–1207.

Núñez-Delicado E, Sojo MM, Sánchez-Ferrer A and García-Carmona F. 1999. Hydroperoxidase activity of lipoxygenase in the presence of cyclodextrins. Arch Biochem Biophys 367(2):274–280.

Orenes-Piñero E, García-Carmona F and Sánchez-Ferrer A. 2005. A kinetic study of p-cresol oxidation by quince fruit polyphenol oxidase. J Agric Food Chem 53(4):1196–1200.

Pérez-Gilabert M and García-Carmona F. 2000. Characterization of catecholase and cresolase activities of eggplant polyphenol oxidase. J Agric Food Chem 48(3):695–700.

Pérez-Gilabert M, Morte A, Honrubia M and García-Carmona F. 2001. Partial purification, characterization, and histochemical localization of fully latent desert truffle (*Terfezia claveryi* Chatin) polyphenol oxidase. J Agric Food Chem 49(4):1922–1927.

Pérez-Gilabert M, Sánchez-Felipe I and García-Carmona F. 2005. Purification and partial characterization of lipoxygenase from desert truffle (*Terfezia claveryi* Chatin) ascocarps. J Agric Food Chem 53(9):3666–3671.

Pérez-Gilabert M, Sánchez-Ferrer A and García-Carmona F. 1994. Enzymatic oxidation of phenothiazines by lipoxygenase/H_2O_2 system. Biochem Pharmacol 47(12):2227–2232.

Porta H and Rocha-Sosa M. 2002. Plant lipoxygenases. Physiological and molecular features. Plant Physiol 130:15–21.

Rodríguez-Couto S and Toca-Herrera JL. 2006. Industrial and biotechnological applications of laccases: a review. Biotechnol Adv 24:500–513.

Sánchez-Ferrer A, Bru R and Garcia-Carmona F. 1989. Novel procedure for extraction of a latent grape polyphenoloxidase using temperature-induced phase separation in Triton X-114. Plant Physiol 91(4):1481–1487.

Sánchez-Ferrer A, Rodríguez-López JN, García-Cánovas F and García-Carmona F. 1995. Tyrosinase: a comprehensive review of its mechanism. Biochim Biophys Acta 1247(1):1–11.

Sojo MM, Núñez-Delicado E, Sánchez-Ferrer A and García-Carmona F. 2000. Oxidation of salsolinol by banana pulp polyphenol oxidase and its kinetic synergism with dopamine. J Agric Food Chem 48(11):5543–5547.

Soler-Rivas C, Moller AC, Arpin N, Olivier JM and Wichers HJ. 2000. Induction of tyrosinase mRNA in Agaricus bisporus upon treatment with a tolaasin preparation from *Pseudomonas tolaasii*. Physiol Mol Plant Pathol 58:95–99.

Tatiana V, Pegasova P, Zwart O, Koroleva V, Stepanova EV, Rebrikovc DV and Lamzinb V. 2003. Crystallization and preliminary X-ray analysis of a four-copper laccase from *Coriolus hirsutus*. Acta Crystallogr D Biol Crystallogr 59(8):1459–1461.

Thipyapong P, Hunt M and Steffens, J. 2004. Antisense down regulation of polyphenol oxidase results in enhanced disease susceptibility. Planta 220:105–117.

Thurston CF. 1994. The structure and function of fungal laccases. Microbiology 140(1):19–26.

Urquiaga I and Leighton F. 2000. Plant polyphenol antioxidants and oxidative stress. Biol Res Biol 33(2):55–64.

Veitch NC. 2004. Horseradish peroxidase: a modern view of a classic enzyme. Phytochemistry 65:249–259.

Wegrzyn TF, Farr JM, Hunter DC, Au J, Wohlers MW, Skinner MA, Stanley RA and Waterhouse DS. 2008. Stability of antioxidants in an apple polyphenol–milk model system. Food Chem 109:310–318.

5 Chemistry of Flavonoids

Rong Tsao* and Jason McCallum

Introduction

Flavonoids are one of the most abundant natural product groups occurring in the plant kingdom. Although the term "flavonoids" has been interchangeably used by many, including researchers, with other terms such as "polyphenols," "polyphenolics," or "phenolics," it is necessary for the readers of this book to make a clear distinction between these terms. Polyphenols or polyphenolics are a broader category of phytochemicals that includes phenolic acids, stilbenes, flavonoids, and other phytochemicals with phenolic features. Most "phenolics" in fruits and vegetables are polyphenolics (with >1 hydroxyl group on a benzene ring).

Flavonoids generally have a common C_6-C_3-C_6 flavone skeleton in which the three-carbon bridge between the phenyl groups is usually enclosed with oxygen. Based on the degree of unsaturation and oxidation of the three-carbon segment (C-ring), flavonoids are further divided into several subclasses (Fig. 5.1). Most flavonoids reported in the literature are glycosides of a relatively small number of flavonoid aglycones, which are generally water-soluble and accumulated in the vacuoles of plant cells (Bohm 1998; Seigler 1998). The structural features of the B-ring and the hydroxylation and glycosylation patterns on all A, B, and C rings of the flavone skeleton have made flavonoids one of the largest and diverse phytochemical groups. The exact number of flavonoids is difficult to know; however, it has been reported to be anywhere from 2,000 to 6,500 (Harborne and Williams 2000; Klejdus and others 2001; Rauha and others 2001, Tsao and Deng 2004; de Rijke and others 2006). These compounds frequently serve as pigments in plants to attract pollinators, or as plants' chemical defense against invading insects and microorganisms. Flavonoids may have been involved in other biological interactions. A significant role that has been under active research in recent years is their possible health beneficial effects to humans. Flavonoids have been found to possess potent antioxidant activities (Pietta 2000). Increasing evidence from epidemiological studies suggests that diets high in flavonoids are contributing significantly to lower risks of cardiovascular diseases and cancer in humans. For this reason, both Health Canada and the US Food and Drug Administration have allowed health claims for fruit and vegetable consumption and lowered risks of heart diseases and certain cancers (Health Canada online; FDA-USA online). More detailed discussions on the health benefit of flavonoids can be found in Chapter 6 of this book.

* Corresponding author.

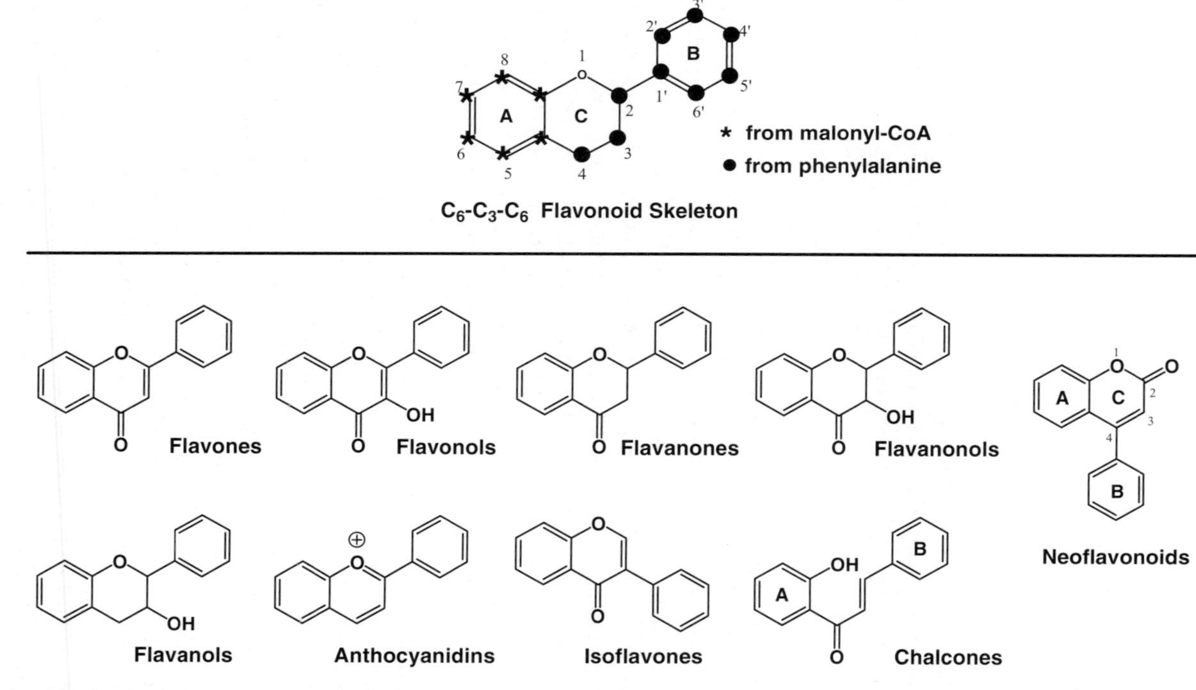

Figure 5.1. The basic C_6-C_3-C_6 flavonoid skeleton (upper panel) and typical flavonoid subgroups (lower panel). The biosynthetic origins of carbon atoms and standard numbering are also illustrated in the upper panel.

Chemical Classification of Flavonoids

Flavonoids can be considered as a subgroup of the polyphenols, and they are plant secondary metabolites with a C_6-C_3-C_6 skeletal system. Most encountered flavonoids contain ring structures as shown in Figure 5.1, although open structures such as the chalcones have traditionally been included in the category because of the similarity in biosynthesis (see "Metabolic pathway and occurrence of flavonoids" in this chapter). The chemical structures of the C_6-C_3-C_6 flavonoids are based on a chromane ring (ring C), which most of the time bears a second aromatic ring B in position 2; however, ring B can also be attached to positions 3 (isoflavones) or 4 (neoflavonoids) of ring C (see Fig. 5.1). Flavonoids shown in Figure 5.1 are aglycones; however these compounds usually exist as glycosides and sometimes acylglycosides in fruits and vegetables. *C*-Glycosylflavonoids, where the sugar unit is attached to the flavonoids through carbon–carbon bonding, are also commonly found in plants, although less in fruits and vegetables. Other flavonoid derivatives such as acylated and methylated flavonoids and flavonoid sulfates have also been found, albeit less frequently and in lower concentrations (Harborne 1994; Anderson and Markham 2006).

Flavonoids with ring B attached to position 2 of ring C can be further categorized into several different groups according to structural features of the chromane ring (ring C): flavones, flavonols, flavanones, flavanonols, flavanols, and anthocyanidins (see Fig. 5.1). Among these subgroups, flavones and flavonols, mostly their *O*-glycosides alone, make up one of the largest classes of flavonoid constituents in plants, with more than 2,000 known structures (Williams 2006). Typical flavonoid subgroups and their distribution in fruits and vegetables are listed in Table 5.1, and their structures in Figures 5.2 and 5.3.

Flavones: Ring C of the flavones contains a double bond between positions 2 and 3, and a ketone on position 4. Most flavones of fruits and vegetables hold a hydroxyl group on position 5 of ring A, whereas hydroxylation on other positions, most often position 7 of ring A or positions 3' and 4' of ring B, can vary depending on the taxonomic classification of a particular fruit or vegetable. Apigenin and luteolin are the most widespread flavone aglycones (see Fig. 5.2); however the diverse substitution patterns make this group the largest, with 309 aglycones in the newest compilation of flavonoids (Valant-Vetschera and Wallenweber 2006). Glycosylation occur mostly at positions 5 and 7, and methylation and acylation on hydroxyl groups of ring B. Some flavones are polymethoxylated, such as tangeretin and nobiletin found in the peels of citrus fruits. Typical flavones and their origins are listed in Table 5.1 and Figure 5.2.

Flavonols: The only difference between flavones and flavonols is the hydroxyl group on position 3 of the latter. This 3-hydroxyl group can be glycosylated as well. This group is perhaps the most common in fruits and vegetables. Valant-Vetschera and Wallenweber (2006) tabulated 393 flavonol aglycones. Similar to the flavones, these aglycones are highly diverse in hydroxylation and methylation patterns; therefore, when the different glycosylation patterns are considered, flavonols are perhaps the largest subgroup. The most common flavonol aglycones, quercetin and kaempferol, alone had at least 279 and 347 different glycosidic combinations, respectively (Williams 2006) (see Table 5.1 and Fig. 5.2).

Table 5.1. Typical flavonoid compounds in major subgroups of selected fruits and vegetables[a]

Flavonoid Subgroups	Major Flavonoids	Fruits/Vegetables
Flavones	Apigenin, luteolin	Beets, bell peppers, Brussels sprouts, cabbage, cauliflower, celery, chives, kale, lettuces, spinach, peppers, tomatoes, watercress
	Tangeretin, nobiletin	Citrus fruits
Flavonols	Quercetin, kaempferol, myricetin, isorhamnetin,	Apples, berries, broccoli, cabbages, chives, cranberries, grapes, kale, onions, peppers, spinach, Swiss chard, tomatoes, watercress
Flavanones	Naringenin, hesperetin,	Oranges and other citrus fruits
Flavanonols	Taxifolin	Citrus fruits*
Anthocyanidins	Cyanidin, delphinidin, malvidin, pelargonidin, peonidin, petunidin,	Blueberries, blackberries, cranberries, egg plants, pomegranates, plums, raspberries, red onions, red potatoes, red grapes, red radishes, strawberries, other red-purple fruits and vegetables
Flavanols/Procyanidins	Catechin, epicatechin, and their gallic acid esters	Apples, grapes, plums, pears, mangoes, okra, peaches, Swiss chard, berry fruits and vegetables in general,
	Monomers, dimers (procyanidin B1, B2), and oligomers	Apples, cherries, berries, grapes, peaches, pears
Chalcones	Xanthohumol	Hops**
	Phloretin	Apple#
Isoflavonoids	Daidzein, genistein	Soybean sprouts##

[a]Source: information was collected from the USDA Flavonoids database Release 2.1; *Kawaii and others 1999, **Zhao and others 2005, #Tsao and others 2003, ##Lee and others 2007.

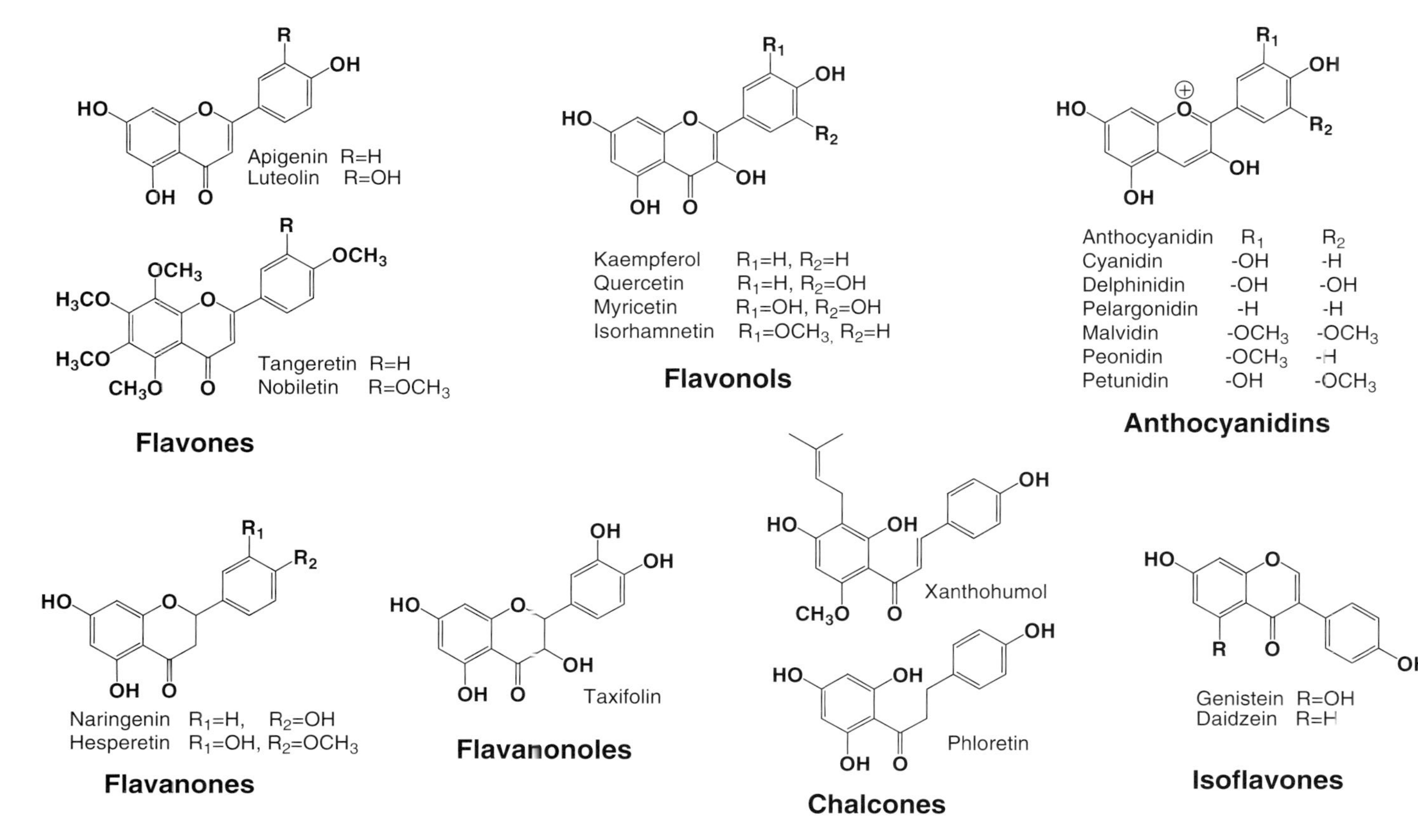

Figure 5.2. Typical flavonoids in the subgroups commonly found in fruits and vegetables.

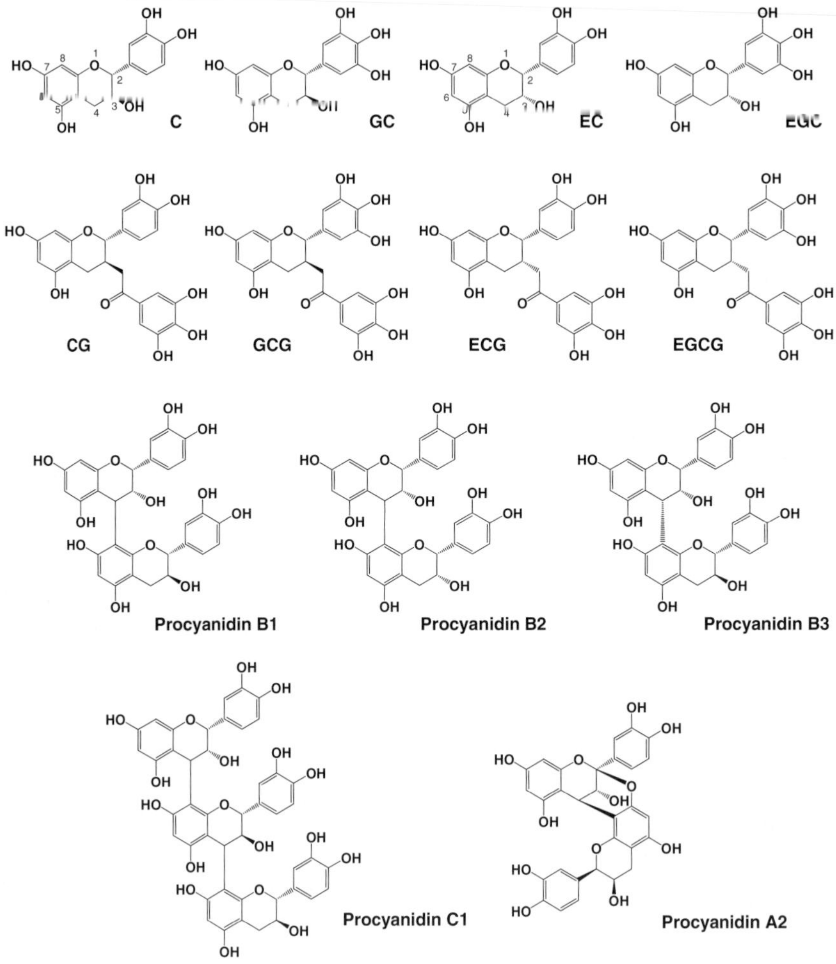

Figure 5.3. Flavanol monomers (catechins, epicatechins, and their gallates) and procyanidins. C, catechin; GC, gallocatechin; EC, epicatechin; EGC, epigallocatechin; CG, catechin gallate; GCG, gallocatechin gallate; ECG, epicatechin gallate, EGCG, epigallocatechin gallate.

Flavanones: The structural features of flavanones are basically the same as those of flavones, except that flavanones lack the double bond between positions 2 and 3. They are also called dihydroflavones. Flavanones were considered to be a relatively minor flavonoid subgroup; however, in the past 15 years, the total number of known compounds in this subgroup has more than doubled. Two representative aglycones of this group are naringenin and hesperetin. However, flavanones can be multihydroxylated, and the various hydroxyl groups can be methylated and/or glycosylated (see Table 5.1 and Fig. 5.2). Some flavanones have unique substitution patterns, such as prenylated flavanones, furanoflavanones, pyranoflavanones, and benzylated flavanones, giving a large number of substituted derivatives (Grayer and Veitch 2006).

Flavanonols: Flavanonols can be considered as flavanones with a hydroxyl group on position 3. They are sometimes referred to as dihydroflavonols. Similar to the situation of flavanones, flavanonols are no longer a minor subgroup of flavonoids, and they are a structurally highly diverse and multisubstituted subgroup (Grayer and Veitch 2006). A well-known flavanonol is taxifolin from citrus fruits (Kawaii and others 1999) (see Table 5.1 and Fig. 5.2).

Flavanols: Flavanols are often referred to as flavan-3-ols, as the hydroxyl group is almost always attached to the position 3 carbon of ring C. Flavanols are also interchangeable with the term catechins. Flavanols do not have the ketone feature as the flavanonols. Catechins have two epimers depending on the stereo configurations of the bond between ring B and the position 2 carbon, and the hydroxyl group on position 3. These two epimers,(−)-epicatechin and (+)-catechin, and their respective derivatives epigallocatechin and gallocatechin are together categorized as catechins (see Fig. 5.3). Gallocatechin and epigallocatechin contain an extra hydroxyl group on ring B. Flavanols or catechins are often found in the skins of fruits and certain vegetables. Many commonly consumed fruits and vegetables are found to contain flavanols and their gallic acid esters (Harnly and others 2006) (see Table 5.1).

Another important feature of the flavanols is their ability to form polymers. Polymeric forms of flavanols are often referred to as procyanidins, which are also called proanthocyanidins or condensed tannins as opposed to hydrolyzable tannins (e.g., gallotannins and ellagitannins, esters of glucose and gallic acid or ellagic acid, respectively). Procyanidins usually contain 2 to 60 units of monomeric flavanols, and the polymerization most often occurs via a carbon–carbon bond between the position 8 carbon (C8) of the terminal unit and C4 of the extender. The four most common dimers are B-type procyanidins (e.g., B1, B2, B3, and B4), and their modes of coupling are shown in Figure 5.3. Further polymerization yield the linear 4,8 polymers. Procyanidins with 2–7 catechin units are oligoprocyanidins (OPC) (e.g., procyanidin C1, Fig. 5.3). Flavanols also form A type procyanidins in which flavanol monomers are doubly linked (e.g., procyanidin A2, Fig. 5.3) (Prior and others 2001; Liu and others 2007). More information on the different types of procyanidins can be found in Prior and others (2001); Xie and Dixon (2005); and Ferreira and others (2006) (see Table 5.1).

Anthocyanidins: Anthocyanidins are a unique subgroup of flavonoids that give plants distinctive colors. The red, blue, and purple colors of small berry fruits, red apples and cherries, red lettuce, and many other fruits and vegetables are from the various anthocyanidins (see Table 5.1). Chemically, anthocyanidins are flavylium cations and most often exist as chloride salts. Depending on the pH values, anthocyanidins show colors from red (in very acidic conditions) to purple-blue (in intermediate pH conditions) and yellow-green (alkaline conditions). The color of anthocyanidins can also be affected by acylation or methylation at the hydroxyl groups of ring A and B. Anthocyanins are glycosides of anthocyanidins, with the sugar group mostly attached to the C3 position of ring C. The sugar units of the anthocyanins often conjugated with some phenolic acids such as ferulic acid. Although more than 500 anthocyanins have been reported, these compounds are based on 31 known monomeric anthocyanidins (Anderson and Jordheim 2006). Moreover, among the 31 anthocyanidins, 30% were from cyanidin, 22% from delphinidin, and 18% from pelargonidin. Methylated derivatives of the aforementioned anthocyanidins, peonidin, malvidin, and petunidin, all together had a 20% share of the anthocyanidins. In other words, 90% of the most frequently encountered anthocyanins are related to cyanidin, delphinidin, and pelargonidin and their methylated derivatives (Anderson and Jordheim 2006) (see Fig. 5.2). The difference in chemical structure that occurs in response to changes in pH is the reason why anthocyanins are often used as pH indicators.

In addition to the flavonoids with ring B attached to the C2 position of ring C, other types of flavonoids, particularly those with ring B attached to C3 (isoflavones), C4 (neoflavonoids), and open C ring (chalcones), have also been commonly found in fruits and vegetables.

Isoflavones: In contrast to most other flavonoids, isoflavonoids have a rather limited taxonomic distribution, mainly within the Leguminosae. Commonly consumed fruits and vegetables are not good sources of isoflavones; however, certain specialty vegetables such as soybean sprouts are becoming increasingly popular because of the health-promoting properties of isoflavones including genistein and daidzein (see Fig. 5.2) (Lee and others 2007).

Chalcones: Chalcones and dihydrochalcones can be considered as flavonoids with an open structure (see Fig. 5.1). Although dihydrochalcones such as phloretin glycosides and 3-hydroxyphloretin glycosides have been found in many fruits and vegetables (Tsao and others 2003; Arabbi and others 2004), the occurrence in general is rare (Tomás-Barberán and Clifford 2000). Prenylated chalcones such as xanthohumol are found in hops (Zhao and others 2005) (see Fig. 5.2).

The total and individual flavonoid contents in commonly consumed fruits and vegetables can be found in several recent surveys of the literature (Arabbi and others 2004; Franke and others 2004; Chun and others 2005; Harnly and others 2006; Sun and Powers 2007). They have been collected and compiled into a database (USDA Flavonoids Database Release 2.1, 2007). Table 5.1 enlists some of the most typical flavonoids, found in the major subgroups just discussed, for selected popular fruits and vegetables based on this database and current literature.

Analytical Methods

A good separation and analytical method is essential to the characterization of flavonoids for both chemical and biological properties. The heightened interest in flavonoids from fruits and vegetables and their potential health benefits has resulted in several good reviews on the separation and analysis of phytochemicals including flavonoids (Sakakibara and others 2003; Tsao and Deng 2004; de Rijke and others 2006), so in this chapter, we briefly summarize and discuss only the most frequently used methods for extraction, separation, and analysis of flavonoids found in commonly consumed fruits and vegetables.

Sample Preparation and the Extraction Process

Plants including fruits and vegetables are a vast reservoir of different phytochemicals. As stated previously, flavonoids are a diverse group of polyphenolic compounds, some of which are relatively stable, whereas others such as anthocyanins are labile under ambient conditions. Sample preparation is of paramount importance in studying flavonoids because a good method prevents compounds of interest from being degraded during sample collection, storage, and any drying processes. Fruit and vegetable samples contain large amounts of water, and any disruption of cells at ambient temperature may trigger enzymatic reactions such as browning caused by the polyphenoloxidase (PPO) in apple (Goupy and others 1995). Degradation of flavonoids can also happen under elevated temperature or exposure to air and light during oven or air drying of samples. It is always advisable to collect fresh samples and extract and analyze immediately; however, when they must be kept for longer time periods, freezing and drying the samples is preferred. Flash freezing with liquid nitrogen and lyophilization (freeze-drying) are the most prudent methods for preserving the identity and concentration of the native flavonoids in fruits and vegetables.

The purpose of extraction is to maximize recovery of the compounds of interest, that is, flavonoids, from the samples. Extraction of flavonoids from fruits and vegetables is generally carried out by using water, organic solvents, or liquefied gas, or combinations of these under various temperature and pressure conditions. The efficiency of solvent extraction, however, depends on several factors, including the physicochemical properties of the solvent (e.g., polarity), temperature, pH, sample/solvent ratio, extraction steps (repeats), and sample particle characteristics (size and shape). The recovery of flavonoids also relies on the sample matrix and the physicochemical properties of the compounds under investigation. Flavonoids are normally stored in the vacuoles of the cell; therefore, solvent molecules must be able to penetrate the cell walls to reach the compounds of interest. Sonication is commonly used in flavonoid solvent extraction. Flavonoid glycosides are more water soluble and thus require the use of relatively polar solvents. Different extraction technologies exist that can enhance the efficiency of flavonoid extraction. To what extent all these factors are considered for a particular method of extraction also depends on the objective of the study, for example, quantitative or qualitative analysis.

Enzyme activity of the plants and the presence of oxygen and light during the extraction also impact the efficiency, therefore extreme care must be taken to avoid hydrolysis, oxidation and/or isomerization (Montedoro and others 1992a, b; Gao and Mazza 1995). Often, noninterfering antioxidants such as ascorbic acid or BHT (*tert*-butylhydroquinone) are added to prevent oxidation (Careri and others 2000; Häkkinen and Törrönen 2000). However, caution must be taken in choosing the antioxidants because some may cause degradation of certain flavonoids (Bradshaw and others 2001). Intentional hydrolysis for obtaining the aglycones of flavonoids with highly complex glycosylation patterns may be incorporated into the extraction process as well. Hydrolysis is usually performed in acidic conditions with 1–2 M HCl at higher temperature (Hertog and others 1992; Franke and others 2004). Sonication and refluxing are commonly used to aid the extraction and hydrolysis (Franke and others 2004; Harnly and others 2006). Pre-extraction hydrolysis can also be done under milder conditions using enzymes such as β-glucuronidase (Pietta and others 1989; de Rijke and others 2006).

Different fruits and vegetables vary significantly in their structural constituents, macronutrients (proteins, oils, and carbohydrates), and micronutrients such as flavonoid profiles. It is almost impossible to develop one optimal method for extraction, separation, and analysis for each and every different fruit or vegetable. However, because of the relatively similar chemistry and biochemistry of flavonoids, some general statements can be abstracted from the existing literature. Flavonoids of fruits and vegetables are usually extracted using water, or more often a mixture of water and miscible polar solvents such as methanol, ethanol, and acetone. Methanol is more frequently used than ethanol and acetone because of its higher extraction efficiency (Marston and Hostettmann 2006). Aqueous methanol between 50% and 80% has been used for extracting flavonoids. Owing to the phenolic nature of most flavonoids, the extraction solvents are often acidified to increase the extraction efficiency, particularly of anthocyanins. However, one must be cautious when choosing the concentration and type of acid, because high concentration can cause hydrolysis. It is reported that solvents containing >0.12 M HCl can cause partial hydrolysis of acylated anthocyanins from red grapes (Revilla and others 1998). On the other hand, higher water percentage in the solvent can aid in the extraction of flavonoids in dry samples; it also helps extracting the glycosides. Relatively lower percentage of water is used for fresh samples. In a recent survey by Harnly and others (2006), all fruit and vegetable samples were freeze-dried. Two extraction methods were used in their study. In the direct extraction method, 0.2–0.5 g freeze-dried sample powder was extracted repeatedly (3 × 5 min homogenization) using 4 mL 90% aqueous methanol in the presence of a synthetic antioxidant TBHQ (*tert*-butylhydroquinone). The indirect extraction (hydrolysis) was done by reflux at 75°C for 5 hr using 1.2 N HCl in methanol in the presence of TBHQ) (0.5–7 g/50 mL methanol) (Harnly and others 2006).

Solvent extraction offers good recovery of flavonoids from fruits and vegetables; however, the use of large amounts of organic solvents poses health and safety risks and is environmentally unfriendly. There are many alternative methods that either eliminate or reduce significantly the use of organic solvents. Some of them offer identical, if not better, extraction efficiency and cost effectiveness. Methods such as solid-phase extraction (SPE) use solid absorbents to extract flavonoids; SPE is perhaps more suitable for sample cleanup, purification, or preconcentration than for extraction because of the selectivity and saturation of the absorbents (Tsao and Deng 2004). Other alternative extraction methods, including microwave-assisted extraction (MAE), supercritical fluid extraction (SFE), and pressurized liquid extraction (PLE), have become increasingly popular in the extraction of flavonoids. However, for reasons of space, readers are referred to Chapter 9 of this book and other reviews for detailed information (Tsao and Deng 2004).

Quantitative and Qualitative Analysis of Flavonoids

Total flavonoid content. Quantitative analysis of flavonoids depends on the objective of the study. Colorimetric estimation of total flavonoid content is measured by the aluminum chloride colorimetric assay (Jia and others 1999; Chang and others 2002). The total flavonoid content measured in this way is normally expressed in equivalent values of a standard flavonoid, often catechin or quercetin equivalents. Not all subgroups of flavonoids can be quantified by colorimetric methods; however, total anthocyanin content is determined using the pH-differentiation method (Boyles and others 1993).

Quantification of individuals: The foregoing quantitative methods provide rapid estimation of the total flavonoids or their subgroups; however, they offer no information on specific flavonoids. To analyze the concentration of individual flavonoids, good separation procedures must first be developed, followed by quantification using various spectrophotometric or other types of detectors. Such a separation and purification procedure is also critical for the identification of unknown compounds. Concentrations of individual flavonoids in fruits and vegetables have been determined using various methods, particularly chromatographic techniques. Conventional chromatographic separations of flavonoids include the use of thin-layer chromatography (TLC) and open column chromatography (CC). These methods are simple and cost less, but are mostly used for preparative separations. For analytical purposes, gas chromatography (GC), capillary electrophoresis (CE), and high-performance liquid chromatography (HPLC) provide high sensitivity and separation efficiency (Tsao and Deng 2004; de Rijke and others 2006). In this chapter, we focus on HPLC because it is often the method of choice for quantitative and qualitative analyses of flavonoids (Tsao and Deng 2004).

HPLC coupled with different detection techniques such as UV-Vis diode-array detection (DAD) and mass spectrometry (LC-MS) has been pivotal in the characterization of different classes of flavonoids, and a vast number of applications can be found in the literature. Many good reviews have been published as well (Merken and Beecher 2000; Sakakibara and others 2003; Tsao and Deng 2004; de Rijke and others 2006). The overwhelming majority of these applications are based on reversed-phase (RP) separation, although normal-phase (NP) HPLC is occasionally used for less polar flavonoid subgroups such as polymethoxylated flavones in citrus fruits (Marston and Hostettmann 2006). The most typical RP-HPLC method usually involves the use of a C18 column (typically 150–250 mm × 4.6 mm, particle size 5 μm), DAD, a binary solvent system containing acidified water (solvent A) and a polar organic solvent (solvent B), and a run time of 1 hr at a flow rate of 1.0–1.5 mL/min (Tsao and Deng 2004; de Rijke and others 2006). Many researchers have attempted to develop a separation method for the quantification and identification of as many compounds as possible in a single analysis; however, the complexity of the flavonoids in fruits and vegetables may predispose such methods to be difficult. Then again, despite the intricacy of such simultaneous separation methods, a few good methods have been recently developed (Merken and Beecher 2000; Tsao and Yang 2003; de Rijke and others 2006). One recent method by Harnly and others (2006) again can be adapted for analysis of multiple fruits and vegetables. They used a C18 column (250 × 4.6 mm, 5 μm) and a C18 guard column (12.5 × 4.6 mm), with the column set at 30°C and the pump at 1.0 mL/min. The sample injection volume was 5 μL. The DAD was set to obtain spectra for the full range with multiple specific monitoring at 210, 260, 278, 370, and 520 nm for the various flavonoid subgroups. More precisely, 260 nm was used to identify the flavones, 278 nm for flavanones and flavonols, 370 nm for flavonols, and 520 nm for anthocyanins. They used the absorbance at 210 nm for the quantification of all flavonoid subgroups except the anthocyanins (520 nm) because of the higher sensitivity (Harnly and others 2006).

Identification of flavonoids: Quantification of individual flavonoids depends heavily on the availability of standard references. Only a limited number of common flavonoids are commercially available as standards. Standard references for flavonoid glycosides are particularly difficult to find; thus direct quantification of the native glycosides is nearly impossible. Analysis of the aglycones after acid or enzymatic hydrolysis is therefore common practice. When standard flavonoids are not available, or when unknown compounds are encountered in a particular fruit or vegetable, use of a DAD and/or MS detector is essential. Identification using technologies such as MS, nuclear magnetic resonance (NMR) spectroscopy, and infrared (IR) spectroscopy is certainly the way to go if such new or unknown compounds are first isolated and purified in enough quantity. To review applications of these technologies is beyond the scope of this chapter; however, the coupling of HPLC with DAD and MS is a great tool in both quantitative and qualitative analysis of fruit- and vegetable-derived flavonoids. This method is worthy of special emphasis.

The UV-Vis absorption spectra and often the subtle differences in the λ_{max} of the different flavonoids can be used to distinguish and identify the subgroup they belong to. Using this technique, Justesen and others successfully separated and analyzed flavonols, flavones, and flavanones in fruits and vegetables (Justesen and others 1998). A searchable in-house UV spectral database was built based on our own research on flavonoids, using the ChemStation software. We found this database highly useful for the confirmation of known and identification of unknown compounds (Tsao and Yang 2003; Tsao and Deng 2004). Harnly and others (2006) have used a similar concept to monitor and identify the various flavonoid subgroups in 60 fruits, vegetables, and nuts.

In the meantime, HPLC coupled with MS is no longer a dream machine to many researchers. Many able and less costly bench top LC-MS instruments are now available for routine applications. Two soft ionization techniques, namely, electrospray ionization (ESI) and atmospheric-pressure chemical ionization (APCI), are frequently used for the identification of flavonoids. Mass spectral data provide structural information on flavonoids and are used to determine molecular masses and to establish the distribution of substituents between the A- and B-rings. A careful study of fragmentation patterns can also be of particular value in the determination of the nature and site of attachment of the sugars in *O*- and *C*-glycosides (Marston and Hostettmann 2006). The combination of LC-DAD-MS provides even more efficient a method for the identification of flavonoids in a mixture. In MS, techniques such as collision-induced dissociation (CID) can be carried out to enhance fragmentation of a flavonoid of interest. A CID experiment generates daughter ions (product ions) from a parent ion MS^2, and this experiment can be done on the daughter ions to produce granddaughter ions MS^3. Such processes can be repeated many times in MS^n, providing sufficient information for the elucidation of flavonoid structure from the fragmentation patterns. There are different mass analyzers and other related techniques used in LC-MS^n, and the reader is directed to in-depth reviews (Tsao and Deng 2004; Marston and Hostettmann 2006; de Rijke and others 2006; Stalikas 2007). Both APCI- and ESI-MS can be done in either positive or negative ionization mode (de Rijke and others 2006).

It must be mentioned that the HPLC technologies, particularly in packing materials and high-pressure pumps, have advanced rapidly in the past few years, and a whole new generation of ultraperformance liquid chromatography (UPLC) is increasingly used for flavonoid separation. UPLC not only presents unprecedented high resolution, it is also highly sensitive (sharper peaks) and fast, and it uses only a fraction of the solvent (mobile phase) used by conventional HPLC (Armenta and others 2008). This is so new a technology that literature related to flavonoid analysis using UPLC is relatively scarce (67 hits using "UPLC and flavonoids" to search the ScienceDirect, for all years, all fields, at 17:54, March 17, 2009). The same search for the *Journal of Agricultural and Food Chemistry* returned 12 articles. Only a few UPLC-related papers described method for flavonoid analysis in fruits and vegetables. Slimestad and others (2008) analyzed flavonoids in tomatoes, and their method was only 16 min long. There are no studies directly comparing HPLC and UPLC for the same flavonoids and same samples; however, naringenin, the aglycone in tomatoes, was detected at a retention time of 15.52 min, whereas the aglycones in Sakakibara and others (2003) were detected after 75 min. Most of the UPLC-related literature was about the use of UPLC-MS for identification purposes (Hosseinian and others 2007; Ortega and others 2008). Furthermore, other so-called hyphenated techniques such as NMR detection have also been adapted to HPLC; therefore powerful systems such as LC-DAD-MS^n-NMR are available. Combinations like this will significantly improve the identification of unknown flavonoids (Wilson 2000).

Metabolic Pathway and Occurrence of Flavonoids

Biosynthetic Pathway and Origins of the A, B, and C Rings

Flavonoid biosynthesis is linked to primary metabolism through both plastid- and mitochondria-derived intermediates, each requiring export to the cytoplasm where they are incorporated into separate halves of the molecule.

Ring B and the central three-carbon bridge forming the C ring (see Fig. 5.1) originate from the amino acid phenylalanine, itself a product of the shikimate pathway, a plastid-based process which generates aromatic amino acids from simple carbohydrate building blocks. Phenylalanine, and to a lesser extent tyrosine, are then fed into flavonoid biosynthesis via phenylpropanoid (C6-C3) metabolism (see Fig. 5.1).

The six-membered aromatic A ring originates from three units of malonyl-CoA, produced from citrate precursors through the activity of a cytosolic acetyl-CoA carboxylase (ACC) (Fatland and others 2004) (see Fig. 5.1). These three malonyl-CoA units are added through sequential decarboxylation condensation reactions and actually represent the first committed step toward flavonoid biosynthesis.

Phenylpropanoid Precursors to Flavonoids

Flavonoid biosynthesis requires *p*-coumaroyl-CoA, itself a product of phenylpropanoid metabolism, synthesized via a core set of three reactions collectively referred to as the group I or early-acting enzymes (Fig. 5.4). The cytosolic-based group I reactions take phenylalanine generated in the plastid by the shikimic pathway and feed it into multiple biosynthetic pathways used to synthesize a wide variety of phenolic compounds including benzoic acids, cinnamic acids, ellagic acids, lignins, lignans, stilbenes, and hydrolyzable tannins, in addition to the flavonoids that are the focus of this chapter.

In the first enzymatic step, phenylalanine ammonia lyase (PAL) converts phenylalanine to *trans* cinnamate, via a deamination reaction liberating ammonia. PAL can also convert tyrosine to *p*-coumarate, albeit at lower efficiency (MacDonald and D'Cunha 2007). PAL functions as a tetramer of identical subunits, with two subunits combining to form one active site (Stafford 1990; MacDonald and D'Cunha 2007).

While PAL itself is recovered from the soluble fractions of cellular lysates via ultracentrifugation purifications, strong experimental evidence suggests it exists as part of a multienzyme complex associated with the endoplasmic reticulum (ER), forming a metabolic channel funneling substrates directly from one enzymatic reactive site to another (Winkel 2004). The enzyme physically anchored to the ER is cinnamate

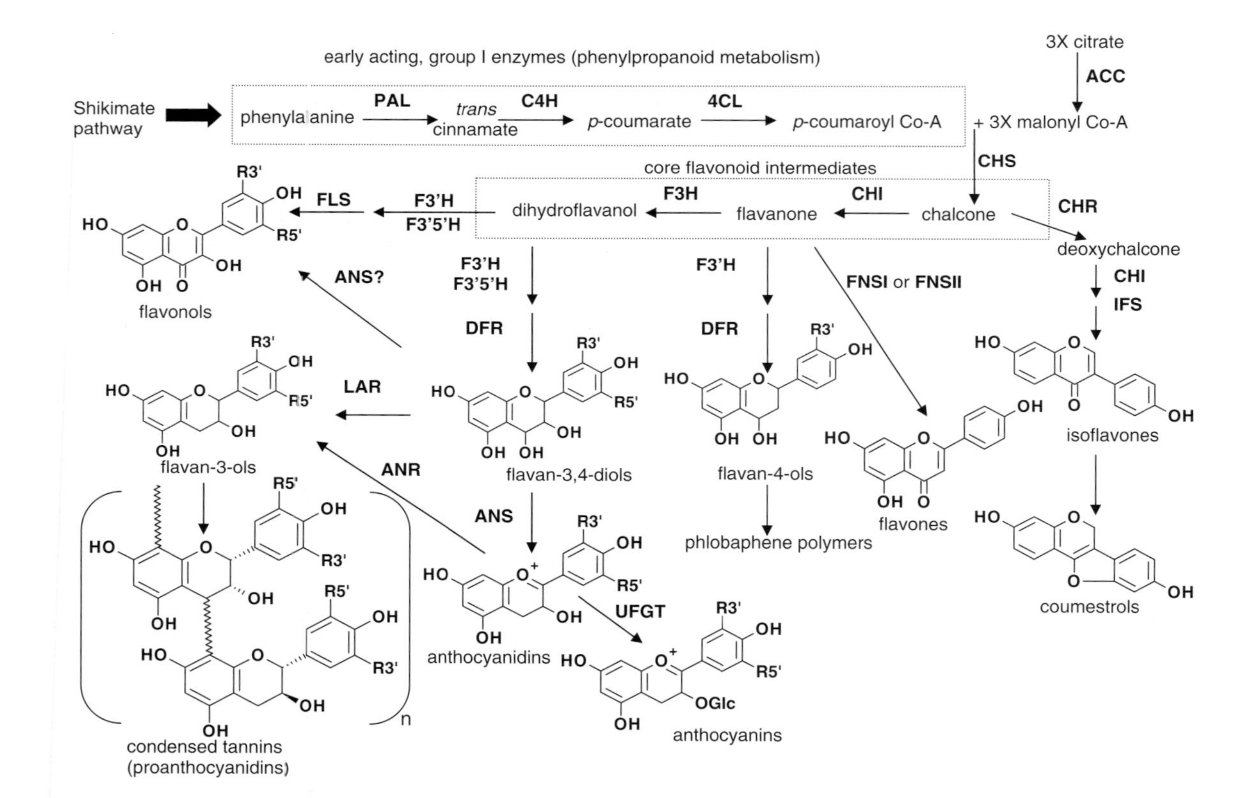

Figure 5.4. Abbreviated scheme for biosynthesis of major flavonoid subclasses, showing the primary enzymes and substrates leading to different subclasses. Bold-faced, uppercase abbreviations refer to enzyme names, whereas substrate names are presented in lowercase letters. PAL, phenylalanine ammonia lyase; C4H, cinnamate 4-hydroxylase; 4CL, 4-coumarate:CoA ligase; CHS, chalcone synthase; CHI, chalcone isomerase; CHR, chalcone reductase; IFS, isoflavone synthase; F3H, flavonone 3-hydroxylase; F3′H, flavonoid 3′-hydroxylase; F3′5′H, flavonoid 3′5′-hydroxylase; FNSI/II, flavone synthase; DFR, dihydroflavonol 4-reductase; FLS, flavonol synthase; ANS, anthocyanidin synthase; LAR, leucoanthocyanidin reductase; ANR, anthocyanidin reductase; UFGT, UDP-glucose:flavonoid 3-*O*-glucosyltransferase. R3′ = H or OH. R5′ = H or OH. Glc = glucose. Please refer to text for more information.

4-hydroxylase (C4H), which catalyzes the synthesis of *p*-coumarate from *trans* cinnamate, introducing the 4′ hydroxyl group seen in nearly all known flavonoids (Stafford 1990; Winkel 2004; Anderson and Markham 2006). C4H is a typical cytochrome P450 monooxygenase enzyme containing a heme cofactor and is dependent on both NADPH and molecular oxygen (Ayabe and Akashi 2006). Also associated with the multienzyme complex is 4-coumarate:CoA ligase (4CL), converting *p*-coumarate to its coenzyme-A ester, a reaction requiring both ATP and Mg^{2+}, and thereby activating it for reactions with malonyl CoA (Stafford 1990; Anderson and Markham 2006).

In most plants, PAL, C4H, and 4CL are encoded by multigene families, with the exact copy number varying widely between species. Individual isozyme forms show distinct temporal, tissue, or elicitor-induced expression patterns, and it seems likely that each individual family member may be primarily utilized for the biosynthesis of one specific type of phenolic compound (Stafford 1990; Anderson and Markham 2006). For instance, different 4CL isoforms show distinct activities toward alternate cinnamic acid substrates such as caffeic and ferulic acids and may function as a control point regulating carbon flux among the lignin, lignan, and flavonoid biosynthetic pathways, which would otherwise compete for a common set of phenylpropanoid enzymes and substrate intermediates (Anderson and Markham 2006). Elucidating the complex web of interacting regulatory mechanisms between these various isozyme variants is no simple task and remains an active area of research.

Core Flavonoid Biosynthesis

The first committed step in flavonoid biosynthesis begins with the condensation of one molecule of 4-coumaroyl-CoA with three molecules of malonyl-CoA, via sequential decarboxylation additions, forming naringenin chalcone (2′,4′,6′,4-tetrahydroxychalcone), in what amounts to an irreversible process (see Fig. 5.4). This flavonoid entry point reaction is cytosol-based and is carried out by chalcone synthase (CHS), again in association with the ER. The A ring hydroxyl groups at positions 5 and 7, observed in most flavonoids, are generated via these CHS reactions as opposed to a cytochrome P450 dependent oxidation. CHS operates as a homodimer, without any known cofactors, and in many plant species including maize, phaseolis, petunia, Arabidopsis, and grape, CHS is present in relatively large, multigene families (Stafford 1990; Austin and Noel 2003). Similar to PAL, these isozymes display different temporal, tissue and elicitor induced expression patterns, apparently under distinct molecular controls, and elucidating the mechanisms of this control is an active area of research. In the next enzymatic step, chalcone isomerase (CHI) converts naringenin chalcone to its (2*S*)-flavanone counterpart, naringenin, via a stereospecific isomerization, thereby closing the C ring. This is a reversible process, but closing the C-ring protects the nascent flavonoid from degradative reactions leading to ring fission. Next, flavanone 3β-hydroxylase (F3H) converts the stereospecific 3β-hydroxylation of (2*S*)-flavanones to dihydroflavonols. F3H is a soluble, nonheme dioxygenase enzyme, which functions as a monomer, making it distinct from most of the other hydroxylase enzymes active within the flavonoid pathways, which instead rely upon cytochrome P450 type enzymes (Turnbull and others 2004; Ayabe and Akashi 2006).

Both (2*S*)-naringenin and its dihydroflavonol derivative are central intermediates in flavonoid biosynthesis (see Fig. 5.4), acting as branch-point metabolites, feeding into several diverging side branches, each resulting in the synthesis of a distinct flavonoid subclass. Not all these side branches are present in every plant species or are even active within every tissue type within a given plant. The regulation of these various side branches is strictly controlled, resulting in distinct, tissue-specific flavonoid metabolite profiles. A striking example occurs in grape (*Vitis* species), where the fruits' skin, flesh (mesocarp), and seeds show distinct flavonol, flavanol (condensed tannin), and anthocyanin profiles, whose accumulation is temporally coordinated throughout development (Grimplet and others 2007; Deluc and others 2008).

Biosynthesis of Specific Metabolites

Flavones: Flavones, like flavonols, are ubiquitous, found in nearly all plant tissues, and are derived from flavanones via desaturation of the C-ring, catalyzed by flavone synthase enzymes (FNSI and FNSII) (see Fig. 5.4). Unlike many other instances of overlapping gene function, FNSI and FNSII are two evolutionarily distinct genes (Martens and Mithofer 2005), despite performing the same biochemical function. FNSI is a soluble dioxygenase enzyme, with rather limited distribution in the plant kingdom, whereas the much more commonly encountered FNSII is anchored to the ER and is a cytochrome P450 enzyme (Ayabe and Akashi 2006). Common flavones include apigenin and luteolin, generally found at high levels in green, vegetative plant tissues such as parsley leaves and celery stalks.

Flavonols: Flavonols are ubiquitous within the plant kingdom, with quercetin likely the most commonly encountered example, being particularly prevalent in onions, apples, and berries (USDA Flavonoids Database Release 2.1, 2007). Flavonols play a role in UV protection and are generally induced in response to light exposure, typically in the epidermal tissues of fruits and leaves. They have been implicated in pollen tube growth and also play a role in the pigmentation of flowers, being visible to insects sensitive to UV radiation and modifying the intensity of anthocyanin coloration via copigmentation effects. Flavonols are primarily synthesized from dihydroflavonols via the activity of flavonol synthase (FLS), a non-heme, 2-oxoglutarate-dependent oxygenase, although emerging evidence suggests that anthocyanidin synthase (ANS) may also generate flavonols via flavan-3,4-diol intermediates (Turnbull and others 2004; Owens and others 2008).

Flavanones: In some cases, flavanones produced by CHI will accumulate to sizeable amounts instead of being diverted away to form flavonols, anthocyanins, and flavanols (see Fig. 5.4). These flavanone products, hesperetin and naringenin being the most common, are frequently encountered in citrus fruits and juices (USDA Flavonoids Database Release 2.1, 2007). In most of these cases, essentially no flavonols or anthocyanins are encountered; the flavonoid pathway is essentially blocked at the F3H step.

Flavanols and procyanidins: Flavanols, or flavan-3-ols, are synthesized via two routes, with (+) catechins formed from flavan-3,4-diols via leucoanthocyanidin reductase (LAR), and (−) epicatechins from anthocyanidins via anthocyanidin reductase (ANR) (see Fig. 5.4). These flavan-3-ol molecules are then polymerized to condensed tannins (proanthocyanidins or procyanidins), widely varying in the number and nature of their component monomers and linkages (Aron and Kennedy 2008; Deluc and others 2008). It is still not known whether these polymerization reactions happen spontaneously, are enzyme catalyzed, or result from a mixture of both.

Anthocyanins: Anthocyanins, the primary red, blue, and purple pigments found in plants, are synthesized from flavan-3,4-diols (leucoanthocyanins) via anthocyaninidin synthase (ANS), occasionally called LDOX (leucoanthocyanidin dioxygenase). Like FLS, to which it shows strong sequence homology, ANS is a nonheme 2-oxoglutarate-dependent oxygenase (Turnbull and others 2004). Anthocyanins are stabilized through a glycosylation reaction at position 3, catalyzed by UDP-glucose:flavonoid 3-*O*-glucosyltransferase (UFGT). In an elegant series of investigations, Kobayashi and others (2004) have shown that green-skinned grapes fail to produce UFGT enzyme, resulting in a lack of anthocyanin accumulation.

Flavan-3,4-diols: Flavan-3,4-diols, also known as leucoanthocyanidins, are not particularly prevalent in the plant kingdom, instead being themselves precursors of flavan-3-ols (catechins), anthocyanidins, and condensed tannins (proanthocyanidins) (see Fig. 5.4). Flavan-3,4-diols are synthesized from dihydroflavonol precursors by the enzyme dihydroflavonol 4-reductase (DFR), through an NADPH-dependent reaction (Anderson and Markham 2006). The substrate binding affinity of DFR is paramount in determining which types of downstream anthocyanins are synthesized, with many fruits and flowers unable to synthesize pelargonidin type anthocyanins, because their particular DFR enzymes cannot accept dihydrokaempferol as a substrate (Anderson and Markham 2006).

Isoflavones: Most flavonoids possess a hydroxyl group at the 5-position, incorporated into the nascent chalcone skeleton by CHS (see Fig. 5.4). A relatively small number of flavonoids lack this hydroxyl group and are collectively called the 5-deoxyflavonoids, with isoflavones being the most conspicuous examples, primarily found in legumes such as soybean, alfalfa, and clover. This alternate biosynthetic route is carried out by chalcone reductase (CHR), in a coupled reaction with CHS (Bomati and others 2005). The resulting deoxychalcone intermediate is then fed into CHI, before the aryl B-ring migration carried out via isoflavone synthase (IFS) (Cheng and others 2008).

Further Structural Modifications: P450 Hydroxylase Enzymes

The addition of new hydroxyl groups to the nascent flavonoid skeleton is performed by cytochrome P450 enzymes (Ayabe and Akashi 2006), the two main activities provided by flavonoid 3′ hydroxylase (F3′H) and flavonoid 3′5′hydroxylase (F3′5′H), at the 3′ or the 3′and 5′ positions, respectively. These modifications increase the water solubility and change the light absorption and bioactivity characteristics of the resulting molecule. F3′H and F3′5′H can act on both flavanones and dihydroflavanols, ultimately generating a variety of flavonol, anthocyanidin, and flavan-3-ol molecules. The bioactivity of various flavonoid molecules has been tied to specific structure-activity relationships associated with changes to the B-ring hydroxylation pattern (Marko and others 2004; Jing and others 2008; Li and others 2008), and therefore characterization of F3′H and F3′5′H has been an important area of research.

Effect of Postharvest Storage and Processing Conditions

Phytochemicals such as flavonoids are a major contributor to the health benefit of fruits and vegetables; however, in a recent study of 43 fruits and vegetables, it was found that the nutritional value of these food plants has declined in the past 50 years (Davis and others 2004). The fact that today's farmers have been planting crops designed to improve traits such as size and yield rather than the nutritional composition has been attributed to this decline. On the other hand, genetic and environmental factors, such as cultivar, maturity, UV light exposure, and postharvest storage and processing procedures, play an important role in retaining the phytochemical compositions of fruits and vegetables (Parr and Bolwell 2000; Tsao and others 2006). Plant secondary metabolites such as flavonoids are found to exist more in the outer layers of apple and other fruits (Tsao and others 2003), to be used by plants as their defense against attacks from invading insects and microorganisms. Conventional growing of fruits and vegetables keeps such insect and microbial pressure low. Interestingly, in recent years we have seen rapid increase in organic production of produce. Plant genetics, farming practices (organic vs. conventional), environmental factors such as geographic location, growing season, soil quality and mineral nutrients, and harvest maturity (many fruits and vegetables are harvested prematurely in order to be shipped long distance and stored for longer periods) also significantly affect the flavonoid content of fruits and vegetables (Tsao and others 2006). In this chapter, we mainly focus on the effect of postharvest storage and processing.

Effect of Postharvest Storage Conditions

Contrary to the negative opinions that most people may hold against the postharvest storage of fruits and vegetables, most literature found postharvest storage conditions may actually not affect the flavonoid content, including anthocyanins, as much as we thought. Phenolics in apple did not seem to change significantly during cold storage for up to 9 months (Burda and others 1990; Golding and others 2001). Similarly, the antioxidant activity and total phenolic and anthocyanin contents in blueberries showed no significant decrease but slight increase depending on cultivar during marketable cold storage period (3–5 weeks) (Connor and others 2002). Zheng and others (2003) found that certain postharvest storage conditions, such as high oxygen treatment (60–100% oxygen) even increased the total phenolic and anthocyanin contents. Similarly, modified-atmosphere packaging (MAP)(7% O_2 and 10% CO_2) of Swiss chard or spinach stored in cold had no effect on total flavonoid content after 8 days (Gil and others 1998, 1999). Storage conditions such as low temperature may act as a physical stress that increases the production of secondary defense metabolites to a certain degree. That may explain these observations to some extent. However, others found that different fruits may vary in flavonoid stability during storage. Häkkinen and others (2000) found that during 9 months of storage at −20°C, quercetin content decreased markedly (40%) in bilberries and lingonberries, but not in black currants or red raspberries. They also found that myricetin and kaempferol were more susceptible than quercetin to losses during storage.

Effect of Processing

Opposite to the effect of postharvest storage, processing of fruits and vegetables normally leads to decreases in the concentrations and compositions of phytochemicals including flavonoids (Tsao and others 2006). Ferreira and others (2002) found that total phenolic content in pears decreased by 64% under sun-drying conditions. Freezing, pasteurization, boiling, and microwave cooking generally reduce the total antioxidant levels (Gil-Izquierdo and others 2002; Guyot and others 2003; Aziz and others 1998). Gil and others (1999) found that although the total flavonoid content remained quite constant during storage in both air and MAP atmospheres, 50% of the content was lost during boiling. Different cooking methods may have different impact on flavonoids. Studies of the effects of domestic processing on the flavonols quercetin, myricetin, and kaempferol in five berries showed that cooking strawberries with sugar to make jam resulted in minor losses of quercetin (15%) and kaempferol (18%) (Häkkinen and others 2000). However, when bilberries were cooked with water and sugar, 40% of quercetin was lost (Häkkinen and others 2000). Lingonberry juice made by the traditional crushing caused a considerable loss (40%) of quercetin; however, only 15% of quercetin and 30% of myricetin present in unprocessed berries were retained in juices made by common domestic methods (steam-extracted black currant juice, unpasteurized lingonberry juice). Cold pressing was superior to steam extraction in extracting flavonols from black currants (Häkkinen and others 2000). In a recent study by Ferracane and others (2008), boiling and steaming artichoke significantly reduced the total apigenin content from 1894 mg/kg DM to 1214 and 1446 mg/kg DM, respectively, but frying had the most severe impact, reducing the concentration to 776 mg/kg DM. Examination of 16 commercial dehydrated onion products sold in the US indicated that these products contained low amounts or no flavonoids (Lee and others 2008). In the same study, the percent losses of onion flavonoids subjected to "cooking" (in percent) were found to be: frying, 33%; sautéing, 21%; boiling, 14–20%; steaming, 14%; microwaving, 4%; baking, 0%. They also found exposure to fluorescent light for 24 and 48 hr induced time-dependent increases in the flavonoid content of fresh onions (Lee and others 2008). Processing of fruits and vegetables without doubt has a significant adverse effect on the flavonoid content.

Summary

Flavonoids are a large and diverse group of phytochemicals in fruits and vegetables. These compounds may be used by plants as their own chemical defense, however, evidence accumulated in the past decade strongly suggests that flavonoid-rich fruits and vegetables may play important roles in human health maintenance and disease prevention. The main flavonoid subgroups, flavones, flavonols, flavanones, flavanonols, flavanols, anthocyanidins, isoflavones, and chalcones, do not always exist as aglycones. In fact, the majority of them exist in fruits and vegetables as glycosides. Some flavonoids are highly methylated; others have different substitutions on all rings A, B, and C. The complexity of the substation patterns makes extraction, separation, and analysis of the flavonoids of fruit and vegetable origin a difficult task. Modern separation technologies such as SPE, SFE, and MAE and analytical instrumentation such as HPLC, UPLC, and LC-MS, particularly various highly hyphenated technologies including LC-DAD-MS^n and LC-MS-NMR, have significantly improved the quantification and identification of flavonoids. Advances in molecular biology and molecular genetics have also helped us better understand the flavonoid biosynthetic pathways, and thus the metabolism of flavonoids in fruits and vegetables pre- and postharvest. This will lead to postharvest storage and processing technologies that can maximize the retention of the health benefits of flavonoids in fruits and vegetables.

References

Anderson OM and Jordheim M. 2006. The anthocyanins. In: Anderson OM, Markham KR, editors, Flavonoids: Chemistry, Biochemistry and Applications. Boca Raton, FL: CRC Press/Taylor & Francis Group, pp. 472–551.

Anderson OM and Markham KR, editors. 2006. Flavonoids: Chemistry, Biochemistry and Applications. Boca Raton, FL: CRC/Taylor & Francis.

Arabbi PR, Genovese MI and Lajolo FM. 2004. Flavonoids in vegetable foods commonly consumed in Brazil and estimated ingestion by the Brazilian population. J Agric Food Chem 52(5):1124–1131.

Armenta S, Garrigues S and de la Guardia M. 2008. Green analytical chemistry. Trends Anal Chem 27(6):497–511.

Aron PM and Kennedy JA. 2008. Flavan-3-ols: nature, occurrence and biological activity. Mol Nutri Food Res 52:79–104.

Austin MB and Noel JP. 2003. The chalcone synthase superfamily of type III polyketide synthases. Nat Prod Rep 20:79–110.

Ayabe SI and Akashi T. 2006. Cytochrome P450s in flavonoid metabolism. Phytochem Rev 5:271–282.

Aziz AA, Edwards CA, Lean ME and Crozier A. 1998. Absorption and excretion of conjugated flavonols, including quercetin-4'-O-β-glucoside and isorhamnetin-4'-O-β-glucoside by human volunteers after the consumption of onions. Free Radic Res 29:257–269.

Bohm BA. 1998. Introduction to Flavonoids. Amsterdam: Harwood Academic Publishers.

Bomati EK, Austin MB, Bowman ME, Dixon RA and Noel JP. 2005. Structural elucidation of chalcone reductase and implications for deoxychalcone biosynthesis. J Biol Chem 280:30496–30503.

Boyles MJ, Matthew J, Wrolstad RE and Ronald E. 1993. Anthocyanin composition of red raspberry juice: Influences of cultivar, processing, and environmental factors. J Food Sci 58(5):1135–1141.

Bradshaw MP, Prenzler PD and Scollary GR. 2001. Ascorbic acid-induced browning of (+)-catechin in a model wine system. J Agric Food Chem 49:934–939.

Burda S, Oleszek W and Lee CY. 1990. Phenolic compounds and their changes in apples during maturation and cold storage. J Agric Food Chem 38:945–948.

Careri M, Elviri L, Mangia A and Musci M. 2000. Spectrophotometric and coulometric detection in the high-performance liquid chromatography of flavonoids and optimization of sample treatment for the determination of quercetin in orange juice. J Chromatogr A 881:449–460.

Chang CC, Yang MH, Wen HM and Chen JC. 2002. Estimation of total flavonoid content in propolis by two complementary colorimetric methods. J Food Drug Anal 10:178–182.

Cheng H, Yu O and Yu D. 2008. Polymorphisms of IFS1 and IFS2 genes are associated with isoflavone concentrations in soybean seeds. Plant Sci 175: 505–512.

Chun OK, Kim DO, Smith N, Schroeder D, Han JT and Lee CY. 2005. Daily consumption of phenolics and total antioxidant capacity from fruit and vegetables in the American diet. J Sci Food Agric 85:1715–1724.

Connor AM, Luby JJ, Hancock JF, Berkheimer S and Hanson EJ. 2002. Changes in fruit antioxidant activity among blueberry cultivars during cold temperature storage. J Agric Food Chem 50:893–898.

Davis DR, Epp MD and Riordan HD. 2004. Changes in USDA food composition data for 43 garden crops, 1950 to 1999. J Am Coll Nutr 23:669–682.

de Rijke E, Out P, Niessen WMA, Ariese F, Gooijer C and Brinkman UATh. 2006. Analytical separation and detection methods for flavonoids. J Chromatogr A. 1112:31–63.

Deluc L, Bogs J, Walker AR, Ferrier T, Decendit A, Merillon JM, Robinson SP and Barrieu F. 2008. The transcription factor VvMYB5b contributes to the regulation of anthocyanin and proanthocyanin biosynthesis in developing grape berries. Plant Physiol 147: 2041–2053.

Fatland BL, Ke J, Anderson MD, Mentzen WI, Cui LW, Allred CC, Johnston JL, Nikolau BJ and Wurtele ES. 2004. Molecular characterization of a heteromeric ATP-citrate lyase that generates cytosolic acetyl-Coenzyme A in Arabidopsis. Plant Plsy 130:740–756.

FDA-USA http://www.cfsan.fda.gov

Ferracane R, Pellegrini N, Visconti A, Graziani G, Chiavaro E, Miglio C and Fogliano V. 2008. Effects of different cooking methods on antioxidant profile, antioxidant capacity, and physical characteristics of artichoke. J Agric Food Chem 56(18):8601–8608.

Ferreira A, Slade D and Marais JPJ. 2006. Flavans and proanthocyanidins. In: Anderson OM, Markham KR, editors. Flavonoids: Chemistry, Biochemistry and Applications. Boca Raton, FL: CRC Press/Taylor & Francis Group, pp. 553–616.

Ferreira D, Guyot S, Marnet N, Delgadillo I, Renard CM and Coimbra MA. 2002. Composition of phenolic compounds in a Portuguese pear (*Pyrus communis* L. var. S. Bartolomeu) and changes after sun-drying. J Agric Food Chem 50:4537–4544.

Franke AA, Custer LJ, Arakaki C and Murphy SP. 2004. Vitamin C and flavonoid levels of fruits and vegetables consumed in Hawaii. J Food Comp Anal 17:1–35.

Gao L and Mazza G. 1995. Characterization, quantitation, and distribution of anthocyanins and phenolics in sweet cherries. J Agric Food Chem 43:343–346.

Gil MI, Ferreres F and Tomás-Barberán FA. 1998. Effect of modified atmosphere packaging on the flavonoids and vitamin C content of minimally processed Swiss chard (*Beta vulgaris* subsp. *cycla*). J Agric Food Chem 46:2007–2012.

Gil MI, Ferreres F and Tomás-Barberán FA. 1999. Effect of postharvest storage and processing on the antioxidant constituents (flavonoids and vitamin C) of fresh-cut spinach. J Agric Food Chem 47(6):2213–2217.

Gil-Izquierdo A, Gil MI and Ferreres F. 2002. Effect of processing techniques at industrial scale on orange juice antioxidant and beneficial health compounds. J Agric Food Chem 50:5107–5114.

Golding JB, McGlasson WB, Wyllie SG and Leach DN. 2001. Fate of apple peel phenolics during cool storage. J Agric Food Chem 49:2283–2289.

Goupy P, Amiot MJ, Richard-Forget F, Duprat F, Aubert S and Nicolas J. 1995. Enzymatic browning of model solutions and apple phenolic extracts by apple polyphenoloxidase. J Food Sci 60:497–501, 505.

Grayer RJ and Veitch NC. Flavanones and dihydroflavonols. In: Anderson OM, Markham KR. editors. Flavonoids: Chemistry, Biochemistry and Applications. Boca Raton, FL: CRC Press/Taylor & Francis Group, pp. 918–1002.

Grimplet J, Deluc LG, Tillett RL, Wheatley MD, Schlauch KA, Cramer GR and Cushman JC. 2007. Tissue-specific mRNA expression profiling in grape berry tissues. BMC Genomics 8:187.

Guyot S, Marnet N, Sanoner P and Drilleau JF. 2003. Variability of the polyphenolic composition of cider apple (*Malus domestica*) fruits and juices. J Agric Food Chem 51:6240–6247.

Häkkinen SH, Kärenlampi SO, Mykkänen HM and Törrönen AR. 2000. Influence of domestic processing and storage on flavonol contents in berries. J Agric Food Chem 48(7):2960–2965.

Häkkinen SH and Törrönen AR. 2000. Content of flavonols and selected phenolic acids in strawberries and Vaccinium species: influence of cultivar, cultivation site and technique. Food Res Int 33:517–524.

Harborne JB. 1994. The Flavonoids: Advances in Research Since 1986. London: Chapman and Hall

Harborne JB and Williams CA. 2000. Advances in flavonoid research since 1992. Phytochemistry 55:481–504.

Harnly JM, Doherty RF, Beecher GR, Holden JM, Haytowitz DB, Bhagwat SA and Gebhardt SE. 2006. Flavonoid content of US fruits, vegetables, and nuts. J Agric Food Chem 54:9966–9977.

Health Canada. http://www.hc-sc.gc.ca

Hertog MGL, Hollman PCH and Katan MB. 1992. Content of potentially anticarcinogenic flavonoids of 28 vegetables and 9 fruits commonly consumed in the Netherlands. J Agric Food Chem 40:2379–2383.

Hosseinian FS, Li W, Hydamaka AW, Tsopmo A, Lowry L, Friel J and Beta T. 2007. Proanthocyanidin profile and ORAC values of Manitoba berries, chokecherries, and seabuckthorn. J Agric Food Chem 55(17):6970–6976.

Jia Z, Tang M and Wu J. 1999. The determination of flavonoid contents in mulberry and their scavenging effects on superoxide radicals. Food Chem 64:555–559.

Jing P, Bomser JA, Schwartz SJ, He J, Magnuson BA and Giusti MM. 2008. Structure-function relationships of anthocyanins from various anthocyanin-rich extracts on the inhibition of colon cancer cell growth. J Agric Food Chem 56:9391–9398.

Justesen U, Knuthsen P and Leth T. 1998. Quantitative analysis of flavonols, flavones, and flavanones in fruits, vegetables and beverages by high-performance liquid chromatography with photo-diode array and mass spectrometric detection. J Chromatogr A 799:101–110.

Kawaii S, Tomono Y, Katase E, Ogawa K and Yano M. 1999. Quantitation of flavonoid constituents in citrus fruits. J Agric Food Chem 47(9):3565–3571.

Klejdus B, Vitamvásová D and Kubán V. 2001. Identification of isoflavone conjugates in red clover (*Trifolium pratense*) by liquid chromatography-mass spectrometry after two-dimensional solid-phase extraction. Anal Chim Acta 450:81–97.

Kobayashi S, Goto-Yamamoto N and Hirochika H. 2004. Retrotransposon-induced mutations in grape skin color. Science 304:982.

Lee SJ, Ahn JK, Khanh TD, Chun SC, Kim SL, Ro HM, Song HK and Chung IM. 2007. Comparison of isoflavone concentrations in soybean (Glycine max (L.) Merrill) sprouts grown under two different light conditions. J Agric Food Chem 55(23):9415–9421.

Lee SU, Lee JH, Choi SH, Lee JS, Ohnisi-Kameyama M, Kozukue N, Levin CE and Friedman M. 2008. Flavonoid content in fresh, home-processed, and light-exposed onions and in dehydrated commercial onion products. J Agric Food Chem 56(18):8541–8548.

Li J, Zhang D, Stoner GD and Huang C. 2008. Differential effects of black raspberry and strawberry extracts on BaPDE-induced activation of transcription factors and their target genes. Mol Carcinogenesis 47:286–294.

Liu L, Xie B, Cao S, Yang E, Xu X and Guo S. 2007. A-type procyanidins from *Litchi chinensis* pericarp with antioxidant activity. Food Chem 105:1446–1451.

MacDonald MJ and D'Cunha GB. 2007. A modern view of phenylalanine ammonia lyase. Biochem Cell Biol 85:273–282.

Marko D, Puppel N, Tjaden Z, Jakobs S and Pahlke G. 2004. The substitution pattern of anthocyanidins affects different cellular signalling cascades regulating cell proliferation. Mol Nutr Food Res 48:318–325.

Marston A and Hostettmann K. 2006. Separation and quantification of flavonoids. In: Anderson OM and Markham KR, editors. Flavonoids: Chemistry, Biochemistry and Applications. Boca Raton, FL: CRC Press/Taylor & Francis Group, pp. 1–36.

Martens S and Mithofer A. 2005. Flavones and flavone synthases. Phytochem 66:2399–2407.

Merken HM and Beecher GR. 2000. Measurement of food flavonoids by high-performance liquid chromatography: a review. J Agric Food Chem 48:577–599.

Montedoro G, Servili M, Baldioli M and Miniati E. 1992a. Simple and hydrolyzable phenolic compounds in virgin olive oil. 1. Their extraction, separation, and quantitative and semiquantitative evaluation by HPLC. J Agric Food Chem 40:1571–1576.

Montedoro G, Servili M, Baldioli M and Miniati E. 1992b. Simple and hydrolyzable phenolic compounds in virgin olive oil. 2. Initial characterization of the hydrolyzable fraction. J Agric Food Chem 40:1577–1580.

Ortega N, Romero MP, Macià A, Reguant J, Anglès N, Morelló JR and Motilva MJ. 2008. Obtention and characterization of phenolic extracts from different cocoa sources. J Agric Food Chem 56(20):9621–9627.

Owens DK, Alerding AB, Crosby KC, Bandara AB, Westwood JH and Winkel BSJ. 2008. Functional analysis of a predicted flavonol synthase gene family in *Arabidopsis*. Plant Physiol 147:1046–1061.

Parr AJ and Bolwell GP. 2000. Phenols in the plant and in man. The potential for possible nutritional enhancement of the diet by modifying the phenols content or profile. J Sci Food Agric 80:985–1012.

Pietta PG. 2000. Flavonoids as Antioxidants. J Nat Prod 63:1035–1042.

Pietta PG, Mauri PL, Manera E and Ceva PL. 1989. HPLC determination of the flavonoid glycosides from *Betulae folium* extracts. Chromatographia 28:311–312.

Prior RL, Lazarus SA, Cao G, Muccitelli H and Hammerstone JF. 2001. Identification of procyanidins and anthocyanins in blueberries and cranberries (*Vaccinium spp.*) using high-performance liquid chromatography/mass spectrometry. J Agric Food Chem 49(3):1270–1276.

Rauha JP, Vuorela H and Kostiainen R. 2001. Effect of eluent on the ionization efficiency of flavonoids by ion spray, atmospheric pressure photoionization mass spectrometry. J Mass Spectrom 36:1269–1280.

Revilla E, Ryan JM and Martin-Ortega G. 1998. Comparison of several procedures used for the extraction of anthocyanins from red grapes. J Agric Food Chem 46:4592–4597.

Sakakibara H, Honda Y, Nakagawa S, Ashida H and Kanazawa K. 2003. Simultaneous determination of all polyphenols in vegetables, fruits, and teas. J Agric Food Chem 51(3):571–581.

Seigler D. (1998) Flavonoids. Plant Secondary Metabolism. Norwell: Kluwer Academic Publishers. pp. 151–192.

Slimestad R, Fossen T and Verheul MJ. 2008. The flavonoids of tomatoes. J Agric Food Chem 56(7):2436–2441.

Stafford HA. 1990. Flavonoid Metabolism. Boca Raton, FL: CRC Press.

Stalikas CD. 2007. Extraction, separation, and detection methods for phenolic acids and flavonoids. J Sep Sci 30:3268–3295.

Sun T and Powers JR. 2007. Antioxidants and antioxidant activities of vegetables. In Shahidi F, Ho C-T, editors. Antioxidant Measurement and Applications. Washington, DC: American Chemical Society, pp. 160–183.

Tomás-Barberán FA and Clifford MN. 2000. Flavanones, chalcones, and dihydrochalcones-nature, occurrence and dietary burden. J Sci Food Agric 80:1073–1080.

Tsao R and Deng Z. 2004. Separation Procedures for naturally occurring antioxidant phytochemicals. J Chromatogr B 812: 85–99.

Tsao R, Khanizadeh S and Dale A. 2006. Designer fruits and vegetables with enriched phytochemicals for human health. Can J Plant Sci 86:773–786.

Tsao R and Yang R. 2003. Optimisation of a new mobile phase to know the complex and real polyphenolic composition: towards a total phenolic index using HPLC. J Chromatogr A 1018:29–40.

Tsao R, Yang R, Young JC and Zhu H. 2003. Polyphenolic profiles in eight apple cultivars using high-performance liquid chromatography (HPLC). J Agric Food Chem 51:6347–6353.

Turnbull JJ, Nakajima J, Welford RW, Yamazaki M, Saito K and Schofield CJ. 2004. Mechanistic studies on three 2-oxoglutarate-dependent oxygenases of flavonoid biosynthesis: anthocyanidin synthase, flavonol synthase, and flavanone 3β-hydroxylase. J Biol Chem 279:1206–1216.

USDA Flavonoids Database, Release 2.1. 2007. http://www.ars.usda.gov/nutrientdata

Valant-Vetschera KM and Wallenweber E. 2006. Flavones and flavonols. In: Anderson OM and Markham KR, editors. Flavonoids: Chemistry, Biochemistry and Applications. Boca Raton, FL: CRC Press/Taylor & Francis Group. pp. 618–748.

Williams CA. 2006. Flavone and flavonol *O*-glycosides. In: Anderson OM and Markham KR. editors. Flavonoids: Chemistry, Biochemistry and Applications. Boca Raton, FL: CRC Press/Taylor & Francis Group, pp. 749–856.

Wilson ID. 2000. Multiple hyphenation of liquid chromatography with nuclear magnetic resonance spectroscopy, mass spectrometry and beyond. J Chromatogr A 892:315–327.

Winkel BSJ. 2004. Metabolic channeling in plants. Annu Rev Plant Biol 55:85–107.

Xie DY and Dixon RA. 2005. Proanthocyanidin biosynthesis—still more questions than answers? Phytochemistry 66: 2127–2144.

Zhao F, Watanabe Y, Nozawa H, Daikonnya A, Kondo K and Kitanaka S. 2005. Prenylflavonoids and phloroglucinol derivatives from hops (*Humulus lupulus*). J Nat Prod 68(1):43–49.

Zheng Y, Wang CY, Wang SY and Zheng W. 2003. Effect of high-oxygen atmospheres on blueberry phenolics, anthocyanins, and antioxidant capacity. J Agric Food Chem 51:7162–7169.

6 Flavonoids and Their Relation to Human Health

Alma E. Robles-Sardin,* Adriana Verónica Bolaños-Villar, Gustavo A. González Aguilar, and Laura A. de la Rosa

Introduction

There is no doubt that the eating patterns and physical activity of the adult have a direct effect on the prevalence of nontransmittable chronic illnesses. In recent decades, the prevalence of cardiovascular disease, obesity, cancer, hypertension, and diabetes, among others, has steadily increased, making these diseases the priority for health care systems in many countries, especially in developed countries.

The development of nontransmittable chronic illnesses is associated with the presence of oxidative agents in the body. These agents are found in air, water, or food or can be produced in the body's cells. Their high content in body cells causes an imbalance that results in oxidative stress damaging proteins, DNA, and others. As a result of this deterioration, an increase in the risk of nontransmittable chronic illnesses has been noted. In order to prevent or decrease oxidative stress, the consumption of foods rich in antioxidants, such as fruits and vegetables, is recommended (van Dokkum and others 2008; Liu 2003).

Nutrition guides, whether pyramids or serving dishes, include a large portion of fruits and vegetables that are rich in natural antioxidants (vitamin C, vitamin E, carotenes, and polyphenols). These compounds have been studied for their protective role in various pathologies such as cardiovascular disease and some types of cancer. In addition to their antioxidant properties, other possible protective health effects of phytochemicals have been studied: modulation of detoxifying enzymes, stimulation of the immune system, decrease in platelet aggregation, changes in cholesterol metabolism, modulation of the concentration and metabolism of steroid hormones, lowering of blood pressure, antibacterial and antiviral activity, and endothelial vascular function.

Therefore, in recent years, consumers have been demanding foods that are beneficial to their health, not only for their nutritional value but also for the prevention of chronic and degenerative diseases. This is due to the fact that many chronic diseases are directly related to nutrition and could be prevented with an appropriate diet rich in fruits and vegetables.

Some of the bioactive phytochemicals found in fruits and vegetables are polyphenols, including flavonoids. This chapter provides a general overview of the relationship between flavonoids and health. The mechanisms of action believed to be behind the healthful effects of some compounds will also be mentioned.

* Corresponding author.

Generalities

Phytochemicals or phytonutrients are bioactive substances that can be found in foods derived from plants and are not essential for life; the human body is not able to produce them. Recently, some of their characteristics, mainly their antioxidant capacity, have given rise to research related to their protective properties on health and the mechanisms of action involved. Flavonoids are a diverse group of phenolic phytochemicals (Fig. 6.1) that are natural pigments. One function of flavonoids is to protect plants from oxidative stress, such as ultraviolet rays, environmental pollution, and chemical substances. Other relevant biological roles of these pigments are discussed in other chapters of this book.

Flavonoids are a complex group of polyphenolic compounds with a basic C_6-C_3-C_6 structure that can be divided in different groups: flavonols, flavones, flavanols (or flavan-3-ols), flavanones, anthocyanidins, and isoflavones. More than 6,000 flavonoids are known; the most widespread are flavonols, such as quercetin; flavones, such as luteolin; and flavanols (flavan-3-ols), such as catechin. Anthocyanidins are also bioactive flavonoids: they are water-soluble vegetable pigments found especially in berries and other red-blue fruits and vegetables.

These structurally diverse compounds exhibit a range of biological activities *in vitro* that may explain their potential health-promoting properties, including antioxidant and anti-inflammatory effects and the induction of apoptosis (Hooper and others 2008). Most of the recent interest in flavonoids as health-promoting compounds is related to their powerful antioxidant properties. The criteria to establish the antioxidant capacity of these compounds is based on several structural characteristics that include (a) the presence of *o*-dihydroxyl substituents in the B-ring; (b) a double bond between positions 2 and 3; and (c) hydroxyl groups in positions 3 and 5.

The health benefits of antioxidants of natural origin are associated with their role in the prevention of several disorders called "oxidative stress pathologies" (Martinez-Flores and others 2002). These are related to the damaging effect of oxygen free radicals, or more generally reactive oxygen species (ROS), products of normal metabolism that become harmful when they cannot be neutralized by the cellular antioxidant defense systems. In this condition of "oxidative stress" an uncontrolled oxidizing process may occur that damages biological molecules, disturbs cellular functions, and can potentially lead to the development of one or more diseases (Valko and others 2007). Thus, brain and cardiovascular diseases have significant components resulting from oxidative stress. Components of oxidative stress are also found in the etiopathogenesis of some types of cancer such as liver, stomach, colon, or prostate. The triggering of oxidative stress in glial and neural cells results in neurodegenerative diseases such as Alzheimer's and Parkinson's. Therefore, the consumption of antioxidants of natural origin is at present a recommendation for all ages and in particular for adults and the elderly (Valenzuela and others 2007).

Although the antioxidant power in flavonoids has been studied for many years, researchers from the Linus Pauling Institute at the University of Oregon (Lotito and Frei, 2006, 2004) suggested that their effect *in vivo* is not significant. Some of the premises that led the researchers to this assumption are the following: (a) it is known

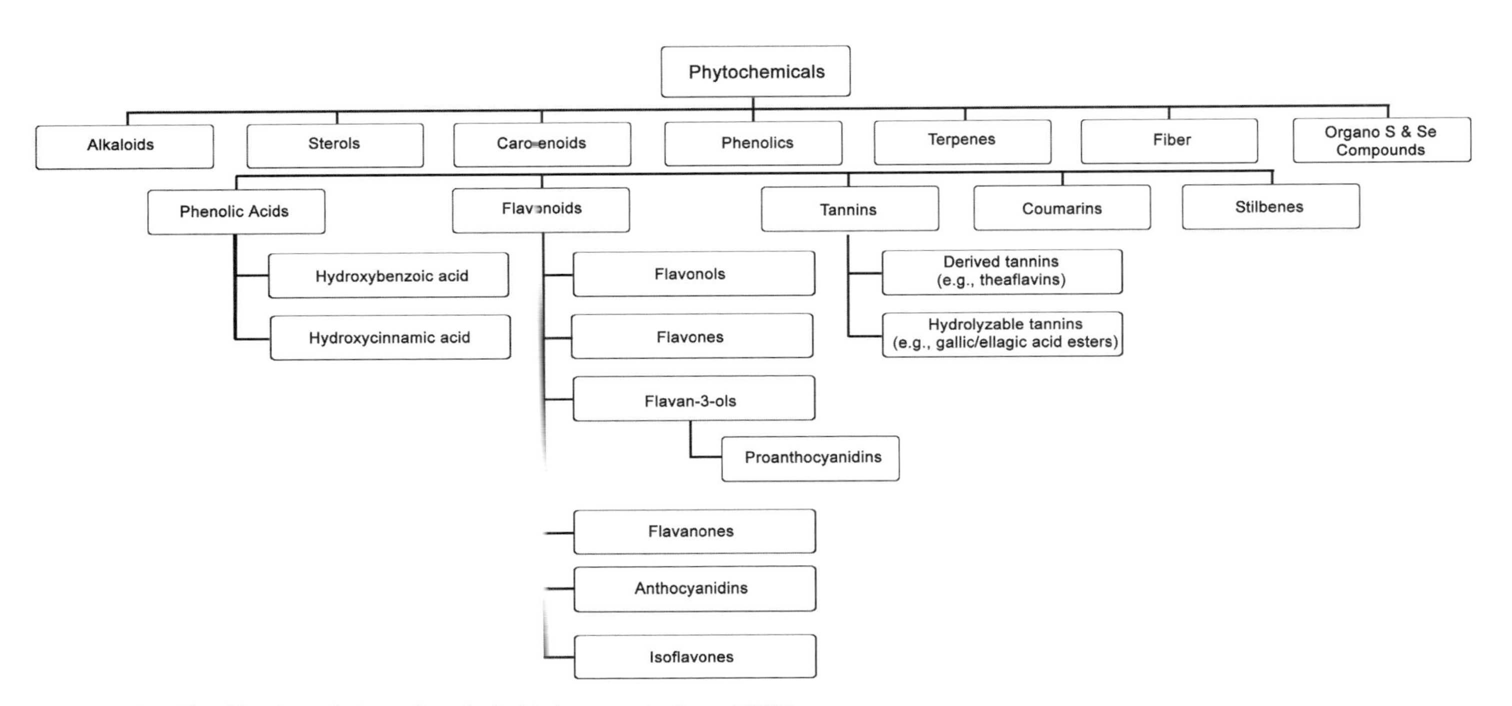

Figure 6.1. Classification of phytochemicals (Erdman and others 2007).

that flavonoids are highly metabolized, which alters their structure and diminishes their antioxidant capacity; (b) as flavonoids enter into the body, they are recognized as strange substances and modified in order to be excreted through the urine or the bile; (c) even though *in vitro* studies show that flavonoids are powerful antioxidants with 3 to 5 times more antioxidant capacity than vitamins C or E, they are less absorbed than these vitamins; and (d) it has been suggested that the increase in antioxidant capacity of plasma observed after eating foods rich in flavonoids is not related to the flavonoids present in plasma but to its high uric acid content.

That flavonoids are not effective antioxidants in the human body does not mean that they have no importance in metabolism. Their action is better explained as that of a vaccine of which only small amounts are needed to trigger an important metabolic response. Thus, they appear to have a strong influence on signaling processes and gene expression, particularly in cancer and coronary diseases. It is thought that human cells recognize flavonoids as xenobiotics (substances that are normally not present in the body), and therefore, their presence activates phase II detoxification enzymes, which also eliminate mutagens and carcinogens. Through this indirect mechanism, flavonoids may help eliminate carcinogenic cells and reduce the risk of cancer development (Lotito and Frei 2006).

Flavonoids are found in fruits, vegetables, seeds, and flowers, as well as in some drinks, such as beer, wine, black and green teas, and soy beverages. Consequently, these kind of compounds are consumed in a normal diet, although they can also be consumed as supplements.

The consumption of flavonoids in different populations is variable, and in many cases the amount is underestimated because calculations are derived from the analysis of few foods or because of the lack of proper food composition tables in the countries (Sarria, 2004; Nijvelt and others 2001). Chun and others (2007) estimated the consumption of flavonoids by adults in the US as 189.7 mg/day. This value was much higher than those reported by other authors for the same population as well as for other countries such as Denmark (23–46 mg/day), Finland (3.4–24 mg/day), Netherlands (23 mg/day), and Japan (63 mg/day)(Chun and others 2007). Johannot and Somerset (2006) estimated the ingestion of flavonoids for an Australian population ($n = 13{,}858$) at around 351 mg/day. They found that tea was typically the main source of flavonoids. Although it was observed that the types of flavonoids and their sources varied according to age, these authors noted that it is necessary to carry out more research with more consistent methodologies to validate the ingestion of specific flavonoids and to facilitate the international comparison.

The diversity of lifestyles in different cultures dictates the kinds of foods and the amount of flavonoids ingested from diet (Sarria, 2004). In Japan, soy and foods that contain soy are consumed in large amounts, resulting in a higher consumption of isoflavones instead of other kinds of flavonoids (Chun and others 2007). Epidemiological studies suggest that in cultures that consume diets rich in soy products, the prevalence of chronic diseases is lower than that in countries with different eating patterns. The message derived from the study of the health benefits of flavonoids is that consumption of plant foods must increase to at least five daily servings of fruit and vegetables (Sarria, 2004).

Flavonoids and Cardiovascular Diseases

Along with cancer, cardiovascular pathology is one of the areas of medicine in which flavonoids have generated major expectations (Johannot and Somerset, 2006; Alvarez Castro and Orallo, 2003). These compounds are important to keep blood vessels healthy, which in turn regulate capillary permeability and allow the flow of oxygen, carbon dioxide, and other nutrients. Many flavonoids increase the resistance of capillaries, preventing them from folding or flattening, probably by enhancing the action of vitamin C. They relax the smooth muscle of the cardiovascular system, which leads to the lowering of blood pressure and improves circulation. They act as antioxidants preventing the oxidation of low-density lipoproteins (LDL), thus preventing formation of atherosclerotic plaque. They can also prevent the excessive accumulation of blood platelets, thus preventing blood coagulation and damage to blood vessels (Mojzisova and Kuchta, 2001).

Vascular Effect

The endothelial vascular cells have an important role in maintaining cardiovascular health, producing nitric oxide (NO), a powerful vasodilator. NO also prevents the adhesion of leukocytes and platelets to the endothelial surface and platelet aggregation (Barringer and others 2008; Erdman and others 2007).

There are several mechanisms involved in the vasodilator effect of flavonoids. The main mechanism seems to be related to the inhibition of protein kinase C or some of the processes activated by this protein. The inhibition of other protein kinases and cyclic nucleotide phosphodiesterase activity and blockage of calcium entry can also contribute to this effect to a greater or lesser extent (Alvarez Castro and Orallo, 2003; Herrera and others 1996). Certain flavonoids, like the flavonol myricetin, have a two-phase action on blood vessels: vasoconstrictor in lowest active concentrations and vasodilator in higher concentrations (Alvarez Castro and Orallo, 2003).

Many studies have been performed in order to evaluate the effect of flavonoids found in different beverages, on the promotion of vascular function. It has been observed that the daily consumption of 4 to 5 cups (900–1250 mL) of black tea for 4 weeks significantly improves vasodilatation in individuals with coronary disease (Duffy and others 2001). This effect is also found with the ingestion of 3 cups (640 mL) of black grape juice (Stein and others 1999) or the consumption of a dark chocolate bar high in flavonoids for 2 weeks (Engler and others 2004). Wang-Polagruto and others (2006) found that the consumption of cocoa for 6 weeks improved endothelial function in postmenopausal women.

Antithrombotic and Antiatherosclerotic Effect

It is well known that high concentrations of LDL, specifically oxidized LDL, are risk factors for coronary artery disease. This fact is explained by the "oxidative hypothesis of atherogenesis." According to this hypothesis, the atheroma is formed by foam cells from the vascular subendothelium that derive from macrophages that have picked up previously oxidized LDL in an uncontrolled manner. These lipoproteins are cytotoxic to the endothelium and, in addition, chemotactic to macrophages and monocytes, promoting their migration to the deep part of the vessel wall (Alvarez Castro and Orallo, 2003; Aviram and Fuhrman, 2003).

Flavonoids protect LDL from oxidation, delaying the onset of lipid peroxidation, however, the prevention of atherosclerosis by flavonoids occurs not only by the inhibition of LDL oxidation, but also by the increase of cellular resistance to harmful effects of the oxidized LDL (de Luis and Aller, 2008). The antioxidant activity of anthocyanidins, as well as their protective role against LDL oxidation, has been well demonstrated in different *in vitro* systems (Aviram and Fuhrman, 2002; Satue-Gracia and others 1997; Teissedre and others 1996).

Flavonoids protect LDL against oxidative changes by directly interacting with this lipoprotein and inhibiting its oxidation and indirectly by accumulating in arterial cells and protecting macrophages against oxidative stress, reducing the formation of foam cells and the development of advanced atherosclerosis (Aviram and Fuhrman, 2003).

The aggregation of platelets contributes to the development of atherosclerosis and to the formation of acute thrombus. The activated platelets that adhere to the vascular endothelium generate lipid peroxides and oxygen free radicals, inhibiting the endothelial formation of prostacyclin and nitric oxide.

Flavonoids, specifically flavonols, inhibit the aggregation of platelets, trapping free radicals directly, thus maintaining adequate concentration of endothelial prostaglandins and nitric oxide (Nijveldt and others 2001; Mojzisova and Kuchta, 2001). It is well known that arachidonic acid (AA) is released when inflammation takes place. AA is metabolized by platelets to form prostaglandins, endoperoxides, and thromboxane A2, causing the activation and aggregation of platelets. It is believed that the inhibition of thromboxane A2 formation, is the main anti-aggregating effect of flavonoids (de Luis and Aller, 2008; Nijveldt and others 2001). Flavonoids can also have an effect on the metabolism of arachidonic acid by inhibition of cyclooxygenase and/or lipoxygenase. One of the most powerful mechanisms by which flavonoids can inhibit platelet aggregation is by induction and increase of intracellular AMPc concentration, through stimulation of adenylcyclase or inhibition of phosphodiesterase activity (Alvarez Castro and Orallo, 2003).

Freedman and others (2001) determined the effects of purple grape juice and its main flavonoids on the functionality of platelets and the production of NO. They observed that incubation of platelets with diluted grape juice resulted in the inhibition of aggregation, increased production of NO, and decreased production of superoxide. To confirm the relevance of these findings, 20 healthy subjects were supplemented with 7 mL of black grape juice/kg/day for 14 days. The inhibition of platelet aggregation was also observed *ex vivo*; there was an increase in the production of NO from 3.5 ± 1.2 to 6.0 ± 1.5 pmol/10^8 platelets and a decrease in the release of superoxide, from 29.5 ± 5.0 to 19.2 ± 3.1 arbitrary units. Under these conditions the antioxidant capacity of protein-free plasma increased by 50% (Freedman and others 2001).

Besides being powerful antioxidants, flavonoids are the most important family of phlebotropic drugs. Their properties of vein protection have proved to be of great help in the treatment of venous insufficiency. Daflon 500 mg (Servier Laboratories, France) is a micronized fraction of flavonoids composed of a mixture of diosmin (90%) and hesperidin (10%), which has become one of the preferred medicines for the treatment of chronic venous deficiency. This drug facilitates the contractility of the venous wall, which activates return flow and reduces venous hypertension. It improves lymphatic drainage because it increases the frequency and intensity of contractions of these ducts and, therefore, the functionality of the lymphatic capillaries is increased (Pascarella 2007). In spite of this, authorization for the marketing of many phlebotonics has been suspended because of insufficient studies that demonstrate their effectiveness (Alvarez Castro and Orallo, 2003).

In Vivo Studies of Flavonoid Effect on Coronary Disease

In vivo studies have assessed the relationship between flavonoids and coronary disease. Flavonoids can contribute to the prevention of the onset of coronary disease mainly because they decrease platelet aggregation and reduce LDL oxidation. In many studies, an inverse correlation between the intake of flavonoids and the risk of death from coronary disease or the risk of stroke has been found. In other studies, like the Caerphilly Study, an increase in the mortality rate of Welshmen due to ischemic cardiac disease was related to a higher intake of flavonols and flavones. It has been suggested that this could be due to their habit of adding milk to tea because milk proteins make the absorption of flavonoids difficult (Hertog and others 1997b). Although some authors have observed the same effect of milk on the absorption of flavonoids (Arts and others 2002b; Langley-Evans 2000), the opposite effect has been found in other studies (Hollman and others 2001; Leenen and others 2000). Kyle and others (2007) demonstrated that it was the infusion time and not the milk addition that influenced the antioxidant capacity of polyphenols in six different brands of black tea. In addition, the consumption of two cups of black tea (400 mL) was associated with a significant increase in the antioxidant capacity of plasma and the total plasma concentrations of phenols, catechins, and flavonols quercetin and kaempferol.

Several studies have been published on the effect of the consumption of flavonols, flavones, and flavan-3-ols on the risk of developing coronary diseases. In some studies, a protective effect of flavonoids has been observed on fatal and nonfatal coronary disease where the risk of death has been reduced by up to 65%. These studies are the following: the Zutphen Elderly Study, which included a group of 805 men from the Netherlands (Hertog and others 1997a, 1993); a later study with a subgroup of 470 men from the Zutphen Elderly Study (Buijsse and others 2006); a 10-year follow-up for the same study (Arts et al., 2001); the Iowa Women's Health Study with a group of 34,500 women from the US (Yochum and others 1999); the α-Tocopherol, β-Carotene Cancer Prevention Study, which included 25,000 male smokers (Hirvonen and others 2001); and the Rotterdam Study, with a group of 4,800 men and women (Geleijnse and others 2002).

A significant protective effect of flavonoid consumption was observed on men, in the Finnish Mobile Clinic Health Examination Survey (Knekt and others 1996). In contrast, in a group of 35,000 men from the USA, coronary death was associated with high intakes of flavonoids in men with a previous history of coronary disease (Rimm and others 1996).

Flavonoids from Different Sources Associated with the Risk of Cardiovascular Disease

The effects of flavonoids derived from soybean, cocoa, wine, and green tea on the reduction of risk for chronic cardiovascular diseases have been studied the most.

Soy has an influence on the risk for developing coronary disease. In cellular and molecular *in vitro* studies, experiments with animal models, and clinical trials with humans, the soy isoflavones (genistein and daidzein) have been related to the prevention of cardiovascular diseases, cancer, osteoporosis, and others. Isoflavones have antioxidant qualities and therefore can reduce the oxidation of LDL cholesterol and inhibit the excessive accumulation of platelets (Geleijnse and Hollman, 2008; Hooper and others 2008). Other potential mechanisms for the prevention of atherosclerosis are the antiproliferative and antimigratory effect on soft muscle cells, the effect on the formation of thrombus, and the maintenance of normal vascular reactivity. It has also been proposed that because of structural similarities to endogenous estrogen, soy isoflavones could interfere with the metabolism of cholesterol by binding to estrogen receptors but with less affinity than estradiol (Alvarez Castro and Orallo, 2003).

In spite of the alleged benefits of isoflavones, some studies have produced inconsistent results in relation to their protective effect against cardiovascular diseases (Lin and others 2007; van der Schouw and others 2005; Engelman and others 2005). Some of the inconsistencies are due to the small sample size in most of the studies and to the very marked differences in metabolism among subjects (Erdman and others 2007; Visioli and others 2000). Therefore, no firm conclusions can be drawn.

Cocoa has the highest content of polyphenols (611 mg/serving) and flavanols (564 mg/serving of epicatechin), higher concentrations than those found in wine and tea (Lee and others 2003). Most of the flavonoids found in cocoa, and therefore in dark chocolate, are flavanols (or flavan-3-ols). Two of their most important monomers are epicatechin and catechin, and among polymeric products, procyanidins are also important (Kris-Etherton and Keen, 2002). Extensive research focuses on the effect of monomeric and polymeric flavanols on the protection of LDL against oxidation. Cocoa procyanidins have shown their capacity to inhibit LDL oxidation and to increase NO production in endothelial tissue, indicating a vasodilator action. It has been found that procyanidins inhibit activation and aggregation of platelets in blood vessels. Another effect of the flavanols derived from cacao is their action on the synthesis of eicosanoids, arachidonic acid–derived molecules, which are directly involved in the inflammatory processes and the regulation of vascular homeostasis (Selmi and others 2006; Brash, 2001). Cocoa also shows hypotensive properties: after continuous consumption of chocolate, a reduction in both systolic and diastolic pressure has been observed, which is expected to reduce the risk of heart attack by 8% and death from coronary disease by 5% (Hooper and others 2008; Ding and others 2006).

It has been reported that a significant portion of epicatechin and catechin contained in chocolate is absorbed, reaching maximum plasma concentrations 3 hours after ingestion (Rein and others 2000): individuals consuming between 35 and 105 g of dark chocolate showed a significant increase of both plasma epicatechin concentration and antioxidant capacity (Wang and others 2000; Rein and others 2000). Although epicatequin has been described as the main contributor to the increase in plasma antioxidant capacity induced by cocoa consumption, the flavonol quercetin is also found in chocolate and could be the main protecting agent against LDL oxidation (Lamuela-Raventós and others 2001).

As in the case of tea flavonoids, there is controversy on the effect of milk proteins on the absorption and beneficial effects of cocoa flavonoids. Serafini and others (2003) concluded that the absorption and *in vivo* plasma antioxidant capacity of epicatechin are notably reduced when chocolate is consumed with milk, because of interactions between milk proteins and flavonoids. In contrast, Keogh and others (2007) found that plasma levels of catechin and epicatechin in healthy men and women who consumed chocolate flavonoids were not affected by addition of milk proteins. Roura and others (2007) also evaluated the possible interactions between milk and absorption of epicatechin in healthy people. They concluded that dilution of cocoa powder in milk seems to have a negative effect on the absorption of polyphenols, if data are analyzed individually; however, because of the high interindividual variation in the absorption, no statistical differences could be observed. These authors also suggest that milk effects could be dependent on fat and not only proteins. If milk products truly affect the health benefits of flavonols, this could have important implications (Schroeter and others 2003).

Consumption of wine has been associated with a reduced risk of coronary diseases, although contradictory results have been reported. Epidemiologic evidences shows that French people have less heart disease than expected, considering their high consumption of saturated fat, high percentage of smokers, and low physical activity (Goldberg 1995). This particular characteristic is known as the French paradox and has been attributed to the high daily intake of red wine. There are several studies that show that daily consumption of one or two glasses of red wine protects against heart attack. In general, greater health benefits are attributed to red wine compared to white wine (Visioli and others 2000).

Red wine contains quercetin, rutin, catechin, and epicatechin, among other flavonoids (Frankel and others 1993). Quercetin and other phenolic compounds isolated from wines were found to be more effective than α-tocopherol in inhibiting copper-catalyzed LDL oxidation. It has been determined that quercetin has also several anti-inflammatory effects: it inhibits inflammatory cytokine production (Boots and others 2008), inducible NO synthase expression and activation of inflammatory transcription factors (Hamalainen and others 2007), and activity of cyclooxygenase and lipooxygenase (Issa 2006), among others.

Tea, like wine, has a protective effect against coronary diseases. In a study by Lagiou and others (2004b) carried out in Greece, they found that the high content of flavan-3-ols in wine and tea could protect against coronary diseases. Therefore, populations that do not include wine in their diet, such as the Japanese, may have a low incidence of coronary diseases due to the high consumption of tea (Kuriyama and others 2006). One of the potential mechanisms of the protective effect of tea could be the antioxidant capacity of tea flavonoids. Other mechanisms include attenuation of inflammatory process in atherosclerosis, reduction of thrombosis, blocking cellular adhesion of molecules, and promoting normal endothelial function (Kris-Etherton and Keen 2002).

Continued intake of green tea could significantly reduce LDL levels and LDL oxidation. In order to obtain this result, drinking two to five glasses of green tea per day is needed (Hooper and others 2008). In the Ohsaki National Health Insurance Group Study, 40,530 Japanese adults aged from 40 to 79 years without a history of strokes or coronary diseases were monitored. The authors found a significant inverse association between the consumption of green tea and death by general causes and with death by cardiovascular disease. The inverse association with cardiovascular disease mortality was stronger than that with all-cause mortality, and in both cases it was stronger for women than for men (Kuriyama and others 2006). In another study with 3,430 subjects in Saudi Arabia, Hakim and others (2003) found that those subjects who consumed more than six cups of black tea per day (>480 mL) had a lower prevalence of coronary disease compared to those who did not drink any tea. Geleijnse and others (2002) found that the higher the tea consumption, the lower the risk of a first myocardial attack. These results agree with those reported previously by Peters and others (2001), who carried out a meta-analysis of ten cohort studies as well as seven case-control studies. They found that no conclusive effects of tea consumption on stroke and coronary heart disease could be observed; however, the incidence of myocardial infarction was estimated to decrease by 11% with an increase in tea consumption of three cups per day. Much of the heterogeneity among studies could be explained by geographical region where they had been performed.

Flavonoids are present in other beverages besides wine and tea. For example, pomegranate and cranberry juice contain high concentrations of polyphenols and a strong antioxidant activity against LDL oxidation. Their antioxidant capacity depends not only on the amount but also on the type of flavonoids present (Aviram and Fuhrman 2003).

Flavonoids and Cancer

In addition to their antioxidant properties, other interesting biological effects of flavonoids include the capacity to scavenge free radicals, regulation of NO production, inhibition of leukocyte adhesion, induction of apoptosis, and inhibition of cell proliferation (Higdon and Frei 2003; Yang and others 2001; Nijveldt and others 2001; Adlercreutz and Mazur 1997). These effects may contribute to the potential protective role of flavonoids in the development of cancer and cardiovascular diseases, among others. However, to date, it is not clear whether these effects, which have been found in experimental animals or *in vitro* studies, are relevant for human health. Although the consumption of flavonoids is frequent, their plasma concentrations are relatively low, depending on their bioavailability and metabolism. After absorption, polyphenols are generally metabolized by phase II enzymes that convert them to methoxylated, glucuronidated, and sulfated derivatives (Hollman and Arts 2000). Additionally, flavonoids can be metabolized by the intestinal flora prior absorption; therefore, the resulting metabolites depend upon the type of colonic microflora normally present on the study subjects. On the other hand, the lifestyles of people and populations play an important role in the consumption of flavonoids.

In the case of a potential benefit of flavonoids in the development of cancer, these compounds may be active in the modulation of cell signaling by:

1. Stimulation of phase II detoxification enzymes (Kong and others 2001; Walle and Walle 2002).
2. Regulation of the normal cell cycle (Chen and others 2004; Wang and others 2004).
3. Inhibition of angiogenesis (Bagli and others 2004; Kim 2003).
4. Decrease of inflammation (O'Leary and others 2004; Sakata and others 2003; Cho and others 2003; Steele and others 2003).
5. Inhibition of cell proliferation and induction of apoptosis (Sah and others 2004; Kavanagh and others 2001; Ramos 2007).

In vivo studies have found that some flavonoids inhibit the development of some kinds of cancer (Yang and others 1998; Balasubramanian and Govindasamy 1996; Li and others 2002; Yamane and others 1996; Guo and others 2004; Huang and others 1997; Haddad and others 2006; Yamagishi and others 2002). However, as stated before, there is still no convincing epidemiological evidence of any inverse relationship between high flavonoid consumption and a reduction of the risk for cancer development in humans. Some prospective studies, which have estimated the consumption of flavonoids through a dietary survey on the frequency of food intake, found no inverse relationship with the risk of cancer (Goldbohm and others 1995, 1996; Hertog and others 1994; Arts and others 2001). In a study of postmenopausal women from the USA, an inverse association between the incidence of rectal cancer and the consumption of tea catechin was found (Arts and others 2002a). Knekt and others (2002) estimated the food intake of volunteers from Finland through a dietary history of 1 year, and they found that men with higher quercetin intakes had a lower incidence of lung cancer, whereas men with higher myricetin intake had a lower risk of prostate cancer. In a previous study, these authors also found an inverse association between the intake of flavonoids and the incidence of lung cancer (Knekt and others 1997). In this case, the beneficial effect was mainly attributed to quercetin, which contributed more than 95% of the total flavonoids ingested. Also, these researchers emphasized that the effect found was not due to the consumption of vitamin C, E or β-carotene because the results were adjusted for these variables.

In Greece, a case-control study was conducted to investigate the incidence of liver cancer by estimating the consumption of six types of flavonoids with a semiquantitative questionnaire on the frequency of foods. The intake of flavones was inversely associated with hepatocellular carcinoma, irrespective of its etiology (viral or nonviral). With respect to cholangiocarcinoma, an inverse association with the consumption of flavan-3-ols, anthocyanidins, and total flavonoids studied was found. However, this last result should be viewed with caution because of the small sample size, due to the fact that this is a rare type of cancer (Lagiou and others 2008).

Lagiou and others (2004a) explored the relationship between the intake of flavonoids and the risk of stomach cancer and found that only the consumption of flavanones showed an inverse relationship with the illness. However, they concluded that it is possible that other factors are part of the inherent protection exerted by the consumption of vegetables. Rossi and others (2007), similarly using a dietary methodology, studied the relationship among several types of flavonoids (flavanones, flavan-3-ols, flavonols, flavones, and anthocyanidins) and esophageal cancer in several regions of northern Italy. The study revealed an inverse relationship between the intake of flavanones and the risk of esophageal cancer (odds ratio = 0.38 for the highest vs. the lowest quintile (95% CI = 0.23–0.66)).

Other studies have estimated the consumption of fruits or vegetables (rich in flavonoids) and have reported an inverse relationship with some types of cancer but in different extent among men and women (Park and others 2007; McCullough and others 2003; Voorrips and others 2000).

In preclinical models of carcinogenesis, which include colorectal, breast, and prostate cancer, a remarkable efficacy of some phytochemicals has been found. Some of these are epigallocatechin from tea; quercetin and genistein from onion and soy, respectively; curcumin from curry; and resveratrol from red grapes. To advance the investigation in this field, it is necessary to perform clinical tests (phase III) on healthy volunteers as well as in patients with premalignant lesions or cancer. With this aim, Thomasset and others (2006) have reviewed some clinical studies performed in past years on healthy volunteers, patients with premalignant lesions, and patients with different types of cancer.

The studies on healthy volunteers have allowed exploration of the pharmacokinetics of these metabolites in humans and have given information on potential novel biomarkers of chemopreventive effectiveness. Oral intake of polyphenols has been studied in relation to an increase of antioxidant activity in plasma or urine (Sung and others 2000; Langley-Evans 2000), to the protection against damage of genetic material induced by the consumption of cigarettes or exposure to radiation (Hakim and others 2003b; Schwartz and others 2005; Klaunig and others 1999), and mammographic density (Boyd and others 1998), among other indicators (Thomasset and others 2006).

The tests performed on patients with premalignant lesions have tried to prove the ability of a compound to interfere with the progression to malignancy. This progression can take many years, and it is important to consider the fact that if a treatment is available, the clinical tests would not be ethically accepted. A premalignant lesion on which clinical tests can be performed is prostatic intraepithelial neoplasia (PIN). One of the studies performed on PIN rendered the result that when the biopsy was repeated after 1 year of treatment with 600 mg of green tea catechins daily, the progression of prostate cancer decreased 10 times (compared with placebo) although the levels of prostate-specific antigen did not change (Bettuzzi and others 2006). In women with chronic cervicitis or cervical intraepithelial neoplasia, the effect of receiving catechins from green tea for two weeks was studied. The results were positive in terms of the histology: patients were without lesions when the biopsy was repeated (Ahn and others 2003).

On patients with cancer, the effects of green tea catechins, soy isoflavones and quercetin as chemoprotective/chemotherapeutic agents have also been studied. Although results have not been entirely satisfactory, a partial response has been achieved in some trials. For example, small decreases in plasma concentration of prostate-specific antigen were observed in prostate cancer patients who consumed soy isoflavones. Nevertheless, results in individuals with premalignant disease who consumed green tea polyphenols support their advancement into phase III clinical intervention trials aimed at the prevention of PIN, leukoplakia, or premalignant cervical disease (Thomasset and others 2006).

Metabolic Syndrome

Most of the studies about the relationship of flavonoids and the regulation of insulin do not provide clear information about their possible direct effects on the pancreas. This organ has the most important function in the regulation of nutrient metabolism, such as control of glucose homeostasis via insulin and glucagon secretion. Experiments using diabetic animal models suggest that flavonoids can ameliorate this disease. Some of the published data suggest that flavonoids may be involved in the secretion of insulin, as well as in preventing apoptosis of beta-cells; this could be due to the antioxidant properties of flavonoids as well as to other modes of action.

The potential beneficial effects of flavonoids on insulin metabolism, diabetes mellitus, and metabolic syndrome have been recently reviewed. They are mainly related to the numerous *in vitro* biological actions of flavonoids: antioxidant, anti-inflamatory, and antiatherogenic activities; stimulation of insulin secretion; regulation of key enzymes in carbohydrate metabolism; promotion of vascular function; and inhibition of angiogenesis (Cazarolli and others 2008; Dembinska-Kiec and others 2008). Several *in vitro studies* have shown that some flavonoids (isoflavones) stimulate insulin secretion through tyrosine kinase inhibition in different insulin-secreting models (Ohno and others 1993; Jonas and others 1995; Sorenson and others 1994; Persaud and others 1999). Anthocyanins and anthocyanidins are also effective at inducing insulin secretion, when tested in pancreatic cell lines. They showed different effectiveness depending on the number of hydroxyl groups in the B-ring of their structures (Jayaprakasam and others 2005).

In a study using healthy rats fed on a diet that included 148 mg catechins from green tea/day for 12 days, it was found that fasting plasma glucose, insulin, tryglyceride, and fatty acid levels were decreased significantly, and insulin-stimulated glucose uptake of adipocytes was increased (Wu and others 2004). On the other hand, in another study on hypertensive rats, improvement of insulin sensitivity was observed when 200 mg of epigallocatechin gallate/kg/day for 3 weeks was fed by gavage (Potenza and others 2007). Venables and others (2008) found that the consumption of green tea extract (GTE) may increase lipid oxidation during moderate exercise and improve insulin sensitivity and glucose tolerance in healthy young men. They suggest that amelioration of glucose tolerance may come from oxidation of lipids instead of their storage, because there was an increase in lipid oxidation during exercise, reducing the accumulation of fatty acid metabolites in muscle when GTE had been administered. It is known that fatty acid metabolites interfere with the cascade of insulin signals via the activation of protein kinase C isoforms (Itani and others 2002; Yu and others 2002).

Effects of Flavonoids in Other Illnesses

Research on the role of nutrition, particularly concerning antioxidants, in illnesses such as dementia, Alzheimer's disease, and Parkinson's disease raises high expectations. Oxidative stress is involved in the pathophysiology of these illnesses and can induce neuronal damage, modulate intracellular signals, and kill these cells by apoptosis or necrosis. In Alzheimer's patients, the accumulation of beta-amyloid protein is associated with the increase of the production of free radicals and lipid peroxidation. Dietary supplementation with fruit and vegetable extracts high in antioxidants can reduce the vulnerability to oxidative stress that occurs in old age. This can occur through the ability of phytochemicals, such as flavonoids, to increase the signals of cell communication, as well as their antioxidant and anti-inflammatory activities. Some investigations indicate that flavonoids can selectively regulate multiple signals at the transcriptional level, especially those that involve protein kinases (Joseph and others 1999; Martin and others 2003; Sato and others 2001).

A number of epidemiological studies have indicated that the individuals consuming diets with great quantities of fruits and vegetables can reduce the risk of developing age-related illnesses like Alzheimer's disease. In a recent study, 1,640 subjects, 65 years or older and free form dementia at baseline, were followed for 10 years for cognitive function. Flavonoid intake was associated with better cognitive performance at baseline and over time, raising the possibility that dietary flavonoid intake is associated with better cognitive evolution (Letenneur and others 2007).

There are few studies involving other illnesses; however, the consumption of flavonoids has been related to the development of asthma and rheumatoid arthritis. In the case of asthma, it has been suggested that high consumption of apples may protect against this and other chronic pulmonary diseases. This effect has been attributed to the high content of flavonoids in apple (mainly catequins and flavonols), but more studies are needed in order to support this evidence (Garcia and others 2005). However, in a study of dietary intake of flavonoids, a low incidence of asthma with a higher consumption of quercetin, naringenin, and hesperetin was observed (Knekt and others 2002). These researchers also found a higher risk of rheumatoid arthritis after consuming more kaempherol and did not find any significant decrease in the incidence of cataracts after a high consumption of flavonoids.

Finally, the role of isoflavones (phytoestrogens) in the prevention of osteoporosis and improvement of bone health must be mentioned. Varied clinical trials suggest that isoflavones reduce bone loss in women in the early period post menopause, but a definitive result requires more investigations that have substantial sample size and are of long duration (Coxam 2008). For example, in two recently published meta-analyses of randomized controlled trials, contradictory results were obtained. Ma and others (2008) found that consumption of 90 mg/day of isoflavones for 6 months increased spine bone mineral density and mineral content. However, Liu and others (2009) found that long-term (1 year or longer) interventions of soy isoflavones had no significant effect on spinal or hip bone mineral density in women, suggesting that soy isoflavone supplementation may have a transient, but not a long-term, beneficial effect on bone health.

Bioavailability of Flavonoids

Bioavailability of flavonoids should be emphasized because the health-promoting effects of dietary flavonoids depend on their absorption and metabolism. Not all flavonoids are absorbed with the same efficiency; therefore, the most common flavonoids in diet need not necessarily be the ones with the main biological activities. Isoflavones and quercetin glucosides (flavonols) are the most absorbed polyphenols followed by catechins and flavanones, with different kinetics. The least absorbed flavonoids are the proanthocyanidins, the galloylated tea catechins, and the anthocyanins (Manach and others 2005). Anthocyanins were thought to be poorly absorbed and metabolized, and therefore it was believed that they were found in plasma only in their intact form (glycosylated) and in concentrations too low to exert biological effects. However, more recent studies have identified metabolites of anthocyanins at higher concentrations that the parent compounds (Kay and others 2005).

Flavonoids in general are extensively metabolized by enterocyte and hepatic cell enzymes and by intestinal microflora. Therefore, it is necessary to explore the biological activity of flavonoids and their metabolites in specific tissues, because the metabolites found in the blood flow or any specific organ may differ among themselves and from the original compounds in terms of biological activity (Sarria 2004). The role of flavonoid metabolism by enteric (colonic) microflora might be especially important in isoflavones: equol produced from soya daidzein appears to have phytoestrogenic properties equivalent to or even greater than those of the original isoflavone (Manach and others 2004). For this reason, in order to fully understand the potential health benefits of flavonoids, precise information about their bioavailability, absorption, and metabolism should be considered.

Final Comments

Flavonoids have received a great deal of attention in literature in the past 10 years, and a great variety of potential beneficial effects have been elucidated. The study of these effects is complex because of their chemical heterogeneity and because information about the exact flavonoid composition of food products (fruits, vegetables, beverages, etc.), metabolism, and bioavailability is incomplete. Enough evidence has been gathered from epidemiological surveys, intervention trials, and *in vitro* studies to allow the conclusion that flavonoids are important for cardiovascular health. Evidence for other types of disease is probably less conclusive, although appealing.

Many studies on the health-promoting properties of flavonoids have dealt mainly with their antioxidant capacity; however, it is unlikely that this unique activity is the cause of all their healthful properties. In fact, although flavonoids have numerous *in vitro* activities, most of them are observed only at concentrations much higher than those that can be achieved through dietary intake. Therefore, these biological actions may have pharmacological significance but small importance in determining the health benefits of fruit and vegetable consumption. More *in vitro* studies of the biological effects of low concentrations of flavonoids and especially their metabolites should be carried out.

Finally, it should also be considered that flavonoid-rich foods contain a great diversity of compounds with bioactive properties (for e.g., carotenoids, other phenolics, fiber, and minerals), and multiple interactions occur among all of them. There is also great diversity in the ingestion, absorption, and metabolism of these compounds in different populations, and all of these circumstances could camouflage any effect of flavonoids on disease prevention or treatment.

Acknowledgments

The authors wish to thank Lic. Aida Espinosa Curiel for preparing the figure.

References

Adlercreutz H and Mazur W. 1997. Phyto-oestrogens and Western diseases. Ann Med 29(2):95–120.

Ahn WS, Yoo J, Huh SW, Kim CK, Lee JM, Namkoong SE, Bae SM and Lee IP. 2003. Protective effects of green tea extracts (polyphenon E and EGCG) on human cervical lesions. Eur J Cancer Prev 12(5):383–390.

Alvarez Castro E and, Orallo Cambeiro F. 2003. Actividad biológica de los flavonoides (II). Acción cardiovascular y sanguínea. OFFARM 22:102–110.

Arts IC, Hollman PC, Bueno De Mesquita HB, Feskens EJ and Kromhout D. 2001. Dietary catechins and epithelial cancer incidence: the Zutphen elderly study. Int J Cancer 92(2):298–302.

Arts IC, Jacobs DR Jr., Gross M, Harnack LJ and Folsom AR. 2002a. Dietary catechins and cancer incidence among postmenopausal women: the Iowa Women's Health Study (United States). Cancer Causes Control 13(4):373–382.

Arts MJ, Haenen GR, Wilms LC, Beetstra SA, Heijnen CG, Voss HP, Bast A. 2002b. Interactions between flavonoids and proteins: effect on the total antioxidant capacity. J Agric Food Chem 50(5):1184–1187.

Aviram M and Fuhrman B. 2003. Effects of flavonoids on the oxidation of LDL and atherosclerosis. In: Rice-Evans C and Packer L, editors. Flavonoids in Health and Disease, 2nd ed. Boca Raton, FL: CRC Press, pp. 165–203.

Aviram M, Fuhrman B. 2002. Wine flavonoids protect against LDL oxidation and atherosclerosis. Ann N Y Acad Sci 957:146–161.

Bagli E, Stefaniotou M, Morbidelli L, Ziche M, Psillas K, Murphy C and Fotsis T. 2004. Luteolin inhibits vascular endothelial growth factor-induced angiogenesis; inhibition of endothelial cell survival and proliferation by targeting phosphatidylinositol 3'-kinase activity. Cancer Res 64(21):7936–7946.

Balasubramanian S, Govindasamy S. 1996. Inhibitory effect of dietary flavonol quercetin on 7,12-dimethylbenz[*a*]anthracene-induced hamster buccal pouch carcinogenesis. Carcinogenesis 17(4):877–879.

Barringer TA, Hatcher L and Sasser HC. 2008. Potential benefits on impairment of endothelial function after a high-fat meal of 4 weeks of flavonoid supplementation. eCAM:1–6.

Bettuzzi S, Brausi M, Rizzi F, Castagnetti G, Peracchia G and Corti A. 2006. Chemoprevention of human prostate cancer by oral administration of green tea catechins in volunteers with high-grade prostate intraepithelial neoplasia: a preliminary report from a one-year proof-of-principle study. Cancer Res 66(2):1234–1240.

Boots AW, Haenen GRMM and Bast A. 2008. Health effects of quercetin: From antioxidant to nutraceutical. Eur J Pharmacol 585:325–337.

Boston Collaborative Drug Surveillance Program. 1972. Coffee drinking and acute myocardial infarction. Report from the Boston Collaborative Drug Surveillance Program. Lancet 2:1278–1281.

Boyd NF, Lockwood GA, Martin LJ, Knight JA, Byng JW, Yaffe MJ and Tritchler DL. 1998. Mammographic densities and breast cancer risk. Breast Dis 10(3–4):113–126.

Brash AR. 2001. Arachidonic acid as a bioactive molecule. J Clin Invest 107(11):1339–1345.

Buijsse B, Feskens EJ, Kok FJ and Kromhout D. 2006. Cocoa intake, blood pressure, and cardiovascular mortality: the Zetphen Elderly Study. Arch Intern Med 166:411–417.

Cazarolli LH, Zanatta L, Alberton EH, Figueredo MS, Folador P, Damazio RG, Pizzolatti MG and Silva FR. 2008. Flavonoids: cellular and molecular mechanisms of action in glucose homeostasis. Mini Rev Med Chem 8(10):1032–1038.

Chen JJ, Ye ZQ and Koo MW. 2004. Growth inhibition and cell cycle arrest effects of epigallocatechin gallate in the NBT-II bladder tumour cell line. BJU Int 93(7):1082–1086.

Cho SY, Park SJ, Kwon MJ, Jeong TS, Bok SH, Choi WY, Jeong WI, Ryu SY, Do SH, Lee CS, Song JC and Jeong KS. 2003. Quercetin suppresses proinflammatory cytokines production through MAP kinases andNF-kappaB pathway in lipopolysaccharide-stimulated macrophage. Mol Cell Biochem 243(1-2):153–160.

Chun OK, Chung SJ and Song WO. 2007. Estimated dietary flavonoid intake and major food sources of U.S. adults. J Nutr 37(5):1244–1252.

Coxam V. Phyto-oestrogens and bone health. 2008. Proc Nutr Soc 67(2):184–195.

de Luis DA and Aller R. 2008. Papel de los flavonoides del té en la protección cardiovascular. An Med Interna 25:105–107.

Dembinska-Kiec A, Mykkanen O, Kiec-Wilk B and Mykkanen H. 2008. Antioxidant phytochemicals against type 2 diabetes. Br J Nut 99:ES109–ES117.

Ding EL, Hutfless SM, Ding X and Girotra S. 2006. Chocolate and prevention of cardiovascular diseases: a systematic review. Nutr Metab 3:2.

Duffy SJ, Keaney JF Jr, Holbrook M, Gokce N, Swerdloff PL, Frei B and Vita JA. 2001. Short- and long-term black tea consumption reverses endothelial dysfunction in patients with coronary artery disease. Circulation 104(2):151–156.

Engelman HM, Alekel DL, Hanson LN, Kanthasamy AG and Reddy MB. 2005. Blood lipid and oxidative stress responses to soy protein with isoflavones and phytic acid in postmenopausal women. Am J Clin Nutr 81:590–596.

Engler MB, Engler MM, Chen CY, Malloy MJ, Browne A, Chiu EY, Kwak HK, Milbury P, Paul SM, Blumberg J and Mietus-Snyder ML. 2004. Flavonoid-rich dark chocolate improves endothelial function and increases plasma epicatechin concentrations in healthy adults. J Am Coll Nutr 23:197–204.

Erdman JW Jr, Balentine D, Arab L, Beecher G, Dwyer JT, Folts J, Harnly J, Hollman P, Keen CL, Mazza G, Messina M, Scalbert A, Vita J, Williamson G and Burrowes J. 2007. Flavonoids and heart health: proceedings of the ILSI North America Flavonoids Workshop, May 31–June 1, 2005, Washington, DC. J Nutr 137:718S–737S.

flavonoid intake and breast cancer risk among women on Long Island. Am J Epidemiol 165(5):514–523.

Frankel EN, Kanner J, German JB, Parks E and Kinsella JE. 1993. Inhibition of oxidation of human low-density lipoprotein by phenolic substances in red wine. Lancet 341:454–457.

Freedman JE, Parker C 3rd, Li L, Perlman JA, Frei B, Ivanov V, Deak LR, Iafrati MD and Folts JD. 2001. Select flavonoids and whole juice from purple grapes inhibit platelet function and enhance nitric oxide release. Circulation 103:2792–2798.

Geleijnse JM and Hollman PCh. 2008. Flavonoids and cardiovascular health: which compounds, what mechanisms? Am J Clin Nutr 88(1):12–13.

Geleijnse JM, Launer LJ, Van Der Kuip DA, Hofman A and Witteman JC. 2002. Inverse association of tea and flavonoid intakes with incident myocardial infarction: the Rotterdam Study. Am J Clin Nutr 75:880–886.

Goldberg DM. 1995. Does wine work?. Clin Chem 1:14–16.

Goldbohm RA, Hertog MG, Brants HA, van Poppel G and van den Brandt PA. 1996. Consumption of black tea and cancer risk: a prospective cohort study. J Natl Cancer Inst 88(2):93–100.

Goldbohm RA, van den Brandt PA, Hertog MG, Brants HA and Van Poppel G. 1995. Flavonoid intake and risk of cancer: a prospective cohort study. Am J Epidemiol 41:s61.

Guo JY, Li X, Browning JD Jr., Rottinghaus GE, Lubahn DB, Constantinou A, Bennink M and MacDonald RS. 2004. Dietary soy isoflavones and estrone protect ovariectomized ERalphaKO and wild-type mice from carcinogen-induced colon cancer. J Nutr 134(1):179–182.

Haddad AQ, Venkateswaran V, Viswanathan L, Teahan SJ, Fleshner NE and Klotz LH. 2006. Novel antiproliferative flavonoids induce cell cycle arrest in human prostate cancer cell lines. Prostate Cancer Prostatic Dis 9(1):68–76.

Hakim IA, Alsaif MA, Alduwaihy M, Al-Rubeaan K, Al-Nuaim AR and Al-Attas OS. 2003. Tea consumption and the prevalence of coronary heart disease in Saudi adults: results from a Saudi national study. Prev Med 36:64–70.

Hamalainen M, Nieminen R, Vuorela P, Heinonen M and Moilanen E. 2007. Anti-inflammatory effects of flavonoids: genistein, kaempferol, quercetin and daidzein inhibit STAT-1 and NF-kB activations whereas flavones, isorhamnetin, naringenins, and pelargonidin inhibit only NF-kB activation along with their inhibitory effect on iNOS expression and NO production in activated macrophagues. Mediators Inflamm 2007:45673.

Herrera MD, Zarzuelo A, Jimenez J, Marhuenda E, Duarte J. 1996. Effects of flavonoids on rat aortic smooth muscle contractility: structure-activity relationships. Gen Pharmacol 27:273–277.

Hertog MG, Feskens EJ, Hollman PC, Katan MB and Kromhout D. 1993. Dietary antioxidant flavonoids and risk of coronary heart disease: the Zutphen Elderly Study. Lancet 342:1007–1011.

Hertog MG, Feskens EJ, Hollman PC, Katan MB and Kromhout D. 1994. Dietary flavonoids and cancer risk in the Zutphen Elderly Study. Nutr Cancer 22(2):175–184.

Hertog MG, Feskens EJ and Kromhout D. 1997a. Antioxidant flavonols and coronary heart disease risk. Lancet 349:699.

Hertog MG, Sweetnam PM, Fehily AM, Elwood PC and Kromhout D. 1997b. Antioxidant flavonols and ischemic heart disease in a Welsh population of men: the Caerphilly Study. Am J Clin Nutr 65:1489–1494.

Higdon JV and Frei B. 2003. Tea catechins and polyphenols: health effects, metabolism, and antioxidant functions. Crit Rev Food Sci Nutr 43(1):89–143. Review.

Hirvonen T, Pietinen P, Virtanen M, Ovaskainen ML, Hakkinen S, Albanes D and Virtamo J. 2001. Intake of flavonols and flavones and risk of coronary heart disease in male smokers. Epidemiology 12:62–67.

Hollman PCH and Arts ICW. 2000. Flavonols, flavones and flavanols: nature, occurrence and dietary burden. J Sci Food Agric 80:1081–1093.

Hollman PC, Van Het Hof KH, Tijburg LB and Katan MB. 2001. Addition of milk does not affect the absorption of flavonols from tea in man. Free Radic Res 34:297–300.

Hooper L, Kroon PA, Rimm EB, Cohn JS, Harvey I, Le Cornu KA, Ryder JJ, Hall WL and Cassidy A. 2008. Flavonoids, flavonoid-rich foods, and cardiovascular risk: a meta-analysis of randomized controlled trials. Am J Clin Nutr 88(1):38–50.

Huang MT, Xie JG, Wang ZY, Ho CT, Lou YR, Wang CX, Hard GC and Conney AH. 1997. Effects of tea, decaffeinated tea, and caffeine on UVB light-induced complete carcinogenesis in SKH-1 mice: demonstration of caffeine as a biologically important constituent of tea. Cancer Res 57(13):2623–2629.

Issa AY, Volate SR and Wargovich MJ. 2006. The role of phytochemicals in inhibition of cancer and inflammation: new directions and perspectives. J Food Comp Anal 19:405–419.

Itani SI, Ruderman NB, Schmieder F and Boden G. 2002. Lipid-induced insulin resistance in human muscle is associated with changes in diacylglycerol, protein kinase C, and IkappaB-alpha. Diabetes 51(7):2005–2011.

Jayaprakasam B, Vareed SK, Olson LK and Nair MG. 2005. Insulin secretion by bioactive anthocyanins and anthocyanidins present in fruits. J Agric Food Chem 53(1):28–31.

Johannot L and Somerset SM. 2006. Age-related variations in flavonoid intake and sources in the Australian population. Public Health Nutr 9:1045–1054.

Jonas JC, Plant TD, Gilon P, Detimary P, Nenquin M and Henquin JC. 1995. Multiple effects and stimulation of insulin secretion by the tyrosine kinase inhibitor genistein in normal mouse islets. Br J Pharmacol 114(4):872–880.

Joseph JA, Shukitt-Hale B, Denisova NA, Bielinski D, Martin A, McEwen JJ and Bickford PC. 1999. Reversals of age-related declines in neuronal signal transduction, cognitive, and motor behavioral deficits with blueberry, spinach, or strawberry dietary supplementation. J Neurosci 19(18):8114–8121.

Kavanagh KT, Hafer LJ, Kim DW, and others. 2001. Green tea extracts decrease carcinogen-induced mammary tumor burden in rats and rate of breast cancer cell proliferation in culture. J Cell Biochem 82(3):387–398.

Kay CD, Mazza G and Holub BJ. 2005. Anthocyanins exist in the circulation primarily as metabolites in adult men. J Nutr 135:2582–2588.

Keli SO, Hertog MG, Feskens EJ, Kromhout D. 1996. Dietary flavonoids, antioxidant vitamins, and incidence of stroke: the Zutphen Study. Arch Intern Med 156:637–642.

Keogh JB, McInerney J and Clifton PM. 2007. The effect of milk protein on the bioavailability of cocoa polyphenols. J Food Sci 72(3):S230–S233.

Kim MH. 2003. Flavonoids inhibit VEGF/bFGF-induced angiogenesis *in vitro* by inhibiting the matrix-degrading proteases. J Cell Biochem 89(3):529–538.

Klaunig JE, Xu Y, Han C, Kamendulis LM, Chen J, Heiser C, Gordon MS and Mohler ER. 1999. The effect of tea consumption on oxidative stress in smokers and nonsmokers. Proc Soc Exp Biol Med 220:249–254.

Knekt P, Jarvinen R, Reunanen A and Maatela J. 1996. Flavonoid intake and coronary mortality in Finland: a cohort study. BMJ 312:478–481.

Knekt P, Järvinen R, Seppänen R, Hellövaara M, Teppo L, Pukkala E and Aromaa A. 1997. Dietary flavonoids and the risk of lung cancer and other malignant neoplasms. Am J Epidemiol 146(3):223–230.

Knekt P, Kumpulainen J, Järvinen R, Rissanen H, Heliövaara M, Reunanen A, Hakulinen T and Aromaa A. 2002. Flavonoid intake and risk of chronic diseases. Am J Clin Nutr 76(3):560–568.

Kong AN, Owuor E, Yu R, Hebbar V, Chen C, Hu R and Mandlekar S. 2001. Induction of xenobiotic enzymes by the MAP kinase pathway and the antioxidant or electrophile response element (ARE/EpRE). Drug Metab Rev 33(3-4):255–271.

Kris-Etherton PM and Keen CL. 2002. Evidence that the antioxidant flavonoids in tea and cocoa are beneficial for cardiovascular health. Curr Opin Lipidol 13(1):41–49.

Kuriyama S, Shimazu T, Ohmori K, Kikuchi N, Nakaya N, Nishino Y, Tsubono Y and Tsuji I. 2006. Green tea consumption and mortality due to cardiovascular disease, cancer, and all causes in Japan: the Ohsaki study. JAMA 296:1255–1265.

Kyle JA, Morrice PC, McNeill G and Duthie GG. 2007. Effects of infusion time and addition of milk on content and absorption of polyphenols from black tea. J Agric Food Chem 55:4889–4894.

Lagiou P, Rossi M, Lagiou A, Tzonou A, La Vecchia C and Trichopoulos D. 2008. Flavonoid intake and liver cancer: a case-control study in Greece. Cancer Causes Control 19(8):813–818.

Lagiou P, Samoli E, Lagiou A, Peterson J, Tzonou A, Dwyer J and Trichopoulos D. 2004a. Flavonoids, vitamin C and adenocarcinoma of the stomach. Cancer Causes Control 15:67–72.

Lagiou P, Samoli E, Lagiou A, Tzonou A, Kalandidi A, Peterson J, Dwyer J and Trichopoulos D. 2004b. Intake of specific flavonoid classes and coronary heart disease. A case-control study in Greece. Eur J Clin Nutr 58:1643–1648.

Lamuela-Raventós RM, Andrés-Lacueva C, Permanyer J and Izquierdo-Pulido M. 2001. More antioxidants in cocoa. J Nutr. 131(3):834–835.

Langley-Evans SC. 2000. Consumption of black tea elicits an increase in plasma antioxidant potential in humans. Int J Food Sci Nutr 51(5):309–315.

Lee KW, Kim YJ, Lee HJ and Lee CY. 2003. Cocoa has more phenolic phytochemicals and a higher antioxidant capacity than teas and red wine. J Agric Food Chem 51:7292–7295.

Leenen R, Roodenburg AJ, Tijburg LB and Wiseman SA. 2000. A single dose of tea with or without milk increases plasma antioxidant activity in humans. Eur J Clin Nutr 54:87–92.

Letenneur L, Proust-Lima C, Le Gouge A, Dartigues JF and Barberger-Gateau P. 2007. Flavonoid intake and cognitive decline over a 10-year period. Am J Epidemiol 165(12):1364–1371.

Li ZG, Shimada Y, Sato F, Maeda M, Itami A, Kaganoi J, Komoto I, Kawabe A and Imamura M. 2002. Inhibitory effects of epigallocatechin-3-gallate on *N*-nitrosomethylbenzylamine-induced esophageal tumorigenesis in F344 rats. Int J Oncol 21(6):1275–1283.

Lin J, Rexrode KM, Hu F, Albert CM, Chae CU, Rimm EB, Stampfer MJ and Manson JE. 2007. Dietary intakes of flavonols and flavones and coronary heart disease in US women. Am J Epidemiol 165:1305–1313.

Liu J, Ho Sc, Su Y, Chen W, Zhang C and Chen Y. 2009. Effect of long-term intervention of soy isoflavones on bone mineral density in women: A meta-analysis of randomized controlled trials. Bone Doi:10.1016/j.bone.2008.12.020.

Liu RH. 2003. Health benefits of fruit and vegetables are from additive and synergistic combinations of phytochemicals. Am J Clin Nutr 78(3 Suppl):517S–520S.

Lotito SB and Frei B. 2004. Relevance of apple polyphenols as antioxidants in human plasma: contrasting *in vitro* and *in vivo* effects. Free Radic Biol Med 36(2):201–211.

Lotito SB and Frei B. 2006. Consumption of flavonoid-rich foods and increased plasma antioxidant capacity in humans: cause, consequence, or epiphenomenon? Free Radic Biol Med 41(12):1727–1746.

Ma DF, Qin LQ, Wang PY and Katoh R. 2008. Soy isoflavone intake increases bone mineral density in the spine of menopausal women: meta-analysis of randomized controlled trials. Clin Nutr 27(1):57–64.

Manach C, Scalbert A, Morand C, Rémésy C and Jimenez L. 2004. Polyphenols: food sources and bioavailability. Am J Clin Nutr 79:727–747.

Manach C, Williamson G, Morand C, Scalbert A and Rémésy C. 2005. Bioavailability and bioefficacy of polyphenols in humans. I. Review of 97 bioavailability studies. Am J Clin Nutr 81(1 Suppl):230S–242S.

Martin S, Favot L, Matz R, Lugnier C and Andriantsitohaina R. 2003. Delphinidin inhibits endothelial cell proliferation and cell cycle progression through a transient activation of ERK-1/-2. Biochem Pharmacol 65(4):669–675.

Martinez-Flores A, Gónzalez-Gallego J, Culebras JM and Tuñón MJ. 2002. Nutr Hosp 17:271–278.

McCullough ML, Robertson AS, Chao A, Jacobs EJ, Stampfer MJ, Jacobs DR, Diver WR, Calle EE and Thun MJ. 2003. A prospective study of whole grains, fruits, vegetables and colon cancer risk. Cancer Causes Control 14(10):959–970.

Mojzisová G and Kuchta M. 2001. Dietary flavonoids and risk of coronary heart disease. Physiol Res 50:529 535.

Nijveldt RJ, van Nood E, van Hoorn DE, Boelens PG, van Norren K and van Leeuwen PA. 2001. Flavonoids: a review of probable mechanisms of action and potential applications. Am J Clin Nutr 74(4):418–425.

Ohno T, Kato N, Ishii C, Shimizu M, Ito Y, Tomono S and Kawazu S. 1993. Genistein augments cyclic adenosine 3'5'-monophosphate(cAMP) accumulation and insulin release in MIN6 cells. Endocr Res 19(4):273–285.

O'Leary KA, de Pascual-Tereasa S, Needs PW, Bao YP, O'Brien NM and Williamson G. 2004. Effect of flavonoids and vitamin E on cyclooxygenase-2 (COX-2) transcription. Mutat Res 551(1–2):245–254.

Park Y, Subar AF, Kipnis V, Thompson FE, Mouw T, Hollenbeck A, Leitzmann MF and Schatzkin A. 2007. Fruit and vegetable intakes and risk of colorectal cancer in the NIH–AARP Diet and Health Study. Am J Epidemiol 166(2):170–180.

Pascarella L. 2007. Essentials of Daflon 500 mg: from early valve protection to long-term benefits in the management of chronic venous disease. Curr Pharm Des 13:431–444.

Persaud SJ, Harris TE, Burns CJ and Jones PM. 1999. Tyrosine kinases play a permissive role in glucose-induced insulin secretion from adult rat islets. J Mol Endocrinol 22(1):19–28.

Peters U, Poole C and Arab L. 2001. Does tea affect cardiovascular disease? A metaanalysis. Am J Epidemiol 154:495–503.

Potenza MA, Marasciulo FL, Tarquinio M, Tiravanti E, Colantuono G, Federici A, Kim JA, Quon MJ and Montagnani M. 2007. EGCG, a green tea polyphenol, improves endothelial function and insulin sensitivity, reduces blood pressure, and protects against myocardial I/R injury in SHR. Am J Physiol Endocrinol Metab 292(5):E1378–E1387.

Ramos S. 2007. Effects of dietary flavonoids on apoptotic pathways related to cancer chemoprevention. J Nutr Biochem 18(7):427–442.

Rein D, Lotito S, Holt R, Keen C, Schmitz H and Fraga C. 2000. Epicatechin in human plasma: *in vivo* determination and effect of chocolate consumption on plasma oxidation status. J Nutr 130:2109S–2114S.

Rimm EB, Katan MB, Ascherio A, Stampfer MJ and Willett WC. 1996. Relation between intake of flavonoids and risk for coronary heart disease in male health professionals. Ann Intern Med 125:384–389.

Rossi M, Garavello W, Talamini R, La Vecchia C, Franceschi S, Lagiou P, Zambon P, Dal Maso L, Bosetti C and Negri E. 2007. Flavonoids and risk of squamous cell esophageal cancer. International Journal of Cancer 120(7):1560–1564.

Roura E, Andrés-Lacueva C, Estruch R, Mata-Bilbao ML, Izquierdo-Pulido M, Waterhouse AL and Lamuela-Raventós RM. 2007. Milk does not affect the bioavailability of cocoa powder flavonoid in healthy human. Ann Nutr Metab 51(6):493–498.

Sah JF, Balasubramanian S, Eckert RL and Rorke EA. 2004. Epigallocatechin-3-gallate inhibits epidermal growth factor receptor signaling pathway. Evidence for direct inhibition of ERK1/2 and AKT kinases. J Biol Chem 279(13):12755–12762.

Sakata K, Hirose Y, Qiao Z, Tanaka T and Mori H. 2003. Inhibition of inducible isoforms of cyclooxygenase and nitric oxide synthase by flavonoid hesperidin in mouse macrophage cell line. Cancer Lett 199(2):139–145.

Sarria A. 2004. Flavonoides: compuestos bioactivos de los alimentos. Boletín Sociedad de Pediatría de Aragón, La Rioja Y Soria 34(3):88–92.

Sato M, Bagchi D, Tosaki A and Das DK. 2001. Grape seed proanthocyanidin reduces cardiomyocyte apoptosis by inhibiting ischemia/reperfusion-induced activation of JNK-1 and C-JUN. Free Radic Biol Med 31(6):729–737.

Satue-Gracia MT, Heinonen M and Frankel EN. 1997. Anthocyanins as antioxidants on human low-density lipoprotein and lecithin-liposome systems. J Agric Food Chem 45:3362–3367.

Schroeter H, Holt RR, Orozco TJ, Schmitz HH and Keen CL. 2003. Nutrition: milk and absorption of dietary flavanols. Nature 426(6968):787–788.

Schwartz JL, Baker V, Larios E and Chung FL. 2005. Molecular and cellular effects of green tea on oral cells of smokers: a pilot study. Mol Nutr Food Res 49(1):43–51.

Serafini M, Bugianesi R, Maiani G, Valtuena S, De Santis S and Crozier A. 2003. Plasma antioxidants from chocolate. Nature 424(6952):1013.

Sorenson RL, Brelje TC and Roth C. 1994. Effect of tyrosine kinase inhibitors on islets of Langerhans: evidence for tyrosine kinases in the regulation of insulin secretion. Endocrinology 134(4):1975–1978.

Steele VE, Hawk ET, Viner JL and Lubet RA. 2003. Mechanisms and applications of nonsteroidal anti-inflammatory drugs in the chemoprevention of cancer. Mutat Res 523-524:137–144.

Stein JH, Keevil JG, Wiebe DA, Aeschlimann S and Folts JD. 1999. Purple grape juice improves endothelial function and reduces the susceptibility of LDL cholesterol to oxidation in patients with coronary artery disease. Circulation 100:1050–1055.

Sung H, Nah J, Chun S, Park H, Yang SE and Min WK. 2000. *In vivo* antioxidant effect of green tea. Eur J Clin Nutr 54(7):527–529.

Teissedre PL, Frankel EN, Waterhouse AL, Peleg H and German JB. 1996. Inhibition of *in vitro* human LDL oxidation by phenolic antioxidants from grapes and wines. J Sci Food Agric 70:55–61.

Thomasset SC, Berry DP, Garcea G, Marczylo T, Steward WP and Gescher AJ. 2006. Dietary polyphenolic phytochemicals—promising cancer chemopreventive agents in humans? A review of their clinical properties. Int J Cancer 120(3):451–458.
Valenzuela A, Arteaga A and Rozowski J. 2007. Rol de la dieta mediterránea en la prevalencia del syndrome metabólico. Revista Chilena de Nutrición 34(3): online version.
Valko M, Leibfritz D, Moncol J, Cronin MTD, Mazur M and Telser J. 2007. Free radicals and antioxidants in normal physiological functions and human disease. Int J Biochem Cell Biol 39:44–84.
van der Schouw YT, Kreijkamp-Kaspers S, Peeters PH, Keinan-Boker L, Rimm EB and Grobbee DE. 2005. Prospective study on usual dietary phytoestrogen intake and cardiovascular disease risk in Western women. Circulation 111(4):465–471.
van Dokkum W, Frølich W, Saltmarsh M, Gee J. 2008. The health effects of bioactive plant components in food: results and opinions of the EU COST 926 action. Nutr Bull 33(2):133–139.
Venables MC, Hulston CJ, Cox HR and Jeukendrup AE. 2008. Green tea extract ingestion, fat oxidation, and glucose tolerance in healthy humans. Am J Clin Nutr 87(3):778–784.
Visioli F, Borsani L and Galli C. 2000. Diet and prevention of coronary heart disease: the potential role of phytochemicals. Cardiovasc Res 47:419–425.
Voorrips LE, Goldbohm RA, van Poppel G, Sturmans F, Hermus RJJ and van den Brandt PA. 2000. Vegetable and fruit consumption and risks of colon and rectal cancer in a prospective cohort study. Am J Epidemiol 152:1081–1092.
Walle UK and Walle T. 2002. Induction of human UDP-glucuronosyltransferase UGT1A1 by flavonoids-structural requirements. Drug Metab Dispos 30(5):564–569.
Wang J, Schramm D and Holt R. 2000. A dose-response effect from chocolate consumption on plasma epicatechin and oxidative damage. J Nutr 130:2115S–2119S.
Wang W, VanAlstyne PC, Irons KA, Chen S, Stewart JW and Birt DF. 2004. Individual and interactive effects of apigenin analogs on G2/M cell-cycle arrest in human colon carcinoma cell lines. Nutr Cancer 48(1):106–114.
Wang-Polagruto JF, Villablanca AC, Polagruto JA, Lee L, Holt RR, Schrader HR, Ensunsa JL, Steinberg FM, Schmitz HH and Keen CL. 2006. Chronic consumption of flavanol-rich cocoa improves endothelial function and decreases vascular cell adhesion molecule in hypercholesterolemic postmenopausal women. J Cardiovasc Pharmacol 47 Suppl 2: S177–186; discussion S206–209.
Wu LY, Juan CC, Ho LT, Hsu YP and Hwang LS. 2004. Effect of green tea supplementation on insulin sensitivity in Sprague-Dawley rats. J Agric Food Chem 52(3):643–648.
Yamagishi M, Natsume M, Osakabe N, and others. 2002. Effects of cacao liquor proanthocyanidins on PhIP-induced mutagenesis *in vitro*, and *in vivo* mammary and pancreatic tumorigenesis in female Sprague-Dawley rats. Cancer Lett 185(2):123–130.
Yamane T, Nakatani H, Kikuoka N, and others. 1996. Inhibitory effects and toxicity of green tea polyphenols for gastrointestinal carcinogenesis. Cancer 77(8 Suppl):1662–1667.
Yang CS, Landau JM, Huang MT and Newmark HL. 2001. Inhibition of carcinogenesis by dietary polyphenolic compounds. Annu Rev Nutr 21:381–406. Review.
Yang CS, Yang GY, Landau JM, Kim S and Liao J. 1998. Tea and tea polyphenols inhibit cell hyperproliferation, lung tumorigenesis, and tumor progression. Exp Lung Res 24(4):629–639.
Yochum L, Kushi LH, Meyer K and Folsom AR. 1999. Dietary flavonoid intake and risk of cardiovascular disease in postmenopausal women. Am J Epidemiol 149:943–949.
Yu C, Chen Y, Cline GW, Zhang D, Zong H, Wang Y, Bergeron R, Kim JK, Cushman SW, Cooney GJ, Atcheson B, White MF, Kraegen EW and Shulman GI. 2002. Mechanism by which fatty acids inhibit insulin activation of insulin receptor substrate-1 (IRS-1)-associated phosphatidylinositol 3-kinase activity in muscle. J Biol Chem 27;277(52):50230–50236.

7 Chemistry, Stability, and Biological Actions of Carotenoids

Elhadi M. Yahia and José de Jesús Ornelas-Paz

Introduction

Carotenoids were discovered during the nineteenth century. Wachen in 1831 proposed the term "carotene" for the hydrocarbon pigment crystallized from carrot roots; Berzelius called the more polar yellow pigments extracted from autumn leaves "xanthophylls"; and Tswett separated many pigments by column chromatography and called the whole group "carotenoids."

Carotenoids are organic pigments naturally occurring in the chromoplasts of plants and some other photosynthetic organisms such as algae, and in some types of fungi and bacteria, where they have diverse and important functions and actions. There are more than 600 known carotenoids, commonly divided into two classes; carotenes (hydrocarbon carotenoids) and xanthophylls (oxygenated carotenoids). Many of these (more than 100) have been reported in fruits and vegetables. Carotenoids can be found in plants along with chlorophylls in leaves and green fruits, and they are widespread in many other parts of the plant such as yellow, orange, and red flowers, fruits, roots (such as carrots), and seeds (such as maize). In some green plants and plant parts, generally the darker the green color, the higher the carotenoid content. For example, carotenoid content in pale green cabbage is less than 1% of that in dark green cabbage. In nongreen tissues, carotenoids are localized in chromoplasts. Fruit carotenoids are very diverse, and those present in ripe fruits can be different from those present in unripe fruits. Although carotenoids are thought of as plant pigments, they also occur extensively in animals and microorganisms.

Fruit and vegetable carotenoids have been classified in different manners such as the scheme shown in Table 7.1 (Goodwin and Britton 1988). Some of the most common carotenes and xanthophylls are shown in Figures 7.1 and 7.2. In plants, carotenoids absorb light energy (light harvesting), serving as accessory pigments, and pass it to chlorophylls to be used for photosynthesis, protect chlorophylls from photodamage, and also act as attractants of insects for pollination. The major light-harvesting carotenoids are lutein, violaxanthin, and neoxanthin. In foods (including fruits and vegetables), carotenoids are important quality parameters, because they impart attractive colors such as yellow, orange, and red. In humans, some carotenoids, such β-carotene, are precursors to vitamin A, and several are potent antioxidants.

Consumer acceptance of fresh and processed fruits and vegetables is influenced by product appearance, flavor, aroma, and textural properties. Color is a key component that influences a consumer's initial perception of fruit and vegetable quality. Lycopene is the principal carotene in tomato fruit that imparts color. Analytical and sensory

Table 7.1. Classification of fruit carotenoids (Goodwin and Britton 1988)

Group	Pigment Characteristics
I	Insignificant amounts
II	Small amounts, generally of chloroplast carotenoids
III	Relatively large amounts of lycopene and its hydroxyl derivatives
IV	Relatively large amounts of β,β-carotene and hydroxyl derivatives
V	Large amounts of epoxides, particularly furanoid epoxides
VI	Unusual carotenoids such as capsanthin
VII	Poly-Z carotenoids such as prolycopene
VIII	Apocarotenoids such as β-citraurin

analyses of fruit quality constituents were conducted to assess real and perceived differences in fruit quality between orange-pigmented, high-β-carotene cherry tomato genotypes and conventional lycopene-rich, red-pigmented cherry tomato cultivars (Stommel and others 2005). Thirteen sensory attributes were evaluated by untrained consumers under red-masking light conditions where differences in fruit color could not be discerned. Panelists preferred the appearance of the red-pigmented cultivars when viewed under white light, but scored many of the other fruit-quality attributes of red- and orange-pigmented genotypes similarly whether they could discern the color or not. The results highlight the influence that color has on the perception of fruit quality.

Carotenoids are implicated as part of the dietary bioactive compounds providing protection against a number of degenerative conditions including cardiovascular diseases, cancer, immunity, and macular degeneration (Mayne 1996; Olson 1999). Eight possible mechanisms were proposed for the health-promoting effects of carotenoids: quenching of singlet oxygen, scavenging of peroxyl radicals, modulation of carcinogen metabolism, inhibition of cell proliferation, enhancement of cell differentiation via retinoids, stimulation of cell-to-cell communication, enhancement of the immune response, and filtering of blue light.

Chemistry

Structure, Nomenclature, and Classification

More than 600 different carotenoids from natural sources have been isolated and characterized. Physical properties and natural functions and actions of carotenoids are determined by their chemical properties, and these properties are defined by their molecular structures. Carotenoids consist of 40 carbon atoms (tetraterpenes) with conjugated double bonds. They consist of eight isoprenoid units joined in such a manner that the arrangement of isoprenoid units is reversed at the center of the molecule so that the two central methyl groups are in a 1,6-position and the remaining nonterminal methyl groups are in a 1,5-position relationship. They can be acyclic or cyclic (mono- or bi-, alicyclic or aryl). Whereas green leaves contain unesterified hydroxy carotenoids, most carotenoids in ripe fruit are esterified with fatty acids. However, those of a few

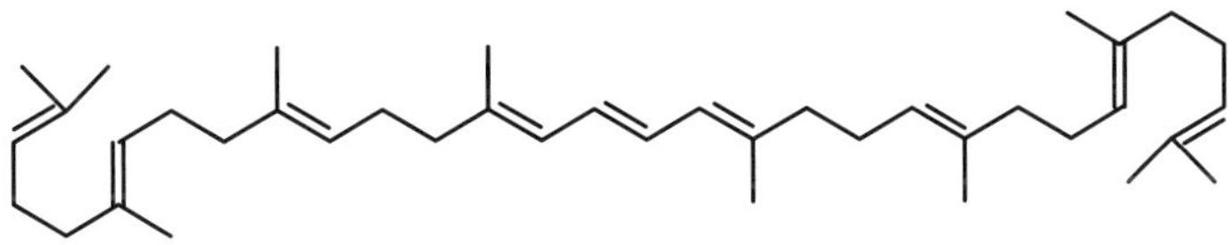

Phytoene (colorless)

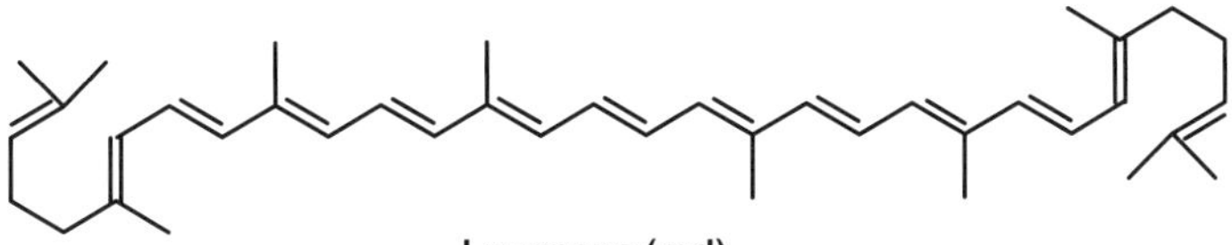

Lycopene (red)

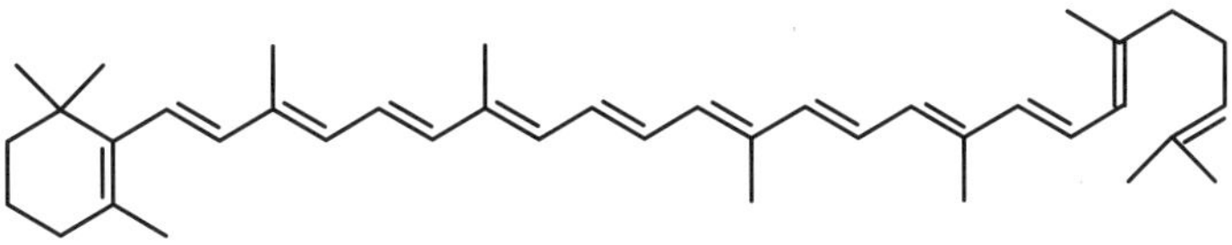

γ-Carotene (orange)

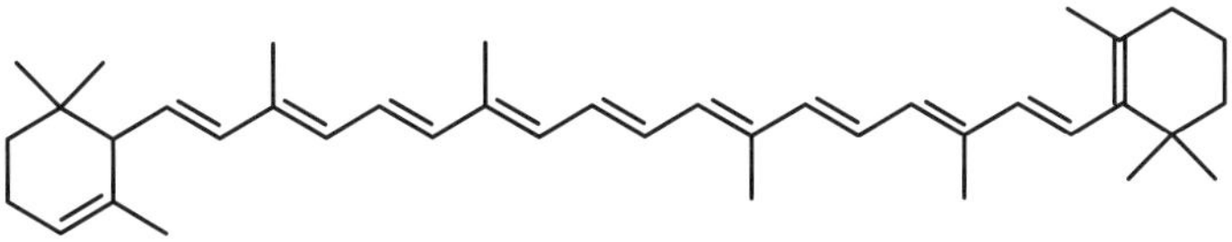

α-Carotene (orange)

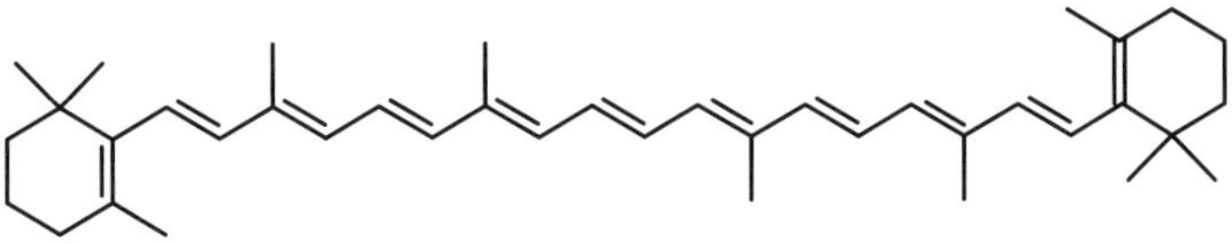

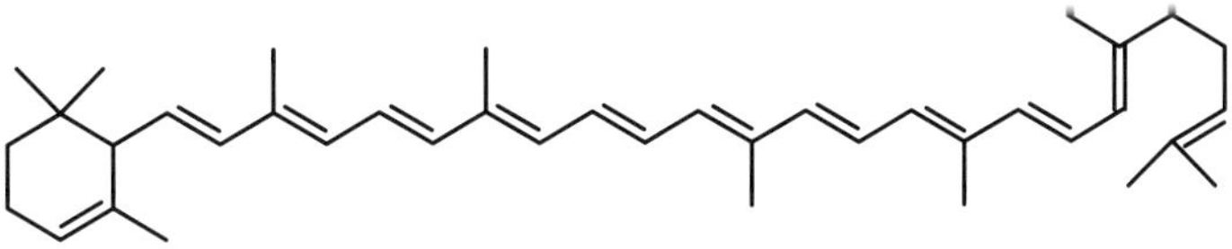

β-Carotene (orange)

δ-Carotene

Figure 7.1. Some examples of carotenes.

fruits, particularly those fruits that remain green when ripe, such as kiwi fruit, undergo limited or no esterification.

Cyclization and other modifications, such as hydrogenation, dehydrogenation, double-bond migration, chain shortening or extension, rearrangement, isomerization, introduction of oxygen functions, or combinations of these processes, result in a myriad of structures. A distinctive characteristic is an extensive conjugated double-bond system, which serves as the light-absorbing chromophore responsible for the yellow, orange, or red color that these compounds impart to many types of foods. As mentioned

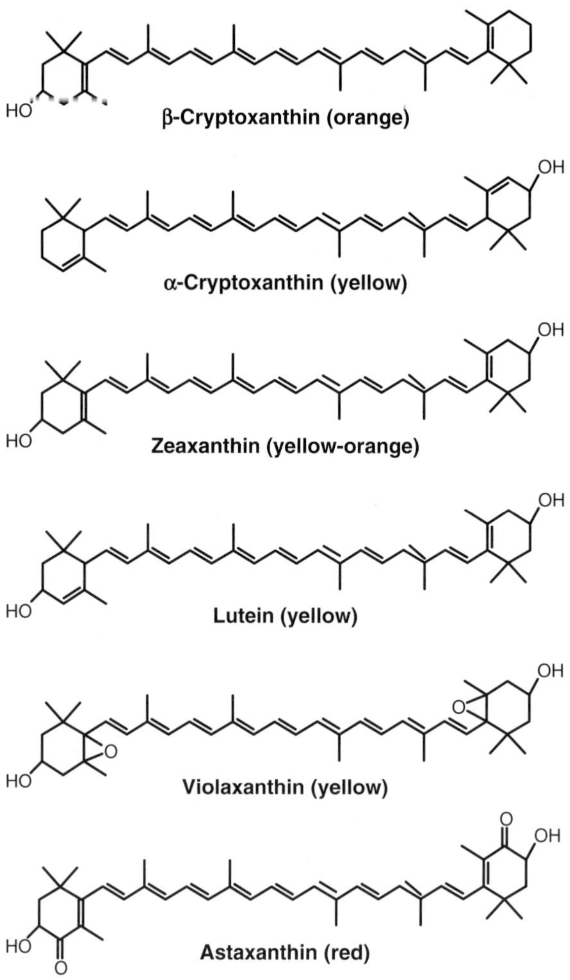

β-Cryptoxanthin (orange)

α-Cryptoxanthin (yellow)

Zeaxanthin (yellow-orange)

Lutein (yellow)

Violaxanthin (yellow)

Astaxanthin (red)

Figure 7.2. Structures of some common food xanthophylls.

before, hydrocarbon carotenoids are collectively called carotenes; those containing oxygen are termed xanthophylls. Carotenoids exist primarily in the more stable all-*trans* isomeric form, but small amounts of *cis* isomers are also present naturally or are transformed from the all-*trans* forms during processing.

Traditionally, carotenoids have been given trivial names derived usually from the biological source from which they are isolated, but a semisystematic scheme has been devised that allows carotenoids to be named unambiguously and in a way that defines and describes their structure (Table 7.2). Specific names are based on the stem name "carotene" preceded by the Greek-letter prefixes that designate the two end groups. For example, β-carotene is correctly referred to as β, β-carotene, and α-carotene as β, ε-carotene.

Table 7.2. Trivial and semisystematic names of common food carotenoids (Rodriguez-Amaya 2001)

Trivial Name	Semisystematic Name
Anteraxanthin	5,6-epoxy-5,6-dihydro-β,β-carotene-3,3′-diol
Astaxanthin	3,3′-dihydroxy-β,β-carotene-4,4′-dione
Auroroxanthin	5,8,5′,8′-diepoxy-5,8,5′,8′-tetrahydro-β,β-carotene-3,3′-diol
Bixin	methyl hydrogen 9′-*cis*-6,6′-diapocarotene-6,6′-dioate
Canthaxanthin	β,β-carotene-4,4′-dione
Capsanthin	3,3′-dihydroxy-β,κ-carotene-6′-one
Capsorubin	3,3′-dihydroxy- κ, κ-carotene-6,6′-dione
α-Carotene	β,ε-carotene
β-Carotene	β,β-carotene
β-Carotene-5,6-epoxide	5,6-epoxy-5,6-dihydro-β,β-carotene
β-Carotene-5,8-epoxide (mutatochrome)	5,8-epoxy-5,8-dihydro-β,β-carotene
β-Carotene-5,6,5′,6′-diepoxide	5,6,5′,6′-diepoxy-5,6,5′,6′-tetrahydro-β,β-carotene
δ-Carotene	ε,ψ-carotene
γ-Carotene	β,ψ-carotene
ζ-Carotene	7,8,7′,8′-tetrahydro-ψ,ψ-carotene
Crocetin	8,8′-diapocarotene-8,8′- dioic acid
α-Cryptoxanthin	β,ε-carotene-3′-ol
β-Cryptoxanthin	β,β-carotene-3-ol
Echineone	β,β-carotene-4-one
Lutein	β,ε-carotene-3,3′-diol
Luteine-5,6-epoxide (taraxanthin)	5,6-epoxy-5,6-dihydro-β,ε-carotene-3,3′-diol
Lycopene	ψ,ψ-carotene
Neoxanthin	5′,6′-epoxy-6,7-didehydro-5,6,5′,6′-tetrahydro-β,β-carotene-3,5,3′-triol
Neurosporene	7,8-dihydro-ψ,ψ-carotene
Phytoene	7,8,11,12,7′,8′,11′12′-octahydro-ψ,ψ-carotene
Phytofluene	7,8,11,12,7′,8′-hexahydro-ψ,ψ-carotene
Rubixanthin	β,ψ-carotene-3-ol
Violaxanthin	5,6,5′,6′-diepoxy-5,6,5′,6′-tetrahydro-β,β-carotene-3,3′-diol
α-Zeacarotene	7′,8′-dihydro-ε,ψ-carotene
β-Zeacarotene	7′,8′-dihydro-β,ψ-carotene
Zeaxanthin	β,β-carotene-3,3′-diol
Zeinoxanthin	β,ε-carotene-3-ol

Some carotenoids have structures containing fewer than 40 carbon atoms and derived formally by loss of part of the C_{40} skeleton. These compounds are referred to as apocarotenoids when carbon atoms have been lost from the ends of the molecule or as norcarotenoids when carbon atoms have been lost formally from within the chain. These modifications are caused by oxidative degradation at the level of the terminal rings either by nonspecific mechanisms (lipoxygenase, photo-oxidation) or by specific mechanisms (dioxygenases). These oxygenated carotenoids have significant roles in developmental and environmental response signaling. They also make important contributions to flavor and nutritional quality of several types of foods such as fruits, tea, wine, and tobacco. Two well-known natural apocarotenoids, bixin and crocetin, have economic importance as pigments and aroma in foods. Apocarotenoids also act as visual or volatile signals to attract pollinating agents and are also important in plant defense mechanisms. Studies have shown that the loss of carotenoid cleavage enzymes induces the development of axillary branches, indicating that apocarotenoids convey signals that regulate plant architecture (Bouvier and others 2005). Another apocarotenal, *trans*-β-apo-8′-carotenal, is found in spinach and citrus fruits; it has a low provitamin A activity, is used in pharmaceuticals and cosmetic products, and is also used as an additive (E160e) legalized by the European Commission for Human Food. Otherwise, a wide range of apocarotenals are produced by oxidative reactions during food processing and are intermediates in the formation of even smaller molecules of significance in food color and flavor.

Chemical and Physical Properties

A striking characteristic of the carotenoid structure is the long system of alternating double and single bonds that forms the central part of the molecule. This constitutes a conjugated system in which the π-electrons are effectively delocalized over the entire length of the polyene chain. This feature gives the carotenoids their distinctive molecular shape, chemical reactivity, and light properties.

With very few exceptions, carotenoids are lipophilic, having little or no solubility in water and good solubility in organic solvents. Therefore, they are expected to be restricted to the hydrophobic areas of the cell, such as the inner core of the membranes, except when association with protein allows them access to an aqueous environment. Polar functional groups alter the polarity of carotenoids and affect their interactions with other molecules. The overall size and shape of the molecule are important in relation to properties and function of the carotenoids. All colored carotenoids in the all-*trans* configuration have an extended conjugated double-bond system and are linear, rigid molecules. The overall shape of the *cis*-isomers differs substantially from that of the all-*trans* forms, and therefore their ability to fit into subcellular structures may be greatly altered. The tendency of the *cis*-isomers to crystallize or aggregate is usually much less, making them more readily solubilized, absorbed, and transported than their all-*trans* counterparts. The shape and size of the end groups are also important. Acyclic carotenoids such as lycopene are essentially long, linear molecules with flexible end groups. Cyclization shortens the overall length of the molecule and increases the effective bulk of the end groups and the space they occupy.

The conjugated double-bond system constitutes the light-absorbing chromophore that gives the carotenoids their attractive color and provides the visible absorption spectrum that serves as a basis for their identification and quantification. Most carotenoids absorb maximally at three wavelengths, resulting in three-peak spectra. At least seven conjugated double bonds are needed for a carotenoid to have perceptible color. The greater the number of conjugated double bonds, the higher the λ_{max} values. Thus, the acyclic carotenoid lycopene, with 11 conjugated double bonds, is red and absorbs at the longest wavelengths (λ_{max} at 444, 470, and 502 nm). Being also acyclic, the spectrum of the ζ-carotene has three well-defined peaks, but these are at wavelengths much lower than those of lycopene (λ_{max} at 378, 400, and 425 nm). Hydroxyl groups exert a great influence on absorption, whereas methylation, acetylation, and silylation markedly reduce this effect. Lutein has its maximum absorption at 450 nm, cryptoxanthin at 453 nm, and zeaxanthin at 454 nm.

Being highly unsaturated, carotenoids are prone to isomerization and oxidation. Heat, light, acids, and adsorption on an active surface such as alumina promote isomerization of all-*trans* carotenoids, their usual configuration, to the *cis* forms. This results in some losses in color and provitamin A activity. Oxidative degradation, the principal cause of extensive losses of carotenoids, depends on the availability of oxygen and is stimulated by light, enzymes, metals, and co-oxidation by lipid hydroperoxides. Carotenoids appear to have different susceptibilities to oxidation; lutein and violaxanthin are thought to be more labile than others. Formation of epoxides and apocarotenoids, that is, carotenoids with shortened carbon skeletons, appears to be the initial step. Subsequent fragmentations yield a series of low-molecular-weight compounds similar to those produced in fatty acid oxidation. Thus, total loss of color and biological activities is the final consequence. Therefore, retention of naturally occurring or added carotenoids in prepared, processed, and stored foods is an important consideration. Carotenoids are also subject to isomerization and oxidation during analysis, and preventative measures must be taken to guarantee the reliability of analytic results.

Biosynthesis

Carotenoid biosynthesis occurs within the chromoplast through the isoprenoid pathway. The carotenoid biosynthetic pathway starts with a condensation of pyruvate and glyceraldehyde 3-phosphate, catalyzed by 1-deoxy-D-xylulose 5-phosphate, which in turn is converted to 2-*C*-methyl-D-erithritol 4-phosphate by the 1-deoxy-D-xylulose-5-phosphate reductoisomerase (Fig. 7.3). This compound is later metabolized to isopentenyl diphosphate through a series of unknown reactions, which may involve stages of reduction, dehydration, and phosphorylation (Lichtenthaler 1999). In the early stages of carotenoid biosynthesis, the C5 primer for chain elongation undergoes successive additions of C_5 units, yielding in sequence C_{10}, C_{15}, and C_{20} compounds. The isoprenoid chain is built up by prenyl transferases to the C_{20} level forming geranylgeranyl diphosphate (Ggps), and two molecules of this are joined tail-to-tail to form phytoene through an intermediate (prephytoene diphosphate), as the first C_{40} carotenoid skeleton from which all the individual variations are derived.

Through the action of several specific enzymes (phytoene desaturase and ζ-carotene desaturase), phytoene is sequentially desaturated to ζ-carotene with seven conjugated double bonds, and lycopene with 11 conjugated double bounds (Cunningham 2002). The latter is the most common substrate for cyclases β and ε, which transform the ψ groups of lycopene into β or ε ionone rings (Britton 1988). The action of cyclase β on both ends of the lycopene moiety leads to the formation of β-carotene. Carotenes are further converted into a diversity of xanthophylls through the action of specific hydroxylases and epoxidases (van den Berg and others 2000).

Phytoene is colorless but undergoes a series of reactions, each of which creates a new double bond and extends the chromophore by two conjugated double bonds. The

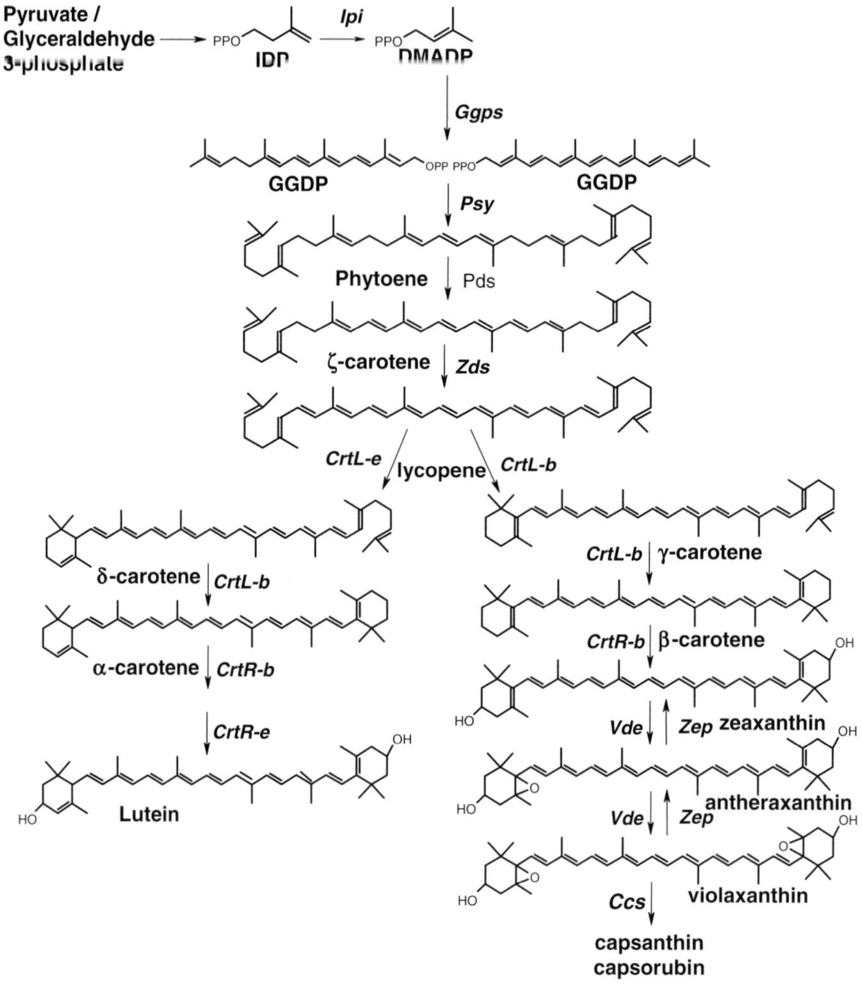

Figure 7.3. The carotenoid biosynthetic pathway. Enzymes are named according to the designation of their genes: Ccs, capsanthin–capsorubin synthase; CrtL-b, lycopene-b-cyclase; CrtL-e, lycopene-e-cyclase; CrtR-b, b-ring hydroxylase, CrtR-e, e-ring hydroxylase; DMADP, dimethylallyl diphosphate; GGDP, geranylgeranyl diphosphate; Ggps, geranylgeranyl-diphosphate synthase; IDP, isopentenyl diphosphate; Ipi, IDP isomerase; Pds, phytoene desaturase; Psy, phytoene synthase; Vde, violaxanthin de-epoxidase; Zds, z-carotene desaturase; Zep, zeaxanthin epoxidase. (From van den Berg and others 2000.)

sequential introduction of double bonds at alternate sides of phytoene (three conjugated double bonds) gives rise to phytofluene. The end product is lycopene produced via the intermediates phytoene, ζ-carotene, and neurosporene. Carotenoids can be considered derivatives of lycopene (found in tomatoes and few other fruits). Its long straight chain is highly unsaturated and composed of two identical units joined by a double bond between carbons 15 and 15′. The lycopene molecule may undergo cyclization to form six-membered rings at one end of the molecule (β-rings). Cyclization reduces λ_{max} and therefore alters the color. For example, lycopene with 11 conjugated double bonds absorbs maximally at 470–500 nm and is strongly colored red. In contrast, β,β-carotene, also with 11 conjugated double bonds, absorbs at 450 nm and is orange-yellow as a consequence of the cyclization. The introduction of oxygen molecules and other structural modifications of end groups such as esterification follow as the final stages of biosynthesis. For example, zeaxanthin and lutein are formed by the introduction of two hydroxy groups at C-3 and C-3′ of β,β-carotene and β,ε-carotene, respectively. The formation of zeaxanthin from lycopene involves two reactions, namely, β-cyclization and hydroxylation at each end group. Reactions are catalyzed by particular enzymes, and these are the product of expression of particular genes (Britton, 1993). The first two C_{40} carotenoids formed in the biosynthetic pathway have the 15-*cis* configuration in plants. The presence of small amounts of *cis* isomers of other carotenoids in natural sources has been increasingly reported.

In tomato (*Solanum lycopersicum*), phytoene synthase-1 (PSY-1) is the key biosynthetic enzyme responsible for the synthesis of fruit carotenoids. Fraser and others (2007) characterized the effect of constitutive expression of an additional tomato Psy-1 gene product. A quantitative data set defining levels of carotenoid/isoprenoid gene expression, enzyme activities, and metabolites was generated from fruit that showed the greatest perturbation in carotenoid content. Transcriptional upregulation, resulting in increased enzyme activities and metabolites, occurred only in the case of Psy-1, Psy-2, and lycopene cyclase B. For reactions involving 1-deoxy-D-xylulose-5-phosphate synthase, geranylgeranyl-diphosphate synthase, phytoene desaturase, X-carotene desaturase, carotene isomerase, and lycopene β-cyclase, there were no correlations among gene expression, enzyme activities, and metabolites. Perturbations in carotenoid composition were associated with changes in plastid type and with chromoplast-like structures arising prematurely during fruit development. The levels of more than 120 known metabolites were determined. Comparison with the wild type illustrated that key metabolites (sucrose, glucose, fructose) and sectors of intermediary metabolism (e.g., trichloroacetic acid cycle intermediates and fatty acids) in the Psy-1 transgenic mature green fruit resembled changes in metabolism associated with fruit ripening. General fruit developmental and ripening properties, such as ethylene production and fruit firmness, were unaffected, allowing Fraser and others (2007) to conclude that the changes in pigmentation, plastid type, and metabolism associated with Psy-1 overexpression do not appear to be connected with the ripening process.

The red color of tomato (*Lycopersicon esculentum*) fruit is provided by the carotenoid lycopene, whose concentration increases dramatically during the ripening process of this fruit (Carrillo-Lopez and Yahia 2009). A single dominant gene, Del, in the tomato mutant Delta changes the fruit color to orange as a result of accumulation of δ-carotene at the expense of lycopene. The cDNA for lycopene ε-cyclase (CRTL-E), which converts lycopene to δ-carotene, was cloned from tomato (Ronen and others 1999). The primary structure of CRTL-E is 71% identical to the homologous polypeptide from *Arabidopsis* and 36% identical to the tomato lycopene β-cyclase, CRTL-B. The CRTL-E gene was mapped to a single locus on chromosome 12 of the tomato linkage map. This locus co-segregated with the Del gene. In the wild-type tomato, the transcript level of CRTL-E decreases at the "breaker" stage of ripening to a nondetectable level in the ripe fruit. In contrast, it increases approximately 30-fold during fruit ripening in the Delta plants. The Delta mutation does not affect carotenoid composition nor the mRNA level of CRTL-E in leaves and flowers. The primary mechanism that controls lycopene accumulation in tomato fruit seem to be based on the differential regulation of expression of carotenoid biosynthetic genes. During fruit development, the mRNA levels for the lycopene-producing enzymes phytoene synthase (Psy) and phytoene desaturase (Pds) increase, whereas the mRNA levels of the genes for the lycopene β- and ε-cyclases, which convert lycopene to either β- or δ-carotene, respectively, decline and completely disappear.

Ripe tomato fruits accumulate significant amounts of lycopene, but only trace amounts of xanthophylls. Dharmapuri and others (2002) overexpressed the lycopene β-cyclase (b-Lcy) and β-carotene hydroxylase (b-Chy) genes under the control of the fruit-specific Pds promoter, and transgene and protein expression was followed through semiquantitative reverse- transcription PCR, Western blotting, and enzyme assays. Fruits of the transformants showed a significant increase of β-carotene, β-cryptoxanthin, and zeaxanthin; the carotenoid composition of leaves remained unaltered, and the transgenes and the phenotype were inherited in a dominant Mendelian fashion.

Because plants are able to synthesize carotenoids *de novo*, the carotenoid composition of plant foods is enriched by the presence of small or trace amounts of biosynthetic precursors, along with derivatives of the main components. Although commonly thought of as plant pigments, carotenoids are also encountered in some animal foods. Animals are incapable of carotenoid biosynthesis; thus their carotenoids need to be derived from the diet. Selectively or unselectively absorbed, carotenoids accumulate in animal tissues unchanged or slightly modified into typical animal carotenoids.

Carotenoids and their biosynthetic precursors can be used as biomarkers of fruit product quality and adulteration of one product with another, such as fraudulent mixing of apricot jams and spreads with pumpkin extracts (Kurz and others 2008), and for differentiation of various pumpkins and squashes (Azevedo-Meleiro and Rodriguez-Amaya 2007).

Analysis

Several analytical approaches and methods are used to analyze carotenoids in foods and biological tissues. A major problem in the analysis of food carotenoids is the sample preparation and extraction. As mentioned previously, the extractability of a carotenoid is very much dependent on its form and on matrix-binding effects. In addition, carotenoids are very sensitive molecules that undergo degradation under the effect of light, heat, and acidification. Preferably, samples are ground under nitrogen in the dark in the presence of a carbonate source, such as sodium carbonate or bicarbonate, and an antioxidant such as pyrogallol, butylated hydroxytoluene, or butylated hydroxyanisole (e.g., Kurz and others 2008). Afterwards, extraction can be performed in amber glass under a nitrogen atmosphere using different proportions of methanol and hexane. Alternatively, chloroform, dichloromethane, *tert*-butyl methyl ether, or diethyl ether may be used. Often, complete extraction of carotenoids is difficult to achieve. For difficult matrices, saponification may be required, but it must be performed under strict conditions and is always accompanied by some losses. Analysis of carotenoids can be performed spectrophotometrically, such as identifying β-carotene using its specific absorption coefficient at 450 nm, or by high performance liquid chromatography (HPLC). Separations are poorly performed on C18 reversed-phase HPLC columns, whereas adequate separations are performed on C30 reversed-phase columns, especially if *cis*-isomers have to be separated. Detection is performed using diode array or mass spectrometric detection (Dugo and others 2008).

Sources of Dietary Carotenoids

Sources in Fruits and Vegetables

About 70–90% of consumed carotenoids originate from fruits and vegetables. Khachik and others (1991) divided foods of plant origin in three groups based on their carotenoid distribution:

1. Green vegetables, such as broccoli, spinach, and green beans, that contain a high diversity of xanthophylls and carotenes
2. Yellow and red fruits and vegetables (plums, carrots, melons, and tomatoes), with a more distinctive carotenoid distribution than the first category (fruits and vegetables belonging to this category contain primarily carotenes)
3. Yellow or orange fruits, including pumpkins, oranges, and peaches, which primarily contain xanthophyll esters

Carotenoids from fruits and vegetables can exist as protein–carotenoid complexes (as in the case of green leaf vegetables), crystals (as in carrots or tomatoes), or in oil solution (as in mango and papaya) (West and Castenmiller 1998). Carotenoids commonly found in human blood are lutein, zeaxanthin, β-cryptoxanthin, lycopene, β-carotene, and α-carotene. The content of some carotenoids in some fruits and vegetables is shown in Table 7.3.

Leaves have a strikingly constant carotenoid pattern, often referred to as the chloroplast carotenoid pattern, the main carotenoids being lutein (about 45%), β-carotene (usually 25–30%), violaxanthin (15%), and neoxanthin (15%) (Britton 1991). α-Carotene, β-cryptoxanthin, α-cryptoxanthin, zeaxanthin, antheraxanthin, and lutein 5,6-epoxide are also reported as minor carotenoids. Lactucaxanthin is a major xanthophyll in a few species, such as lettuce. Other green vegetables, such as broccoli, follow the same pattern as green leafy vegetables. In contrast to leafy and other green vegetables, fruits, including those used as vegetables, are known for their complex and variable carotenoid composition. Some palm fruits are especially rich in carotenoids, particularly provitamin A carotenes. Carotenoid mono- and diesters were identified in mandarin essential oil (Dugo and others 2008), in various vegetables and fruits (Breithaupt and Bamedi 2001), in shrimps and *Haematococcus pluvialis* (Breithaupt 2004), and in the Antarctic krill *Euphausia superba* (Takaichi and others 2003).

Of the acyclic carotenes, lycopene and ζ-carotene are the most common. Lycopene is the principal pigment of some red-fleshed fruits and fruit vegetables, such as tomato, watermelon, red-fleshed papaya and guava, and red or pink grapefruit (see Table 7.3). ζ-Carotene is more ubiquitous, but it is usually present at low levels except in Brazilian passion fruit (Mercadante and others 1998) and in carambola (Gross and others

Table 7.3. The content of some carotenoids in some fruits and vegetables

	Carotenoids (μg/100g) fresh weight basis				
Produce	**Lutein**	**β-Cryptoxanthin**	**Lycopene**	**α-Carotene**	**β-Carotene**
Apples (with skin)	100–840			30	
Apricot	0–141	28–231	0.5	0–37	140–6939
Asparagus	610–750	–		12	493
Avocados		36		28	53
Banana[1]	0–37			0–157	0–92
Black currant	210				62
Blackberry	270			9	100
Blueberry	230				49
Broccoli	830–4300	0		1	414–2760
Carrot	110–2097			530–35833	1161–64350
Cherries, sweet	10				28
Cucumbers (with skin)	160				138
Grapefruit (red)		12		5	603
Guava	270		769–1816		102–2669
Jackfruit			37–111		40–772
Lettuce	73–4537				48–3120
Mango	100	0–1640			300–4200
Melon, cantaloupe		0		27	1,595
Nectarines	20	59		0	101
Orange flesh sweet potatoes (raw)					736–6600
Orange juice	67	34		8	13
Orange	64–350	14–1395		0–400	0–500
Papaya	20–820	60–1483	2080–4750	0–60	71–1210
Peach	9–120	12–510		0–9	30–1480
Pepper, bell, green	340–660	1		22	198
Pineapple					171–476
Raspberry	320			24	9
Red currant	28				13
Spinach	2047–20300				840–24070
Strawberry	6–21				5
Sweet potatoes, white flesh (cooked)					25–157
Sweet potatoes, white flesh (raw)					6–249
Tomato	40–1300		21–62273		36–2232
Watermelon	0–40	62–457	2300–7200	0–1	44–324
White flesh sweet potatoes (cooked)					25–157
White flesh sweet potatoes (raw)					6–249

Sources: van den Berg and others (2000); Ameny and Wilson (1997); Marín and others (2004); Setiawan and others (2001); Lee and Coates (2003); Marinova and Ribarova (2007); Calva (2005)

1983), in which it occurs as a major pigment. Phytoene and phytofluene are probably more widely distributed than reported; because they are both colorless and vitamin A–inactive, their presence may often be overlooked. Neurosporene has limited occurrence and is normally found in small amounts. The bicyclic β-carotene is the most widespread of all carotenoids in foods, either as a minor or as the major constituent. The bicyclic α-carotene and the monocyclic γ-carotene sometimes accompany β-carotene, generally at much lower concentrations. Substantial amounts of α-carotene are found in carrot and some varieties of squash and pumpkin (Arima and Rodriguez-Amaya 1988, 1990), and substantial amounts of γ-carotene are found in rose hips and *Eugenia uniflora* (Cavalcante and Rodriguez-Amaya 1992). Although less frequently encountered, δ-carotene is the principal carotenoid of the high delta strain of tomato and the peach palm fruit. The hydroxy derivatives of lycopene, lycoxanthin, and lycophyll, are rarely encountered; they are found in trace amounts in tomato.

The xanthophyll cryptoxanthin can be present as three isomers: β-cryptoxanthin, α-cryptoxanthin, and zeinoxanthin. α-Cryptoxanthin is rarely found in fruits, whereas zeinoxanthin is widely distributed in citrus fruits and maize, but generally at low levels (Schlatterer and Breithaupt 2005). β-Cryptoxanthin is the main pigment of many orange-fleshed fruits such as peach, nectarine, orange-fleshed papaya, persimmon, fruit of the tree tomato, and *Spondias lutea*, but occurs rarely as a secondary pigment. In contrast to the relative abundance of the parent carotenes, α- and β-carotene, respectively, lutein is normally present in plant tissues at considerably higher levels than zeaxanthin. Lutein is the predominant carotenoid in leaves, green vegetables, and yellow flowers (e.g., marigold, *Tagetes erecta*). Zeaxanthin is usually a minor food carotenoid, except for yellow corn, the Brazilian fruit *Caryocar villosum*, sweet orange-fleshed pepper, and *Lycium barbarum* (wolfberry), in which it is the major pigment. The epoxycarotenoid violaxanthin may be underestimated in foods because it is easily degraded (Mercadante and Rodriguez-Amaya 1998).

Some of the unique carotenoids include capsanthin and capsorubin, predominant in red pepper, bixin as the major pigment of the food colorant annatto; and crocetin as the main coloring component of saffron. Although green leaves contain unesterified hydroxy carotenoids, most carotenols in ripe fruits are esterified with fatty acids. However, the carotenols of a few fruits, particularly those that remain green when ripe, such as kiwi fruit (Gross 1982b), undergo limited or no esterification. A more detailed discussion of the carotenoids present in leafy vegetables, roots, and fruits is given next.

Leafy vegetables. Leafy vegetables are excellent sources of carotenoids, especially β-carotene and lutein, which represent about 80% of the total carotenoids, with very low content of α-carotene. In these commodities, β-carotene is practically the only source of vitamin A (Takyi 2001). Bhaskarachary and others (1995) reported a range of 1,400 to 19,700 μg/100 g β-carotene in 38 leafy vegetables; the lowest content was in lettuce (*Lactuca sativa*) and the highest value in leaves of *Moringa oleifera*. Some members of *Brassica oleracea* are reported to rank highest for reported levels of lutein and β-carotene. Twenty-three leafy *B. oleracea* cultigens were field grown under similar fertility over two separate years and evaluated for leaf lutein and β-carotene accumulation (Kopsell and others 2004). Choice of *B. oleracea* cultigen and year significantly affected carotenoid levels. Lutein concentrations ranged from a high of 13,430 μg/100 g fresh weight (FW) for *B. oleracea* var. Acephala Toscano to a low of

4,840 μg/100 g FW for *B. oleracea* var. Acephala 343-93G1. β-Carotene accumulation ranged from a high of 10,000 μg/100 g FW for *B. oleracea* var. Acephala Toscano to a low of 3,820 μg/100 g FW for *B. oleracea* var. Acephala 3343-93G1.

Roots. Some roots, such as carrots and sweet potatoes, are important sources of carotenoids. Carrots, from which carotenes derive their name, are a traditional carotenoid-rich food source. The level of α-carotene in carrots is high, but usually about half the content of β-carotene. β-Carotene was reported to be 6,500 μg/100 g by Müller (1997) and 8,840 μg/100 g by Murkovic and others (2000). It was reported to be 20 μg/100 g in white-fleshed sweet potato (Rodriguez-Amaya 1997) and from 1,870 to 21,800 in yellow-fleshed sweet potato (Bhaskarachary and others 1995; van Jaarsveld and others 2006). The carotenoid levels are heterogeneous in different yellow, orange, red, and purple carrot cultivars with the level of β-carotene increasing gradually from the periderm toward the core, the levels of α-carotene and lutein being higher than those of β-carotene in younger cells and with lycopene accumulating throughout the whole secondary phloem in red cultivars (Baranska and others 2006). Hagenimana and Low (2000) evaluated 17 cultivars of sweet potato of different colored flesh and reported values ranging from 100 to 8,000 μg/100 g β-carotene.

Fruits. Pumpkins are excellent sources of carotenoids and vitamin A. The content of β-carotene and α-carotene in different species (*Cucurbita pepo*, *C. maxima*, and *C. moschata*) was found to be 60–7,400 and 0–7,500 μg/100 g, respectively (Murkovic and others 2000). Several other fruits are important sources of carotenoids, although there is great diversity between them. The most important carotenoids with vitamin A activity found in fruits include β-carotene and β-cryptoxanthin. Tropical and subtropical fruits are thought to contain more carotenoids compared to temperate fruits, which contain more anthocyanins. Mango and papaya are among the tropical fruits rich in carotenoids (Pott and others 2003b; Yahia and others 2006; Corral-Aguayo and others 2008; Ornelas-Paz and others 2008a; Rivera-Pastarna and others 2009). β-Carotene in mango has been reported as 60 μg/100 g in Thailand, 2,900 μg/100 g in India, and 6,700 μg/100 g in "Keitt" and 5,800 μg/100 g in "Tommy Atkins" mangoes from Brazil (Mercadante and Rodriguez-Amaya 1998). β-Cryptoxanthin was reported as half the content of β-carotene in mango (Rodriguez-Amaya 1997) and 288 to 1,034 μg/100 g in papaya (Wall 2006). The content of carotenoids in banana is low, but the high amount of this fruit consumed makes it an important carotenoid source. Recently, a total of 37 varieties of fresh fruits obtained from six representative markets in Bangkok (Thailand) were screened for their β-carotene and lycopene contents (Charoensiri and others 2009). β-Carotene content ranged from undetectable up to 616 μg/100 g of edible portion, lycopene content from undetectable up to 6,693 μg/100 g of edible portion, and red watermelon, *Citrullus vulgaris* (variety "jin-trarah") was the richest source of both carotenoids. Several carotenoids have been identified and quantified in the fruit of *Lycium barbarum*, a traditional Chinese herb containing functional components such as carotenoids, flavonoids, and polysaccharides, widely used in the health food industry because of its possible role in the prevention of chronic diseases such as age-related macular degeneration (Weller and Breithaupt 2003). Recently, Inbaraj and others (2008) found that zeaxanthin dipalmitate (1,143.7 μg/g) was present in a large amount, followed by β-cryptoxanthin monopalmitate and its two isomers (32.9–68.5 μg/g), zeaxanthin monopalmitate and its two isomers (11.3–62.8 μg/g), all-*trans*-β-carotene (23.7 μg/g), and all-*trans*-zeaxanthin (1.4 μg/g).

Factors Influencing Carotenoid Content in Fruits and Vegetables

Many factors can affect carotenoid content in fruits and vegetables. Type of product, cultivar or variety, climate, geographic site of production, maturation and ripening stages, temperature of ripening, exposure to sunlight, harvesting and postharvest handling and technologies (such as storage temperature, controlled/modified atmospheres, heat treatments, etc), processing and agrotechnological conditions, and method of analysis are some of the many factors that can alter carotenoid concentration in fruits and vegetables (Gross 1987, 1991; Rodriguez-Amaya 1993; Wright and Kader 1997; Tran and Raymundo 1999; Solovchenko and others 2006; Yahia and others 2007). Type of product can affect type and quantities of carotenoids, but cultivar differences may be related mostly to quantitative differences, although differences in type of carotenoids can also be found.

Maturation and ripening. The main factor affecting carotenoid content in fruits and vegetables is the ripening process. The content of some carotenoids can increase from zero to high levels in a few days as a consequence of maturation and ripening (Carrillo-Lopez and Yahia 2009). This increase in carotenoid content is triggered by the metabolic activity of ethylene. Carotenoid content in mangoes increases steadily during ripening (Vázquez-Caicedo and others 2005). Ornelas-Paz and others (2008a) demonstrated that the content of the main carotene and xanthophyll esters in mango fruit exponentially increased during ripening. Similar findings have been reported in several other fruits and vegetables such as apricots (Dragovic-Uzelac and others 2007), acerola (Lima and others 2005), and tomatoes (Abushita and others 1997; Carrillo-Lopez and Yahia 2009). Immature pepper (*Capsicum* spp.) fruit generally contained lower levels of lutein and xeaxanthin than mature, colored fruit (Lee and others 2005a). In tomatoes, carotenoids, especially lycopene, significantly increase during maturation and ripening on or off the plant (Carrillo-Lopez and Yahia 2009), and the magnitude of carotenoid accumulation depends on various factors such as temperature and light intensity (Von Elbe and Shwartz 1996). However, there are exceptions. For example, Boudries and others (2007) found that carotenoid content in three date cultivars decreased during ripening and was highest in the "Khalal" stage and lowest in "Rutab" stage.

During ripening, chlorophylls are commonly degraded or disappear, chloroplasts are degraded or transformed into chromoplasts, and carotenoids are synthesized or appear. Green, unripe fruits commonly contain chloroplast carotenoids, and when the fruit ripens, chromoplasts develop and carotenoids are produced on a large scale, usually not the same ones as those found in the chloroplast. During ripening, many genes are "switched on," including those related to carotenoid biosynthesis or chlorophyll degradation (the latter ones cause the disappearance of chlorophylls and the appearance of carotenoids), thus leading to increased accumulation or increased appearance of carotenoids. Several enzymes have been detected such as a ripening-specific phytoene synthase in tomato (Fray and Grierson 1993). Molecular genetic manipulation has been employed in fruits such tomatoes, where transgenic tomato plants have been developed containing increased amounts of phytoene synthase through the introduction of multiple copies of the phytoene synthase gene, thus substantially increasing lycopene content (Fray and Grierson 1993). Several examples of increased carotenoids during maturation and ripening of several fruits and vegetables have been reported, such as in *Momordica charantia* (Rodriguez-Amaya and others 1976), yellow Lauffener gooseberry (Gross 1982b), red pepper (Rahman and Buckle 1980), Badami mango fruit (John and others 1970), and leafy vegetables (Hulshof and others 1997; Ramos and Rodriguez-Amaya 1987). In fruits in which the color at the ripe stage is due to anthocyanins, such as yellow cherry (Gross 1985), red currant (Gross 1982c), strawberry (Gross 1982a), and olive fruit (Mínguez-Mosquera and others 1989) and in fruits that retain their green color when ripe, such as kiwi fruit (Gross 1982b), the carotenoid concentrations decrease with ripening. The same trend is seen with some fruits that undergo yellowing simply by unmasking the carotenoids through chlorophyll degradation (Gross 1987). Carotenoid content decreased in senescent green vegetables (Kopsell and Kopsell 2006). Ripening conditions can also affect carotenoid content. For example, carotenoid biosynthesis in the flesh of ripening Alphonso mango was highest at tropical ambient temperature (28–32°C) (Thomas and Janave 1975). "Keitt" and "Tommy Atkins" mangoes harvested at the mature-green stage showed a marked increase in all-*trans*-β-carotene, all-*trans*-violaxanthin, and 9-*cis*-violaxanthin during ripening after harvest.

Degreening with ethylene is a common postharvest practice in citrus fruit. The effect of ethylene treatment on carotenoid content and composition and on the expression of carotenoid biosynthetic genes in the flavedo of "Navelate" orange (*Citrus sinensis* L.) harvested at two ripening stages has been investigated by Rodrigo and Zacarias (2007). The ethylene-induced fruit coloration and carotenoid content in the flavedo increased with the ripening stage of the fruit. Ethylene stimulated an increase in phytoene; phytofluene; 9-*cis*-violaxanthin, which is the main carotenoid in fully ripened orange peel; and the apocarotenoid β-citraurin. Ethylene decreased the concentration of chloroplastic carotenoids. These changes are consistent with the effect of ethylene on the expression of carotenoid biosynthetic genes, because it upregulated the expression of phytoene synthase, ζ-carotene desaturase, and β-carotene hydroxylase genes; sustained or transiently increased accumulation of phytoene desaturase, plastid terminal oxidase, β-lycopene cyclase, and zeaxanthin epoxidase mRNAs; and decreased the expression of the ε-lycopene cyclase gene (Rodrigo and Zacarias 2007). These data indicate that exogenous ethylene reproduces and accelerates the physiological and molecular changes in the carotenoid biosynthesis naturally occurring during maturation of citrus fruit. On the other hand, gibberellic acid, which delays fruit degreening, reduced the ethylene-induced expression of early carotenoid biosynthetic genes and the accumulation of phytoene, phytofluene, and β-citraurin (Rodrigo and Zacarias 2007).

Genotypes. Genotype is another factor influencing carotenoid content in fruits and vegetables. Ornelas-Paz and others (2007) found marked differences in the content of the main carotenes and xanthophyll esters in several mango cultivars, and in a further study (Ornelas-Paz and others 2008a) they proposed a carotenoid index (a relationship between the content of the two most abundant carotenoids) as a valuable tool to identify mango fruit genotypes (Fig. 7.4). The effect of genotype on carotenoid content has also been reported in several other fruits and vegetables such as tomatoes (Abushita and others 1997), acerola (Lima and others 2005), dates (Boudries and

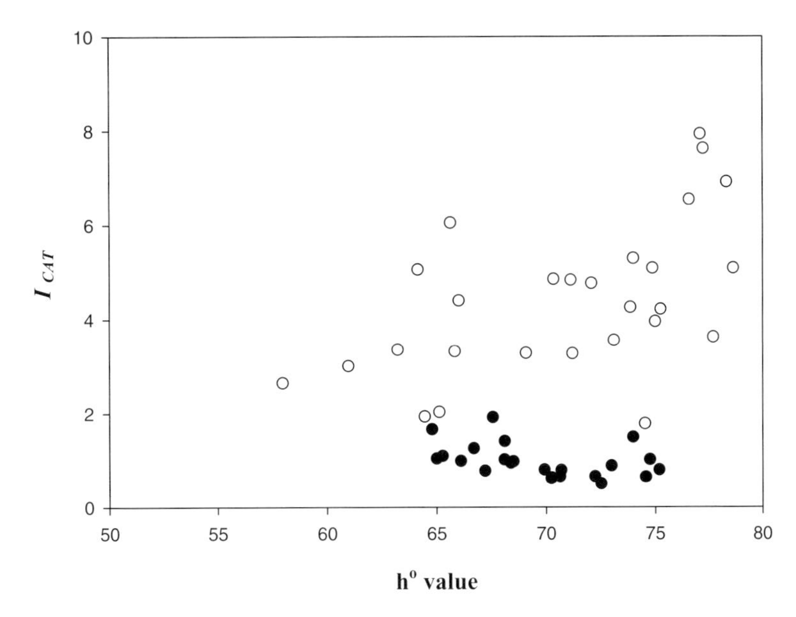

Figure 7.4. Carotenoid index (I_{CAT}) measured during the postharvest ripening of Manila (•) and Ataulfo (○) mango fruit. I_{CAT} represents the proportion of all-*trans*-β-carotene to all-*trans*-violaxanthin (as dibutyrate) in the mesocarp plotted against mesocarp h° values, which decreases during ripening of mango fruit. (Adapted from Ornelas-Paz and others 2008a.)

others 2007), apricots (Dragovic-Uzelac and others 2007), sweet potatoes (Ameny and Wilson 1997), and cassava (Thakkar and others 2007). The carotenoid contents of 52 lettuce genotypes, including crisphead, leaf, romaine, butterhead, primitive, Latin, and stem lettuces, and wild species were investigated by Beiquan (2005). Wild accessions (*L. serriola*, *L. saligna*, *L. virosa*, and primitive form) had higher β-carotene and lutein concentrations than cultivated lettuces. Among the major types of cultivated lettuce, carotenoid concentration followed the order of green leaf or romaine > red leaf > butterhead > crisphead. There was significant genetic variation in carotenoid concentration within each of these lettuce types. Crisphead lettuce accumulated more lutein than β-carotene, whereas other lettuce types had more β-carotene than lutein. Pepper (*Capsicum* spp.) cultivar, as well as location of cultivation, significantly affected carotenoid contents (Lee and others 2005a). The best sources of β-carotene were mature greenhouse-grown fruit of Fidel (23.7 μg/g) and C 127 (22.3 μg/g). Mature greenhouse fruit of Tropic Bell (10.1 μg/g) and PI 357509 (9.2 μg/g) had high lutein, but field-grown mature fruit of these lines were low in this compound (1.4 and 0.5 μg/g, respectively).

Geographical origin. Geographical origin of fruits and vegetables affects their carotenoid content, probably because of differences in climatic and cultural variables. Dragovic-Uzelac and others (2007) reported higher carotenoid content in apricots grown in the Mediterranean region than in apricots from other geographical origins. Mercadante and Rodríguez-Amaya (1998) found big differences in carotenoid content in the same mango genotype at the same ripening stage but grown in two different regions of Brazil. All-*trans*-β-carotene was twice as high in "Keitt" mango from Bahia as in "Keitt" mango from São Paulo, and all-*trans*-violaxanthin and 9-*cis*-violaxanthin were also higher in the Bahian mangoes. These differences were greater than those between "Keitt" and "Tommy Atkins" mangoes from São Paulo, indicating that climatic effects could have the same or even greater influence on carotenoid composition as cultivar differences, with fruits from hot regions having generally higher carotenoid content. Similar data have been reported for other fruits and vegetables such as papaya (Kimura and others 1991). Examples of effects due to geographical differences were also shown by Formosa papayas produced in two Brazilian states with different climates. Those from the temperate São Paulo had lower β-carotene, β-cryptoxanthin, and lycopene concentrations than did papayas from the hot state of Bahia (Rodriguez-Amaya 1993). Similarly, β-carotene content of West Indian cherry from the hot Northeastern states of Pernambuco and Ceará was found to be 5–6 times greater than that of the same fruit from São Paulo (Rodriguez-Amaya 1993).

Fruit and vegetable structure. Generally, carotenoids have been reported to be concentrated more in the peel than in the pulp of fruits and vegetables (Carrillo-Lopez and Yahia 2009), with some exceptions. In the *Cucurbita* hybrid *tetsukabuto*, the pulp and the peel had 56 and 642 μg/g total carotenoids, respectively (Arima and Rodriguez-Amaya 1988). This distribution pattern was also noted in tomato fruit (Carrillo-Lopez and Yahia 2009), papaya (Rivera-Pastarna and others 2009), kumquat (Huyskens and others 1985), mandarin hybrid (Gross 1987), muskmelon (Flügel and Gross 1982), and persimmon (Gross 1987). An exception to the usual pattern is seen in pink-fleshed guava (Padula and Rodriguez-Amaya 1986) and red pomelo (Gross 1987), in which the high lycopene concentration in the pulp compensates for the greater amounts of other carotenoids in the peel.

Temperature. Temperature during fruit development significantly influenced the carotenoid concentration of tomato produced in a controlled-environment greenhouse (Koskitalo and Ormrod 1972). At diurnal 17.8/25.6°C minimum-maximum temperatures, the β-carotene concentration was 2.97, 2.18, and 2.19 μg/g, respectively, in fruits harvested 7, 14, and 21 days following the onset of initial coloration. The corresponding levels for lycopene were 43.5, 57.7, and 64.8 μg/g. At 2.8/13.9°C, β-carotene content was 3.56, 3.73, and 3.67 μg/g and lycopene content was 9.30, 20.5, and 24.2 μg/g in fruit collected 7, 14, and 21 days following color break, respectively.

Processing. Processing of fruits and vegetables alters their carotenoid content, which can be positive or negative depending on the type and intensity of food processing (see Processing section).

Other factors. Greater exposure to sunlight and elevated temperatures were reported to increase carotenoid biosynthesis in fruits and vegetables. Carotenoid content was significantly higher in kale leaves harvested from an organic farm compared to those harvested from a neighboring farm that used agrochemicals, both collected at the same stage of maturity (Young and Britton 1990). The β-carotene, lutein, and total carotenoids were significantly higher in the winter than in the summer in "Manteiga," which may be due to greater destruction of leaf carotenoids with higher temperature and greater sunlight, but neoxanthin content was significantly higher in the summer for the "Tronchuda" cultivar. Carotenoid concentration in lettuce was higher in summer than in the fall, but was not affected by the position of the plant on the raised bed (Beiquan 2005). Greenhouse-grown peppers contained more carotenoids than the field-grown peppers (Lee and others 2005a). The location of cultivation of pepper (*Capsicum* spp.) significantly affected carotenoid content (Lee and others 2005a).

Postharvest and Processing Effects

Carotenoids are susceptible to heat, light, and oxygen. Isomerization and oxidation, such as during processing and storage, result in loss of color and biological activity and the formation of volatile compounds that impart desirable or undesirable flavor in some foods. The oxidation process depends on the presence of oxygen, metals, enzymes, unsaturated lipids, prooxidants, or antioxidants; exposure to light; type and physical state of carotenoid present; severity of the treatment; packaging material; and storage conditions such as temperature and relative humidity. Heating promotes *trans–cis* isomerization. Isomerization leads to increased amounts of *cis* isomers, 5,6-epoxide to 5,8-furanoid oxide rearrangement, structural modification by released enzymes, and oxidative cleavage of the polyene chain by free radical reactions to give shorter-chain apocarotenals and ketones as products (Simpson and others 1976), a process that can lead to bleaching of the tissue. Oxidative breakdown of carotenoids can lead to the production of flavor (aroma) products such as β-ionone and dihydroactinodiolide. Carotenoids are known to be stable during freezing and freeze-drying when oxygen is excluded.

Storage Conditions

Great variations in the stability of carotenoids in fruits and vegetables have been reported during storage. Generally, losses or changes in carotenoids in mature fruits and vegetables that can be stored for long periods of time are small and occur slowly compared to those in ripe fruits and vegetables. Carotenoid accumulation can occur after harvest and during maturation and ripening in storage or during transport. Rapid losses in carotenoids and changes in their composition can occur as a result of inappropriate storage conditions. Carotenoid esters are generally more stable than free carotenoids. Carotenoid losses during postharvest storage were reported in some vegetables, particularly leaves (Ezell and Wilcox 1962; Takama and Saito 1974; Simonetti and others 1991; Kopas-Lane and Warthesen 1995), especially under conditions favorable to wilting, high temperature, and light exposure. The effect of different postharvest treatments (film wrapping in conventional and biodegradable materials, foodtainer, surface coating) were evaluated in respect to their suitability to prevent loss of bioactive compounds in highly perishable fruits and vegetables. In lettuce, film packaging did not show any beneficial effects on pectic substances and pigments, but in pepino dulce fruits the use of foodtainer maintained the β-carotene and chlorophyll contents.

Storage temperature influences phytochemical content in fruits and vegetables. Carotenoid changes during storage depend on the ripening stage of fruit and vegetables (Gross 1991). Carotenoid content usually increases during maturation and usually decreases during senescence, and these changes can be promoted or suppressed by temperature. β-Carotene content of tomatoes increased during storage when fruit were still in the ripening process, and this increase was more pronounced at higher temperatures of up to 25°C, whereas in some sweet potato cultivars no increase in β-carotene during storage occurred because the synthesis of carotenoids was already completed by the time of harvest (Watada 1987). This maturity–temperature interaction was also observed in pepino (Prono-Widayat and others 2003). Premature and mature pepino fruit stored at 18°C showed a stronger increase in β-carotene compared to those stored at 5°C; however, in ripe pepinos, β-carotene was unaffected by temperature. Storage at 7–20°C for 16–43 days caused a substantial decrease in total carotenoid content even when fruits were subsequently ripened at optimal conditions. Losses in α-carotene, β-carotene, and lutein increased in carrots as storage temperature increased above 4°C and as illumination increased (Meléndez-Martínez and others 2004a,b). Lycopene content in several fruits and vegetables has been reported to increase at a temperature of about 25°C, but no synthesis occurs above 30°C (Goodwin and Jamikorn 1952; Lurie and others 1996). In chilling sensitive crops such as tomato, lycopene was reported to decrease at low (chilling) temperatures (Ajlouni and others 2001; Yahia and others 2007). However, this effect was not reported in hydroponically grown tomato fruit (Talcott and others 2005). Contents of lycopene and β-carotene precursors, phytoene, phytofluene, and ζ–carotene were highest in tomatoes stored at 20°C when compared with those stored at 30°C (Hamauzu and Chachin 1995). This indicates that the activity of the key enzymes catalyzing the synthesis of β-carotene and lycopene, phytoene synthase and phytoene desaturase, is mainly triggered at the maturation process and can be additionally influenced by temperature. For example, temperatures above 30°C limit carotenoid production (Goodwin and Jamikorn 1952; Lurie and others 1996). This phenomenon may be due to inhibition of ethylene biosynthesis and/or to reduced levels of phytoene synthase as reported for tomatoes (Lurie and others 1996). However, peaches held at 41°C for 24 hr prior to postharvest ripening at 27°C had a higher carotenoid content than tree-ripened fruit (Dekazos 1983).

Adequate modified atmospheres (MA) and controlled atmospheres (CA), especially atmospheres with low concentrations of oxygen, are known to maintain carotenoids and reduce their losses (Yahia 2009). Very high concentrations of carbon dioxide have been shown to result in some losses in carotenoids. The use of MA storage for 6 days at 5°C preserved carotenoid content in broccoli compared to those stored in normal air at the same temperature, which lost about half of their carotenoid content (Kalt 2005). CA storage of mature pepino fruit for a total of 21 days with varying high CO_2 concentrations (15 kPa for 2 days followed by 5 kPa for 19 days) maintained β-carotene content, but a high-CO_2 atmosphere (15 kPa) caused a strong reduction within 14 days (Huyskens-Keil and others 2005). High CO_2 also reduced β-carotene content in spinach, melon, and kale (Kader, 2002). "Golden Delicious" apple fruits held in 15 kPa CO_2 retained their green color due to inhibition of carotenoid production (Hribar and others 1994). A controlled atmosphere of 3 kPa O_2 and 20 kPa CO_2 inhibited lycopene accumulation in tomato fruit, but accumulation increased slightly once the fruit were replaced in air (Sozzi and others 1999). In addition, modified atmosphere packaging (MAP) with 10 kPa CO_2 and 10 kPa O_2 resulted in a decline in lycopene content in watermelon stored at 2°C for 7 or 10 days (Perkins-Veazie and Collins 2004). Low-oxygen atmospheres enhanced the retention of carotenes in carrots; air + 5 kPa CO_2 caused a loss, whereas air + 7.5 kPa CO_2 or higher appeared to cause *de novo* synthesis of carotene (Weichmann 1986). The carotene content of leeks was found to be higher after storage in 1 kPa O_2 + 10 kPa CO_2 than after storage in air (Weichmann, 1986). Barth and Zhuang (1996) found that MAP retained carotenoids in broccoli florets. Howard and Hernandez-Brenes (1998) found that the retention of β-carotene in jalapeno pepper rings after 12 days at 4.4°C plus 3 days at 13°C was 87% in MAP (5 kPa O_2 + 4 kPa CO_2) and 68% in air, and retention of α-carotene was 92% in MAP and 52% in air after 15 days. Wright and Kader (1997) reported that peach slices kept in air + 12 kPa CO_2 had a lower content of β-carotene and β-cryptoxanthin (retinol equivalent) than slices kept in air, 2 kPa O_2, or 2 kPa O_2 + 12 kPa CO_2 for 8 days at 5°C. Storage of persimmon slices in 2 kPa O_2 or air + 12 kPa CO_2 resulted in slightly lower retinol equivalent after 8 days at 5°C, but the loss was insignificant in slices kept under 2 kPa O_2 + 12 kPa CO_2. For sliced peaches and persimmons, the limit of shelf life based on sensory quality was reached before major losses of carotenoids occurred.

Ethylene treatment was reported to be associated with carotenogenic gene expression and the corresponding carotenoid accumulation in apricot (Marty and others 2005). Generally, postharvest ethylene applications accelerate tomato fruit ripening and the accumulation of lycopene, whereas ethylene inhibitors inhibit lycopene accumulation (Edwards and others 1983). However, this effect can be influenced by stage of ripening. For example, tomato fruit harvested at the breaker stage accumulated more lycopene at the same temperature than those harvested green and treated with ethylene (Thompson and others 2000).

The use of 1-methylcyclopropene (1-MCP) to extend the shelf life of cherry tomato (*Lycopersicon esculentum* var. Cerasiforme) delayed the ethylene-induced climacteric peaks in mature green and breaker fruits, chlorophyll degradation, and accumulation of lycopene and carotenoids (Opiyo and Tie-Jin 2005). Higher 1-MCP concentrations inhibited the accumulation of lycopene and carotene such that the color of the fruit did not reach that of control fruit.

Lower doses (at or below 200 Gy) of irradiation coupled with 35 days of storage at 10°C were not harmful in the retention of lycopene and other health-promoting compounds in early-season grapefruit, but higher doses (400 and 700 Gy) and 35 days of storage had detrimental effects. However, no significant effect of radiation and storage was observed in late-season fruit (Patil and others 2004).

Hot water treatment was reported to delay carotenoid synthesis and thus yellowing of broccoli florets (at 40°C for 60 min) and kale (at 45°C for 30 min), but did not affect Brussels sprouts (Wang 2000). Hot air treatment (38°C and 95% RH for 24 hr) slightly decreased lycopene and β-carotene content in tomato fruit (Yahia and others 2007); however, fruit heated at 34°C for 24 hr and stored 20°C developed higher lycopene and β-carotene than nonheated fruit (Soto-Zamora and others 2005). Moist (100% RH) hot air (48.5 or 50°C) for 4 hr caused injury to papaya and losses in lycopene and β-carotene, but similar treatment with dry air (50% RH), alone or in combination with thiabendazole, had no effect on lycopene and β-carotene (Perez-Carrillo and Yahia 2004). High-temperature treatment also suppressed 1-aminocyclopropane-1-carboxylic acid oxidase activity and thus indirectly prevented carotenoid synthesis (Suzuki and others 2005).

Processing

Processing of fruits and vegetables exerts important and diverse effects on carotenoids depending on type and conditions of processing method. Tissue disruption, such as by chopping or homogenizing, can lead to substantial losses in carotenoids, especially through the action of oxidation processes. For example, up to 30% of β-carotene can be lost within a few minutes when green leaves are macerated. Carotenoids are generally stable during heat processing and cooking of vegetables and fruits, but can increase amounts of *cis*-isomers (Khachik and others 1992). A significant increase in *cis*-lycopene and a slight, statistically insignificant decrease in all-*trans*-lycopene were observed in guava juice after processing (Padula and Rodriguez-Amaya 1987). Both isomers decreased during 10 months of storage. The small amount of β-carotene was retained during processing and storage. Carotenoids were essentially retained during the processing of mango slices (Godoy and Rodriguez-Amaya 1987). The only significant change was the increase in luteoxanthin, compatible with the conversion of 5,6- to 5,8-epoxide. More evident changes occurred on processing mango puree, where the principal pigment β-carotene decreased 13%, auroxanthin appeared, and violaxanthin and luteoxanthin decreased. During storage of mango slices in lacquered or plain tinplate cans, no appreciable loss of β-carotene was noted for 10 months, but between the 10th and 14th month a 50% reduction occurred. Violaxanthin tended to decrease and auroxanthin to increase during storage. β-Carotene showed a greater susceptibility to degrade in bottled mango puree (18% loss after 10 months) than in the canned product. Both bottled and canned mango puree suffered a 50% loss of β-carotene after the 14th month. Violaxanthin and luteoxanthin tended to decrease whereas auroxanthin maintained a comparatively high level throughout storage (Godoy and Rodriguez-Amaya 1987). Substantial differences have been noted in commercially processed mango juice (Mercadante and Rodriguez-Amaya 1998). Violaxanthin, the principal carotenoid of the fresh mango, was not detected; auroxanthin appeared at an appreciable level; and β-carotene became the principal carotenoid. Both lycopene (the major pigment) and β-carotene showed no significant changes during the processing of papaya puree (Godoy and Rodriguez-Amaya 1991). *cis*-Lycopene increased sevenfold and β-cryptoxanthin decreased 34%. During 14 months of storage, β-carotene, lycopene, and *cis*-lycopene remained practically constant. β-Cryptoxanthin did not change significantly during the first 10 months but decreased 27% after 14 months. Auroxanthin and flavoxanthin appeared during storage. In olives, only β-carotene and lutein resisted the fermentation and brine storage (Mínguez-Mosquera and others 1989). Phytofluene and ζ-carotene disappeared. Violaxanthin, luteoxanthin, and neoxanthin gave rise to auroxanthin and neochrome; the total pigment content did not change.

Canning increased the percentage of total *cis* isomers of provitamin A carotenoids in several fruits and vegetables (Lessin and others 1997). Canning of sweet potatoes caused the largest increase (39%), followed by processing of carrots (33%), tomato juice (20%), collards (19%), tomatoes (18%), spinach (13%), and peaches (10%). Heat induced the formation of 13-*cis*-β-carotene in sweet potatoes, and the quantity formed was related to the severity and length of the heat treatment (Chandler and Schwartz 1988). However, carotene content was reduced 20% by canning, 21% by dehydration, 23% by microwaving, and 31% by baking. *Cis* isomers have also been reported to increase during heating of carrot juice (Chen and others 1995), 13-*cis*-β-carotene being formed in the largest amount, followed by 13-*cis*-β-lutein and 15-*cis*-α-carotene. However, canning (121°C, 30 min) resulted in the greatest loss of carotenoids, followed by high-temperature short-term heating (120°C for 30 sec, 110°C for 30 sec, acidification plus 105°C for 25 sec, and acidification) (Chen and others 1996). Carrot juice color turned from orange to yellow with intensive treatment. When this processed carrot juice was stored, the concentrations of lutein, α-carotene, and β-carotene decreased with increasing storage temperature and illumination. The formation of 13-*cis*-isomers increased under light and 9-*cis*-isomers increased under dark storage. Storage of tomato and carrot juices under light (230 ft-c at 4°C) caused a 75% reduction in α- and β-carotene and 25% reduction in lycopene, but storage in darkness showed only negligible losses (Rodriguez-Amaya 1999).

Dehydration (hot air drying at 65°C), freezing (at −30°C), and freeze-drying of spinach previously immersed in salt and bicarbonate solution induced a 12% loss of β-carotene, but caused no significant losses of lutein, violaxanthin, and zeaxanthin (Rodriguez-Amaya 1999). Losses in all-*trans*-β-carotene were 33% in freeze-dried Italian spinach, and 43% to 38% in solar-dried Italian spinach, spring cabbage, and cowpea leaves (Nyambaka and Ryley 1996). In winter squash no losses in carotenoids were reported during blanching and freeze-drying; lutein decreased slightly and β-carotene was stable during freezing (Kon and Shimba 1989). However, losses in β-carotene were 15%, 20%, and 53% in freeze-dried squash stored at 30°C for 1, 2, and 3 months, respectively. Storage at lower temperature (3°C) for 3 months caused only 10% losses in β-carotene.

Thermal processing is common for fruits and vegetables. Thermal sensitivity of carotenoids is commonly assumed; however, the effect of thermal treatment of fruits and vegetables is not clear. Bengtsson and others (2008) found that boiling (for 20 min), frying (for 10 min), or drying (at 57°C for 10 hr) of sweet potatoes reduced the content of all-*trans*-β-carotene by 77–88%. These types of processing induced β-carotene isomerization (from all-*trans* to 9-*cis*). Similarly, Ade-Omowaye and others (2002) found that carotenoid content in bell peppers decreased by 55–80% as temperature of the osmotic dehydration process increased from 25 to 55°C. Pott and others (2003a) demonstrated that conventional drying of mango fruit induced isomerization of all-*trans*-β-carotene to 13-*cis*-β-carotene, whereas solar drying favored formation of 9-*cis*-β-carotene. In contrast, Mínguez-Mosquera and Hornero-Méndez (1994) found that thermal processing (drying) of paprika increased the content of red carotenoids by 40%. Similarly, Wen-ping and others (2008) showed that zeaxanthin and β-carotene content in fruits increased significantly by drying: 2 to 22 times that of fresh fruits of *Lycium barbarum* at the beginning of the drying period. Odriozola-Serrano and others (2009) found that heat pasteurization enhanced the content of some carotenoids (lycopene, β-carotene, and phytofluene) and the red color of juices. On the other hand, freezing can decrease, increase, or maintain carotenoid levels of fruits and vegetables (Scott and Eldridge 2005; Gębczyński and Kmiecik 2007). Minimal processing and milling of fruits and vegetables induce carotenoid losses (Mínguez-Mosquera and Hornero-Méndez 1994; Ruiz-Cruz and others 2007). In contrast, pulsed electric fields (Cortés and others 2006), osmotic dehydration (Tonon and others 2007), radiation

(Hajare and others 2007), and high-pressure processing induce small or no losses of carotenoids in intact fruits and vegetables and their juices and purees.

The evaluation of different cooking forms on the content of α- and β-carotene in carrots has shown that lower temperature and less cooking time resulted in the most retention (Meléndez-Martínez and others 2004a). Scalding of carrots in water at different temperatures (50, 70, 90°C) for 15 min did not result in significant losses in lycopene, and only slight losses in β-carotene at 90°C (Meléndez-Martínez and others 2004a). Sun drying has been reported to result in significant losses in carotenoids (Rodriguez-Amaya and Kimura 2004). However, fresh mangoes dried using a solar dryer to a final moisture content of 10–12% had 4,000 ± 500 μg/100 g and 3,680 ± 150 μg/100 g of β-carotene after 2 and 6 months of storage, respectively (Rankins and others 2008).

Therefore, in most processing methods, carotenoids retention seem to be reduced and losses seem to be increased at higher temperature and longer exposure/processing time. Reducing temperature, reducing processing time, reducing the presence of oxygen, and incorporating antioxidants can greatly reduce the losses in carotenoids.

Absorption and Metabolism

Carotenoid metabolism is expressed in different forms (Perera and Yen 2007). On one hand, their absorption and transport in the body would determine, to a certain extent, their biological actions. On the other hand, their chemical changes in foods or in the body would determine the extent of their losses and/or conversion to other compounds. In this section, we emphasize their absorption and transport in the body. Their chemical changes are equally important and were discussed in other sections of this chapter. It is generally accepted that oxidation of carotenoids begins with epoxidation and cleavage to apocarotenals prior to transformation into other derivatives. Several epoxycarotenoids and apocarotenals were observed in experimental oxidation models, but some were also identified in processed foods (Rodriguez and Rodriguez-Amaya 2007). A peroxidase from edible fungi was shown to be a key enzyme able to degrade carotenoids into important flavor compounds (such as ionone, cyclocitral, and terpineol) (Zorn and others 2003).

Absorption and Transport

The absorption efficiency of the different carotenoids is variable. For example, β-cryptoxanthin has been reported to have higher absorption efficiency than α-cryptoxanthin in rats (Breithaupt and others 2007). Carotenoids must be liberated from the food before they can be absorbed by intestinal cells (Faulks and Southon 2005). Mechanical disruption of the food by mastication, ingestion, and mixing leads to carotenoid liberation (Guyton and Hall 2001). The enzymatic and acid-mediated hydrolysis of carbohydrates, lipids, and proteins (chemical breaking of the food) also contributes to carotenoids liberation from the food matrix (Faulks and Southon 2005). Once released, carotenoids must be dissolved in oil droplets, which are emulsified with the aqueous components of the chyme. When these oil droplets are mixed with bile in the small intestine, their size is reduced, facilitating the hydrolytic processing of lipids by the pancreatic enzymes (Pasquier and others 1996; Furr and Clark 1997;

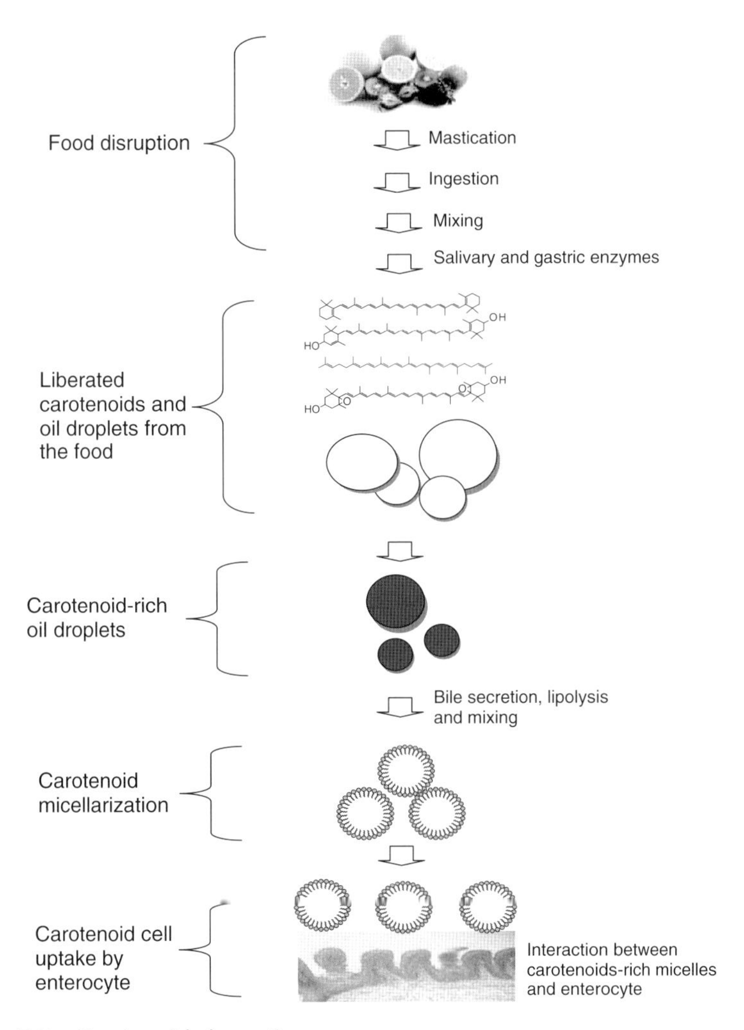

Figure 7.5. Carotenoid absorption process.

Guyton and Hall 2001; Faulks and Southon 2005). Only micellarized carotenoids can be absorbed by intestinal cells (Furr and Clark 1997). The products of the lipid digestion and carotenoids are then transferred to mixed micelles (Pasquier and others 1996) through a mechanism that is not yet fully understood (Fig. 7.5). Cholesterol membrane transporters scavenger receptor class B type I (SR-BI) and cluster determinant 36 (Cd36), but not Niemann-Pick C1-like 1 transporter (NPC1L1), were found to be involved in intestinal uptake of lutein and β-carotene (Moussa and others 2008). The small intestine bioaccessibility of lutein (79%) was found to be higher than that of lycopene (40%) and β-carotene (27%), and it was suggested that about 90% of the β-carotene, lutein, and lycopene contained in fruits and vegetables is available in the gut during the entire digestion process (Goñi and others 2006).

In the enterocyte, provitamin A carotenoids are immediately converted to vitamin A esters. Carotenoids, vitamin A esters, and other lipophilic compounds are packaged into chylomicrons, which are secreted into lymph and then into the bloodstream. Chylomicrons are attacked by endothelial lipoprotein lipases in the bloodstream, leading to chylomicron remnants, which are taken up by the liver (van den Berg and others 2000). Carotenoids are exported from liver to various tissues by lipoproteins. Carotenes (such as β-carotene and lycopene) are transported by low-density lipoproteins (LDL) and very low-density lipoproteins (VLDL), whereas xanthophylls (such as lutein, zeaxanthin, and β-cryptoxanthin) are transported by high-density lipoproteins (HDL) and LDL (Furr and Clark 1997).

Limiting Factors for Absorption

Effects of Food Matrix

Food matrix plays an important role in carotenoid bioavailability. Carotenoids from supplements are absorbed more easily than those from fruits and vegetables (West and Castenmiller 1998). On the other hand, carotenoids from fruits seem to be four times more bioavailable than those from vegetables (De Pee and others 1998). The carotenoids from carrots and green leaf vegetables are poorly absorbed, presumably because they are united to proteins. This union is not only physical (effect of the matrix) but also chemical, which reduces carotenoid bioavailability (West and Castenmiller 1998). β-Carotene from carrots and lycopene from tomatoes are also poorly bioavailable because they are in the form of large crystals, which are not completely dissolved during their passage through the gastrointestinal track (De Pee and others 1998). In contrast, carotenoids from papaya and mango are readily absorbed because they are dissolved in oil droplets (West and Castenmiller 1998).

Recently, Ornelas-Paz and others (2008b) demonstrated that fruit ripening stage affects the bioavailability of carotenoids. They hypothesized that the changes caused by fruit ripening (mainly the cleavage of cell-wall polymers, especially pectic polysaccharides) cause a kind of "biological processing" that favors carotenoids bioavailability. The effect of this biological processing is similar to thermal or mechanical processing of foods, which also causes losses in cell integrity. Food processing increases carotenoid bioavailability because it breaks food cells and protein carotenoid complexes. Food processing diminishes the negative effect of food matrix on carotenoid bioavailability. van het Hof and others (2000) demonstrated that homogenization and thermal processing of tomatoes significantly increased the bioavailability of β-carotene and lycopene. Similarly, the bioavailability of β-carotene of puree from cooked carrots was found to be higher than that of puree from raw carrots (Livny and others 2003).

Effects of Dietary Fats

Dietary fats are required for carotenoid uptake by intestinal cells. Fats have an important role in the continuation of the process of carotenoid absorption, because the human intestine is incapable of secreting significant quantities of chylomicrons into the bloodstream in the absence of fats (Ornelas-Paz and others 2008b). Some studies have suggested that at least 5 g/day of dietary fat are required for suitable β-carotene absorption (West and Castenmiller 1998), whereas others suggested the consumption of 3–5 g of fat in the meal (Roodenburg and others 2000; Brown and others 2004). On the other hand, some human studies have demonstrated that carotenoid bioavailability increases as the amount of dietary fat increases, whereas other studies have demonstrated the opposite. Unlu and others (2005) observed a gradual increase in β-carotene bioavailability when a salad containing 75 and 150 g of avocado pulp (as a fat source) was consumed. In contrast, Roodenburg and others (2000) found only small increases in β-carotene bioavailability when a meal containing 3.1 or 36 g of fat was consumed. Ornelas-Paz and others (2008b) demonstrated that dietary fat increased the carotenoid micellarization (potential bioavailability) of mango carotenoids by 25–231%, depending on the stage of ripening of mango fruit. Besides the amount of dietary fat, the type of fat consumed seems to have an important effect on carotenoid absorption. Hu and others (2000) found that β-carotene absorption was lower when a meal was consumed in the presence of sunflower oil than in the presence of beef tallow, suggesting that polyunsaturated fatty acids and carotenoids compete for absorption.

Effects of Dietary Fiber

Dietary fiber plays an important role in carotenoid bioavailability. Rock and Swendseid (1992) demonstrated in humans that pectins reduced β-carotene bioavailability considerably. Several fibers (pectin, guar gum, cellulose, and wheat bran) reduced β-carotene bioavailability by 33–43% (Riedl and others 1999). Horvitz and others (2004) demonstrated that carrot fiber reduces the bioavailability of carotenoids of this vegetable. This negative effect of dietary fiber can be attributed to fiber's ability to alter the process of mixed micelle formation. Pasquier and others (1996) demonstrated that several dietary fibers, including pectins, increased the viscosity of reconstituted duodenal medium and affected emulsification and lipolysis of fat, two indispensable steps for carotenoid micellarization (Borel and others 1996). Fiber can also decrease pancreatic lipase activity (Koseki and others 1989) and bind bile acids (Wang and others 2001), thus reducing the formation of micelles and micellarization of carotenoids. Recently, Ornelas-Paz and others (2008b) found that the amount and the molecular weight, and possibly other physicochemical characteristics, of pectins affect the bioavailability of carotenoids from supplements.

Carotenoid Quality and Quantity

The type and amount of carotenoids consumed affects carotenoid absorption. Absorption of β-carotene from large doses is independent of dose size (Tanumihardjo 2002). The response of β-carotene in serum and milk was similar in women supplemented with 60 or 210 mg of β-carotene (Canfield and others 1997). In contrast, small carotenoid doses are more efficiently absorbed than large ones (West and Castenmiller 1998; Furusho and others 2000; Tanumihardjo 2002).

Chemical properties of carotenoids play an important role in carotenoid micellarization and, therefore, bioavailability. Apolar carotenoids (carotenes) are generally incorporated in the central region, which is highly hydrophobic, of the oil droplets, whereas polar carotenoids (xanthophylls) are localized on the surface, and therefore xanthophylls are more easily micellarized and absorbed than carotenes (Borel and others 1996). van het Hof and others (2000) found in humans that lutein is five times more bioavailable than β-carotene.

Carotenoids in foods are mostly found in the all-*trans* configuration. In the case of β-carotene the 9-*cis* isomer is 7–16% more bioavailable in humans than the all-*trans* isomer (West and Castenmiller 1998), presumably because 9-*cis*-β-carotene is a more soluble molecule (van den Berg and others 2000), as a consequence of its bent backbone. However, an important 9-*cis* to all-*trans* isomerization occurs before entering the bloodstream (You and others 1996). Bioavailability of 9-*cis*-β-carotene was also found to be higher than that of its all-*trans* counterpart in rats (Ben-Amotz and Fishler 1998) and chicks (Mokady and others 1990), whereas the opposite has been reported for ferrets (Erdman and others 1988) and gerbils (Deming and others 2002). Lycopene *cis* isomers from tomato are also more bioavailable than all-*trans*-lycopene (Stahl and Sies 1992).

Xanthophyll esters are common in fruits and vegetables. Few data exist regarding the effect of carotenoid esterification on carotenoid bioavailability. Xanthophyll esters are readily broken in the human intestine (West and Castenmiller 1998; Breithaupt and others 2003; Faulks and Southon 2005). Chitchumroonchokchai and Failla (2006) demonstrated that hydrolysis of zeaxanthin esters increases zeaxanthin bioavailability. Wingerath and others (1995) did not find β-cryptoxanthin esters in chylomicrons from humans fed with tangerine juice. Herbst and others (1997) demonstrated that lutein diesters are more bioavailable than free lutein. However, the question of whether the free or the esterified form is more bioavailable to humans is still an ongoing discussion.

Effects of the Interaction between Carotenoids

In systems where several carotenoids are involved, the absorption of each carotenoid is governed by interactions among them; carotenoids compete for absorption (Furr and Clark 1997). For example, β-carotene supplementation reduced absorption of dietary lutein and lycopene in humans (Micozzi and others 1992). Tyssandier and others (2002) found that the absorption of dietary lycopene was reduced when a portion of spinach or pills of lutein were additionally administered to the volunteers. Similarly, the absorption of dietary lutein was reduced by consumption of tomato puree or lycopene pills (Tyssandier and others 2002). Furusho and others (2000) demonstrated that liver retinol accumulation in Wistar rats was significantly reduced when a fixed amount of β-carotene was replaced by a mixture of β- and α-carotene, suggesting that each one of these carotenoids mutually inhibits the utilization of the other. The proportion of β- and α-carotene in the mixture used in that study (Furusho and others 2000) simulated that of carrots.

Subject-related Factors Limiting Carotenoid Absorption

Studies in humans and animals suggest that carotenoid absorption depends on several factors including vitamin A status. Sklan and others (1989) demonstrated that vitamin A supplementation reduced β-carotene and canthaxanthin absorption in chickens. Dietary carotenoids absorption and bioconversion to vitamin A varied inversely with the vitamin A status of Philippine children (Ribaya-Mercado and others 2000). Some studies (Lecomte and others 1994; Albanes and others 1997) have suggested a possible negative effect of alcohol consumption on carotenoid absorption; however, it has not been clearly demonstrated. Nierenberg and others (1991) suggested that carotenoid absorption was higher in men than in women, which could be caused by the differences in weight and body composition between them (West and Castenmiller 1998).

There is no solid evidence that relates human aging and reduction of carotenoid absorption. In some studies, old people have shown a lower β-carotene absorption than that of young people (Madani and others 1989), whereas the opposite has also been reported by other studies (Sugarman and others 1991). The absorption of lipid-soluble substances, including carotenoids, is affected by any disease related to the digestion and absorption of fats (West and Castenmiller 1998). Inadequate production of lipase and bile as well as an inadequate neutralization of the chyme in the duodenum affect carotenoid bioavailability (Guyton and Hall 2001).

In the stomach, carotenoids are exposed to acid environments. This can lead to carotenoid isomerization, which can change carotenoid antioxidant properties, solubility, and absorption. In humans, β-carotene absorption is reduced when the pH of the gastric fluids is below 4.5 (Tang and others 1995). Vitamin E consumption seems to reduce carotenoid absorption in animals, presumably because vitamin E and carotenoids compete for absorption (Furr and Clark 1997). Dietary sterols, such as those in sterol-supplemented functional foods, are also known to decrease carotenoid absorption.

Biological Actions and Disease Prevention

Carotenoids have shown several interesting biological properties, such as anticarcinogenic effects, anti-inflammatory effects, and radical scavenging activity. It has been shown that fucoxanthin has an antiobesity effect in connection with the expression of the uncoupling protein UCP1 in white adipose tissue (Maeda and others 2005). Lutein supplements have been suggested to improve visual function and optical density in aged people (Olmedilla and others 2003).

Structurally, vitamin A (retinol) is essentially one half of the molecule of β-carotene. Thus, β-carotene is a potent provitamin A to which 100% activity is assigned. An unsubstituted β ring with a C_{11} polyene chain is the minimum requirement for vitamin A activity. γ-Carotene, α-carotene, β-cryptoxanthin, α-cryptoxanthin, and β-carotene 5,6-epoxide, all having one unsubstituted ring, have about half the bioactivity of β-carotene (Table 7.4) On the other hand, the acyclic carotenoids, devoid of β-rings, and the xanthophylls, in which the β-rings have hydroxy, epoxy, and carbonyl substituents, are not provitamin A–active for humans.

Other biological functions or actions attributed to carotenoids include influencing gene expression and immune function as well as prevention of some diseases such as certain types of cancer, cardiovascular disease, and age-related macular degeneration (Heinen and others 2007; Kim and others 2008; Murr and others 2009). These functions are independent of the provitamin A activity and have been attributed to an antioxidant property of carotenoids through singlet oxygen quenching and deactivation of free radicals (Stahl and others 1998; Palozza and Krinsky 1992a; Burton 1989; Krinsky 1989).

Table 7.4. Relative vitamin A activity of some carotenoids (Gross 1991)

Carotenoid	Relative Activity (%)
all-*trans*-β-carotene	100
9-*cis*-β-carotene	38
13-*cis*-β-carotene	53
all-*trans*-α-carotene	53
9-*cis*-α-carotene	13
13-*cis*-α-carotene	16
all-*trans*-cryptoxanthin	57
9-*cis*-cryptoxanthin	27
15-*cis*-cryptoxanthin	42
β-carotene 5,6 epoxide	21
β-carotene 5,8-epoxide (mutatochrome)	50
γ-carotene	42–50
β-zeacarotene	20–40

Antioxidant Properties

Carotenoids are well known as effective deactivators of electronically excited molecules involved in the generation of singlet oxygen and other radicals (Young and Lowe 2001). The stabilization of singlet oxygen by carotenoids is of both a physical and chemical nature (Stahl and Sies 2003). Chemical stabilization involves the union between the carotenoid and the free radical (van den Berg and others 2000). In physical stabilization, singlet oxygen transfers its excitation energy to the carotenoid, leading to a singlet oxygen radical of low energy and an excited carotenoid. Later, the carotenoid releases the acquired energy, as heat, to the surrounding environment (El-Agamey and others 2004). In physical stabilization, the carotenoid remains intact and can carry out other cycles of stabilization of singlet oxygen (van den Berg and others 2000). The ability of carotenoids to quench singlet oxygen is related to the conjugated double-bond system, and maximum protection is given by those having nine or more double bonds (Foote and others 1970). The acyclic lycopene was observed to be more effective than the bicyclic β-carotene (Di Mascio and others 1989), and therefore studies related to human health have focused on lycopene in recent years.

In addition to singlet oxygen quenching, carotenoids may be involved in scavenging several reactive oxygen and nitrogen species that are generated during the aerobic metabolism and pathological processes. These species are able to damage membrane lipids, DNA, and other biologically important molecules, leading to degenerative diseases in humans. Carotenoids are involved in the elimination of the most harmful oxygen-reactive species (van den Berg and others 2000). Carotenoids efficiently eliminate peroxyl radicals, especially at low oxygen tensions, contributing to a reduction in lipid peroxidation (Burton and Ingold 1984).

The antioxidant activity of carotenoids depends on the number of conjugated double bonds and possibly the presence of oxygenated functions in the molecule (Schmidt 2004). The high antioxidant activity of lycopene has been identified against singlet oxygen and other oxygen-reactive species. However, this activity is synergized by other antioxidants such as vitamin E, vitamin C, and β-carotene (Liu and others 2008). Lycopene also acts as antioxidant by repairing the radicals of α-tocopherol and vitamin C generated during the antioxidant processes that these other antioxidants undergo (Bast and others 1998). Results obtained with a free radical–initiated system also indicated that canthaxanthin and astaxanthin, in both of which the conjugated double-bond system is extended with carbonyl groups, were better antioxidants than β-carotene and zeaxanthin (Terão 1989). A mixture of β-carotene, vitamin E, and vitamin C neutralized nitrogen-reactive species more efficiently than each individual compound separately; similarly, lipid peroxidation was more efficiently reduced by a mixture of antioxidant compounds (β-carotene and α-tocopherol) than by each individual compound separately (Palozza and Krinsky 1992b).

The antioxidant system in humans is a complex network composed by several enzymatic and nonenzymatic antioxidants. In addition to being an antioxidant, lycopene also exerts indirect antioxidant properties by inducing the production of cellular enzymes such as superoxide dismutase, glutathione S-transferase, and quinone reductase that also protect cells from reactive oxygen species and other electrophilic molecules (Goo and others 2007).

Cell Signaling

Intercellular signaling is a requirement for biochemical functions and coordination in multicellular organisms. Carotenoids exhibit several biological actions that include modulation of cellular signaling pathways, gene expression, or enzyme activities. Carotenoids are able to stimulate the genetic expression of gap junctions, conduits that allow interchange of small molecules between neighboring cells. Gap junctions are composed of several proteins called connexins (Stahl and Sies 2005).

Cancer Prevention

In vitro and animal studies have demonstrated that carotenoids can protect against several forms of cancer. Similar results have been obtained from epidemiological data in humans. For example, high levels of β-carotene (from dietary sources) in human serum have been successfully associated with low risk of suffering from lung cancer (van den Berg and others 2000). However, some human intervention trials have demonstrated that supplements of carotenoids do not exert preventive effects against cancer (Omaye and others 1997), or that they increase the risk for some forms of cancer in populations already at high risk, such as smokers and those exposed to asbestos (Albanes and others 1996; Omenn and others 1996). Concentration of lycopene in blood was reported to correlate with reduced risk of prostate cancer (van Breemen and Pajkovic 2008), digestive tract cancers (Tang and others 2008), pancreatic cancer (Burney and others 1989), and cervical intraepithelial neoplasia (van Eenwyk and others 1991).

Cardiovascular Disease Prevention

Existing data regarding the prevention of cardiovascular diseases and consumption of carotenoids, especially β-carotene, are not clear. Even though β-carotene is able to reduce lipid peroxidation in the LDL, a process probably involved in the pathogenesis of arteriosclerosis, there is still no strong evidence that carotenoids exert protective effects against cardiovascular diseases (Palace and others 1999).

Skin Protection

Several studies in humans have demonstrated that carotenoid levels in plasma and skin decrease after exposure to UV radiation. Thus, a protection mechanism involving carotenoids has been proposed. The protective effects of β-carotene might be related to its antioxidant properties. Photo-oxidative damage occurs in the skin after exposure to the sun, leading to oxygen reactive species production. Photo-oxidative damage seems to be involved in the pathobiochemistry of erythema, photodermatosis, and skin cancer (Berneburg and Krutmann 2000; Krutmann 2000). Even though β-carotene supplements, called oral sun protectants, are widely used, there is little information about their effects on skin protection against the sun. Stahl and others (2000) demonstrated that supplements of β-carotene (24 mg/day), alone or mixed with vitamin E, were able to reduce erythema formation in humans. Similar results have been found by using supplements of mixtures of β-carotene, lutein, and lycopene (Heinrich and others 2003) and carotenoids from fruits and vegetables (Stahl and others 2001).

Provitamin A Activity

Vitamin A deficiency affects more than 100 million children around the world (Miller and others 2002) and thus remains an important public health problem in many countries. Vitamin A is essential for vision, reproduction, growth, immune function, and general health of humans (van Lieshout and others 2001). The major sources of vitamin A in the human diet are retinyl esters (preformed vitamin A) found in foods of animal origin and provitamin A carotenoids from fruits and vegetables. Unfortunately, foods containing preformed vitamin A (meat, milk, eggs, etc.) are frequently too expensive for some economically deprived developing countries, and therefore dietary carotenoids are the main source of vitamin A in these countries.

Of the 600 different carotenoids that have been identified in natural sources, only about 50 have provitamin A activity, with all-*trans*-β-carotene being the main precursor of this vitamin (Wolf 1984). Other carotenoids possessing at least one β-ionone ring without oxygenated groups, such as β-cryptoxanthin and β-carotene, are precursors of vitamin A (see Table 7.4). Carrots, sweet potatoes (yellow), pumpkins, and spinach are good sources of β-carotene and other provitamin A carotenoids (Harrison 2005). The provitamin A value of fruits and vegetables is frequently calculated based on their β-carotene and other provitamin A carotenoid content using the recommendations of FAO/WHO (1967) and NRC (1989), which consider approximately 33% absorption and 50% conversion to vitamin A. This method is still widely used in spite of the fact that it overestimates the vitamin A value in foods (Scott and Rodriguez-Amaya 2000).

Bioconversion of Carotenoids to Vitamin A

After cell uptake, provitamin A carotenoids are readily bioconverted to retinal by intestinal cells, although some bioconversion of carotenoids can take place in the liver (West and Castenmiller 1998). The retinal can be reversibly reduced to retinol or irreversibly oxidized to retinoic acid, before being esterified with several fatty acids, mainly palmitic acid, and transported via the bloodstream to the liver (Faulks and Southon 2005).

There are two pathways for carotenoid bioconversion to vitamin A: (1) central cleavage, in the 15:15′ double bond of the carotenoid backbone, leading to two molecules of retinal; and (2) asymmetric cleavage, in steps, producing a series of apocarotenals, which in turn can lead to one molecule of retinal and several small fragments (Nagao and others 1996) (Fig. 7.6). The central cleavage is carried out by the enzyme β-carotene-15,15′-dioxygenase, which is a cytosolic enzyme that has been only partially purified (Nagao and others 1996). This enzyme cleaves several apocarotenals and 9-*cis*-β-carotene to retinal at the 15:15′ double bond; however, its capacity to cleave other double bonds remains unknown (Nagao and others 1996). The stoichiometry of the central cleavage of β-carotene to retinal gives a molar ratio of 1:2 (1:1 w/w), whereas the asymmetric cleavage gives a molar ratio of 1:1 (2:1 w/w); thus, for practical purposes an absorption of 30% and a conversion factor of 50% (w/w) of dietary β-carotene was assumed (NRC 1989). However, this assumption is far from reality, as mentioned earlier. After oral supplementation, a significant proportion of β-carotene appears in serum, indicating clearly that β-carotene bioconversion in the enterocyte is incomplete (van Vliet and others 1995; Borel and others 1998). Bioconversion of β-carotene to vitamin A can only reach a molar ratio of 1:2 in vitamin A–deficient individuals fed with low amounts of β-carotene (van Lieshout and others 2001). Efficiency of bioconversion of carotenoids to retinal depends on carotenoids bioavailability, the reducing power of intestinal cells, activity of β-carotene 15:15′ dioxygenase, and vitamin A status of the individual (van Vliet and others 1995). Efficiencies of bioconversion of β-carotene to retinal vary widely: 100% (Nagao and others 1996); 40–60%, 82%, and 54–100% (Wang and others 1991); and 35–71% and 52–83% (van Vliet and others 1995).

Biological Actions of Carotenoids as Vitamin A

Carotenoids, as vitamin A, can exert several biological functions. Vitamin A regulates the expression of several genes (McGrane 2007), and it has protective effects against some forms of cancer at the stages of initiation (Sampaio and others 2007), promotion (Rizzi and others 1997), and progression (Moreno and others 2002). Retinoids induce connexin formation (Bertram and Vine 2005) and tissue differentiation (Wolf 1984). Vitamin A is involved in human growth (Mogi and others 2005), male and female reproduction (Clagett-Dame and DeLuca 2002), vision and eye health (Lee and others 2005b), immune function (Humphrey and others 2006), facial nerve palsy in babies (Cameron and others 2007), lung function in asthmatic patients (McGowan 2007), and general health of humans (van Lieshout and others 2001).

Age-related Macular Degeneration

Age-related macular degeneration (AMD) is the main cause of irreversible blindness in the Western world and affects 20% of people older than 65 years (Krinsky and others 2003). The macula, or yellow spot, is part of the retina and is the area of maximum visual

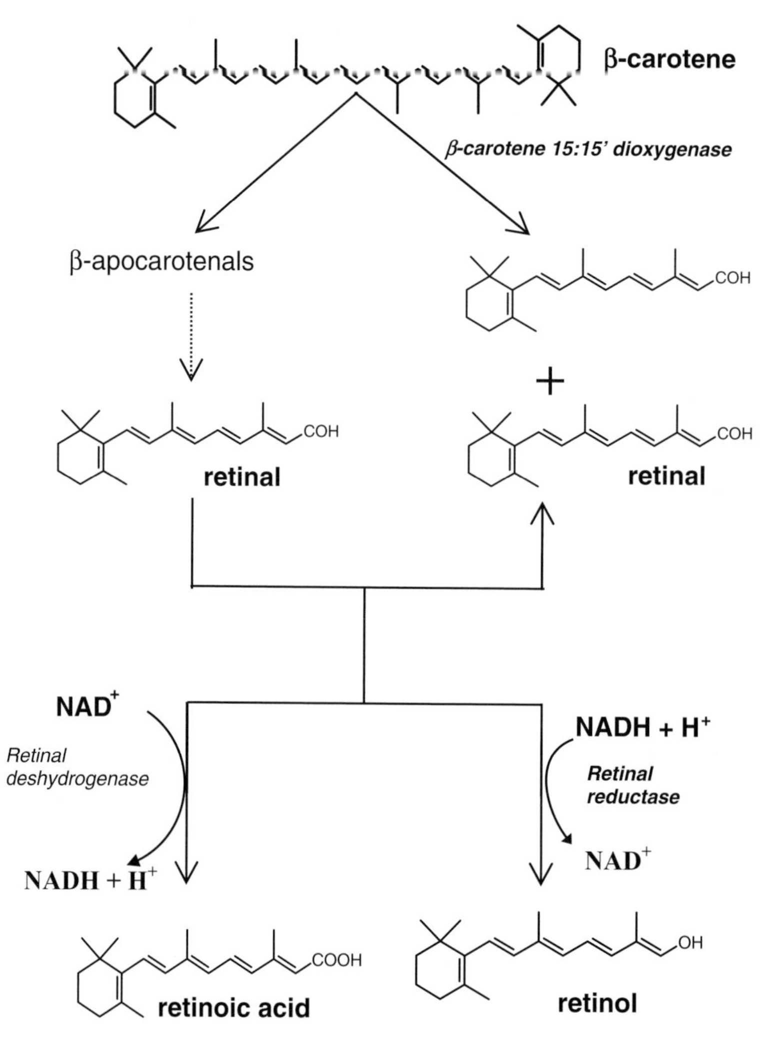

Figure 7.6. A schematic simplification of the bioconversion of β-carotene to vitamin A.

acuity. Lutein and zeaxanthin are the pigments responsible for the yellow pigmentation of the macula. Lutein and zeaxanthin are also the main carotenoids in the whole retina (Landrum and others 1999). The incorporation and transport of xanthophylls in the retina is carried out by several specific proteins (Stahl and Sies 2005). Lutein exerts protective effects on the human macula and prevents cataract development (Schalch 1992). Several studies have demonstrated that lutein levels in patients suffering from AMD are quite low and that the visual function in such patients improves substantially after lutein supplementation (Richer and others 2004). The protective effects of lutein on AMD are based on its capacity for absorbing blue light (harmful) and quenching singlet oxygen and other free radicals (Krinsky and others 2003; Leung 2008).

Conclusions

Carotenoids are very important natural organic molecules with diverse and important biological actions and functions. In plants they participate in the process of photosynthesis as accessory pigments and also in protecting chlorophylls from photodamage. In addition, they impart important diverse colors that are essential not only for their role in pollination, but also as a major quality component of foods. Several carotenoids are provitamin A–active, many act as anti-oxidants, and several others have been associated with the prevention of several types of illnesses including cancer and other chronic diseases. Fruits and vegetables are important sources of carotenoids for the human diet; they provide about 70–90% of consumed carotenoids. Many factors influence the accumulation and also the interconversion, degradation, and loss of carotenoids. Research is still needed in the effort to increase carotenoid contents and reduce their losses in fruits and vegetables.

Acknowledgment

We would like to thank Dr. Afaf Kamal-Eldin and Dr. Dietmar Breithaupt for reviewing this chapter.

References

Abushita AA, Hebshi EA, Daood HG and Biacs PA. 1997. Determination of antioxidant vitamins in tomatoes. Food Chem 60:207–212.

Ade-Omowaye BIO, Rastogi NK, Angersbach A and Knorr D. 2002. Osmotic dehydration of bell peppers: influence of high intensity electric field pulses and elevated temperature treatment. J Food Eng 54:35–43.

Ajlouni S, Kremer S and Masih L. 2001. Lycopene contents in two different tomato cultivars. Food Aust 53(5):195.

Albanes D, Heinonen OP, Taylor PR, Virtamo J, Edwards BK, Rautalahti M, Hartman AM, Palmgren J, Freedman LS, Haapakoski, J, Barrett MJ, Pietinen P, Malila N, Tala E, Liippo K, Salomaa ER, Tangrea JA, Teppo L, Askin FB, Taskinen E, Erozan Y, Greenwald P and Huttunen JK. 1996. Alpha-tocopherol and beta-carotene supplements and lung cancer incidence in the alpha-tocopherol, beta-carotene cancer prevention study: effects of base-line characteristics and study compliance. J Natl Cancer Inst 88:1560–1570.

Albanes D, Virtamo J, Taylor PR, Rautalahti M, Pietinen P and Heinonen OP. 1997. Effects of supplemental β-carotene, cigarette smoking, and alcohol consumption on serum carotenoids in the Alpha-Tocopherol, Beta-Carotene Cancer Prevention Study. Am J Clin Nutr 66:366–372.

Ameny MA and Wilson PW. 1997. Relationship between hunter color values and β-carotene contents in white-fleshed African sweetpotatoes (*Ipomoea batatas* Lam). J Sci Food Agric 73:301–306.

Arima HK and Rodriguez-Amaya DB. 1988. Carotenoid composition and vitamin A value of commercial Brazilian squashes and pumpkins. J Micronutr Anal 4:177–191.

Arima HK and Rodriguez-Amaya DB. 1990. Carotenoid composition and vitamin A value of a squash and a pumpkin from Northeastern Brasil. Arch Latinoam Nutr 40:284–292.

Azevedo-Meleiro CH and Rodriguez-Amaya DB. 2007. Qualitative and quantitative differences in carotenoid composition among *Cucurbita moschata*, *Cucurbita maxima*, and *Cucurbita pepo*. J Agric Food Chem 55:4027–4033.

Baranska M, Baranski R, Schulz H and Nothnagel T. 2006. Tissue-specific accumulation of carotenoids in carrot roots. Planta 224:1028–1037.

Barth MM and Zhuang H. 1996. Packaging design affects antioxidant vitamin retention and quality of broccoli florets during postharvest storage. Postharv Biol Technol 9:141–150.

Bast A, Haenen GR, van den Berg R and van den Berg H. 1998. Antioxidant effects of carotenoids. Int J Vitam Nutr Res 68:399–403.

Beiquan M. 2005. Genetic variation of β-carotene and lutein contents in lettuce. J Amer Soc Hort Sci 130:870–876.

Ben-Amotz A and Fishler R. 1998. Analysis of carotenoids with emphasis on 9-*cis*-β-carotene in vegetables and fruits commonly consumed in Israel. Food Chem 62:515–520.

Bengtsson A, Namutebi A, Alminger ML and Svanberg U. 2008. Effects of various traditional processing methods on the all-*trans*-β-carotene content of orange-fleshed sweet potato. J Food Compost Anal 21:134–143.

Berneburg M and Krutmann J. 2000. Photoimmunology, DNA repair and photocarcinogenesis. J Photochem Photobiol B 54:87–93.

Bertram JS and Vine AL. 2005. Cancer prevention by retinoids and carotenoids: independent action on a common target. Biochim Biophys Acta 1740:170–178.

Bhaskarachary K, Sankar-Rao DS, Deosthale Y and Vinodini-Reddy G. 1995. Carotene content of some common and less familiar foods of plant origin. Food Chem 54:189–193.

Borel P, Grolier P, Armand M, Partier A, Lafont H, Lairon D and Azais-Braesco V. 1996. Carotenoids in biological emulsions: solubility, surface-to-core distribution, and release from lipid droplets. J Lipid Res 37:250–260.

Borel P, Grolier P, Mekki N, Boirie Y, Rochette Y, Le Roy B, Alexandre-Gouabau MC, Lairon D and Azais-Braesco V. 1998. Low and high responders to pharmacological doses of β-carotene: proportion in the population, mechanisms involved and consequences on β-carotene metabolism. J Lipid Res 39:2250–2260.

Boudries H, Kefalas P and Hornero-Méndez D. 2007. Carotenoid composition of Algerian date varieties (*Phoenix dactylifera*) at different edible maturation stages. Food Chem 101:1372–1377.

Bouvier F, Rahier A and Camara, B. 2005. Biogenesis, molecular regulation and function of plant isoprenoids. Progr Lipid Res 44(6):357–429.

Breithaupt DE. 2004. Identification and quantification of astaxanthin esters in shrimp (*Pandalus borealis*) and in a microalga (*Haematococcus pluvialis*) by liquid chromatography–mass spectrometry using negative ion atmospheric pressure chemical ionization. J Agric Food Chem 52:3870–3875.

Breithaupt DE and Bamedi A. 2001. Carotenoid esters in vegetables and fruits: a screening with emphasis on β-cryptoxanthin esters. J Agric Food Chem 49:2064–2070.

Breithaupt DE, Weller P, Wolters M and Hahn A. 2003. Plasma response to a single dose of dietary β-cryptoxanthin esters from papaya (*Carica papaya* L.) or non-esterified β-cryptoxanthin in adult human subjects: a comparative study. Br J Nutr 90:795–801.

Breithaupt DE, Yahia EM and Valdez F. 2007. Comparison of the absorption efficiency of α- and β-cryptoxanthin in female Wistar rats. Br J Nutr 97:329–336.

Britton G. 1988. Biosynthesis of carotenoids. In: Goodwin TR, editor. Plant Pigments. London: Academic Press, pp. 133–182.

Britton G. 1991. Carotenoids. Methods Plant Biochem 7:473–518.

Britton G. 1993. In: Young AJ and Britton G, editors. Carotenoids in Photosynthesis. London: Chapman & Hall, pp. 96–126.

Brown MJ, Ferruzzi MG, Nguyen ML, Cooper DA, Eldridge AL, Schwartz SJ and White WS. 2004. Carotenoid bioavailability is higher from salads ingested with full-fat than with fat-reduced salad dressings as measured with electrochemical detection. Am J Clin Nutr 80:396–403.

Burney PG, Comstock GW and Morris JS. 1989. Serologic precursors of cancer: serum micronutrients and the subsequent risk of pancreatic cancer. Am J Clin Nutr 49:895–900.

Burton GW. 1989. Antioxidant action of carotenoids. J Nutr 119:109–111.

Burton GW and Ingold KU. 1984. β-Carotene: an unusual type of lipid antioxidant. Science 224:569–573.

Calva MM. 2005. Lutein: A valuable ingredient of fruit and vegetables. Crit Rev Food Sci Nutr 45:671–696.

Cameron C, Lodes MW and Gershan WM. 2007. Facial nerve palsy associated with serum vitamin A level in an infant with cystic fibrosis. J Cyst Fibros 6:241–243.

Canfield LM, Guillado AR, Neilson EM, Yap HH, Graver FJ, Cui HA and Blashill BM. 1997. Carotene in breast milk and serum is increased after a single β-carotene dose. Am J Clin Nutr 66:52–61.

Carrillo-Lopez A and Yahia EM. 2009. Qualitative and quantitative changes in carotenoids and phenolic compounds in tomato fruit during ripening. In preparation.

Cavalcante ML and Rodriguez-Amaya DB. 1992. Carotenoid composition of the tropical fruits *Eugenia uniflora and Malpighia glabra*. In Charalambous G, editor. Food Science and Human Nutrition. Amsterdam: Elsevier Science, pp. 643–650.

Chandler LA and Schwartz SJ. 1988. Isomerization and losses of trans-β-carotene in sweet potatoes as affected by processing treatments. J Agric Food Chem 36:129–133.

Charoensiri R, Kongkachuichai R, Suknicom S and Sungpuag P. 2009. Beta-carotene, lycopene, and alpha-tocopherol contents of selected Thai fruits. Food Chem 113:202–207.

Chen BH, Peng HY and Chen HE. 1995. Changes of carotenoids, color, and vitamin A contents during processing of carrot juice. J Agric Food Chem 43:1912–1918.

Chen BH, Peng HY and Chen HE. 1996. Stability of carotenoids and vitamin A during storage of carrot juice. Food Chem 57:497–503.

Chitchumroonchokchai C and Failla ML. 2006. Hydrolysis of zeaxanthin esters by carboxyl ester lipase during digestion facilitates micellarization and uptake of the xanthophyll by Caco-2 human intestinal cells. J Nutr 136:588–594.

Clagett-Dame M and DeLuca HF. 2002. The role of vitamin A in mammalian reproduction and embryonic development. Annu Rev Nutr 22:347–381.

Corral-Aguayo R, Yahia EM, Carrillo-Lopez A and Gonzalez-Aguilar G. 2008. Correlation between some nutritional components and the total antioxidant capacity measured with six different assays in eight horticultural crops. J Agric Food Chem 56:10498–10504.

Cortés C, Esteve MJ, Rodrigo D, Torregrosa F and Frígola A. 2006. Changes of colour and carotenoid contents during high intensity pulsed electric field treatment in orange juices. Food Chem Toxicol 44:1932–1939.

Cunningham Jr FX. 2002. Regulation of carotenoid synthesis and accumulation in plants. Pure Appl Chem 74:1409–1417.

De Pee S, West CE, Permaesih D, Martuti S, Muhilal and Hautvast JGAJ. 1998. Increasing intake of orange fruits is more effective than increasing intake of dark-green leafy vegetables in increasing serum concentrations of retinol and β-carotene in schoolchildren in Indonesia. Am J Clin Nutr 68:1058–1067.

Dekazos ED. 1983. Effects of postharvest treatments on ripening, carotenoids and quality of canned "Babygold 7" peaches. Proc Fla State Hort Soc 96:235–238.

Deming DM, Teixeira SR, Erdman JW. 2002. All-*trans*-β-carotene appears to be more bioavailable that 9-*cis* or 13-*cis*-β-carotene in gerbils given single oral doses of each isomer. J Nutr 132:2700–2708.

Dharmapuri S, Rosati C, Pallara P, Aquilani R, Bouvier F, Camara B and Giuliano G. 2002. Metabolic engineering of xanthophyll content in tomato fruits. FEBS Lett 519:30–34.

Di Mascio P, Kaiser S and Sies H. 1989. Lycopene as the most efficient biological carotenoid singlet oxygen quencher. Arch Biochem Biophys 274:532–538.

Dragovic-Uzelac V, Levaj B, Mrkic V, Bursac D and Boras M. 2007. The content of polyphenols and carotenoids in three apricot cultivars depending on stage of maturity and geographical region. Food Chem 102:966–975.

Dugo P, Herrero M, Giuffrida D, Ragonese C, Dugo G and Mondello L. 2008. Analysis of native carotenoid composition in orange juice using C30 columns in tandem. J Sep Sci 31:2151–2160.

Edwards JI, Saltveit ME and Henderson WR. 1983. Inhibition of lycopene synthesis in tomato pericarp tissue by inhibitors of ethylene biosynthesis and reversal with applied ethylene. J Am Soc Hort Sci 108:512–514.

El-Agamey A, Lowe GM, McGarvey DJ, Mortensen A, Phillip DM, Truscott TG and Young AJ. 2004. Carotenoid radical chemistry and antioxidant/pro-oxidant properties. Arch Biochem Biophys 430:37–48.

Erdman JW, Thatcher AJ, Hofmann NE, Lederman JD, Block SS, Lee CM and Mokady S. 1988. All-*trans*-β-carotene is absorbed preferentially to 9-*cis*-β-carotene but the latter accumulates in the tissues of domestic ferrets (*Mustela putorius* furo). J Nutr 128:2009–2013.

Ezell BD and Wilcox MS. 1962. Loss of carotene in fresh vegetables as related to wilting and temperature. J Agric Food Chem 10:124–126.

FAO/WHO. 1967. Requirements of Vitamin A, Thiamine, Riboflavine and Niacin. WHO Technical Report Series No. 362. Geneva: World Health Organization.

Faulks RM and Southon S. 2005. Challenges to understanding and measuring carotenoid bioavailability. Biochim Biophys Acta 1740:95–100.

Flügel M and Gross J. 1982. Pigment and plastid changes in mesocarp and exocarp of ripening muskmelon *Cucumis melo* cv. *Galia*. Angew Botanik 56:393–406.

Foote CS, Chang YC and Denny RW. 1970. Chemistry of singlet oxygen X. Carotenoid quenching parallels biological protection. J Am Oil Chem Soc 92:5216–5218.

Fraser PD, Enfissi EMA, Halket JM, Truesdale MR, Dongmei Y, Gerrish C and Bramley PM. 2007. Manipulation of phytoene levels in tomato fruit: effects on isoprenoids, plastids, and intermediary metabolism. Plant Cell 10:3194–3211.

Fray RG and Grierson D. 1993. Identification and genetic analysis of normal and mutant phytoene synthase genes of tomato by sequencing, complementation and co-suppression. Plant Mol Biol 22:589–602.

Furr HC and Clark RM. 1997. Intestinal absorption and tissue distribution of carotenoids. J Nutr Biochem 8:364–377.

Furusho T, Kataoka E, Yasuhara T, Wada M and Masushige S. 2000. Retinol equivalence of carotenoids can be evaluated by hepatic vitamin A content. Int J Vitam Nutr Res 70:43–47.

Gębczyński P and Kmiecik W. 2007. Effects of traditional and modified technology, in the production of frozen cauliflower, on the contents of selected antioxidative compounds. Food Chem 101:229–235.

Godoy HT and Rodriguez-Amaya DB. 1987. Changes in individual carotenoids on processing and storage of mango (*Mangifera indica*) slices and puree. Int J Food Sci Technol 22:451–460.

Godoy HT and Rodriguez-Amaya DB. 1991. Comportamento dos carotenóides de purê de mamão (*Carica papaya*) sob processamento e estocagem. Cienc Tecnol Aliment 11:210–220.

Goñi I, Serrano J and Saura-Calixto F. 2006. Bioaccessibility of beta-carotene, lutein, and lycopene from fruits and vegetables. J Agric Food Chem 54:5382–5387.

Goo YA, Li Z, Pajkovic N, Shaffer S, Taylor G, Chen J, Campbell D, Arnstein L, Goodlett DR and van Breemen R. 2007. Systematic investigation of lycopene effects in LNCaP cells by use of novel large-scale proteomic analysis software. Proteomics Clin Appl 1:513–523.

Goodwin TW and Britton G. 1988. In Goodwin TW, editor. Plant Pigments. London: Academic Press, pp. 61–134.

Goodwin T and Jamikorn M. 1952. Biosynthesis of carotenes in ripening tomatoes. Nature 170:104–105.

Gross J. 1982a. Changes of chlorophylls and carotenoids in developing strawberry fruits (*Fragaria anonassa*) cv. Tenira. Gartenbauwiss 47:142–144.

Gross J. 1982b. Pigment changes in the pericarp of the Chinese goosebery or kiwi fruit (*Actinidia chinensis*) cv. Bruno during ripening. Gartenbauwiss 47:162–167.

Gross J. 1982c. Chlorophyll and carotenoid pigments in *Ribes* fruits. Sci Hort 18:131–136.

Gross J. 1985. Carotenoid pigments in the developing cherry (*Prunus avium*) cv. "Dönissen"s Gelbe". Gartenbauwiss 50:88–90.

Gross J. 1987. Pigments in Fruits. London: Academic Press.

Gross J. 1991. Pigments in Vegetables. New York: Van Nostrand Reinhold.

Gross J, Ikan R and Eckhardt G. 1983. Carotenoids of the fruit of *Overrhoa carambola*. Phytochemistry 22:1479–1481.

Guyton AC, Hall JE. 2001. Digestión y absorción en el tubo digestivo. In: Tratado de Fisiología Médica, 10th ed. New York: McGraw-Hill, pp. 909–920.

Hagenimana V and Low J. 2000. Potential of orange-fleshed sweet potatoes for raising vitamin A intake in Africa. Food Nutr Bull 21(4):414–418.

Hajare SN, Saroj SD, Dhokane VS, Shashidhar R and Bandekar JR. 2007. Effect of radiation processing on nutritional and sensory quality of minimally processed green gram and garden pea sprouts. Radiat Phys Chem 76:1642–1649.

Hamauzu Y and Chachin K. 1995. Effect of high-temperature on the postharvest biosynthesis of carotenes and alpha-tocopherol in tomato fruit. J Jap Soc Hort Sci 63:879–886.

Harrison EH. 2005. Mechanisms of digestion and absorption of dietary vitamin A. Annu Rev Nutr 25:87–103.

Heinen MM, Hughes MC, Ibiebele TI, Marks GC, Green AC and Van der Pols JC. 2007. Intake of antioxidant nutrients and the risk of skin cancer. Eur J Cancer 43:2707–2716.

Heinrich U, Gartner C, Wiebusch M, Eichler O, Sies H, Tronnier H and Stahl W. 2003. Supplementation with beta-carotene or a similar amount of mixed carotenoids protects humans from UV-induced erythema. J Nutr 133:98–101.

Herbst S, Bowen P, Hussain E, Stacewicz-Sapuntzakis M, Damayanti B and Burns J. 1997. Evaluation of the bioavailability of lutein (L) and lutein diesters (LD) in humans. FASEB J 11:A447.

Horvitz MA, Simon PW and Tanumihardjo SA. 2004. Lycopene and beta-carotene are bioavailable from lycopene "red" carrots in humans. Eur J Clin Nutr 58:803–811.

Howard LR and Hernandez-Brenes C. 1998. Antioxidant content and market quality of jalapeno pepper rings as affected by minimal processing and modified atmosphere packaging. J Food Qual 21:317–327.

Hribar J, Plestenjal A, Vidrih R and Simcic M. 1994. Influence of CO_2 shock treatment and ULO storage on apple quality. Acta Hort 368:634.

Hu X, Jandacek RJ and White WS. 2000. Intestinal absorption of β-carotene ingested with a meal rich in sunflower oil or beef tallow: postprandial appearance in triacylglycerol-rich lipoproteins in women. Am J Clin Nutr 71:1170–1180.

Hulshof PJM, Xu C, van de Bovenkamp P, Muhilal and West CE. 1997. Application of a validated method for the determination of provitamin A carotenoids in Indonesian foods of different maturity and origin. J Agric Food Chem 45:1174–1179.

Humphrey JH, Iliff PJ, Marinda ET, Mutasa K, Moulton LH, Chidawanyika H, Ward BJ, Nathoo KJ, Malaba LC, Zijenah LS, Zvandasara P, Ntozini R, Mzengeza F, Mahomva AI, Ruff AJ, Mbizvo MT and Zunguza CD. 2006. Effects of a single large dose of vitamin A, given during the postpartum period to HIV-positive women and their infants, on child HIV infection, HIV-free survival, and mortality. J Infect Dis 193:860–871.

Huyskens S, Timberg R and Gross J. 1985. Pigment and plastid ultra-structural changes in kumquat (*Fortunella margarita*) "Nagami" during ripening. J Plant Physiol 118:61–72.

Huyskens-Keil S, Prono-Widayat H, Lüdders P and Schreiner M. 2005. Postharvest quality of pepino (*Solanum muricatum* Ait.) fruit in controlled atmosphere storage. J Food Eng 77(3):628–634.

Inbaraj BS, Lu H, Hung CF, Wu WB, Lin CL and Chen BH. 2008. Determination of carotenoids and their esters in fruits of *Lycium barbarum* Linnaeus by HPLC-DAD-APCI-MS. J Pharm & Biomed Anal 47:812–818.

John J, Subbarayan C and Cama HR. 1970. Carotenoids in 3 stages of ripening of mango. J Food Sci 35:262–265.

Kader A. 2002. Postharvest Technology of Horticultural Crops. University of California, Division of Agriculture and Natural Resources, Publication 3311.

Kalt W. 2005. Effects of production and processing factors on major fruit and vegetable antioxidants. J Food Sci 70:R9–R11.

Khachik F, Beecher GR and Goli MG. 1991. Separation, identification, and quantification of carotenoids in fruits, vegetables and human plasma by high performance liquid chromatography. Pure Appl Chem 63:71–80.

Khachik F, Goli MB, Beecher, GR, Holden J, Lusby, WR, Tenorio, MD and Barrera MR. 1992. Effect of food preparation on qualitative and quantitative distribution of major carotenoid constituents of tomatoes and several green vegetables. J Agric Food Chem 40:390–398.

Kim JH, Na HJ, Kim CK, Kim JY, Ha KS, Lee H, Chung HT, Kwon HJ, Kwon YG and Kim YM. 2008. The non-provitamin A carotenoid, lutein, inhibits NF-κB-dependent gene expression through redox-based regulation of the phosphatidylinositol 3-kinase/PTEN/Akt and NF-κB-inducing kinase pathways: role of H_2O_2 in NF-κB activation. Free Radic Biol Med 45(6):885–896.

Kimura M, Rodríguez-Amaya DB and Yokoyama SM. 1991. Cultivar differences and geographic effects on the carotenoid composition and vitamin A value of papaya. Lebens Wiss Technol 24:415–418.

Kon M and Shimba R. 1989. Changes in carotenoid composition during preparation and storage of frozen and freeze-dried squash. Nippon Shokuhin Kogyo Gakkaishi 36:619–624.

Kopas-Lane LM and Warthesen JJ. 1995. Carotenoid photostability in raw spinach and carrots during cold storage. J Food Sci 60:773–776.

Kopsell DA and Kopsell DE. 2006. Accumulation and bioavailability of dietary carotenoids in vegetable crops. Trends Plant Sci 11(10):499–507.

Kopsell DA, Kopsell DE, Lefsrud MG, Curran-Celentano J and Dukach LE. 2004. Variation in lutein, β-carotene, and chlorophyll concentrations among *Brassica oleracea* cultigens and seasons. HortSci 39:361–364.

Koseki M, Tsuji K, Nakagawa Y, Kawamura M, Ichikawa T, Kazama M, Kitabatake N and Doi E. 1989. Effects of gum Arabic and pectin on the emulsification, the lipase reaction and the plasma cholesterol levels in rats. Agric Biol Chem 53:3127–3132.

Koskitalo LN and Ormrod DP. 1972. Effects of sub-optimal ripening temperatures on the color quality and pigment composition of tomato fruit. J Food Sci 37:56–59.

Krinsky NI. 1989. Antioxidant functions of carotenoids. Free Radical Biol Med 7:617–635.

Krinsky NI, Landrum JT and Bone RA. 2003. Biologic mechanisms of the protective role of lutein and zeaxanthin in the eye. Annu Rev Nutr 23:171–201.

Krutmann J. 2000. Ultraviolet A radiation-induced biological effects in human skin: relevance for photoaging and photodermatosis. J Dermatol Sci 23:S22–S26.

Kurz C, Carle R and Schieber A. 2008. HPLC-DAD-MS^n characterization of carotenoids from apricots and pumpkins for the evaluation of fruit product authenticity. Food Chem 110:522–530.

Landrum JT, Bone RA, Chen Y, Herrero C, Llerena M and Twarowska E. 1999. Carotenoids in the human retina. Pure Appl Chem 71:2237–2244.

Lecomte E, Herberth B, Pirollet P, Chancerelle Y and Arnaud J. 1994. Effect of alcohol consumption on blood antioxidant nutrients and oxidative stress indicators. Am J Clin Nutr 60:255–261.

Lee HS and Coates GA. 2003. Effect of thermal pasteurization on Valencia orange juice color and pigments. Lebensm Wiss Technol 36:153–156.

Lee JJ, Crosby KM, Pike LM, Yoo KS and Leskovar DI. 2005a. Impact of genetic and environmental variation on development of flavonoids and carotenoids in pepper (*Capsicum* spp.). Scientia Hortic 106:341–352.

Lee WB, Hamilton SM, Harris JP and Schwab IR. 2005b. Ocular complications of hypovitaminosis A after bariatric surgery. Ophthalmology 112:1031–1034.

Lessin WJ, Catigani GL and Schwartz SJ. 1997. Quantification of *cis-trans* isomers of provitamin A carotenoids in fresh and processed fruits and vegetables. J Agric Food Chem 45:3728–3732.

Leung IYF. 2008. Macular pigment: new clinical methods of detection and the role of carotenoids in age-related macular degeneration. Optometry 79:266–272.

Lichtenthaler HK. 1999. The 1-deoxy-D-xylulose-5-phosphate pathway of isoprenoid biosynthesis in plants. Ann Rev Plant Physiol Mol Biol 50:47–65.

Lima VLAG, Mélo EA, Maciel MIS, Prazeres FG, Musser RS and Lima DES. 2005. Total phenolic and carotenoid contents in acerola genotypes harvested at three ripening stages. Food Chem 90:565–568.

Liu D, Shi J, Ibarra AC, Kakuda Y and Xue SJ. 2008. The scavenging capacity and synergistic effects of lycopene, vitamin E, vitamin C, and β-carotene mixtures on the DPPH free radical. Food Sci Technol 41:1344–1349.

Livny O, Reifen R, Levy I, Madar Z, Faulks R, Southon S and Schwartz SJ. 2003. β-Carotene bioavailability from differently processed carrot meals in human ileostomy volunteers. Eur J Nutr 42:338–345.

Lurie S, Handros A, Fallik E and Shapira R. 1996. Reversible inhibition of tomato fruit gene expression at high temperature. Effects on tomato fruit ripening. Plant Physiol 110:1207–1214.

Madani G, Mobarhan S, Ceccanti M, Ranaldi L and Gettner S. 1989. Beta-carotene serum response in young and elderly females. Eur J Clin Nutr 43:749–761.

Maeda H, Hosokawa M, Sashima T, Funayama K and Miyashita K. 2005. Fucoxanthin from edible seaweed, *Undaria pinnatifida*, shows antiobesity effect through UCP1 expression in white adipose tissues. Biochem Biophys Res Commun 332(2):392–397.

Marín A, Ferreres F and Tomás-Barberán FA, Gil MI. 2004. Characterization and quantitation of antioxidant constituents of sweet pepper (*Capsicum annuum* L.). J Agric Food Chem 52:3861–3869.

Marinova D and Ribarova F. 2007. HPLC determination of carotenoids in Bulgarian berries. J Food Comp Anal 20:370–374.

Marty I, Bureau S, Sarkissian G, Gouble B, Audergon JM and Albagnac G. 2005. Ethylene regulation of carotenoid accumulation and carotenogenic gene expression in colour-contrasted apricot varieties (*Prunus armeniaca*). J Exp Bot 56(417):1877–1886.

Mayne ST. 1996. β-Carotene, carotenoids, and disease prevention in humans. FASEB J 10:690–701.

McGowan SE. 2007. Vitamin A deficiency increases airway resistance following C-fiber stimulation. Respir Physiol Neurobiol 157:281–289.

McGrane MM. 2007. Vitamin A regulation gene expression: molecular mechanism of a prototype gene. J Nutr Biochem 18:497–508.

Meléndez-Martínez AJ, Vicario IM and Heredia FJ. 2004a. Estabilidad de los pigmentos carotenoides en los alimentos. ALAN 50(2):209–215.

Meléndez-Martínez AJ, Vicario IM and Heredia FJ. 2004b. Importancia nutricional de los pigmentos carotenoides. ALAN 54(2):149–155.

Mercadante AZ, Britton G and Rodriguez-Amaya DB. 1998. Carotenoids from yellow passion fruit (*Passiflora edulis*). J Agric Food Chem 46:4102–4106.

Mercadante AZ and Rodriguez-Amaya DB. 1998. Effects of ripening, cultivar differences, and processing on the carotenoid composition of mango. J Agric Food Chem 46:128–130.

Micozzi MS, Brown ED, Edwards BK, Bieri JG, Taylor PR, Khachik F, Beecher GR and Stimth JC. 1992. Plasma carotenoid response to chronic intake of selected foods and beta-carotene supplements in men. Am J Clin Nutr 55:1120–1125.

Miller M, Humphrey J, Johnson E, Marinda E, Brookmeyer R and Katz J. 2002. Why do children become vitamin A deficient? J Nutr 132:2867–2880.

Mínguez-Mosquera MI, Garrido-Fernández J and Gandul-Rojas B. 1989. Pigment changes in olives during fermentation and brine storage. J Agric Food Chem 37:8–11.

Mínguez-Mosquera MI and Hornero-Méndez D. 1994. Comparative study of the effect of páprika processing on the carotenoids in peppers (*Capsicum annuum*) of the Bola and Agridulce varieties. J Agric Food Chem 42:1555–1560.

Mogi C, Goda H, Mogi K, Takaki A, Yokohama K and Tomida M. 2005. Multistep differentiation of GH-producing cells from their immature cells. J Endocrinol 184:41–50.

Mokady S, Avron M and Ben-Amotz A. 1990. Accumulation in chick livers of 9-*cis* versus all-*trans*-beta-carotene. J Nutr 120:889–892.

Moreno FS, S-Wu T, Naves MM, Silveira ER, Oloris SC, Costa MA, Dagli ML and Ong TP. 2002. Inhibitory effects of beta-carotene and vitamin A during progression phase of hepatocarcinogenesis involve inhibition of cell proliferation but not alterations in DNA methylation. Nutr Cancer 44:80–88.

Moussa M, Landrier JF, Reboul E, Ghiringhelli O, Coméra C, Collet X, Fröhlich K, Böhm V and Borel P. 2008. Lycopene absorption in human intestinal cells and in mice involves scavenger receptor class B type I but not Niemann-Pick C1-like 1. J Nutr 138:1432–1436.

Murkovic M, Gams K, Draxl S and Pfannhauser W. 2000. Development of an Austrian carotenoid database. J Food Comp Anal 13:435–444.

Murr C, Winklhofer-Roob BM, Schroecksnadel K, Maritschnegg M, Mangge H, Böhm BO, Winkelmann BR, März M and Fuchs D. 2009. Inverse association between serum concentrations of neopterin and antioxidants in patients with and without angiographic coronary artery disease. Atherosclerosis. In press.

Müller H. 1997. Determination of the carotenoid content in selected vegetables and fruit by HPLC and photodiode array detection. Z Lebensm Unters Forsch A 204:88–94.

Nagao A, During A, Hoshino C, Terao J and Olson JA. 1996. Stoichiometric conversion of all-*trans*-β-carotene to retinal by pig intestinal extract. Arch Biochem Biophys 328:57–63.

Nierenberg DW, Stukel TA, Baron JA, Dain BJ and Greenberg ER. 1991. Determinants of increase in plasma concentration of β-carotene after chronic oral supplementation. Am J Clin Nutr 53:1443–1449.

NRC (National Research Council). 1989. Diet and Health: Implications for Reducing Chronic Disease Risk. Washington, DC: National Academies Press.

Nyambaka H and Ryley J. 1996. An isocratic reversed-phase HPLC separation of stereoisomers of the provitamin A carotenoids (α- and β-carotene) in dark green vegetables. Food Chem 55:63–72.

Odriozola-Serrano I, Soliva-Fortuny R, Hernández-Jover T and Martín-Belloso O. 2009. Carotenoid and phenolic profile of tomato juices processed by high intensity pulsed electric fields compared with conventional thermal treatments. Food Chem 112:258–266.

Olmedilla B, Granado F, Blanco I and Vaquero M. 2003. Lutein, but not α-tocopherol, supplementation improves visual function in patients with age-related cataracts: a 2-y double-blind, placebo-controlled pilot study. Nutrition 19(1):21–24.

Olson JA. 1999. Carotenoids. In: Shills ME, Olson JA, Shike M and Ross AC, editors. Modern Nutrition in Health and Disease, 9th ed. Baltimore: Williams & Wilkins, pp. 525–541.

Omaye ST, Krinsky NI, Kagan VE, Mayne ST, Liebler DC and Bidlack WR. 1997. β-Carotene: friend or foe? Fundam Appl Toxicol 40:163–174.

Omenn GS, Goodman GE, Thornquist MD, Balmes J, Cullen MR, Glass A, Keogh JP, Meyskens FL, Valanis B, Williams JH, Barnhart S, Cherniack MG, Brodkin CA and Hammar S. 1996. Risk factors for lung cancer and for intervention effects in CARET, the beta-Carotene and Retinol Efficacy Trial. J Natl Cancer Inst 88:1550–1559.

Opiyo AM and Tie-Jin Y. 2005. The effects of 1-methylcyclopropene treatment on the shelf life and quality of cherry tomato (*Lycopersicon esculentum* var. cerasiforme) fruit. Int J Food Sci Technol 40:665–673.

Ornelas-Paz JJ, Yahia EM and Gardea AA. 2007. Identification and quantification of xanthophyll esters, carotenes and tocopherols in the fruit of seven Mexican mango cultivars by liquid chromatography–APcI$^+$-time of flight mass spectrometry. J Agric Food Chem 55:6628–6635.

Ornelas-Paz JJ, Yahia EM and Gardea-Béjar AA. 2008a. Changes in external and internal color during postharvest ripening of "Manila" and "Ataulfo" mango fruit and relationship with carotenoid content determined by liquid chromatography-APcI$^+$-time-of-flight mass spectrometry. Postharvest Biol Technol 50:145–152.

Ornelas-Paz JJ, Failla ML, Yahia EM and Gardea-Béjar AA. 2008b. Impact of the stage of ripening and dietary fat on *in vitro* bioaccessibility of β-carotene in "Ataulfo" mango. J Agric Food Chem 56:1511–1516.

Padula M and Rodriguez-Amaya DB. 1986. Characterization of the carotenoids and assessment of the vitamin A value of Brazilian guavas (*Psidium guajava* L.). Food Chem 20:11–19.

Padula M and Rodriguez-Amaya DB. 1987. Changes in individual carotenoids and vitamin C on processing and storage of guava juice. Acta Aliment 16:209–216.

Palace VP, Khaper N, Qin Q and Singal PK. 1999. Antioxidant potentials of vitamin A and carotenoids and their relevance to heart disease. Free Rad Biol Med 26:746–761.

Palozza P and Krinsky NI. 1992a. Antioxidant effects of carotenoids *in vivo* and *in vitro*: an overview. Methods Enzymol 213:403–420.

Palozza P and Krinsky NI. 1992b. Beta-carotene and alpha-tocopherol are synergistic antioxidants. Arch Biochem Biophys 297:184–187.

Pasquier B, Armand M, Guillon F, Castelain C, Borel P, Barry JL, Pieroni G and Lairon D. 1996. Viscous soluble dietary fibers alter emulsification and lipolysis of triacylglycerols in duodenal medium *in vitro*. J Nutr Biochem 7:293–302.

Patil BS, Vanamala J and Hallman G. 2004. Irradiation and storage influence on bioactive components and quality of early and late season "Rio Red" grapefruit (*Citrus paradisi* Macf.). Postharvest Biol Technol 34:53–64.

Perera CO and Yen GM. 2007. Functional properties of carotenoids in human health. Int J Food Prop 10:201–230.

Perez-Carrillo E and Yahia EM. 2004. Effect of postharvest hot air and fungicide treatments on the quality of "Maradol" papaya (*Carica papaya*). J Food Qual 27:127–139.

Perkins-Veazie PM and Collins JK. 2004. Flesh quality and lycopene stability of minimally processed watermelon. Postharv Biol Technol 31(2):159–166.

Pott I, Breithaupt DE and Carle R. 2003b. Detection of unusual carotenoid esters in fresh mango (*Mangifera indica* L. Cv. Kent). Phytochemistry 64:825–829.

Pott I, Marx M, Neidhart S, Mühlbauer W and Carle R. 2003a. Quantitative determination of β-carotene stereoisomers in fresh, dried, and solar-dried mango (*Mangifera indica* L.). J Agric Food Chem 51:4527–4531.

Prono-Widayat H, Schreiner M, Huyskens-Keil S and Lüdders, P. 2003. Effect of ripening stage and storage temperature on postharvest quality of pepino (*Solanum muricatum* Ait.). J Food Agric Environ 1:35–41.

Rahman FMM and Buckle KA. 1980. Pigment changes in capsicum cultivars during maturation and ripening. J Food Technol 15:241–249.

Ramos DMR and Rodriguez-Amaya DB. 1987. Determination of the vitamin A value of common Brazilian leafy vegetables. J Micronutr Anal 3:147–155.

Rankins J, Sathe SK and Spicer MT. 2008. Solar drying of mangoes: preservation of an important source of vitamin A in French-speaking West Africa. J Am Diet Assoc 108:986–990.

Ribaya-Mercado JD, Solon FS, Solon MA, Cabal-Barza MA, Perfecto CS, Tang G, Solon JA, Fjeld CR and Rusell RM. 2000. Bioconversion of plant carotenoids to vitamin A in Filipino school-aged children varies inversely with vitamin A status. Am J Clin Nutr 72:455–465.

Richer S, Stiles W, Statkute L, Pulido J, Frankowski J, Rudy D, Pei K, Tsipursky M and Nyland J. 2004. Double-masked, placebo-controlled, randomized trial of lutein and antioxidant supplementation in the intervention of atrophic age-related macular degeneration: the Veterans LAST study (Lutein Antioxidant Supplementation Trial). Optometry 75:216–230.

Riedl J, Linseisen J, Hoffmann J and Wolfram G. 1999. Some dietary fibers reduce the absorption of carotenoids in women. J Nutr 129:2170–2176.

Rivera-Pastarna D, Yahia EM and Gonzalez-Aguilar G. 2009. Identification and quantification of carotenoids and phenolic compounds in papaya using mass spectroscopy. In preparation.

Rizzi MB, Dagli ML, Jordao AA, Penteado MV and Moreno FS. 1997. Beta-carotene inhibits persistent and stimulates remodeling c-GT-positive preneoplastic lesions during early promotion of hepatocarcinogenesis. Int J Vitam Nutr Res 67:415–422.

Rock CL and Swendseid ME. 1992. Plasma β-carotene response in humans after meals supplemented with dietary pectin. Am J Clin Nutr 55:96–99.

Rodrigo M and Zacarias L. 2007. Effect of postharvest ethylene treatment on carotenoid accumulation and the expression of carotenoid biosynthetic genes in the flavedo of orange (*Citrus sinensis* L. Osbeck) fruit. Postharvest Biol Technol 43:14–22.

Rodriguez EB and Rodriguez-Amaya DB. 2007. Formation of apocarotenals and epoxycarotenoids from β-carotene by chemical reactions and by autoxidation in model systems and processed foods. Food Chem 101(2):563–572.

Rodriguez-Amaya D. 2001. A Guide to Carotenoid Analysis in Foods. Washington, DC: ILSI Press, International Life Science Institute.

Rodriguez-Amaya DB. 1999. Changes in carotenoids during processing and storage of foods. ALAN 49(1):38–47.

Rodriguez-Amaya DB. 1997. Carotenoids and Food Preparation: The Retention of Provitamin A Carotenoids in Prepared, Processed, and Stored Foods. Washington, DC: OMNI/USAID.

Rodriguez-Amaya DB. 1993. Nature and distribution of carotenoids in foods. In: Charalambous G, editor. Shelflife Studies of Foods and Beverages. Chemical, Biological, Physical and Nutritional Aspects. Amsterdam: Elsevier Science, pp. 547–589.

Rodriguez-Amaya DB and Kimura M. 2004. HarvestPlus Handbook for Carotenoid Analysis. HarvestPlus Technical Monograph 2. Washington, DC and Cali: International Food Policy Research Institute (IFPRI) and International Center for Tropical Agriculture (CIAT).

Rodriguez DB, Tanaka Y, Katayama T, Simpson KL, Lee T-C and Chichester CO. 1976. Hydroxylation of β-carotene on micro-cel C. J Agric Food Chem 24:819–822.

Ronen G, Cohen M, Zamir D and Hirschberg J. 1999. Regulation of carotenoid biosynthesis during tomato fruit development: expression of the gene for lycopene epsilon-cyclase is down-regulated during ripening and is elevated in the mutant Delta. Plant J 17:341–351.

Roodenburg AJ, Leenen R, van het Hof KH, Weststrate JA and Tijburg LB. 2000. Amount of fat in the diet affects bioavailability of lutein esters but not of alpha–carotene, beta-carotene and vitamin E in humans. Am J Clin Nutr 71:1187–1193.

Ruiz Cruz S, Islas Osuna M, Sotelo Mundo RR, Vázquez-Ortiz F and González-Aguilar G. 2007. Sanitation procedure affects biochemical and nutritional changes of shredded carrots. J Food Sci 72:S146–S152.

Sampaio ARD, Chagas CEA, Ong TP and Moreno FS. 2007. Vitamin A and β-carotene inhibitory effect during 1,2-dimethylhydrazine induced hepatocarcinogenesis potentiated by 5-azacytidine. Food Chem Tox 45:563–567.

Schalch W. 1992. Carotenoids in the retina—a review of their possible role in preventing or limiting damage caused by light and oxygen. In Emerit I and Chance B, editors. Free Radicals and Aging. Basel: Birkhauser Verlag.

Schlatterer J and Breithaupt DE. 2005. Cryptoxanthin structural isomers in oranges, orange juice and other fruits. J Agric Food Chem 53:6355–6361.

Schmidt R. 2004. Deactivation of singlet oxygen by carotenoids: internal conversion of excited encounter complexes. J Phys Chem 108:5509–5513.

Scott KJ and Rodríguez-Amaya D. 2000. Pro-vitamin A carotenoids conversion factors: retinol equivalents–fact or fiction? Food Chem 69:125–127.

Scott CE and Eldridge AL. 2005. Comparison of carotenoid content in fresh, frozen and canned corn. J Food Compost Anal 18:551–559.

Setiawan B, Sulaeman A, Giraud DW and Driskell JA. 2001. Carotenoid content of selected Indonesian fruits. J Food Compost Anal 14:169–176.

Simonetti P, Porrini M and Testolin G. 1991. Effect of environmental factors and storage on vitamin content of *Pisum sativum* and *Spinacea oleracea*. Ital J Food Sci 3:187–196.

Simpson KL, Rodriguez DB and Chichester CO. 1976. In Goodwin TW, editor. Chemistry and Biochemistry of Plant Pigments, 2nd ed. London: Academic Press, pp. 780-842.

Sklan D, Yosefov T and Friedman A. 1989. The effects of vitamin A, betacarotene and canthaxanthin on vitamin A metabolism and immune responses in the chick. Int J Vitam Nutr Res 59:245–250.
Solovchenko AE, Avertcheva OV and Merzlyak MN. 2006. Elevated sunlight promotes ripening-associated pigment changes in apple fruit. Postharvest Biol Technol 40:183–189.
Soto-Zamora G, Yahia EM, Brecht JK and Gardea A. 2005. Effect of postharvest hot air treatment on the quality of "Rhapsody" tomato fruit. J Food Qual 28:492–504.
Sozzi GO, Trinchero GD and Fraschina AA. 1999. Controlled-atmosphere storage of tomato fruit: low oxygen or elevated carbon dioxide levels alter galactosidase activity and inhibit exogenous ethylene action. J Sci Food Agric 79(8):1056–1070.
Stahl W and Sies H. 1992. Uptake of lycopene and its geometrical isomers is greater from heat-processed than from unprocessed tomato juice in humans. J Nutr 122:2161–2166.
Stahl W and Sies H. 2003. Antioxidant activity of carotenoids. Mol Aspects Med 24:345–351.
Stahl W and Sies H. 2005. Bioactivity and protective effects of natural carotenoids. Biochim Biophys Acta 1740:101–107.
Stahl W, Heinrich U, Jungmann H, Sies H and Tronnier H. 2000. Carotenoids and carotenoids plus vitamin E protect against ultraviolet light-induced erythema in humans. Am J Clin Nutr 71:795–798.
Stahl W, Heinrich U, Wiseman S, Eichler O, Sies H and Tronnier H. 2001. Dietary tomato paste protects against ultraviolet light-induced erythema in humans. J Nutr 131:1449–1451.
Stahl W, Junghans A, de Boer B, Driomina E, Briviba K and Sies H. 1998. Carotenoid mixtures protect multilamellar liposomes against oxidative damage: synergistic effects of lycopene and lutein. FEBS Lett 427:305–308.
Stommel J, Abbott JA, Saftner RA and Camp MJ. 2005. Sensory and objective quality attributes of β-carotene and lycopene-rich tomato fruit. J Am Soc Hort Sci 130:244–251.
Sugarman SB, Mobarhan S, Bowen PE, Stacewicz-Sapuntzakis M and Langenberg P. 1991. Serum time curve characteristics of a fixed dose of β-carotene in young and old men. J Am Coll Nutr 10:297–307.
Suzuki Y, Toshiya A, Yukie M, Hirofumi T and Masaya K. 2005. Suppression of the expression of genes encoding ethylene biosynthetic enzymes in harvested broccoli with high temperature treatment. Postharv Biol Technol 36:265–271.
Takaichi S, Matsui K, Nakamura M, Muramatsu M and Hanada S. 2003. Fatty acids of astaxanthin esters in krill determined by mild mass spectrometry. Comp Biochem Physiol 136(2):317–322.
Takama F and Saito S. 1974. Studies on the storage of vegetables and fruits II. Total carotene content of sweet pepper, carrot, leek and parsley during storage. J Agric Sci (Japan) 19:11.
Takyi EK. 2001. Bioavailability of carotenoids from vegetables versus supplements. In: Watson RR, editor. Vegetables, Fruits, and Herbs in Health Promotion. Boca Raton, FL: CRC Press, pp. 19–34.
Talcott ST, Moore J, Lounds-Singleton A and Percival SS. 2005. Ripening associated phytochemical changes in mangos (*Mangifera indica*) following thermal quarantine and low-temperature storage. J Food Sci 70(5):337–341.
Tang FY, Cho HJ, Pai MH and Chen YH. 2008. Concomitant supplementation of lycopene and eicosapentaenoic acid inhibits the proliferation of human colon cancer cells. J Nutr Biochem. In press.
Tang G, Serfaty C and Rusell RM. 1995. Effects of gastric acidity on intestinal absorption of beta carotene in humans. FASEB J 9:A442.
Tanumihardjo SA. 2002. Factors influencing the conversion of carotenoids to retinol: bioavailability to bioconversion to bioefficacy. Int J Vitam Nutr Res 72:40-5.
Terão J. 1989. Antioxidant activity of β-carotene-related carotenoids in solution. Lipids 24:659–661.
Thakkar SK, Maziya-Dixon B, Dixon AG and Failla ML. 2007. Beta-carotene micellarization during *in vitro* digestion and uptake by Caco-2 cells is directly proportional to beta-carotene content in different genotypes of cassava. J Nutr 137:2229–2233.
Thomas P and Janave MT. 1975. Effects of gamma irradiation and storage temperature on arotenoids and ascorbic acid content of mangoes on ripening. J Sci Food Agric 26:1503–1512.
Thompson KA, Marshall MR, Sims CA, Wei CI, Sargent SA and Scott JW. 2000. Cultivar, maturity, and heat treatment on lycopene content in tomatoes. J Food Sci 65(5):791–795.
Tonon RV, Baroni AF and Hubinger MD. 2007. Osmotic dehydration of tomato in ternary solutions: influence of process variables on mass transfer kinetics and an evaluation of the retention of carotenoids. J Food Eng 82:509–817.

Tran TLH and Raymundo LC. 1999. Biosynthesis of carotenoids in bittermelon at high temperature. Phytochemistry 52:275–280.

Tyssandier V, Cardinault N, Caris-Veyrat C, Amiot M-J, Grolier P, Bouteloup C, Azais-Braesco V and Borel P. 2002. Vegetable-borne lutein, lycopene, and β-carotene compete for incorporation into chylomicrons, with no adverse effect on the medium-term (3-wk) plasma status of carotenoids in humans. Am J Clin Nutr 75:526–534.

Unlu NZ, Bohn T, Clinton SK and Schwartz SJ. 2005. Carotenoid absorption from salad and salsa by humans is enhanced by the addition of avocado or avocado oil. J Nutr 135:431–436.

van Breemen RB and Pajkovic N. 2008. Multitargeted therapy of cancer by lycopene. Cancer Lett 269(2):339–351.

van den Berg H, Faulks R, Granado HF, Hirschberg J, Olmedilla B, Sandmann G, Southon S and Stahl W. 2000. The potential for the improvement of carotenoid levels in foods and the likely systemic effects. J Sci Food Agric 80:880–912.

van Eenwyk J, Davis FG and Bowen PE. 1991. Dietary and serum carotenoids and cervical intraepithelial neoplasia. Int J Cancer 48:34–38.

van het Hof KH, de Boer BCJ, Tijburg LBM, Lucius BRHM, Zijp I, West CE, Hautvast JGAJ and Weststrate JA. 2000. Carotenoid bioavailability in humans from tomatoes processed in different ways determined from the carotenoid response in the triglyceride-rich lipoprotein fraction of plasma after a single consumption and in the plasma after four days of consumption. J Nutr 130:1189–1196.

van Jaarsveld PJ, Marais D, Harmse E, Nestel P, Rodriguez-Amaya DB. 2006. Retention of β-carotene in boiled, mashed orange-fleshed sweet potato. J Food Comp Anal 19(4):321–329.

van Lieshout M, West CE, Permaesih D, Wang Y, Xu X, van Breeman RB, Creemers AF, Verhoeven MA, Lugtenburg J. 2001. Bioefficacy of β-carotene dissolved in oil studied in children in Indonesia. Am J Clin Nutr 73:949–958.

van Vliet T, Schereurs WHP, van den Berg H. 1995. Intestinal β-carotene absorption and cleavage in men: response of β-carotene and retinyl esters in the triglyceride-rich lipoprotein fraction after a single oral dose of β-carotene. Am J Clin Nutr 62:110–116.

Vázquez-Caicedo AL, Sruamsiri P, Carle R, Neidhart S. 2005. Accumulation of all-*trans*-β-carotene and its 9-*cis* and 13-*cis* stereoisomers during postharvest ripening of nine Thai mango cultivars. J Agric Food Chem 53:4827–4835.

Von Elbe JH, Shwartz SJ. 1996. Colorants. In: Fennema OR, editor. Food Chemistry, 3rd ed. New York: Marcel Dekker, pp. 651–722.

Wall MM. 2006. Ascorbic acid, vitamin A, and mineral composition of banana (*Musa* sp.) and papaya (*Carica papaya*) cultivars grown in Hawaii. J Food Comp Anal 19:434–443.

Wang CY. 2000. Effect of heat treatment on postharvest quality of kale, collard and Brussels sprouts. Acta Hort 518:71–78.

Wang W, Onnagawa M, Yoshie Y and Suzuki T. 2001. Binding of bile salts to soluble and insoluble dietary fibers of seaweeds. Fish Sci 67:1169–1173.

Wang XD, Tang G, Fox JG, Krinsky NI and Rusell RM. 1991. Enzymatic conversion of β-carotene into β-apo-carotenals and retinoids by human, monkey, ferret and rat tissues. Arch Biochem Biophys 258:8–16.

Watada A. 1987. Vitamins. In: Weichmann J, editor. Postharvest Physiology of Vegetables. New York: Marcel Dekker, pp. 455–468.

Weichmann J. 1986. The effect of controlled-atmosphere storage on the sensory and nutritional quality of fruits and vegetables. Hort Rev 8:101–127.

Weller P and Breithaupt DE. 2003. Identification and quantification of zeaxanthin esters in plants using liquid chromatography-mass spectrometry. J Agric Food Chem 51:7044–7049.

Wen-ping MA, Zhi-jing NI, He LI and Min C. 2008. Changes of the main carotenoid pigment contents during the drying processes of the different harvest stage fruits of *Lycium barbarum* L. Agric Sci China 7:363–369.

West CE and Castenmiller JJM. 1998. Quantification of the "SLAMENGHI" factor for carotenoid bioavailability and bioconversion. Int J Vit Nutr Res 68:371–377.

Wingerath T, Stahl W and Sies H. 1995. β-Cryptoxanthin selectively increases in human chylomicrons upon ingestion of tangerine concentrate rich in β-cryptoxanthin esters. Arch Biochem Biophys 324:385–390.

Wolf G. 1984. Multiple functions of vitamin A. Physiol Rev 64:873–937.

Wright KP and Kader AA. 1997. Effect of controlled-atmosphere storage on the quality and carotenoid content of sliced persimmons and peaches. Postharvest Biol Technol 10:89–97.

Yahia EM, editor. 2009. Modified and Controlled Atmospheres for Transportation, Storage and Packaging of Horticultural Commodities. Boca Raton, FL: CRC Press/Taylor & Francis. In press.

Yahia EM, Ornelas-Paz JJ and Gardea A. 2006. Extraction, separation and partial identification of "Ataulfo" mango fruit carotenoid. Acta Hort 712:333–338.

Yahia EM, Soto-Zamora G, Brecht JK and Gardea A. 2007. Postharvest hot air treatment effects on the antioxidant system in stored mature-green tomatoes. Postharvest Biol Technol 44:107–115.

You CS, Parker RS, Goodman KJ, Swanson JE and Corso TN. 1996. Evidence of *cis-trans* isomerization of 9-*cis*-β-carotene during absorption in humans. Am J Clin Nutr 64:177–183.

Young A and Britton G. 1990. Carotenoids and stress. In: Alscher RG and Cumming JR, editors. Stress Responses in Plants: Adaptation and Acclimation Mechanisms. Plant Biology, Volume 12. New York: Wiley-Liss.

Young AJ and Lowe GM. 2001. Antioxidant and prooxidant properties of carotenoids. Arch Biochem Biophys 385:20–27.

Zorn H, Langhoff S, Scheibner M, Nimtz M and Berger RG. 2003. A peroxidase from *Lepista irina* cleaves β,β-carotene to flavor compounds. Biol Chem 384(7):1049–1056.

8 Dietary Fiber and Associated Antioxidants in Fruit and Vegetables

Fulgencio Saura-Calixto, Jara Pérez-Jiménez, and Isabel Goñi

Introduction

The health benefits of fruit and vegetables in the human diet are well known and are supported by strong scientific evidence. A systematic review of studies reveals favorable effects on risk factors for chronic diseases (Dragsted and others 2006; Esmaillzadeh and others 2006; Stea and others 2008) and a significant association between high consumption of fruit and vegetables and lower total mortality, mortality from coronary heart disease, and mortality from cancer (Bazzano and others 2003; Heidemann and others 2008). A daily consumption of at least 400 g of fruit and vegetables is therefore generally recommended by national and international health institutions, including the World Health Organization, for adults in Western societies. This is taken to equate to five (3 vegetable and 2 fruit portions) or more (5 vegetable and 4 fruit portions) servings (FAO 2003; Lin and others 2003).

Fruits and vegetables are generally high in water and low in fat, and, in addition to vitamins and minerals, they contain significant amounts of dietary fiber (DF) and phytochemicals—mainly polyphenols and carotenoids—with significant biological properties, including antioxidant activity.

The composition and physicochemical structure of DF and phytochemicals in fruit and vegetables have specific characteristics that lend this food group significant nutrition- and health-related properties. Indeed, the potential health benefits of fruit and vegetables are mainly attributed to the effects of DF and antioxidants.

Fruits and vegetables generally possess a high soluble DF/insoluble DF ratio that is associated with slow glucose absorption, high colonic fermentability, lower serum cholesterol levels, and enhancement of immune functions (McCleary and Prosky 2001).

Vitamins (C and E), polyphenolic compounds, and carotenoids are the main groups of antioxidants present in fruits and vegetables. Vitamins are single molecules, but polyphenols and carotenoids are made up of hundreds of compounds with a wide range of structures and molecular masses. The intake of these antioxidants can lead to sustained reduction of the kind of oxidative damage to lipids, proteins, and DNA that is associated with the development of chronic diseases (Evans and Halliwell 2001).

DF and antioxidants are generally addressed separately as groups of food constituents in both chemical and nutritional studies. However, it is a little-known fact that a substantial proportion of the antioxidant polyphenols and carotenoids contained in fruit and vegetables are linked to DF (Saura-Calixto and others 2007), and some of the postulated benefits of fiber intake can be attributed to these associated compounds. These compounds are not bioaccessible in the human small intestine, but they may exert significant health effects when they reach the colon (Gonthier and others 2006).

With regard to chemical analysis, literature data may underestimate phytochemical antioxidant contents, because studies generally focus only on antioxidants extracted by aqueous-organic solvents without taking antioxidants linked to DF into account (Pérez-Jiménez and Saura-Calixto 2005; Saura-Calixto and others 2007).

This chapter focuses on the DF in fruit and vegetables, with special attention to the phytochemicals associated with them. It also discusses the contribution of fruit and vegetables to the intake of DF and antioxidants in the diet.

Dietary Fiber in Fruits and Vegetables

Dietary Fiber

DF is a major constituent of plant foods, and its importance in nutrition and health is widely recognized. Numerous clinical and epidemiological studies have addressed the role of DF in intestinal health and in the prevention of cardiovascular disease and cancer, obesity, and diabetes (Sungsoo Cho and Dreher 2001; Spiller 2005). The recommended daily intake of DF is 25–30 g/person (Lunn and Buttriss 2007).

DF has been defined as plant polysaccharides and lignin which are resistant to hydrolysis by the digestive enzymes of man (Trowell 1976). These DF components are neither degraded nor absorbed during their passage through the upper part of the gastrointestinal tract, and they can exert nutritionally important effects by slowing down gastric emptying and affecting nutrient assimilation in the small intestine. When nondigestible compounds pass into the large intestine, they are degraded by bacterial enzymes and completely or partially fermented to produce short-chain fatty acids as well as gases and water (Brownlee and others 2006).

Total dietary fiber (TDF) includes many components that are categorized according to whether or not they are soluble in the human digestive system. Insoluble dietary fiber (IDF) includes cellulose and other polysaccharides along with noncarbohydrate compounds such as lignin and cutin and other cell-wall constituents. Soluble dietary fiber (SDF) includes pectins, beta-glucans, arabinoxylans, galactomannans, and other indigestible polysaccharides and oligosaccharides.

SDF content is high in fruits, vegetables, and legumes and is associated with colonic degradation and high fermentability, slow glucose absorption, enhanced immune functions, and lower serum cholesterol levels. IDF is predominant in cereals and legumes; fermentation is slow and incomplete, and they have more pronounced effects on bowel habit.

Nowadays there is scientific evidence that, besides plant polysaccharides and lignin, other indigestible compounds such as resistant starch, oligosaccharides, Maillard compounds, and phytochemicals—mainly polyphenols—can be considered DF constituents (Saura-Calixto and others 2000). Of these substances, resistant starch is a major constituent in cereals, whereas phytochemicals are the most important such substance in fruits and vegetables. Here, we address mainly polyphenols and carotenoids associated with DF in fruits and vegetables because of the important biological properties derived from them.

Table 8.1. Dietary fiber content (% edible fresh matter) in common fruits and vegetables

Fruit	Moisture (%)	Soluble Dietary Fiber	Insoluble Dietary Fiber	Total Dietary Fiber
Apple	85.2	0.70	1.86	2.56
Avocado	64.6	2.03	3.51	5.53
Banana	77.2	0.64	1.16	1.80
Broccoli	89.1	0.44	3.06	3.50
Cabbage	90.8	0.46	1.79	2.24
Carrot	88.0	0.49	2.39	2.88
Cauliflower	90.7	0.47	2.15	2.62
Cucumber	95.6	0.20	0.94	1.14
Cherry	92.2	0.6	0.9	1.50
Custard apple	73.2	1.46	2.36	3.82
Fig	78.6	2.33	2.98	5.31
Grape	81.9	0.24	1.08	1.32
Guava	80.3	1.47	9.45	10.61
Kiwifruit	83.3	1.01	2.27	3.28
Lettuce	95.5	0.10	0.88	0.98
Mango	81.8	0.85	1.04	1.93
Medlar	87.7	0.86	1.29	2.15
Melon	91.4	0.20	0.91	1.11
Nectarine	85.3	0.98	1.06	2.04
Onion	85.6	0.71	1.22	1.93
Orange	87.2	0.95	0.9	1.85
Papaya	84.4	1.24	1.80	3.04
Peach	85.8	0.67	1.13	1.80
Pear	83.3	0.75	3.02	3.77
Pepper, green	94.4	0.53	0.99	1.52
Persimmon	80.9	0.27	2.53	2.80
Pineapple	85.7	0.27	1.86	2.13
Plum	86.9	0.79	1.28	2.07
Pomegranate	81.3	0.50	2.30	2.80
Spinach	90.3	0.77	2.43	3.20
Strawberry	92.9	0.70	1.60	2.30
Tomato	94.4	0.15	1.19	1.34
Watermelon	91.4	0.17	0.27	0.44

Sources: Li and Andrews 2002; Pak 2003; Ramalu and Rao 2003.

Dietary Fiber in Fruits and Vegetables

Fruits and vegetables are a major source of DF in the diet. Table 8.1 shows the SDF, IDF, and TDF contents of common fruits and vegetables, expressed in edible fresh weight. TDF content ranges from around 1 g/100 g fresh matter for grape or lettuce to more than 10 g/100 g fresh matter for guava. When results are expressed as dry matter,

DF is one of their major constituents, ranging from 5% in watermelon and 33% in spinach to 54% in guava.

As well as the TDF content, the proportion between SDF and IDF is an important nutritional parameter because of the different physiological effects that these exert. Proportions of SDF are characteristically high in fruits and vegetables, although they can differ considerably. SDF is less than 20% of TDF in grape but more than 30% in onion and higher than 50% in fig.

DF content of fruits and vegetables may be affected by several factors, such as variety, agronomic and climatic conditions of cultivation, stage of ripening of the fruit, postharvest processing, and cooking procedure. For instance, DF content in several grape varieties ranges from 0.89 to 2.20 g/100 g fresh matter (Pak 2003). Although traditional vs. organic growing does not seem to affect DF content in fruits such as banana or plum, these values may be affected by the use of different types of organic cultivation (Forster and others 2002; Lombardi-Boccia 2004). Long-term storage of fruits produces an increase in Klason lignin content and changes in pectin composition and solubility, although these changes are reduced if storage is controlled (Marlett 2000). Cooking produces a significant increase in SDF and a concomitant decrease in IDF in vegetables such as carrot or cabbage in relation to the raw samples (Khanum and others 2000). DF content also varies depending on the fraction of the fruit or vegetable considered; peeled apple and pear contain 11 and 23% DF, respectively, less than the whole fruit (Holland and others 1991).

The properties of DF in fruits and vegetables depend not only on the content, but also on the composition (neutral sugars and uronic acids in SDF; neutral sugars, uronic acids and Klason lignin in IDF). Table 8.2 shows the different DF compositions of some fruits and vegetables. Banana and orange, for example, have a similar TDF content, but the uronic acid content in SDF of orange is three times that of banana. There are also significant differences between the samples in terms of neutral sugars; mannose content in IDF, for example, ranges from 0.3 to 1.1 g/100 g dry weight in carrot and aubergine, respectively. The monomeric composition of DF corresponds to the most common polymers in DF of fruits and vegetables: pectins, arabinoxylans, arabinogalactans, etc.

Klason lignin content is lower in fruits than in vegetables, ranging from 0.7 to 6.8 g/200 g dry weight in orange and broccoli, respectively. Fruits and vegetables also present differences in lignin composition; some, like carrot or kiwi, are rich in guaiacyl units, and others, like pear or apple, have a balanced content of guaiacyl and syringyl units (Bunzel and others 2005).

It has recently been reported that a fraction of food starch, named resistant starch (Asp and others 1996), is not digestible by humans. Nowadays resistant starch is considered a DF constituent and is a major constituent of IDF in starchy foods. However, resistant starch is absent in most fruits and vegetables, with the exception of bananas, which contain more than 15% dry weight when the fruit is unripe (Goñi and others 1996).

Nutrition and Health Claims

The European Union recently approved the *Regulation on nutrition and health claims made on foods* (European Parliament 2006). Regarding DF, it establishes that a food

Table 8.2. Composition of dietary fiber in several fruits and vegetables (g/100 g edible dry weight)

				Noncellulosic Polysaccharides							
Sample		**KL**	**Cel**	**Rha**	**Fuc**	**Ara**	**Xyl**	**Man**	**Gal**	**Glu**	**U.ac.**
Apple	Soluble DF	n.d.	n.d.	0.2	0.1	0.8	0.1	t	0.5	0.2	2.8
	Insoluble DF	4.9	4.0	0.1	0.1	0.7	0.7	0.2	0.5	0.1	0.3
Apricot	Soluble DF	n.d.	n.d.	0.5	t	1.5	0.2	0.1	0.5	0.1	5.1
	Insoluble DF	1.6	4.0	t	t	0.2	0.6	0.3	0.3	0.1	0.2
Aubergine	Soluble DF	n.d.	n.d.	0.7	t	0.9	t	t	2.8	t	11.2
	Insoluble DF	4.22	10.4	t	t	0.4	1.8	1.1	0.7	0.5	0.5
Banana	Soluble DF	n.d.	n.d.	t	t	0.1	t	0.6	0.1	t	2.0
	Insoluble DF	8.03	1.0	t	t	0.2	0.2	t	0.1	0.1	0.1
Broccoli	Soluble DF	n.d.	n.d.	1.0	0.1	3.0	0.1	0.2	1.8	0.3	7.9
	Insoluble DF	6.8	10.0	t	0.1	0.6	1.5	0.7	0.9	0.1	0.5
Cabbage	Soluble DF	n.d.	n.d.	0.7	0.1	3.7	0.2	0.1	2.7	0.2	10.1
	Insoluble DF	5.0	13.9	t	0.1	1.1	1.6	0.9	1.6	t	0.1
Carrot	Soluble DF	n.d.	n.d.	0.7	t	1.7	t	0.1	3.0	t	5.9
	Insoluble DF	2.0	6.4	t	t	0.3	0.3	0.3	0.4	0.1	0.3
Orange	Soluble DF	n.d.	n.d.	0.3	t	1.9	0.1	0.1	1.4	0.1	5.9
	Insoluble DF	0.7	3.4	t	t	0.3	0.5	0.4	0.4	t	0.3

DF: dietary fiber; Cel: cellulose; Rha: rhamnose; Fuc: fucose; Ara: arabinose; Xyl: xylose; Man: mannose; Gal: galactose; Glu: glucose; U. ac.: uronic acids; KL: Klason lignin; n.d.: not determined; t: traces
Source: Englyst and Cummings 1988; Englyst and others 1988; Marlett and Vollendorf 1994.

can be categorized as a "source of fibre" when the food contains at least 3 g of DF per 100 g or at least 1.5 g of DF per 100 kcal, and "high fibre" when the food contains at least 6 g of DF per 100g or at least 3 g of DF per 100 kcal.

Considering TDF content in fruits and vegetables (see Table 8.1), the "source of fibre" claim could be applied to broccoli and spinach among vegetables, and kiwi, pear, and cherimoya among fruits, and the "high fibre" claim to guava. Moreover, considering the content of DF in relation to the caloric value provided by the food, all vegetables and most fruits included in Table 8.1 could be classified as "high fibre." Avocado, cherry, grape, melon, and watermelon can be considered "sources of fibre."

Phytochemicals and Antioxidant Capacity Associated with Dietary Fiber

DF does not constitute a defined chemical group but a combination of chemically heterogeneous substances. The procedures most widely used today to determine DF in foods were developed to determine nonstarch polysaccharides (NSP)(Englyst and Cummings 1988; Englyst and others 1988) or NSP plus lignin (Prosky and others 1992), which are the compounds included in the original physiological definition of DF. The DF data shown previously for fruits and vegetables (see Tables 8.1 and 8.2) correspond to nonstarch polysaccharides plus lignin.

Table 8.3. Dietary fiber content and phenolic compounds associated with insoluble dietary fiber (% on dry basis; mean values ± SD; $n \geq 1$)

	Insoluble Dietary Fiber	**Soluble Dietary Fiber**	**Condensed Tannins**	**Hydrolyzable Tannins**
Onion	18.83	0.53	not detected	0.41
Green beans	27.51	4.79	not detected	0.80
Lettuce	19.76	1.20	not detected	0.67
Tomato	36.88	1.26	not detected	0.66
Cherry	13.75	1.40	0.11	0.18
Plum	14.60	3.60	0.11	0.23
Strawberry	21.51	3.39	0.12	0.44
Tangerine	23.60	1.20	not detected	0.50
Orange	23.52	2.98	not detected	0.34
Apple with peel	14.89	1.73	0.20	0.16
Pear with peel	16.63	1.20	0.29	0.49

However, DF of fruits and vegetables transports a significant amount of polyphenols and carotenoids linked to the fiber matrix through the human gut (Goñi and others 2006; Saura-Calixto and others 2007). Therefore, associated phytochemicals can make a significant contribution to the health benefits attributed to the DF of fruits and vegetables.

Phytochemicals

Appreciable amounts of polyphenols (PPs) associated with both IDF and SDF have been reported in fruit, vegetables, and beverages (Tables 8.3 and 8.4). These compounds may be considered DF constituents in view of the similarity of their properties in terms of resistance to digestive enzymes and colonic fermentability.

Polyphenols are significant constituents in the IDF of the samples analyzed, accounting for 1.4% to 4.7% (average, 2.5%). PPs are therefore an important constituent of DF in fruits and vegetables; they are mainly condensed tannins (proanthocyanidins) and hydrolyzable tannins.

Beverages contain appreciable amounts of SDF given that a certain amount of soluble indigestible polysaccharides may pass from the solid material to the beverages during processing.

Table 8.4. Components of beverage dietary fiber (g/liter; mean values ± SD; $n \geq 4$)

	Soluble Dietary Fiber	**Associated Polyphenols**
Orange juice	0.79	0.16
Apple juice	1.67	0.05
Red wine	1.40	0.89
Cider	0.17	0.13

DF content in the beverages listed in Table 8.4 ranges from 0.2 to 1.7 g/liter. These DFs contain an appreciable amount of associated phenolic compounds (from 0.05 to 0.89 g/liter). The main polyphenols associated with DF in wine are flavan-3-ols and benzoic acids (Saura-Calixto and Díaz-Rubio 2007), whereas in beer there would be flavonoids, followed by hydroxycinnamic acids linked to arabinoxylans (Díaz-Rubio 2008).

Polyphenols bound to DF can account for a substantial part of total PPs in foods and beverages. These polyphenols are not bioavailable in the human upper intestine and reach the colon, where they become fermentable substrates for bacterial microflora, along with the DF.

The fermentation of polyphenols in the colon improves antioxidant status and yields different metabolites with potential systemic effects.

As in the case of polyphenols, some of the carotenoids contained in fruits and vegetables are bound to the DF matrix. For instance, between 20% and 70% of the β-carotene and lutein in green leafy vegetables was found to be associated with the DF matrix. The bulk of the unreleased carotenoids was associated with the IDF, and a very small proportion was associated with the SDF (Serrano and others 2005; Goñi and others 2006).

Antioxidant Capacity

Chronic intake of dietary antioxidants from fruits and vegetables would lead to a sustained decrease in oxidative damage to key structures in the body, including lipids, proteins, and DNA. Vitamins, polyphenolic compounds, and carotenoids are recognized as the main groups of antioxidants present in fruits and vegetables. Vitamins are single molecules, but polyphenols (flavonoids, phenolic acids, stilbenes, tannins, etc.) and carotenoids (carotenes and xanthophylls) are made up of hundreds of compounds with a wide range of structures and molecular masses.

The antioxidant capacity of a food is derived from the accumulative and synergistic antioxidant power of vitamins, polyphenols, carotenoids, and other minor constituents (Liu and others 2008).

Because part of the total content of antioxidant polyphenols and carotenoids is linked to DF as noted previously, an appreciable proportion of the total antioxidant capacity in fruits and vegetables is associated with DF.

After screening a large number of fruit products, DF with exceptional antioxidant capacity was found in mango peel (Larraruri and others 1996), pineapple shell (Larraruri and others 1997), guava pulp (Jiménez-Escrig and others 2001), and grape pomace (Saura-Calixto 1998). These fibers combine the physiological effects of both DF and antioxidants in a single material. The concept of antioxidant DF was defined on these bases (Saura-Calixto 1998).

Grape antioxidant dietary fiber (GADF), a natural product obtained from grape pomace, is an example of antioxidant DF. Its DF content and composition, including associated polyphenols, are shown in Table 8.5. GADF has a high DF content (more than 70%), with a significant proportion of SDF. Also, GADF presents 0.64% and 18.15% (in dry weight) of polyphenols associated with SDF and IDF, respectively. The most abundant of these polyphenols are high-molecular-weight proanthocyanidins, followed by epicatechin, several anthocyanidins, and benzoic acid; overall, more than

Table 8.5. Dietary fiber content and associated polyphenols of grape antioxidant dietary fiber (g/100 g dry weight)

	Soluble Fiber		Insoluble Fiber
Total content	15.53		57.95
Glucose	0.26		8.28
Galactose	0.45		4.75
Mannose	1.45		3.12
Xylose	9.70		25.84
Arabinose	2.50		10.89
Rhamnose	0.01		0.06
Uronic acids	1.04		4.40
Associated polyphenols	0.64		18.15
Extractable polyphenols	0.64		2.65
Condensed tannins	n.d.		15.50
Antioxidant capacity			
ABTS (μmol Trolox/g dm)		124.40	
ORAC (μmol Trolox/g dm)		214.20	

ABTS: 2,2′-Azino-bis(3-ethylbenzthiazoline-6-sulfonic acid); dm: dry matter; n.d.: not detected; ORAC: oxygen radical absorbance capacity.
Source: Pérez-Jiménez and others 2008.

100 polyphenols have been reported in this matrix (Touriño and others 2008). The synergistic combination of so large a number of polyphenols lends GADF a high antioxidant capacity.

The combination of high DF and associated polyphenol contents in a single matrix endows GADF with specific properties as a dietary supplement and food ingredient. Dietary supplementation of rats with GADF produced hypocholesterolemic effects and increased fat excretion (Martín-Carrión and others 1999, 2000), as well as increased antioxidant status in cecum (Goñi and Serrano 2005) and antiproliferative capacity in colonic epithelium through a decrease in the total number of crypts per milliliter (López-Oliva and others 2006). In humans, supplementation to normo- and hypercholesterolemic subjects for 3.5 months had a reducing effect on several cardiovascular disease risk factors, such as blood pressure or plasma lipid profile (Pérez-Jiménez and others 2008). As a food ingredient, GADF prevents lipid peroxidation in a number of fish and meat products (Sánchez-Alonso and others 2006; Sáyago-Ayerdi and others 2009).

The exceptional properties of GADF are probably also present in some tropical fruit materials.

Contribution of Fruits and Vegetables to the Intake of Dietary Fiber and Antioxidants in the Diet

The daily intake of DF is quantitatively similar in Mediterranean and non-Mediterranean European countries (around 20 g per capita) (Elmadfa and Weichselbaum 2005). However, there are qualitative differences arising from the fact that a

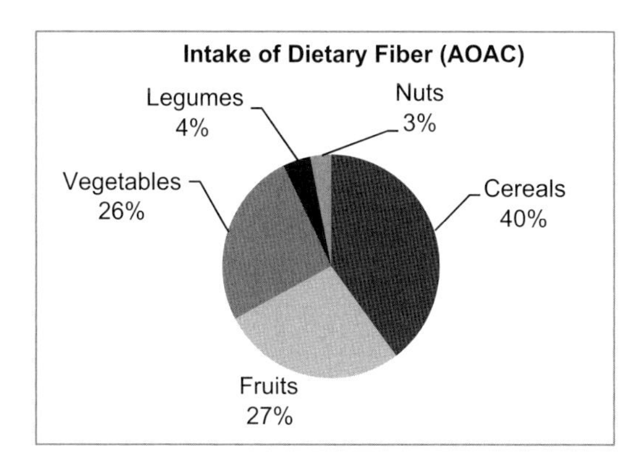

Figure 8.1. Contribution of the various plant food groups to dietary fiber intake in the Spanish diet, according to AOAC method. Sources: Saura-Calixto & Goñi 1993; Goñi 2001.

large proportion of the DF intake in Mediterranean countries comes from fresh fruits and vegetables, whereas in Northern European countries it comes more from cereals.

The intake of DF in the Spanish diet, a Mediterranean dietary pattern, has been estimated at 18.30 g/person/day, which is equivalent to 7 g/1000 kcal. Figure 8.1 shows that the consumption of fruits and vegetables (266 g of fruits and 331 g of vegetables) accounts for 54% of TDF, with a high SDF/IDF ratio (Saura-Calixto and Goñi 1993; Goñi 2001). These figures are based on a definition of DF as plant polysaccharides and lignin. The inclusion of other indigestible compounds in a wider concept of DF, as mentioned previously, would produce higher intake figures, estimated at 45 to 60 g/person/day (Saura-Calixto and Goñi 2004; Tabernero and others 2007).

Regarding antioxidants, the parameter total dietary antioxidant capacity (TDAC) can be taken to reflect antioxidant intake; it is defined as the antioxidant capacity of all plant foods and beverages (alcoholic and nonalcoholic) consumed daily in a diet and may represent the amount of antioxidant units (Trolox equivalents) present daily in the human gut (Saura-Calixto and Goñi 2006).

TDAC in the Spanish diet has been estimated at 3,500 μmol Trolox equivalents by the ABTS method. The contribution of each specific food to the TDAC was dependent on both food intake and food antioxidant capacity. The largest contributors to the TDAC were beverages (about 68%) and fruits and vegetables (about 20%).

Apart from vitamins, polyphenols and carotenoids are the main antioxidants. Polyphenols are the main contributors to TDAC (90%). Vitamins and carotenoids account for 10%.

It is estimated that the mean total intake of polyphenols in the Spanish diet ranges from 2,590 to 3,016 mg/person/day including polyphenols soluble in aqueous–organic solvents, plus insoluble condensed and hydrolyzable tannins. Fruits and vegetables provide a daily intake of 700–1,000 mg of polyphenols/person/diet; a major fraction of this (600 mg/person/day) is associated with DF (Saura-Calixto and others 2007).

A minor proportion of TDAC comes from carotenoids. Daily intake of carotenoids in the Spanish diet is estimated at between 9 and 13 mg/person (O'Neill and others 2001). Lutein, β-carotene, and lycopene account for over 80% of the total carotenoid intake in the Spanish diet. Like polyphenols, a large proportion of the carotenoids contained in the original sample remained in the food matrix after the digestive enzyme action (estimated at around 75% in fruit and 50% in vegetables). This means that around 7 mg of the total dietary intake of carotenoids is associated with DF and hence is not bioaccessible in the small intestine (Goñi and others 2006).

In short, the DF supplied by fruits and vegetable in the Spanish diet (10 g) transports around 600 mg of polyphenols and 7 mg of carotenoids.

References

Asp NG, Amelsvoort JM and Hautvast JG. 1996. Nutritional implications of resistant starch. Nutr Res Rev 9:1–31.

Bazzano LA, Serdula MK and Liu S. 2003. Dietary intake of fruits and vegetables and risk of cardiovascular disease. Curr Atheroscler Rep 5:492–499.

Brownlee IA, Dettmar PW, Strugala V and Pearson JP. 2006. The interaction of dietary fibres with the colon. Curr Nutr Food Sci 2:243–264.

Bunzel M, Seiler A and Steinhart H. 2005. Characterization of dietary fiber lignins from fruits and vegetables using the DCFR method. J Agric Food Chem 53:9553–9559.

Díaz-Rubio ME. 2008. Fibra dietética en bebidas de la dieta. Determinación, composición y contribución a la ingesta de fibra. Ph.D. thesis. Department of Applied Physical Chemistry, Universidad Autónoma de Madrid, Spain.

Dragsted LO, Krath B, Ravn-Haren G, Vogel UB, Vinggaard AM, Jensen PB, Loft S, Rasmussen SE, Sandstrom B and Pedersen A. 2006. Biological effects of fruits and vegetables. Proc Nutr Soc 65:61–67.

Elmadfa I and Weichselbaum E. 2005. On the nutrition and health situation in the European Union. J Publ Health 13:62–68.

Englyst HN, Bingham SA, Runswick SA, Collinson E and Cummings JH. 1988. Dietary fibre (non-starch polysaccharides) in fruit, vegetables and nuts. J Hum Nutr Diet 1:247–286.

Englyst HN and Cummings JH. 1988. Improved method for measurement of dietary fiber as non-starch polysaccharides in plant foods. JAOAC 71:808–814.

Esmaillzadeh A, Kimiagar M, Mehrabi Y, Azadbakht L, Hu FB and Willett WC. 2006. Fruit and vegetables intakes, C-reactive protein and the metabolic syndrome. Am J Clin Nutr 84:1489–1497.

European Parliament. 2006. Regulation on nutrition and health claims made on foods. Regulation 1994/2006 of the European Parliament and of the Council.

Evans P and Halliwell B. 2001. Micronutrients: oxidant/antioxidant status. Br J Nutr 85:s67–s74.

FAO. 2003. Increasing fruit and vegetable consumption becomes a global priority. http://www.fao.org/english/newsroom/focus/2003/fruitveg1.htm.

Forster MP, Rodríguez Rodríguez E and Díaz Romero C. Differential characteristics in the chemical composition of bananas from Tenerife (Canary Islands) and Ecuador. 2002. J Agric Food Chem 50:7586–7592.

Goñi I. 2001. Dietary fiber intake in Spain: recommendations and actual consumption patterns. In: Sungsoo Cho S and Dreher ML, editors. Handbook of Dietary Fiber. New York: Marcel Dekker.

Goñi I, García-Diz L, Mañas E and Saura-Calixto F. 1996. Analysis of resistant starch: A method for foods and food products. Food Chem 56:445–449.

Goñi I and Serrano J. 2005. The intake of dietary fiber from grape seeds modifies the antioxidant status in rat cecum. J Sci Food Agric 85:1877–1881.

Goñi I, Serrano J and Saura-Calixto F. 2006. Bioaccessibility of β-carotene, lutein and lycopene from fruits and vegetables. J Agric Food Chem 54:5382–5387.

Gonthier MP, Remesy C, Scalbert A, Cheynier V, Souquet JM, Poutanen K and Aura AM. 2006. Microbial metabolism of caffeic acid and its sters chlorogenic and caftaric acid by human faecal microbiota *in vitro*. Biomed Pharmacother 60:536–540.

Heidemann C, Schulza MB, Franco OH, Van Dam RM, Mantzoros CS and Hu FB. 2008. Dietary patterns and risk of mortality from cardiovascular disease, cancer and all causes in a prospective cohort of women. Circulation 118:230–237.

Holland G, Welch AA, Unwin DD, Paul AA and Southgate DAT. 1991. In: McCance and Widdowson's The Composition of Foods, 5th ed. Cambridge, UK: Royal Society of Chemistry and Ministry of Agriculture, Fisheries and Food.

Jiménez-Escrig A, Rincón M, Pulido R and Saura-Calixto F. 2001. Guava fruit (*Psidium guajava* L.) as a new source of antioxidant dietary fiber. J Agric Food Chem 49:5489–5493.

Khanum F, Swamy MS, Sudarshana Krishna KR, Santhanam K and Viswanathan KR. 2000. Dietary fiber content of commonly fresh and cooked vegetables consumed in India. Plant Foods Hum Nutr 55:207–218.

Larraruri JA, Rupérez P, Borroto B and Saura-Calixto F. 1996. Mango peel as a new tropical fibre: preparation and characterization. LWT Food Sci Technol 29:729–733.

Larraruri JA, Rupérez P and Saura-Calixto F. 1997. Pineapple shell as a source of dietary fiber with associated polyphenols. J Agric Food Chem 45:428–431.

Li BW and Andrews KW. 2002. Individual sugars, soluble, and insoluble dietary fiber contents of 70 high consumption foods. J Food Compos Anal 15:1401–1411.

Lin PH, Aickin M, Champagne C, Craddick S, Sacks FM, McCarron P, Most-Windhauser MM, Rukenbrod F and Haworth L. 2003. Food group sources of nutrients in the dietary patterns of the DASH-Sodium trial. J Am Diet Assoc 103:488–496.

Liu D, Shi J, Colina Ibarra A, Kakuda Y and Hue SJ. 2008. The scavenging capacity and synergistic effects of lycopene, vitamin E, vitamin C and β-carotene mixtures on the DPPH free radical. LWT Food Sci Technol 41:1344–1349.

Lombardi-Boccia G, Lucarini M, Lanzi S, Aguzzi A and Cappelloni M. 2004. Nutrients and antioxidant molecules in yellow plums (*Prunus domestica* L.) from conventional and organic productions: a comparative study. J Agric Food Chem 52:90–94.

López-Oliva ME, Agis-Torres A, García-Palencia P, Goñi I and Muñoz-Martínez E. 2006. Induction of epithelial hypoplasia in cecal and distal colonic mucosa by grape antioxidant dietary fiber. Nutr Res 26:651–658.

Lunn J and Buttriss JL. 2007. Carbohydrates and dietary fiber. Nutr Bull 32:61–64.

Marlett JA. 2000. Changes in content and composition of dietary fiber in yellow onions and Red Delicious apples during commercial storage. J AOAC Int 83:992–996.

Marlett JA and Vollendorf NW. 1994. Dietary fiber content and composition of different forms of fruits. Food Chem 51:39–44.

Martín-Carrón N, Goñi I, Larrauri JA, Garcia-Alonso A and Saura-Calixto F. 1999. Reduction in serum total and LDL cholesterol concentrations by a dietary fiber and polyphenol-rich grape products in hypercholesterolemic rats. Nutr Res 19:1371–1381.

Martín-Carrón N, Saura-Calixto F and Goñi I. 2000. Effects of dietary fibre and polyphenol-rich grape products on lipidaemia and nutritional parameters in rats. J Sci Food Agric 80:1183–1188.

McCleary BW and Prosky L, editors. 2001. Advanced Dietary Fibre Technology. Oxford: Blackwell Sciences.

O'Neill ME, Carroll Y, Corridan B, Olmedilla B, Grandao F, Blanco I, Van den Berg H, Hininger I, Rousell AM, Chopra M, Southon S and Thurnham DI. 2001. European carotenoid intakes and its use: a five-country comparative study. Br J Nutr 85:499–507.

Pak ND. Dietary fiber in fruits cultivated in Chile. 2003. Arch Latinoam Nutr 53:413–417.

Pérez-Jiménez J and Saura-Calixto F. 2005. Literature data may underestimate the actual antioxidant capacity of cereals. J Agric Food Chem 53:5036–5040.

Pérez-Jiménez J, Serrano J, Tabernero M, Arranz S, Díaz-Rubio ME, García-Diz L, Goñi I and Saura-Calixto F. 2008. Effects of grape antioxidant dietary fiber on cardiovascular disease risk factors. Nutrition 24:646–653.

Prosky L, Asp NG, Schweizer TF, Devries JW and Furda IJ. 1992. Determination of insoluble, soluble and total dietary fiber in foods and food products: collaborative study. JAOAC 75:360–367.

Ramalu P and Rao PU. 2003. Total, insoluble and soluble dietary fiber contents of Indian fruits. J Food Compos Anal 16:677–685.

Sánchez-Alonso I, Jiménez-Escrig A, Saura-Calixto F and Borderías AJ. 2006. Effect of grape antioxidant dietary fibre on the prevention of lipid oxidation in minced fish: evaluation by different methodologies. Food Chem 101:372–378.

Saura-Calixto F. 1998. Antioxidant dietary fiber product: a new concept and a potential food ingredient. J Agric Food Chem 46:4303–4306.

Saura-Calixto F, Garcia-Alonso A, Goñi I and Bravo L. 2000. In vitro determination of the indigestible fraction in foods: an alternative to dietary fiber analysis. J Agric Food Chem 48:3342–3347.

Saura-Calixto F and Goñi I. 1993. In: Cummings JH and Frolich W, editors. Dietary Fibre Intakes in Europe—COST 92. Luxembourg: Office for Official Publications of the European Communities.

Saura-Calixto F and Goñi I. 2004. The intake of dietary indigestible fraction in the Spanish diet shows the limitations of dietary fibre data for nutritional studies. Eur J Clin Nutr 58:1078–1082.

Saura-Calixto F and Goñi I. 2006. Antioxidant capacity of the Spanish Mediterranean diet. Food Chem 94:442–447.

Saura-Calixto F and Díaz-Rubio ME. 2007. Polyphenols associated with dietary fiber in wine. A wine polyphenols gap? Food Res Int 40:613–419.

Saura-Calixto F, Serrano J and Goñi I. 2007. Intake and bioaccessibility of total polyphenols in a whole diet. Food Chem 101:492–501.

Sáyago-Ayerdi SG, Brenes A and Goñi I. 2009. Effect of grape antioxidant dietary fiber on the lipid oxidation on raw and cooking chicken hamburgers. LWT Food Sci Technol. 42:971–976.

Serrano J, Goñi I and Saura-Calixto F. 2005. Determination of β-carotene and lutein available from green leafy vegetables by an in vitro digestion and colonic fermentation method. J Agric Food Chem 53:2936–2940.

Spiller GA. 2005. Dietary fiber in the prevention and treatment of disease. In: Handbook of Dietary Fiber in Human Nutrition. Boca Raton, FL: CRC Press.

Stea TH, Mansoor MA, Wandel M, Uglem S and Frolich W. 2008. Changes in predictors and status of homocysteine in young male adults after a dietary intervention with vegetables, fruit and bread. Eur J Nutr 47:201–209.

Sungsoo Cho S and Dreher ML, editors. 2001. Handbook of Dietary Fiber. New York: Marcel Dekker.

Tabernero M, Serrano J and Saura-Calixto F. 2007. Dietary fiber intake in two European diets with high (Copenhagen, Denmark) and low (Murcia, Spain) colorectal cancer incidence. J Agric Food Chem 55:94443–94449.

Touriño S, Fuguet E, Jáuregui O, Saura-Calixto F, Cascante M and Torres JL. 2008. High-resolution liquid chromatography/electrospray ionization tandem mass spectrometry to identify polyphenols from grape antioxidant dietary fiber. Rapid Commun Mass Spectrom 22:3489–3500.

Trowell H. 1976. Definition of dietary fiber and hypotheses that it is a protective factor in certain diseases. Am J Clin Nutr 29:417–427.

9 Emerging Technologies Used for the Extraction of Phytochemicals from Fruits, Vegetables, and Other Natural Sources

Marleny D. A. Saldaña*, Felix M. C. Gamarra, and Rodrigo M. P. Siloto

Abstract

Phytochemicals such as phenolics, carotenoids, sterols, and alkaloids are receiving increasing attention due to their demonstrated health benefits, but conventional technologies used to extract some of these phytochemicals may have limitations. Therefore, there is a need for emerging efficient technologies that are environmentally friendly, that are cost-effective, and that can guarantee the sustainability of the food chain and product development. Within such emerging technologies are supercritical fluid extraction (SFE), microwave extraction (MWE), pulsed electric field (PEF), high-pressure processing (HPP), ultrasonic extraction (UE), and ohmic heating (OH). Supercritical fluid extraction, which has already reached commercial application for some products, is one of the most widely studied techniques. This technology, which uses a supercritical fluid as the solvent, is simple and fast and allows selective extraction. The most common supercritical fluid is carbon dioxide for agricultural commodities because of its innumerable advantages, operating at moderate pressures and low temperatures (7.4 MPa and 31°C). Microwave extraction is also emerging and has already reached commercial application. This technology uses electromagnetic radiation that transfers rapid and uniform heat to the matrix, resulting in extractions with high concentrations of phytochemicals and a minimum of impurities. Although there are some commercial facilities using this technology, there is a great interest for further research to optimize processing conditions. In spite of the recent attention to pulsed electric field due to its many advantages in comparison with thermal treatments, its use is still circumscribed to a few industrial plants operating in Europe and in the US. This technology uses electric pulses in direct contact with the matrix, reaching moderate temperatures.

This chapter reviews recent findings about the health benefits of phytochemicals present in fruits, vegetables, nuts, seeds, and herbs, including phenolics, carotenoids, sterols, and alkaloids. These phytochemicals are extracted using emerging technologies such as supercritical carbon dioxide (SC-CO_2) extraction, PEF, MWE, HPP, UE, and OH. The impact of important parameters related to sample preparation (particle size and moisture content) and extraction process (temperature, pressure, solvent flow rate, extraction time, and the use of a cosolvent) on the efficiency of extraction and on the characteristics of the extracted products is evaluated based on an extensive review of recent literature. The future of extraction of phytochemicals is certainly bright with the

* Corresponding author: Dr. Marleny D. A. Saldaña, University of Alberta, marleny@valberta.ca

use of combined emerging technologies that will allow high-quality phytochemicals to be produced with minimal degradation.

Introduction

Phytochemicals comprise a number of substances such as phenolics (flavonols, anthocyanins, catechins, etc.), carotenoids (β-carotene, lutein, lycopene), sterols (campesterol, sitosterol, stigmasterol), and alkaloids (caffeine, theobromine, theophylline, etc). They are generally commercialized as nutraceuticals and as ingredients for functional foods. Phytochemicals are found in fruits (e.g., buriti, cloudberry, elderberry, rose hip, and lemon) and vegetables (e.g., beetroot, carrot, olives, and tomatoes), nuts (e.g., acorn, almond, cocoa beans, hazelnut, and walnut), seeds (e.g., guarana, grape, munch, pumpkin, and rose hip), and herbs (e.g., cocoa leaves, maté tea leaves, oregano, epimedii). Phytochemicals have been commonly obtained with conventional extraction methods using petrochemical solvents (e.g., methanol, ethyl acetate, chloroform), requiring their complete removal via subsequent evaporation. But, the heat applied for solvent removal may be detrimental to heat-labile phytochemicals such as lycopene. In addition, government regulations on the use of organic solvents are becoming more strict, and the safety of residual organic solvents in the final product is being questioned. The use of emerging technologies such as SC-CO_2 extraction, MWE, PEF, HPP, UE, and OH is growing because of the capability of these technologies to overcome many of the disadvantages associated with conventional technologies.

One of the most studied technologies is supercritical fluid extraction with SC-CO_2. The advantages of SC-CO_2 include its low processing temperature, which minimizes thermal degradation; the ease of separation with no solvent residue left in the final product; and minimization of undesirable oxidation reactions.

Recently, the use of microwaves (MW) is growing considerably because of its minor effect on the final product quality. This technology uses radiation that causes cell membrane rupture within a short processing time, at low temperature, and using little solvent. The MW energy is delivered directly to the matrix through molecular interaction with the electromagnetic field, facilitating a rapid and uniform heating of relatively thicker matrices. Promising results have been reported through the combination of MW and extraction, referred as microwave-assisted extraction (MAE), pressurized microwave-assisted extraction (PMAE), dynamic microwave-assisted extraction (DMAE), and vacuum microwave-assisted extraction (VMAE).

Pulsed electric field is another alternative to conventional methods of extraction. PEF enhances mass transfer rates using an external electrical field, which results in an electric potential across the membranes of matrix cells that minimizes thermal degradation and changes textural properties. PEF has been considered as a nonthermal pretreatment stage used to increase the extraction efficiency, increasing also permeability throughout the cell membranes.

High-pressure processing shows great potential in preserving food quality and increasing mass transfer and cell permeability, as well as enhancing diffusion. HPP eliminates the adverse effects of heat and significantly improves texture. Compared to conventional extraction methods, HPP uses high pressures, moderate temperature, and short processing time.

Ultrasonic extraction is a well-known commercial method to increase mass transfer rate by cavitation forces. Bubbles in the liquid–solid extraction using UE can explosively collapse and produce localized pressure, improving the interaction between the intracellular substances and the solvent to facilitate the extraction of the phytochemical.

Ohmic heating, despite its great potential, is the emerging technology least known in the literature. OH occurs when alternating current is passed through the matrix and heat is generated by discharge of the sample's electrical resistance. OH allows better contact of the matrix with the electrodes because the matrix is placed between the electrodes. Electrical energy is dissipated into heat, which results in rapid and uniform heating. OH has been shown to increase the extraction of phytochemicals at low frequencies. This might be attributed to the electroporation effect, which allows cell walls to build up charges and form pores. Ohmic heating, also known as Joule heating, electroheating, direct heating, and electroconductive heating, has demonstrated high efficiency for extracting sucrose from sugar beet and oil from rice.

The objectives of this chapter are to review some of the recent literature findings related to the health benefits of phytochemicals present mainly in fruits, vegetables, nuts, seeds, and herbs and the use of emerging extraction technologies such as SC-CO_2 extraction, microwave extraction, pulsed electric field, high-pressure processing, ultrasonic extraction, and ohmic heating for the recovery of phytochemicals from different plant sources. Factors affecting extraction yield such as sample preparation (moisture content and particle size) and extraction parameters (temperature, pressure, flow rate) will be discussed and compared with conventional extraction methods.

Phytochemicals

A large variety of phytochemicals are found within agricultural commodities. This chapter focuses on four main groups: phenolics, carotenoids, sterols, and alkaloids. In addition, recent research related to the health benefits of these phytochemicals will be briefly reviewed. Table 9.1 summarizes the main chemical structure and solubility in organic solvents of phytochemicals such as phenolics (flavonoids), carotenoids, sterols, and alkaloids.

Phenolics

Phenolic compounds are an heterogeneous class of secondary plant metabolites that can be divided into two main groups: flavonoid compounds, with a C_6-C_3-C_6 structure (anthocyanins, flavan-3-ols, flavonols, flavones, and flavanones) and nonflavonoid compounds, with C_6-C_1 and C_6-C_3 structures, (hydroxybenzoic and hydroxycinnamic acids and stilbenes). Anthocyanins are a class of flavonoids that are usually found as natural pigments responsible for the orange, red, blue, and violet colors of many fruits and flowers. The limitation in the use of anthocyanins as natural food colorants is their color instability. The major degradation factors of the anthocyanins are the temperature, the presence of oxygen and light, copigmentation, metal ions, enzymes, and the pH value (Francis 1989; Bridle and Timberlake 1997). However, other studies show that acylation, glycosylation, and condensation with different flavonoids improves the stability of the anthocyanins (Mazza and Brouillard 1987).

Table 9.1. Physico-chemical properties of bioactive compounds

Bioactive Compound	Formula	Molecular Weight (g/mol)	Solubility	Structure
Phenolics				
Anthocyanins			vs EtOH, chl, s eth., ace, H_2O, MeOH, n-He	
Pelargonidin (R_1:H; R_2:H)	$C_{15}H_{11}O_5$	271		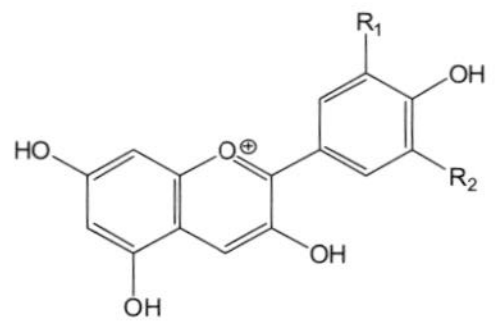
Cyanidin (R_1:OH; R_2:H)	$C_{15}H_{11}O_6$	287		
Delphinidin (R_1:OH; R_2:OH)	$C_{15}H_{11}O_7$	303		
Peonidin (R_1:OCH_3; R_2:H)	$C_{16}H_{13}O_6$	301		
Petunidin (R_1:OCH_3; R_2:OH)	$C_{16}H_{13}O_7$	317		
Malvidin (R_1:OCH_3; R_2:OCH3)	$C_{17}H_{15}O_7$	331		
Flavanones, flavonols and flavones			s H_2O, EtOH, MeOH, eth, n-He	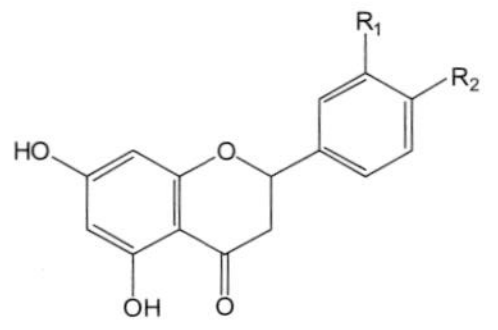
Naringenin (R_1:H; R_2:OH)	$C_{15}H_{12}O_5$	272		
Eriodictyol (R_1:OH; R_2:OH)	$C_{15}H_{12}O_6$	288		
Hesperetin (R_1:OH; R_2:OCH3)	$C_{16}H_{14}O_6$	302		
Kaempferol (R_1:H; R_2:H)	$C_{15}H_{10}O_6$	286		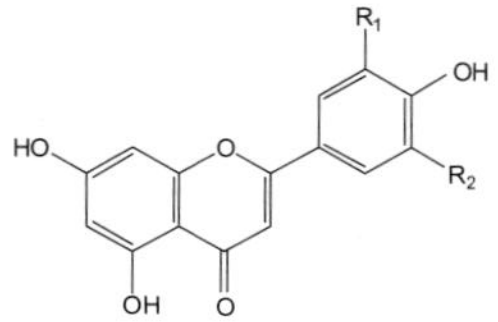
Quercetin (R_1:OH; R_2:H)	$C_{15}H_{10}O_7$	302		
Myricetin (R_1:OH; R_2:H)	$C_{15}H_{10}O_8$	318		
Isorhamnetin (R_1:OCH_3; R_2:H)	$C_{17}H_{14}O_7$	330		

Flavanones, flavonols and flavones (continued)

				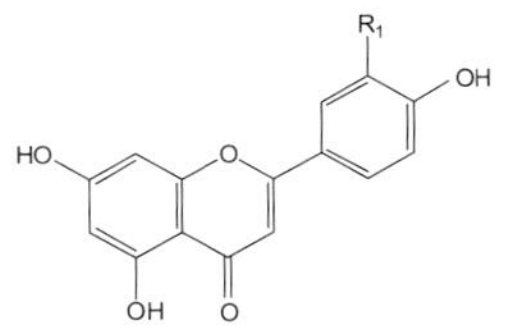
Apigenin (R_1:H)	$C_{15}H_{11}O_4$	255		
Luteolin (R_1:OH)	$C_{15}H_{10}O_6$	286		
Isoflavones			s H_2O, EtOH, MeOH, eth, n-He	
Daidzein	$C_{15}H_{10}O_5$	270		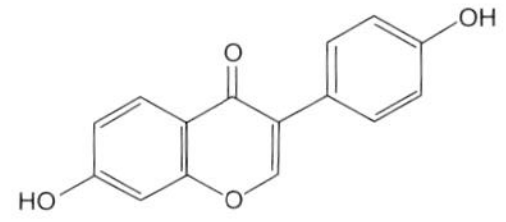
Genistein	$C_{15}H_{10}O_5$	270		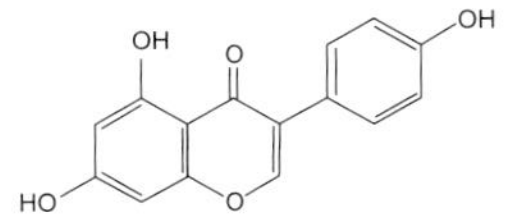
Glycitein	$C_{16}H_{12}O_5$	284		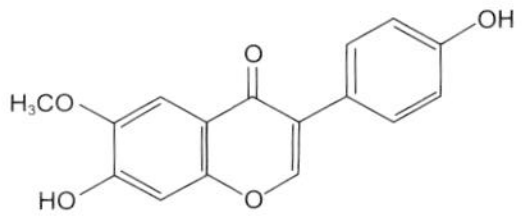

Table 9.1. *Continued*

Bioactive Compound	Formula	Molecular Weight (g/mol)	Solubility	Structure
Catechins			s H_2O, EtOH, MeOH, eth, n-He	
(+)-Catechin R_1:H	$C_{15}H_{14}O_6$	290		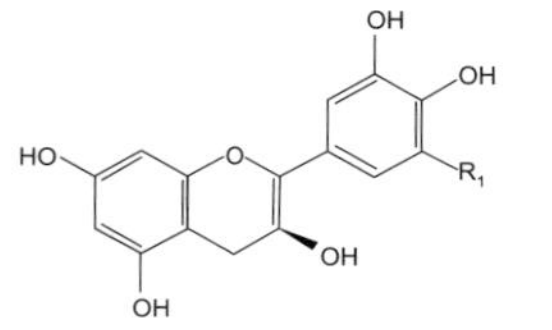
(+)-Gallocatechin R_1:OH	$C_{16}H_{21}ABO_7$	336		
(−)-Epicatechin R_1=H	$C_{15}H_{14}O_6$	290		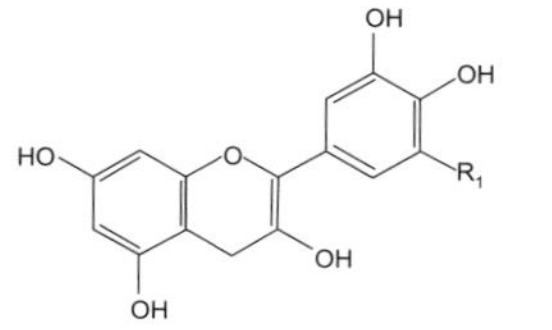
Epigallocatechin R_1:OH	$C_{15}H_{14}O_7$	306		
(+)-Epicatechin gallate R_1:H	$C_{22}H_{18}O_{10}$	442		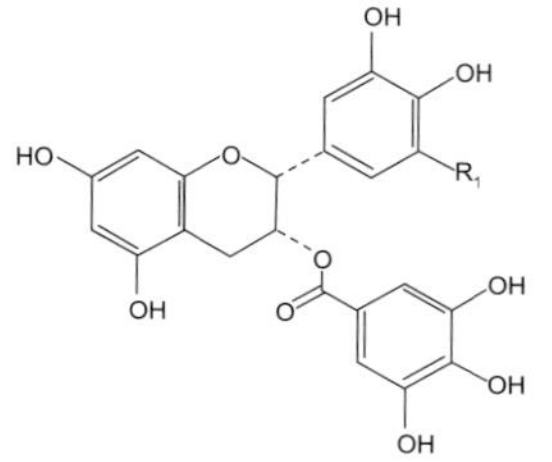
(+)-Epigallocatechin gallate R_1:OH	$C_{22}H_{18}O_{11}$	458		

Bioactive Compound	Formula	Molecular Weight (g/mol)	Solubility	Structure
Carotenoids				
β-Carotene	$C_{40}H_{56}$	536.87	sl EtOH, chl; s eth., ace, bz	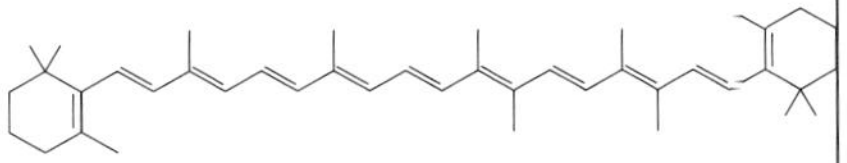
Lycopene	$C_{40}H_{56}$	536.87	sl EtOH, peth; s eth; vs bz, chl, CS_2	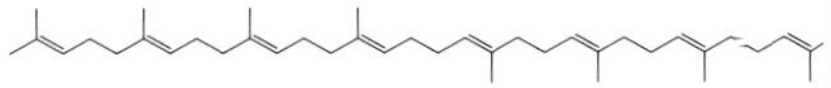
Lutein	$C_{40}H_{56}O_2$	568.87	vs bz, eth, EtOH, peth	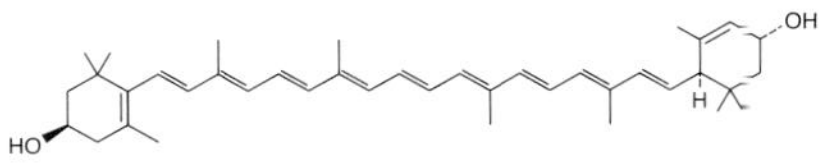

Table 9.1. *Continued*

Bioactive Compound	Formula	Molecular Weight (g/mol)	Solubility	Structure
Sterols				
Campesterol	$C_{28}H_{48}O$	400.68	—	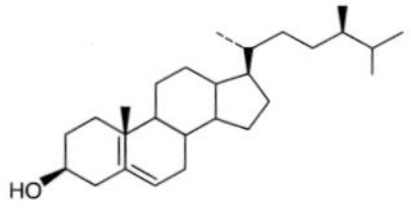
Stigmasterol	$C_{29}H_{48}O$	412.69	vs bz, eth, EtOH	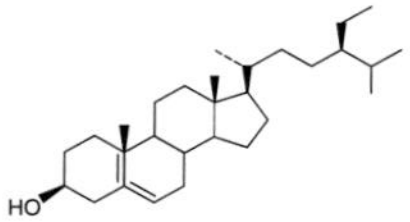
β-Sitosterol	$C_{28}H_{50}O$	414.71	s EtOH, eth, HOAc	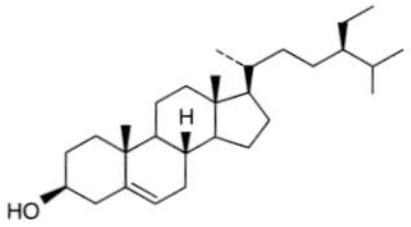

Bioactive Compound	Formula	Molecular Weight (g/mol)	Solubility	Structure
Alkaloids				
Benzoylecgonine	$C_{16}D_3H_{16}NO_4$	292.34	s eth	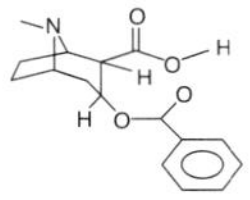
Caffeine	$C_8H_{10}N_4O_2$	194	s H_2O, EtOH	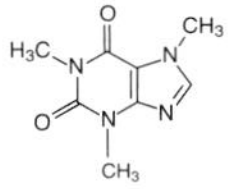
Cocaine	$C_{17}H_{21}NO_5$	319	s H_2O	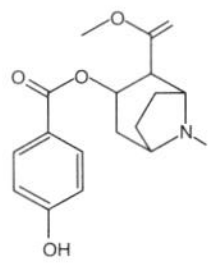
Emodin	$C_{15}H_{10}O_5$	270	s H_2O, EtOH	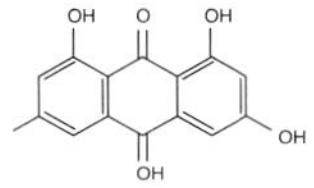

Table 9.1. *Continued*

Bioactive Compound	Formula	Molecular Weight (g/mol)	Solubility	Structure
Imperatorin	$C_{16}H_{14}O_4$	270	s DMSO, sl H_2O, MeOH	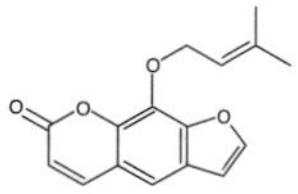
Isoimperatorin	$C_{16}H_{14}O_4$	270.28	s EtOH, H_2O	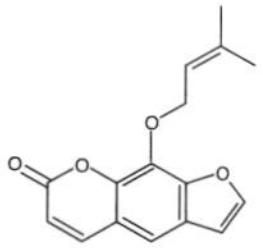
Resveratrol	$C_{14}H_{12}O_3$	228	s MeOH, EtOH, H_2O, DMSO,	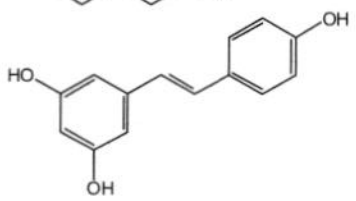

Bioactive Compound	Formula	Molecular Weight (g/mol)	Solubility	Structure
Safflomin A	$C_{27}H_{32}O_{16}$	612.53	s EtOH, eth	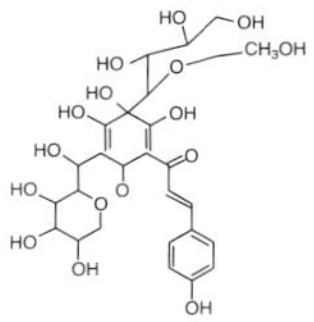
Tetrahydropalmatine	$C_{21}H_{25}NO_4$	355.43	s chl	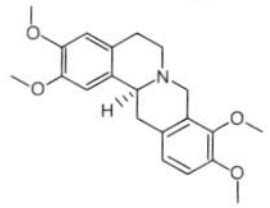
Theobromine	$C_7H_8N_4O_2$	180.16	s H_2O	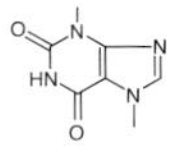
Theophylline	$C_7H_8N_4O_2$	180.16	s H_2O, chl, EtOH	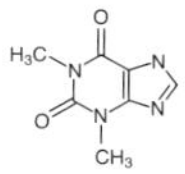

sl: slightly soluble, s: soluble, vs: very soluble, ace: acetone, bz: benzene, chl: chloroform, EtOH: ethanol, eth: diethyl ether, HOAc: acetic acid, peth: petroleum ether, MeOH: methanol, n-He: *n*-hexane, DMSO: dimethyl sulfoxide. CS_2: carbon disulfide, H_2O: water

Phenolic phytochemicals are mainly present in fruits, seeds, and herbs such as elderberry, grape, maté tea leaves, rose hip, *Rosmarinus officinalis, Origanum dictamnus, Teucrium polium, and Styrax officinalis*. The amount and type of phenolics and their conjugates differ markedly even in different tissues of the same species. For example, anthocyanins are the main polyphenols in the red grape skin, whereas flavan-3-ols are the major polyphenols in the grape seeds (Makris and others 2006). It has been proven that a diet rich in various classes of polyphenols (phenolic acids, flavonols, catechin monomers, proanthocyanidins, flavones, flavanones, anthocyanins) decreases the risk of premature mortality from major clinical conditions, including cancer and heart disease (Duthie and others 2003). Phenolic phytochemicals are known to exhibit several activities that are beneficial to health, such as antioxidant, anti-inflammatory, antihepatotoxic, antitumor, and antimicrobial activities (Middleton and others 2000; Rice-Evans and others 1996; Podsedek 2007; Kong and others 2003; Revilla and Ryan 2000). The consumption of polyphenols present in tea leaves, for example, can inhibit the formation and growth of tumors (Wang and others 1994; Inagake and others 1995; Hasegawa and others 1995). These effects are due to the properties of antioxidants to act as reducing agents by donating hydrogen, quenching singlet oxygen, acting as chelators, and trapping free radicals.

Carotenoids

Carotenoids are formed by C_{40} polyunsaturated hydrocarbons, which could be considered the backbone of the molecule. This chain may be terminated by cyclic end groups (rings) and may be complemented with oxygen-containing functional groups. The hydrocarbon carotenoids are known as carotenes, whereas oxygenated derivatives of these hydrocarbons are known as xanthophylls. β-Carotene, the principal carotenoid in carrots, is a familiar carotene. Lutein, the major yellow pigment of marigold petals, is a common xanthophyll (Britton and others 2004; Giovannucci 2002). Carotenoids are responsible for the bright yellow, orange, and red colors of fruits, roots, flowers, algae, bacteria, mold, and yeast. By 2004, more than 750 naturally occurring carotenoids had been reported (Britton and others 2004). Carotenoids are quite unstable because of their many conjugated double bonds, which make them prone to undergo isomerization to produce *trans/cis* isomers, mainly during food processing and storage (Rao and Agarwal, 2000). Carotenoids are mainly present in hiprose and buriti fruit, carrot, rose hip, and tomato. Numerous studies have shown an association between consumption of carotenoids and reduced risk of several chronic diseases. The role of dietary carotenoids in enhancing immune function has been reviewed by Hughes (2001) and Fullmer and Shao (2001). β-Carotene is important because of its provitamin A activity, lutein is associated with reduced incidence of cataract, and lycopene is linked with reduced risk of prostate cancer (Goodman and others 2004; Clinton 1998; Granado and others 2003). Lycopene is known as one of the most potent antioxidants, although it lacks provitamin A activity. Its highly conjugated molecular structure is responsible for the bright red color of tomato, watermelon, pink grapefruit, and apricot (Giovannucci 2002; Shi and Le Maguer 2000). Lycopene might play a protective role in the development of atherosclerosis (Klipstein-Grobusch and others 2000; Rissanen and others 2003) and in the reduction of prostate cancer (Giovannucci 2002) and stomach cancer (Omoni and Aluko 2005). The recommended daily intake of lycopene is 5–10 mg/day

(Rao and Shen 2002). Approximately 80% of dietary lycopene comes from tomatoes. Processed tomatoes have a higher level of lycopene compared to fresh tomatoes. This could be attributed to the heat treatment and homogenization of tomatoes that enhance the lycopene availability (van het Hof and others 2000; Porrini and others 1998).

Sterols

Plant sterols are isoprenoid compounds with an sterol nucleus and an alkyl chain. Most plant sterols have a double bond in position C5 in the nucleus, whereas the stanols are saturated (Moreau and others 2002). The main sterols in plant materials are sitosterol, campesterol, and stigmasterol (Heinemann and others 1993). They are mainly found in acorn, hazelnut, walnut, grape, pumpkin, and rose hip. Plant sterols are of interest because of their potential to lower both total serum cholesterol and low-density lipoprotein (LDL) cholesterol in humans by inhibiting the absorption of dietary cholesterol as well as the reabsorption of cholesterol excreted into the bile in the course of the enterohepatic cycle (Piironen and others 2000, 2003). A detailed review of plant sterols and their role in health and disease can be found elsewhere (Patel 2008). Consumption of sitosterol was shown to reduce colon cancer in rats (Raicht and others 1980). Some studies indicated a beneficial effect of these sterols on immune function (Bouic and Lamprecht 1999). However, there is still controversy in the literature about the daily dietary intake of phytosterols. Some researchers reported that a high dietary intake of phytosterols lowers blood cholesterol levels by competing with dietary and biliary cholesterol during intestinal absorption (Piironen and others 2000; Normen and others 2000; Jones and others 1999). But, it has been also hypothesized that because phytosterols are more readily oxidized by free radicals than cholesterol, they could increase the level of oxidized LDL, which is responsible for formation of atherosclerotic plaques in arteries (Plat and Mensink 2005). According to the Scientific Committee on Food, the average amount of phytosterols in the Western diet is 150–400 mg/day (Scientific Committee on Food 2002). However, the recommended dose of phytosterols to reduce LDL–plasma cholesterol level by 5–15% is 1.3–2 g/day (Hallikainen and others 2000; Law 2000). To achieve such high levels and comply with the approved health claim on the role of plant sterols or stanol esters in reducing the risk of coronary heart disease, the food industry has introduced various functional food products with added phytosterols. However, another study demonstrated that daily consumption of 3.8–4.0 g/day of plant sterol esters might significantly lower serum concentrations of carotenoids and tocopherols (Plat and Mensink 2001). Therefore, more research is needed to determine the best dose recommended for daily consumption of plant sterols.

Alkaloids

Alkaloids are compounds that contain nitrogen in a heterocyclic ring and are commonly found in about 15–20% of all vascular plants. Alkaloids are subclassified on the basis of the chemical type of their nitrogen-containing ring. They are formed as secondary metabolites from amino acids and usually present a bitter taste accompanied by toxicity that should help to repel insects and herbivores. Alkaloids are found in seeds, leaves, and roots of plants such as coffee beans, guarana seeds, cocoa beans, maté tea leaves, peppermint leaves, coca leaves, and many other plant sources. The most common alkaloids are caffeine, theophylline, nicotine, codeine, and indole alkaloids. Research has demonstrated that the consumption of some alkaloids provides health benefits. For example, theobromine has strong diuretic, stimulant, and arterial dilating effects (Li and others 1991; Sotelo and Alvarez 1991). In the case of indole alkaloids, many studies have shown to have biological activity such as antitumoral, anti-inflammatory, analgesic, antioxidant, and antimycobacterial effects (Rates and others 1993; Delorenzi and others 2001; Pereira and others 2005). Alkaloids can be potentially used in the pharmaceutical industry as drugs, or in a few cases such as caffeine, they find application in beverages.

Extraction of Phytochemicals from Different Sources

Various methods are used to extract phytochemicals from different natural sources. One of the methods widely used is the supercritical fluid extraction (SFE). A compound is in its supercritical state when it is heated and compressed above its critical temperature (T_c) and critical pressure (P_c). Figure 9.1 shows the pressure–temperature (P-T) diagram with the supercritical fluid (SCF) region for carbon dioxide. In the supercritical state, the substance exists as a single fluid phase with properties intermediate between those of liquids and gases: the densities are liquidlike, whereas the diffusivities and viscosities are gaslike (McHugh and Krukonis, 1994). Moreover, SCFs have zero surface tension, which allows easy penetration into most matrices including fruits and vegetables. In addition, in the supercritical state, SCFs are extremely sensitive to small changes in temperature and pressure such that a compound may be extracted from a matrix at one set of conditions and then separated from the SCF in a downstream operation under a slightly different set of conditions. Most SFE processes use SC-CO_2 because CO_2 is nontoxic, nonflammable, and chemically inert and has a moderate and relatively easily achievable critical point of 31°C and 7.4 MPa. Some of the other advantages of using CO_2 as the SCF are that CO_2 is available in high purity at relatively

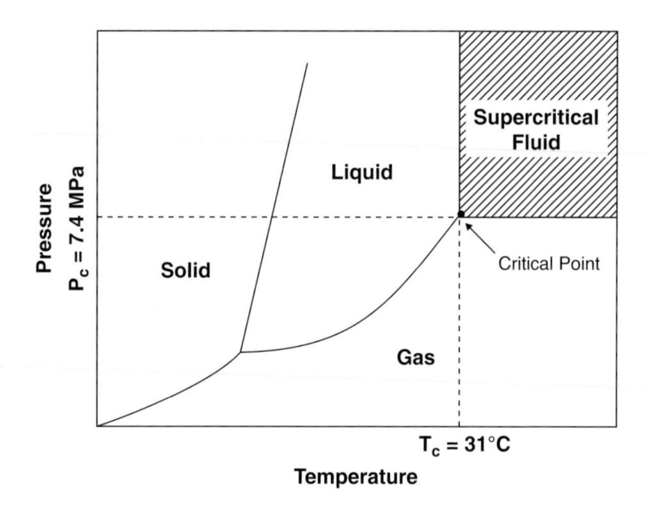

Figure 9.1. Carbon dioxide pressure–temperature phase diagram adapted from McHugh and Krukonis (1994).

low cost, it can be easily removed from the matrix after the SFE process, and it can be easily separated from the extracted compounds. It is well documented that CO_2, a nonpolar solvent, is best suited for the extraction of nonpolar organic compounds. Therefore, for the extraction of more polar compounds, the polarity of SC-CO_2 can be increased by adding modifiers such as ethanol and water, which in turn increases the solubility of more polar compounds in the SCF. Other solvents such as nitrous oxide, ethane, and propane have been used as SCFs in various applications (Reid and others 1987). However, issues such as safety, disposal, toxic emissions, and flammability have limited their use.

One important consideration for the extraction with any solvent is the solubility of the phytochemical in the solvent. For example, when using supercritical fluid extraction, solubility is a strong function of SC-CO_2 density and the properties of the solute such as molecular weight, polarity and vapor pressure. Phytochemicals are soluble in SC-CO_2 to different extents depending on the temperature and pressure conditions. Solubility behavior of phenolics, carotenoids, sterols, and alkaloids in SC-CO_2 has been previously reviewed (Choi and others 1998; Murga and others 2002, 2003; Gomez-Prieto and others 2002; Güçlü-Üstündag and Temelli 2004; Saldaña and others 1999, 2006, 2007). Other potential emerging methods to extract phytochemicals are PEF, MWE, UE, and OH. Emerging extraction methods have been studied for a number of agricultural commodities (Saldaña and others 2002a,b; Lianfu and Zelong 2008; Lopes and Bernardo-Gil 2005; Vatai and others 2009; Corrales and others 2008; Fincan and others 2004; Bernardo-Gil and others 2001; and many others) as reported in Tables 9.2 through 9.4. The extraction efficiency of these different extraction techniques is affected by several parameters such as particle size and moisture content of the feed, extraction temperature and pressure, power, solvent flow rate, extraction time, frequency, and the use of a cosolvent or a mixture of solvents. Therefore, the following discussion will focus on the impact of these processing parameters on the yield of phytochemicals obtained from fruits and vegetables, nuts and seeds, and herbs.

Fruits and Vegetables

Table 9.2 shows a number of fruits and vegetables from which phytochemicals have been extracted using emerging and conventional technologies. For example, carotenoids were obtained from cloudberry (Manninen and others 1997), buriti fruit (De Franca and others 1999), and hiprose fruit (Illés and others 1997) using SC-CO_2 at temperatures varying from 35 to 60°C and pressures varying from 8 to 30 MPa. Using SC-CO_2, carotenoids were also obtained from tomato at temperatures of 32–110°C and pressures of 8–48 MPa (Vasapollo and others 2004; Lianfu and Zelong 2008; Ruiz del Castillo and others 2003; Ollanketo and others 2001; Shi 2001; Rozzi and others 2002; Gomez-Prieto and others 2003), and from carrots at temperatures of 40–70°C and pressures of 7.8–55 MPa (Goto and others 1994; Saldaña and others 2006; Sun and Temelli 2006; El-Belghiti and others 2007). In these studies, it was observed that lycopene was the predominant carotenoid (85–90% of total carotenoids) extracted from tomato (Vasapollo and others 2004), whereas β-carotene (60–80%) was the major carotenoid extracted from carrot (Saldaña and others 2006). Sterols were mainly found in olives (Esquivel and others 1999), whereas phenolic compounds were found in red beetroot (Fincan and others 2004) and in elderberry and grapes (Vatai and others 2009).

Table 9.2. Extraction of phytochemicals from fruits and vegetables using emerging technologies.

Raw Material	Extraction Methods	Feed (g) (i.c. %)	Sample Preparation		Bioactive Compound	Extraction Conditions					Recovery (%)	Ref.
			p.s.	m		T	P	f.r.	t	c-s.		
Buriti fruit	SC-CO_2	n.i.	n.i.	11	carotenoids	40, 55	20, 30	0.31, 0.43 g/sec	n.i.	no	7.8**	De Franca and others (1999)
Cloudberry	SC-CO_2	42	n.i.	n.i.	unsaturated fatty acids, β-carotene	40, 60	9, 10, 12, 15, 30	n.i.	n.i	no	n.i.	Manninen (1997)
Elderberry	CEM	n.i.	n.i.	n.i.	anthocyanins, phenol	20, 40, 60	0.1	20 mL/g	2	no	n.i.	Vatai and others (2009)
Elderberry	SC-CO_2	n.i.	n.i.	n.i.	anthocyanins, phenol	40	15, 30	0.2 liter/min	n.i.	no	100	Vatai and others (2009)
Grape	HPP	n.i.	6–8	76	anthocyanins	70	600	n.i.	1	ethanol/water 50:50 v/v	n.i.	Corrales and others (2008)
Grape	[I]UE	n.i.	6–8	76	anthocyanins	70	n.i.	n.i.	1	no	n.i.	Corrales and others (2008)
Grape	[I]PEF	n.i.	6–8	76	anthocyanins	25–28	no	n.i.	0.004–0.01	no	75	Corrales and others (2008)
Grape marc	CEM	n.i.	n.i.	n.i.	anthocyanins, phenol	20, 40, 60	0.1	20 mL/g	2	no	n.i.	Vatai and others (2009)

Grape marc	SC-CO_2	n.i.	n.i.	n.i.	anthocyanins, phenol	40	15, 30	0.2 mL/min	n.i.	no	100	Vatai and others (2009)
Hiprose fruit	SC-CO_2	n.i.	0.36	n.i.	carotenoids	35	8–25	1–1.5 liter/min	n.i.	no	100	Illés and others (1997)
Lemon	Steam Distillation	n.i.	n.i.	n.i.	terpenes, sesquiterpenes	25-110	0.1	n.i.	7-10	no	1.1	Gamarra and others (2006)
Carrot	[II]PEF + CA	n.i.	1.5	88-92	aqueous solution	18, 25, 35	n.i.	n.i.	0.25	no	96	El-Belghiti and others (2007)
Carrot	SC-CO_2	2	0.5–1	0.8	α-,β-carotene, lutein	40, 50	12–33	1.2	8	no	n.i.	Saldaña and others (2006)
Carrot	SC-CO_2	n.i.	0.26, 0.47, 1.12	n.i.	carotenoids	40, 50, 60	7.8–29.4	n.i.	n.i.	1, 3, 5% ethanol	n.i.	Goto and others (1994)
Carrot	SC-CO_2	2000	n.i.	n.i.	carotenes, phenolics, phytosterols, linolenic acid, trilinolein	45-50	35–38	n.i.	2–3	no	n.i.	Ranalli others (2004)

Table 9.2. *Continued*

Raw Material	Extraction Methods	Feed (g) (i.c. %)	Sample Preparation		Bioactive Compound	Extraction Conditions					Recovery (%)	Ref.
Carrot	SC-CO_2	2	0.25–0.5, 0.5–1, 1–2	0.8, 17.5, 48.7, 84.6	α-,β-carotene, lutein	40, 55, 70	27.6, 41.3, 55.1	0.5, 1, 2	4	0, 2.5, 5% canola oil	0.19**	Sun and Temelli (2006)
Olive husks	SC-CO_2	n.i.	0.4	n.i.	unsaturated fatty acids sterols	35–57	10.4–18	n.i.	n.i.	no	n.i.	Esquivel and others (1999)
Red beetroot	[III]PEF	n.i.	1, 4	n.i.	betanin pigment	25	no	n.i.	0.05-0.06	no	90	Fincan and others (2004)
Tomato skin	SC-CO_2	0.5	0.05–0.25	n.i.	phytoene, phytofluene, ξ-carotene, β-carotene, lycopene	40, 50, 60	8–26	4 mL/min	0.5	no	n.i.	Gomez-Prieto and others (2002)
Tomato	[III]UMAE	n.i.	n.i.	78.8	lycopene	no	no	n.i.	0.05–0.14	no	n.i	Lianfu and Zelong (2008)
Tomato	[IV]UAE	n.i.	n.i.	78.8	lycopene	53, 60, 70, 80, 86	no	n.i.	0.22–0.78	no	n.i	Lianfu and Zelong (2008)

Tomato	SC-CO_2	0.3	n.i.	n.i.	lycopene	60, 85, 110	40.5	1.5 mL/min	0.83	acetone, methanol, ethanol, hexane, dichloromethane, water+	100	Ollanketo and others (2001)
		n.i.	n.i.	n.i.	lycopene	45–80	35–38	n.i.	2–3	no	55	Shi (2001)
		3	n.i.	n.i.	lycopene	32–86	13.8–48.3	2.5 mL/min	n.i.	no	61	Rozzi and others (2002)
		0.5	n.i.	n.i.	lycopene	40	8–28	4 mL/min	n.i.	no	n.i.	Gomez-Prieto and others (2003)
		20	n.i.	n.i.	lycopene	40	32	n.i.	n.i.	no	n.i.	Ruiz del Castillo and others (2003)
		3,000	1	n.i.	lycopene	45–70	33.5–45	8–20 kg/hr	2–8	1–20% hazelnut oil	60	Vasapollo and others (2004)

i.c.: initial content (%), p.s.: particle size (mm), m: moisture (%), T: temperature (°C), P: pressure (MPa), f.r.: flow rate (liter/min), t: time (hr), c-s.: cosolvent (%), n.i.: not indicated, **yield (g/100 g initial material), CEM: conventional extraction method, SC-CO_2: supercritical carbon dioxide extraction, UE: ultrasonic extraction, HHP: high-pressure processing, PEF: pulsed electric field, CA: centrifugal acceleration, UMAE: ultrasound/microwave assisted extraction, UAE: ultrasonic assisted extraction. +500 μL of cosolvent was added to the sample before extraction; [I]UE: a frequency of 35 kHz was used; [I]PEF: a power of 8 kW was used; [II]PEF + CA extraction: electric pulses varying from 270 to 670 V/cm were used; [III]PEF: constant electric pulses of 1 kV/cm were used; [III]UMAE: powers of 60, 70, 85, 100, and 110 W were used; [IV]UAE: powers ranging from 50 to 300 W were used.

Table 9.3. Extraction of phytochemicals from nuts and seeds using emerging technologies

Raw Material	Extraction Methods	Feed (g) (i.c. %)	Sample Preparation		Bioactive Compound	Extraction Conditions					Recovery (%)	Ref.
			p.s.	m		T	P	f.r.	t	c-s.		
Acorns	SC-CO_2	n.i.	0.27	n.i.	β-sitosterol, stigmasterol, campesterol	40	18	2.5×10^{-4} m/sec*	n.i.	no	n.i.	Lopes and Bernardo-Gil (2005)
Almond	SC-CO_2	3,000–4,000	milled, broken, whole	n.i.	oleic acid, linoleic acid	35, 40, 50	35, 45, 55	10, 20, 30 kg/hr	n.i.	no	n.i.	Leo and others (2005)
Cocoa beans	SC-CO_2	1–2	0.35–2 mm	n.i.	theobromine	50, 70	20, 40	5 liter/min	6.3	no	29	Saldaña and others (2002c)
Grape	SC-CO_2	176	<1.25	n.i.	phytosterols (β-sitosterol, campesterol, stigmasterol, squalene)	65	37	60 g/min	6	no	13.6**	Beveridge and others (2005)
Grape	SC-CO_2	20	n.i.	6.8	n.i.	40	28	0.5–1 liter/min	n.i.	no	n.i.	Sovova and others (1994)
Grape	SC-CO_2	3	n.i.	n.i.	phenolics	40	20, 30	n.i.	n.i.	no	n.i.	Murga and others (2000)
Guaraná seeds	SC-CO_2	3–6	0.6–1 mm	n.i.	caffeine	40, 70	10, 20, 40	5 liter/min	3.5	no	97	Saldaña and others (2002a,b)
Hazelnut	[I]MW	n.i.	n.i.	7.9	fatty acids, sterols	n.i.	no	n.i.	0.05-0.06	no	45.3	Uquiche and others (2008)
Hazelnut	SC-CO_2	50	0.7	n.i.	β-sitosterol	35–48	8–23.4	4.42–7.1 $\times 10^{-4}$ m/s	4	no	95	Bernardo-Gil others (2002)

Hybrid Hibiscus	$SC-CO_2$	5	0.1	n.i.	phytosterols	80	53.7	2 mL/min	0.83	no	20**	Holser and Bost (2004)
Pumpkin	$SC-CO_2$	20	0.36	n.i.	sterols	35–45	18–20	0.05–0.082 cm/s*	2	no	n.i.	Bernardo-Gil others (2001)
Rose hip	$SC-CO_2$	10	n.i.	n.i.	carotenoids	28, 35	10, 25	1–1.5 liter/min	n.i.	propane	n.i.	Szentmihalyi and others (2002)
		100	n.i.	n.i.	flavonoids, carotenoids	40, 50, 60	30, 40, 50	21 g/min	n.i.	no	7.1**	Del Valle. and others (2000)
		300	n.i.	10.1	sterols, oryzanol	0, 20, 40, 60	17, 24, 31	2.5 kg/hr	6	no	96.8	Shen others (1997;1996)
		300	n.i.	8.48	sterols, oryzanol	40, 45, 50	8.6, 9.9, 11	3.5 kg/hr	4	no	4.1**	
		30	0.5	n.i.	squalene, sterol	40, 50, 70	20.7, 27.6, 34.5, 41.3	n.i.	8	no	80	Kim and others (1999)
		20	n.i.	n.i.	γ-oryzanol, stero	40, 60, 80	34.5, 51.7, 68.9	1.1 g/min	n.i.	no	24.65**	Perretti and others (2003)
		800	0.25, 0.5, 0.75	n.i.	sitosterol and campesterol	50, 60	10, 20, 30, 40	300 g/min	3	no	20**	Danielski and others (2005)
		600–700	0.35–0.5	n.i.	β-carotene, lutein, and zeaxanthin	55	25, 38	0.5 liter/min	3	no	92	Panfili and others (2003)
Walnut	$SC-CO_2$	n.i.	0.01, 0.05, 0.1, 0.5	n.i.	β-sitosterol, campesterol	35, 40, 45, 48	18, 20, 22, 23.4	n.i.	n.i.	no	95	Oliveira and others (2002)

i.c.: initial content (%), p.s.: particle size (mm), m: moisture (%), T: emperature (°C), P: pressure (MPa), f.r.: flow rate (liter/min), t: time (hr), c-s.: cosolvent (%), n.i.: not indicated, *superficial velocity, **yield (g/100 g initial material), $SC-CO_2$: supercritical carbon dioxide extraction, MW: Microwave, [I]MW: power of 400 W and 600 W were used.

Table 9.4. Extraction of phytochemicals from herbs using emerging technologies.

Raw Material	Extraction Methods	Feed (g) (i.c. %)	Sample Preparation		Bioactive Compound	Extraction Conditions					Recovery (%)	Ref.
			p.s.	m		T	P	f.r.	t	c-s.		
Coca leaves	[I]MAE	n.i.	0.09–0.47	n.i.	benzoylecgonine, cocaine	25	no	n.i.	0.4	methanol/water (71/29), methanol/ ethanol, water	39	Brachet others (2002)
Flos carthami	VMAE	n.i.	n.i.	n.i.	safflomin A	40–80	0.04–0–06	n.i.	0.08–3	no	n.i.	Wang others (2008)
Dandelion leaves	SC-CO_2	n.i.	0.2–0.8	12.47	β-amyrin, β-sitosterol	35–65	15–45	7.4 kg/hr	n.i.	no	3.2–4	Simándi others (2002)
Epimedii	CME	n.i.	4	n.i.	flavonoids	80	0.1	1–10	4	Ethanol/water	n.i.	Chen others (2008)
Epimedii	[II]DMAE	n.i.	4	n.i.	flavonoids	n.i.	no	0.6–1.0 mL/min	0.1	Ethanol/water	n.i.	
Epimedii	PMAE	n.i.	4	n.i.	flavonoids	n.i.	0.04	n.i	0.23	Ethanol/water	n.i.	
Epimedii	RE	n.i.	4	n.i.	flavonoids	80	0.1	n.i	1	Ethanol/water	n.i.	
Epimedii	UE	n.i.	4	n.i.	flavonoids	n.i.	n.i.	n.i	1	Ethanol/water	n.i.	
Ginkgo leaves	SC-CO_2	n.i.	0.105	n.i.	flavonoids	59–120	24.2–31.2	2.5, 5 and 10 mL/min	n.i.	CO_2/Ethanol (90/10)	4–8	Chiu others (2002)
Maté tea leaves	SC-CO_2	3–6	0.046 mm	10	caffeine, theobromine, theophylline	40, 50, 70	14–40	1 mL/min	7	no	99, 96, 95	Saldaña others (1999)

Myrica rubra leaves	VMAE	n.i.	n.i.	n.i.	myricetin, quercetin	40–80	0.04–0.06	n.i.	0.08–3	no	n.i.	Wang others (2008)
[I]Various herbs	CEM	n.i.	n.i.	11.9–13.1	phenol c	90	0.1	n.i.	2	acetone/water 60:40 v/v, methanol/water 60:40 v/v, water, ethyl acetate-water 60:30 v/v	88	Proestos and Komaitis (2008)
[I]Various herbs	[III]MAE	n.i.	n.i.	11.9–13.1	phenolic	n.i.	no	n.i.	0.01–0.06	acetate–water 60:40 v/v, ethyl acetate–water 60:30 v/v	92	Proestos and Komaitis (2008)
Rhizoma polygoni Cuspidati	VMAE	n.i.	n.i.	n.i.	resveratrol, emodin	40–80	0.04–0.06	n.i.	0.08–3	no	n.i.	Wang others (2008)
Yuanhu zhitong	[IV]MAE	n.i.	2	n.i.	tetrahydropalmatine imperatorin isoimperatorin	n.i.	no	n.i.	0.16–0.66	no	100	Liao others (2008)

i.c.: initial content (%), p.s.: particle size (mm), m: moisture (%), T: temperature (°C), P: pressure (MPa), f.r.: flow rate (liter/min), t: time (hr), c-s.: cosolvent (%), n.i.: not indicated, **yield (g/100 g initial material), [I]Various herbs: *Rosmarinus officinalis*, *Origanum dictamnus*, *Origanum majorana*, *Teucrium polium*, *Vitex agnus–cactus*, and *Styrax officinalis*, MAE: Microwave-assisted extraction, VMAE: Vacuum microwave-assisted extraction, CEM: Conventional extraction method, DMAE: Dynamic microwave-assisted extraction, PMAE: Pressurized microwave-assisted extraction, RE: Heat reflux extraction, UE: Ultrasonic extraction, [I]MAE: powers ranging from 25 to 250 W were used; [II]DMAE: 20 and 100 W were used, [III]MAE: powers up to 750 W were used; [IV]MAE: powers from 500 to 850 W were used, SC-CO_2 supercritical carbon dioxide extraction.

Factors Affecting Extraction Yield in SC-CO_2 Extraction

Sample Preparation

Drying and grinding the samples are two important steps prior to extraction, because many fruits and vegetables contain 80–95% moisture and grinding to achieve small particle size will favor the high contact surface area necessary for efficient extraction.

Particle size. For the extraction of carotenoids from freeze-dried carrot using SC-CO_2, the studies of Goto and others (1994) and Sun and Temelli (2006) show that a higher extraction yield was obtained with small carrot particles. In the study of Sun and Temelli (2006), the total carotenoid yield increased from 1,110 to 1,370 and 1,504 μg/g dry carrot with particle sizes of 1–2 mm to 0.5–1 mm and 0.3–0.5 mm, respectively.

Moisture. Moisture had different effects on the extraction yield of phytochemicals. For example, the α- and β-carotene extraction yields using SC-CO_2 increased from 184 to 599 μg/g dry carrot and from 354 to 892 μg/g, respectively, with decreasing the moisture in the feed material from 84.6 to 0.8%. The lutein yield decreased from 55.3 to 13 μg/g dry carrot with a decrease in moisture from 84.6 to 0.8% (Sun and Temelli 2006). For the extraction of lycopene from tomato with 50–60% moisture content, only trace amounts of lycopene were reported (Vasapollo and others 2004).

Extraction Parameters

Temperature and pressure. Most of the fruit and vegetables studies on extraction of phytochemicals using SC-CO_2 were performed at a temperature range of 35–110°C and a pressure range of 8–55 MPa (see Table 9.2). In general, extraction yield of carotenoids, such as α- and β-carotene, lutein, and lycopene, using SC-CO_2 increased with temperature and pressure (Ollanketo and others 2001; Gomez-Prieto and others 2003; Vasapollo and others 2004; Saldaña and others 2006). For the extraction of lycopene at 60°C and 85°C, about 20% and 30% recovery was achieved in 80 min, respectively (Vasapollo and others 2004). The amounts of β-carotene extracted from carrots using SC-CO_2 at 40°C and 50°C and 120–327 bar for 8 hr extraction increased with temperature and pressure (Saldaña and others 2006). Higher amounts of β-carotene followed by α-carotene and lutein were obtained. Furthermore, the quantities of β- and α-carotene extracted by SC-CO_2 were one order of magnitude higher than those of lutein. This difference might facilitate the separation of lutein from β- and α-carotene by changing the operating conditions.

Flow rate and extraction time. Dynamic techniques for the extraction of carotenoids with SC-CO_2 use flow rates that vary from 0.5 to 15 mL/min (measured at extraction temperature and pressure) with different effects depending on the matrix (Rozzi and others 2002; Subra and others 1998; Saldaña and others 2006). Subra and others (1998) extracted β-carotene from 1 to 2.5 g freeze-dried carrots and studied the effect of flow rates (0.4 and 1.2 liter/min); they obtained higher yields of β-carotene at a flow rate of 1.2 liter/min. Sun and Temelli (2006) also evaluated the effect of flow rate (0.5 and 1.0 liter/min) on the extraction of β-carotene with SC-CO_2 + canola oil. The total carotenoids yield increased with flow rate, ranging from 934.8 to 1,973.6 μg/g dry carrot at CO_2 flow rates from 0.5 to 2 liter/min (measured at STP), respectively (Sun and Temelli, 2006). However, the lycopene yield decreased from 38.8% to 8% as flow rate was increased from 2.5 to 15 mL/min (measured at extraction temperature and pressure) (Rozzi and others 2002).

Use of cosolvent. Various cosolvents, such as acetone, ethanol, methanol, hexane, dichloromethane, and water, have been used for the removal of carotenoids using SC-CO_2 extraction (Ollanketo and others 2001). All these cosolvents except water (only 2% of recovery) increased the carotenoid recovery. The use of vegetable oils such as hazelnut and canola oil as a cosolvent for the recovery of carotenoids from carrots and tomatoes have been reported (Sun and Temelli, 2006; Shi, 2001; Vasapollo and others 2004). For the extraction without cosolvent addition, the lycopene yield was below 10% for 2- to 5-hr extraction time, whereas in the presence of hazelnut oil, the lycopene yield increased to about 20% and 30% in 5 and 8 hr, respectively. The advantages of using vegetable oils as cosolvents are the higher extraction yield; the elimination of organic solvent addition, which needs to be removed later; and the enrichment of the oil with carotenoids that can be potentially used in a variety of product applications.

Comparison of Emerging Extraction Methods

Phenolic compounds were extracted from elderberry and grape marc using a conventional extraction method (CEM) working at temperatures of 20, 40, and 60°C for 2 hr, and a SC-CO_2 extraction method working at a temperature of 40°C and pressures of 15 and 30 MPa (Vatai and others 2009). Comparing these two methods, CEM was better than SC-CO_2 in terms of yield of phenolic compounds. The highest yields of phenolic compounds and anthocyanins were found using a mixture of solvents (50% ethanol–acetone–water for phenolics and 50% ethanol or ethyl acetate–water for anthocyanins). Although CEM lead to a higher yield of phenolic compounds, this method has disadvantages such as the solvent residue left in the final extract and degradation during storage (Vatai and others 2009). Anthocyanins were extracted from grapes using three methods: HPP, UE, and PEF (Corrales and others 2008). However, the use of different process conditions (HPP: 70°C, 600 MPa for 1 hr, UE: 70°C, 1 hr, and PEF: 25–28°C, 15–60 sec) makes it difficult to compare the various extraction methods. At these conditions, PEF and UE increased the anthocyanin extraction yield up to 10% in comparison with HPP, and up to 17% when compared to CEM. Furthermore, the extracts obtained using the three different methods were free of organic solvents, demonstrating the potential of these methods as extraction techniques. Ranalli and others (2004) showed that carrot oil extracted using SC-CO_2 had higher carotenes (1,850 mg/kg) and sterols (30,248.4 mg/kg) compared with conventional commercial carrot oil (170 mg carotenes/kg and 1,726.2 mg sterols/kg). The extraction of phytochemicals from vegetables using emerging technologies combined with conventional technologies is also promising. For example, a combination of PEF and aqueous centrifugal extraction methods was used to obtain carotenoids from carrots at 18–35°C within 15 min (El-Belghiti and others 2007). Using the combined methods, a yield of 96% was achieved, probably due to the softening of carrot tissue during the PEF treatment (intensity of 300 V/cm and extent of 30 × 100 μs or 3 ms). PEF was also used to extract betanin pigment from red beetroot at 1 kV/cm, 25°C within 3–4 min (Fincan and others 2004). In this study, PEF technology was shown to be an effective method of permeabilization for the extraction of betanin pigments, consuming very low energy (approximately 7 kJ/kg).

The extraction of lycopene from tomato using ultrasonic assisted extraction (UAE) and ultrasound/microwave assisted extraction (UMAE) was also reported (Lianfu and

Zelong 2008). UAE used temperatures of 53–86°C and processing times of 13–47 min, whereas UMAE used shorter processing times of 3–9 min (Lianfu and Zelong 2008). For UMAE, the optimal processing conditions were obtained at 98 W microwave power with 40 kHz ultrasonic processing. For UAE, the optimal processing temperature was 86°C in 29 min. Comparing UMAE to PEF, lycopene yield was higher using UMAE (97.4% of lycopene) than using PEF (89.4%).

Nuts and Seeds

Nuts and seeds are natural sources of phytochemicals. Table 9.3 summarizes studies on the extraction of phytochemicals from nuts. Examples include sterols from acorn with SC-CO_2 (Lopes and Bernardo-Gil 2005), from hazelnut using microwave extraction (Uquiche and others 2008; Bernardo-Gil and others 2002), from walnut using SC-CO_2 (Oliveira and others 2002), from grape seeds using SC-CO_2 (Beveridge and others 2005; Sovova and others 1994; Murga and others 2000), from hibiscus using SC-CO_2 (Holser and Bost 2004), from pumpkin using SC-CO_2 (Bernardo-Gil and others 2001), and from rosehip using SC-CO_2 (Panfili and others 2003; Shen and others 1997, 1996; Danielski and others 2005). Phenolics and carotenoids were also extracted from rosehip and grapes using SC-CO_2 (Del Valle and others 2000; Szentmihalyi and others 2002). Alkaloids such as caffeine and theobromine were obtained from cocoa beans and guarana seeds (Saldaña and others 2002a,b).

Factors Affecting Extraction Yield in SC-CO_2 Extraction

Sample Preparation

In the case of seeds such as guarana seeds, it is necessary to grind-them because of their hard structure, but this might not be the case for some nuts (Marrone and others 1998; Saldaña and others 2002a,b).

Particle size. Phytochemical yield increased with decreased particle size, and therefore nuts and seeds are generally ground to small particles to increase the surface area and facilitate extraction.

Moisture. Moisture content is an important parameter in extraction of phytochemicals, because the nut or seed might expand upon moisture absorption, resulting in a more permeable cell membrane, allowing both phytochemical and CO_2 to flow easily as in the case of coffee beans (Saldaña 1997). However, excess water in some other matrices can also impede the diffusion of solute and have a negative effect on the extraction. For example, the extraction of phytochemicals from grape seed with 6.3, 2.4, 1.1, and 0.3% moisture content was modified by drying the ground sample for different lengths of time: 0, 2, 4, and 6.3 hr, respectively (Gomez and others 1996). The extraction yield was not significantly affected by the moisture content of the grape seeds. However, the sample exposed to the longest drying time (6.3 hr) had a slightly lower yield, which might be due to the evaporation of volatile constituents (Gomez and others 1996).

Extraction Parameters

Temperature and pressure. Most of the nut and seed studies on extractions of phytochemicals using SC-CO_2 were performed at a temperature range of 35–80°C and a pressure range of 10–70 MPa (see Table 9.3). The effects of pressure and temperature were studied in the removal of caffeine from wet ground guarana seeds at 40 and

70°C and pressures ranging from 10 to 40 MPa. The quantity of caffeine extracted increased with an increase in pressure, with this effect being more pronounced at 70°C than at 40°C (Saldaña and others 2002a,b). One important aspect to consider while depressurizing the system is that this step might affect the breakage of the coffee beans (Saldaña 1997).

Flow rate and extraction time. Decreasing solvent flow rate results in an increased of extraction yield using SC-CO_2. The extraction time is a function of the matrix structure, differing with the type of material. For example, diffusion through a nut is faster than that through a seed. Time is inversely related to the particle size, and many other process parameters can influence this variable, such as temperature, pressure, flow rate, and cosolvent addition (Saldaña 1997; Saldaña and others 2002a,b; Mohamed and others 2002).

Use of cosolvent. Few studies were found in the literature for the extraction of phytochemicals from nuts and seeds using a cosolvent. Alkaloids, caffeine, and theobromine were extracted from guarana seeds and cocoa beans using two different carbon dioxide/ethanol mixtures at 20 MPa and 70°C. The quantities of caffeine and theobromine extracted at these conditions increased with quantity of ethanol added, with this effect being more pronounced at higher ethanol concentrations. Within 115 min of extraction, approximately 54%, 58%, and 74% of the caffeine was removed from guarana seeds using 215 kg of water-saturated CO_2, 5% ethanol/95% CO_2, and 10% ethanol/90% CO_2 mixtures, respectively (Saldaña and others 2002b). Theobromine was extracted from ground cocoa beans using dry, water-saturated, and ethanol/CO_2 mixtures containing 5% and 10% ethanol. Within 145 min, the amounts of theobromine extracted when using dry CO_2 and water-saturated CO_2 were about 10%, in agreement with the findings of Li and Hartland (1992). The addition of ethanol to CO_2 resulted in a substantial increase in the extraction of theobromine from ground cocoa beans (42% and 90% of theobromine was removed using 5% ethanol/95% CO_2 and 10% ethanol/90% CO_2 mixtures, respectively). In addition, a study of extraction of caffeine and theobromine from cocoa beans using both supercritical CO_2 and supercritical ethane at pressures from 15 to 40 MPa and at 70°C was reported (Mohamed and others 2002). For either solvent, the results revealed a large influence of pressure on the extraction yields of cocoa butter and alkaloids from cocoa beans due to the relatively large changes in solvent isothermal compressibility with pressure at the conditions investigated. Relatively shorter extraction times were observed for the removal of caffeine than for theobromine, probably because of the formation of theobromine complexes through strong hydrogen bonding, as demonstrated by the higher melting temperature and enthalpy of theobromine in comparison with those of caffeine.

Comparison of Emerging Extraction Methods

Nuts and bean phytochemicals are extracted by conventional methods using organic solvents. But, a considerable amount of fat-soluble vitamins and other valuable constituents such as phospholipids are removed during hexane extraction (Crowe and White 2003; Zhang and others 1995). Phospholipids are important for lipid stability. Furthermore, organic solvent extraction also leaves undesirable solvent residue in the final product. Alkaloids such as caffeine and theobromine were extracted from guarana seeds and cocoa beans using SC-CO_2 (Saldaña and others 2002a). In this research, higher selectivity of SC-CO_2 for caffeine was obtained in comparison with theobromine. Using MWE, better results were obtained for fatty acids and sterols in hazelnut than when using CEM (Uquiche and others 2008). Microwave (MW) was also used to extract fatty acids from hazelnut with a power of 400 and 600 W and a radiation time of 3–4 min (Uquiche and others 2008). An oil extraction yield up to 45% was obtained using 400 W, 4 min, and 2,450 MHz, indicating that MW may facilitate the rupture of cell membrane to improve the efficiency of the extraction process.

Herbs

Herbs are plants that have nitrogen organic bases. A few studies on the extraction of phytochemicals from herbs are reported in the literature using conventional and emerging extraction technologies. Table 9.4 summarizes studies on the extraction of phytochemicals from herbs, such as flavonoids from epimedii using CEM, DMAE, PMAE, heat reflux extraction (RE), and UE. Of all these methods, DMAE used the least extraction time (approximately 6 min) to obtain flavonoids (Chen and others 2008). Other phytochemicals such as gonine, cocaine, sitosterol, flavonoids, alkaloids, and quercetin were also extracted from natural sources such as coca leaves, dandelion, ginkgo, maté, and *Myrica rubra* leaves (Chen and others 2008; Liao and others 2008; Proestos and Komaitis, 2008; Simándi and others 2002; Chiu and others 2002; Saldaña and others 1999, 2002a; Brachet and others 2002).

Factors Affecting Extraction Yield in SC-CO_2 Extraction

Sample Preparation

Leaves such as maté tea, laurel, and coca contain approximately 80–95% moisture content, so drying is needed. Dry leaves are then ground to reduce their particle size and facilitate the extraction.

Particle size. Phytochemical yield extraction using SC-CO_2 increased with a decreased in particle size. In a study by Simándi and others (2002), the extraction of β-amyrin and β-sitosterol was evaluated at different temperatures and pressures as well as with different particle sizes of 0.2–0.8 mm. In a study by Saldaña and others (2000), the extraction of caffeine and theobromine were compared using whole and ground leaves at pressures of 13.8 and 25.5 MPa and temperatures of 40°C and 70°C. The results clearly revealed higher caffeine extraction rates for ground commercial maté tea at the early stages of extraction as expected due to the absence of mass transfer resistance and plant matrix interference. All caffeine was removed from the ground leaves using only 0.5 kg of SC-CO_2. But, more than 2.5 kg of SC-CO_2 was required to achieve the same level of removal using whole leaves (Saldaña and others 2000).

Moisture. Moisture increased the extraction yield of some alkaloids as demonstrated for black tea and maté tea leaves (Saldaña and others 1999). This can be explained considering that water can act as a cosolvent for the extraction of slightly polar compounds, whereas the presence of water is favorable at about 10–15% to increase extraction yield.

Extraction Parameters

Temperature and pressure. Most of the studies for the extraction of phytochemicals from herbs using SC-CO_2 were performed at temperatures of 40-70°C and pressures of 14–40 MPa (see Table 9.4). The influence of temperature and pressure using

SC-CO_2 was reported for the extraction of sitosterol from dandelion leaves (Simándi and others 2002); flavonoids from ginkgo leaves (Chiu and others 2002); and alkaloids from maté tea leaves (Saldaña and others 1999, 2000, 2002a).

Flow rate and extraction time. No report was found for the effect of flow rate in the extraction of phytochemicals from herbs. Extractions at lower pressures and/or temperatures required prolonged time and large amounts of CO_2 to achieve the same yield as reported for the extraction of caffeine, theophylline, and theobromine from maté tea leaves (Saldaña and others 1999, 2002a).

Use of cosolvent. The use of small amounts of ethanol as a cosolvent could have significant positive impact on the supercritical fluid extraction of alkaloids from agricultural commodities. The use of ethanol as a cosolvent with SC-CO_2 resulted in an increased extraction yield of flavonoids from ginkgo leaves (Chiu and others 2002) and of alkaloids from various sources (Kim and Choi 2000; Lack and Seidlitz 1993; Sanders 1993; Katz 1998; Saldaña and others 2002b; Mohamed and others 2002). Modification of the solvent with ethanol resulted in an increase in the quantity of caffeine extracted, from approximately 41% to 75% of caffeine using 448 kg of solvent/kg maté leaves. Whereas only 30% of the original caffeine was extracted using water-saturated SC-CO_2, doubling the ethanol concentration resulted in an increase in caffeine extraction that is almost proportional to the increase in the ethanol concentration (Mohamed and others 2002).

Comparison of Emerging Extraction Methods

Among the emerging technologies to extract phytochemicals from herbs, MAE, CEM, VMAE, PMAE, and DMAE are reported. For example, cocaine and benzoylecgonine were extracted using MAE from coca leaves at 25°C within 24 min in solvents such as ethanol, methanol, hexane, toluene, and water (Brachet and others 2002). The results indicated that the extracts obtained using MAE were similar to the extracts obtained using CEM. However, MAE (0.5 min) is more efficient than CEM in terms of processing time. Microwave energy causes molecular motion by migration of ions and rotation of dipoles; therefore, microwave heating depends on the presence of polar molecules or ionic species. However, a drawback of this technology is that it requires filtration to separate the raw extract from the solid material. Another study using VMAE, MAE, and CEM for the removal of polyphenolic compounds and pigments from Chinese herbs reveals that VMAE is more efficient than MAE and CEM (Wang and others 2008). The extraction of flavonoids from epimedii herb was also evaluated using methods such as PMAE, DMAE, RE, UE, and CEM (Chen and others 2008). The processing conditions were 80°C and 1–10 hr for CEM, 0.6–1 mL/min and 1–10 min for DMAE, 400 kPa and 14 min for PMAE, 80°C and 1 hr for RE, and 1 hr for UE (Chen and others 2008). Similar extraction yield of flavonoids was obtained using DMAE and PMAE. However, DMAE was more efficient than CEM, RE, and UE. In addition, DMAE also reduced the energy consumption, amount of solvent, and processing time. In the study of Proestos and Komaitis (2008), phenolic compounds were extracted from few plants using two methods: CEM (90°C and 2 hr) and MAE (1–4 min). MAE (only 4 min) was faster than CEM (2 hr) and used less solvent. Phytochemicals from yuanhu zhitong were also extracted using MAE and CEM (Liao and others 2008). MAE had several advantages (saving energy, reducing extraction time, and increasing extraction yield) compared to CEM. The optimal processing conditions for MAE were 500 W (0–850 W), 70% of ethanol, 2,450 MHz, and 27 min of extraction time (10–40 min).

Future Trends

The literature reviewed in this chapter demonstrates the feasibility of using emerging technologies such as SC-CO_2, PEF, HPP, UE, OH, MW, and MAE to recover phytochemicals from a variety of biological plant sources. However, it is essential to study each plant composition individually because the pretreatment of material and optimum extraction conditions will depend on the structure and on the composition of specific plant source. Nowadays, the majority of studies are carried out at laboratory scale. There are few pilot-scale studies using SC-CO_2, OH, MW, and PEF extraction, and some of these studies have already reached commercial application using mainly SC-CO_2, and MW. For a technology such as PEF or MW to be adopted more widely, its economic viability and advantages over conventional techniques must be proven. In addition, supercritical fluid technology allows combination of extraction with fractionation to further separate bioactive components of interest. In the case of MW, a combination of this technology with methods such as PMAE, VMAE, and MAE will increase the extraction yield and efficiency. However, more research is needed to investigate the quality attributes of extracted phytochemicals, such as oxidative stability, chemical composition, and stability of phytochemicals throughout extraction and storage by using the emerging extraction technologies. Much more study is needed of PEF, UE, HPP, MW, and OH as well as their combinations. Finally, it is also necessary to better communicate to consumers the advantages of these technologies compared to conventional extraction technologies.

Conclusions

Although for many years very few new methods to extract phytochemicals were developed, recently there has been a boom in emerging extraction technologies. For a long time traditional methods using organic solvents were utilized to extract phytochemicals. These methods present some disadvantages such as residues of solvent that are left in the final product, high processing temperatures, and emissions of volatile organic compounds into the atmosphere. Some promising emerging extraction technologies that can overcome these disadvantages are SC-CO_2, OH, HPP, PEF, and MW extractions. These methods are able to recover phytochemicals such as phenolics, carotenoids, sterols, and alkaloids from a wide variety of agricultural plant sources. Extensive research within a large variety of plant materials such as fruits, vegetables, nuts and seeds, and herbs have shown that SC-CO_2 is the most utilized emerging extraction method for the recovery of phytochemicals. There are many factors that affect extraction efficiency, such as sample preparation, moisture content, and the extraction parameters of temperature, pressure, solvent flow rate, extraction time, and use of a cosolvent. These parameters also have an impact on the various quality attributes such as color characteristics, flavor, and oxidative stability of the extracted phytochemicals and residual product. Extensive research using other emerging technologies such as OH, HPP, PEF, MW, and PMAE alone or in combination is urgently needed. The optimization of these emerging extraction technologies is also necessary to improve the extraction methods.

Acknowledgments

The authors would like to acknowledge the Natural Sciences and Engineering Research Council of Canada (NSERC) and Alberta Agricultural Research Institute (AARI) for financial support of our research program on sub-/supercritical fluid technology applied to the extraction of phytochemicals from biomass.

REFERENCES

Bernardo-Gil G, Oneto C, Antunes P, Rodrigues MF and Empis JM. 2001. Extraction of lipids from cherry seed oil using supercritical carbon dioxide. Eur Food Res Technol 212(2):170–174.

Bernardo-Gil MG, Grenha J, Santos J and Cardoso P. 2002. Supercritical fluid extraction and characterization of oil from hazelnut. Eur J Lipid Sci Technol 104(7):402–409.

Beveridge THJ, Girard B, Kopp T and Drover JCG. 2005. Yield and composition of grape seed oils extracted by supercritical carbon dioxide and petroleum ether: varietal effects. J Agric Food Chem 53(5):1799–1804.

Bouic PJD and Lamprecht JH. 1999. Plant sterols and sterolins: a review of their immune-modulating properties. Altern Med Rev 4(3):170–177.

Brachet A, Christen P and Veuthey J. 2002. Focused microwave-assisted extraction of cocaine and benzoylecgonine from coca leaves. Phytochem Anal 13(3):162–169.

Bridle P and Timberlake CF. 1997. Anthocyanins as natural food colours–selected aspects. Food Chem 58(1–2):103–109.

Britton G, Liaaen-Jensen S and Pfander H, editors. 2004. Carotenoids Handbook. Basel: Birkhäuser.

Chen L, Jin H, Ding L, Zhang H, Li J, Qu C and Zhang H 2008. Dynamic microwave-assisted extraction of flavonoids from Herba Epimedii. Sep Purif Technol 59(1):50–57.

Chiu K-L, Cheng Y-C, Chen J-H, Chang CJ and Yang P-W. 2002. Supercritical fluids extraction of *Ginkgo ginkgoloides* and flavonoids. J Supercrit Fluids 24:77–87.

Choi ES, Noh MJ and Yoo KP. 1998. Solubilities of *o*-, *m*- and *p*-coumaric acid isomers in carbon dioxide at 308.15–323.15K and 8.5–25 MPa. J Chem Eng Data 43(1):6–8.

Clinton SK. 1998. Lycopene: chemistry, biology and implication for human health and disease. Nutr Rev 56(2 Pt 1):35–51.

Corrales M, Toepfl S, Butz P, Knorr D and Tauscher B. 2008. Extraction of anthocyanins from grape by-products assisted by ultrasonics, high hydrostatic pressure or pulsed electric fields: a comparison. Innov Food Sci Emerg Technol 9(1):85–91.

Crowe TD and White PJ. 2003. Oxidation, flavor, and texture of walnuts reduced in fat content by supercritical carbon dioxide. J Am Oil Chem Soc 80(6):569–574.

Danielski L, Zetzl C, Hense H and Brunner G. 2005. A process line for the production of raffinated rice oil from rice bran. J Supercrit Fluids 34(2):133–141.

De Franca LF, Reber G, Meireles MAA, Machado NT and Brunner G. 1999. Supercritical extraction of carotenoids and lipids from buriti (*Mauriti flexuosa*), a fruit from the Amazon region. J Supercrit Fluids 14(3):247–256.

Del Valle J M, Bello S, Thiel J, Allen A and Chordia L. 2000. Comparison of conventional and supercritical CO_2-extracted rosehip oil. Braz J Chem Eng 17(3):335–348.

Delorenzi JC, Attias M, Gattass CR, Andrade M, Rezende C, Pinto AC, Henriques AT, Bou-Habib DC and Saraiva EM. 2001. Antileishmanial activity of indole alkaloid from *Peschiera australis*. Antimicrob Agents Chemother 45(5):1349–1354.

Duthie GG, Gardner PT and Kyle JAM. 2003. Plant polyphenols: are they the new magic bullet? Rowett Research Institute. Proc Nutr Soc 62(3):599–603.

El-Belghiti K, Rabhi Z and Vorobiev E. 2007. Effect of process parameters on solute centrifugal extraction from electropermeabilized carrot gratings. Food Bioprod Process 85(1):24–28.
Esquivel MM, Bernardo-Gil MG and King MB. 1999. Mathematical models for supercritical extraction of olive husk oil. J Supercrit Fluids 16(1):43–58.
Fincan M, DeVito F and Dejmek P 2004. Pulsed electric field treatment for solid–liquid extraction of red beetroot pigment. J Food Eng 64(3):381–388.
Francis FJ. 1989. Food colorants: anthocyanins. Crit Rev Food Sci Nutr 28 (4): 273–314.
Fullmer LA and Shao A 2001. The role of lutein in eye health and nutrition. Am Assoc Cereal Chem 46(9):408–413.
Gamarra FM.C, Sakanaka LS, Tambourgi EB and Cabral FA. 2006. Influence on the quality of essential lemon (*Citrus aurantifolia*) oil by distillation process. Braz J Chem Eng 23(1):147–151.
Giovannucci E. 2002. A review of epidemiologic studies of tomatoes, lycopene, and prostate cancer. Exp Biol Med 227(10):852–859.
Gomez AM, Lopez CP and de la Ossa EM. 1996. Recovery of grape seed oil by liquid and supercritical carbon dioxide extraction: a comparison with conventional solvent extraction. Chem Eng J 61(3):227–231.
Gomez-Prieto MS, Caja MM and Santa-Maria G. 2002. Solubility in supercritical carbon dioxide of the predominant carotenes in tomato skin. J Am Oil Chem Soc 79(9):897–902.
Gomez-Prieto MS, Caja MM, Herraiz M and Santa-Maria G. 2003. Supercritical fluid extraction of all trans-lycopene from tomato. J Agric Food Chem 51(1):3–7.
Goodman GE, Thornquist MD, Balmes J, Cullen MR, Meyskens FL, Omenn GS, Valanis B and Williams JH. 2004. The β-carotene and retinol efficacy trial: incidence of lung cancer and cardiovascular disease mortality during 6-year follow-up after stopping β-carotene and retinol supplements. J Natl Cancer Inst 96(23):1743–1750.
Goto M, Sato M and Hirose T. 1994. Supercritical carbon dioxide extraction of carotenoids from carrots, in Yano T, Matsuno R and Nakamura K, Eds. Developments in Food Engineering, Proc. Sixth International Congress on Engineering and Food, Blackie Academic & Professional, London, 835–837.
Granado F, Olmedilla B and Blanco I. 2003. Nutritional and clinical relevance of lutein in human health. Br J Nutr 90(3):487–502.
Güçlü-Üstündag O and Temelli F. 2004. Correlating the solubility behavior of minor lipid components in supercritical carbon dioxide. J Supercrit Fluids 31(3):227–234.
Hallikainen MA, Sarkkinen ES and Uusitupa MIJ. 2000. Plant stanol esters affect serum cholesterol concentrations of hypercholesterolemic men and women in a dose-dependent manner. J Nutr 130:767–776.
Hasegawa R, Chujo T, Sai-Kato K, Umemura T, Tanimura A and Kurokawa Y. 1995. Preventive effects of green tea against liver oxidative DNA damage and hepatotoxicity in rats treated with 2-nitropropane. Food Chem Toxicol 33(11):961–970.
Heinemann T, Axtmann G and Von Bergmann. K 1993. Comparison of intestinal absorption of cholesterol with different plant sterols in man. Eur J Clin Invest 23(12):827–831.
Holser RA and Bost G. 2004. Hybrid hibiscus seed oil compositions. J Am Oil Chem Soc 81(8):795–797.
Hughes DA. 2001. Dietary carotenoids and human immune function. Nutrition 17(10):823–827.
Illés V, Szalai O, Then M, Daood H and Perneczki S. 1997. Extraction of hiprose fruit by supercritical CO_2 and propane. J Supercrit Fluids 10(3):209–218.
Inagake M, Yamane T, Kitao Y, Oya K, Matsumoto H, Kikuoka N, Nakatani H, Takahashi T, Nishimura H and Iwashima A. 1995. Inhibition of 1,2-dimethylhydrazine-induced oxidative DNA damage by green tea extract in rat. Jpn J Cancer Res 86(11):1106–1111.
Jones PJH, Ntanios FY, Raeini-Sarjaz M and Vanstone CA. 1999. Cholesterol-lowering efficacy of a sitostanol-containing phytosterol mixture with a prudent diet in hyperlipidemic men. Am J Clin Nutr 69(6):1144–1150.
Katz SN. 1998. Decaffeinitation of coffee. In: Clarke RJ, Macrae R, editors. Coffee. Technology, Volume 2. London: Elsevier Applied Science, p. 59.
Kim HJ, Lee SB, Park KA and Hong IK. 1999. Characterization of extraction and separation of rice bran oil rich in EFA using SFE process. Sep Purif Technol 15(1):1–8.
Kim J and Choi YH. 2000. SPE of ephedrine derivates from *Ephedra sinica*. Proceedings of 5th International Symposium on Supercritical Fluids, Atlanta, GA, pp. 1–12.
Klipstein-Grobusch K, Launer LJ, Geleijnse JM, Boeing H, Hofman A and Witteman JCM. 2000. Serum carotenoids and atherosclerosis: The Rotterdam Study. Atherosclerosis 148:49–56.

Kong J-M, Chia L-S, Goh N-K, Chia T-F and Brouillard R. 2003. Review. Analysis and biological activities of anthocyanins. Phytochemistry 64:923–933.

Lack A and Seidlitz H. 1993. Commercial scale decaffeination of coffee and tea using supercritical CO_2. In: King MB and Bott TR, editors. Extraction of natural products using near-critical solvents. Glasgow: Blackie Academic, p. 101–139.

Law M. 2000. Plant sterol and stanol margarines and health. BMJ 320:861–864.

Li S and Hartland S. 1992. Influence of cosolvents on solubility and selectivity in extraction of xanthines and cocoa butter from cocoa beans with supercritical carbon CO_2. J Supercrit Fluids 5: 7–12.

Li S, Varadarajan GS and Stanley H. 1991. Solubilities of theobromine and caffeine in supercritical carbon dioxide: correlation with density-based models. Fluid Phase Equilib 68:263–280.

Lianfu Z and Zelong L. 2008. Optimization and comparison of ultrasound/microwave assisted extraction (UMAE) and ultrasonic assisted extraction (UAE) of lycopene from tomatoes. Ultrason Sonochem 15(5):731–737.

Liao Z, Wang G, Liang X, Zhao G and Jiang Q. 2008. Optimization of microwave-assisted extraction of active components from Yuanhu Zhitong prescription. Sep Purif Technol 63(2):424–433.

Lopes IMG and Bernardo-Gil MG. 2005. Characterization of acorn oils extracted by hexane and by supercritical carbon dioxide. Eur J Lipid Sci Technol 107:12–19.

Makris DP, Kallithtaka S and Kefalas P. 2006. Flavonols in grapes, grape products and wine: burden, profile and influential parameters. J Food Compos Anal 19: 396–404.

Manninen P, Pakarinen J and Kallio H 1997. Large-scale supercritical carbon dioxide extraction and supercritical carbon dioxide countercurrent extraction of cloudberry seed oil. J Agric Food Chem 45(7):2533–2538.

Marrone C, Poletto M, Reverchon E and Stassi A. 1998. Almond oil extraction by supercritical CO_2: Experiments and modeling. Chem Eng Sci 53(21):3711–3718.

Mazza G and Brouillard R. 1987. Recent developments in the stabilization of anthocyanins in food-products. Food Chem 25:207–225.

McHugh MA and Krukonis VJ. 1994. Supercritical Fluid Extraction: Practice and Principles, 2nd ed. Boston: Butterworth-Heinemann, pp. 14–16.

Middleton E, Kandaswami C and Theoharides TC. 2000. The effects of plant flavonoids on mammalian cells: implications for inflammations, heart disease, and cancer. Pharmacol Rev 52(4):673–751.

Mohamed RS, Saldaña MDA, Zetzl C, Mazzafera P and Brunner G. 2002. Extraction of caffeine, theobromine, and cocoa butter from Brazilian cocoa beans using supercritical CO_2 and ethane. Ind Eng Chem Res 41(26):6751–6758.

Moreau RA, Whitaker BD and Hicks KB. 2002. Phytosterols, phytostanols, and their conjugates in foods: structural diversity, quantitative analysis, and health-promoting uses. Prog Lipid Res 41: 457–500.

Murga R, Ruiz R, Beltran S and Cabezas JL. 2000. Extraction of natural complex phenols and tannins from grape seeds by using supercritical mixtures of carbon dioxide and alcohol. J Agric Food Chem 48(8):3408–3412.

Murga R, Sanz MT, Beltran S and Cabezas JL. 2002. Solubility of some phenolic compounds contained in grape seeds in supercritical carbon dioxide. J Supercrit Fluids 23(2):113–121.

Murga R, Sanz MT, Beltran S and Cabezas JL. 2003. Solubility of three hydroxycinnamic acids in supercritical carbon dioxide. J Supercrit Fluids 27(3):239–245.

Normen L, Dutta P, Lia A and Andersson H. 2000. Soy sterol esters and β-sitostanol ester as inhibitors of cholesterol absorption in human small bowel. Am J Clin Nutr 71(4):908–913.

Oliveira R, Rodrigues MF and Bernardo-Gil MG. 2002. Characterization and supercritical carbon dioxide extraction of walnut oil. J Am Oil Chem Soc 79(3):225–230.

Ollanketo M, Hartonen K, Riekkola ML, Holm Y and Hiltunen R. 2001. Supercritical carbon dioxide extraction of lycopene in tomato skins. Eur Food Res Technol 212(5):561–565.

Omoni AO and Aluko RE. 2005. The anti-carcinogenic and anti-atherogenic effects of lycopene: a review. Trends Food Sci Technol 16(8):344–350.

Panfili G, Cinquanta L, Fratianni A and Cubadda R. 2003. Extraction of wheat germ oil by supercritical CO_2: oil and defatted cake characterization. J Am Oil Chem Soc 80(2):157–161.

Patel SB. 2008. Plant sterols and stanols: their role in health and disease. J Clin Lipidol 2:S11–S19.

Pereira CG, Leal PF, Sato DN and Meireles MAA. 2005. Antioxidant and antimycobacterial activities of Tabernaemontana catharinensis extracts obtained by Supercritical CO_2 +cosolvent. J Med Food 8(4):533–538.

Perretti G, Miniati E, Montanari L and Fantozzi P. 2003. Improving the value of rice by-products by SFE. J Supercrit Fluids 26(1):63–71.

Piironen V, Lindsay DG, Miettinen TA, Toivo J and Lampi AM. 2000. Plant sterols: biosynthesis, biological function and their importance to human nutrition. J Sci Food Agric 80(7):939–966.

Piironen V, Toivo J, Puupponen-Pimia R and Lampi A-M. 2003. Plant sterols in vegetables, fruits and berries. J Sci Food Agric 83(4):330–337.

Plat J and Mensink RP. 2001. Effects of diets enriched with two different plant stanol ester mixtures on plasma ubiquinol-10 and fat-soluble antioxidant concentrations. Metab Clin Exp 50(5): 520–529.

Plat J and Mensink RP. 2005. Plant stanol and sterol esters in the control of blood cholesterol levels: mechanism and safety aspects. Am J Cardiol 96(1):15D–22D.

Podsedek A. 2007. Natural antioxidant capacity of *Brassica* vegetables: a review. LWT Food Sci Technol 40(1):1–11.

Porrini M, Riso P and Testolin G. 1998. Absorption of lycopene from single or daily portions of raw and processed tomato. Br J Nutr 80(4):353–361.

Proestos C and Komaitis M 2008. Application of microwave-assisted extraction to the fast extraction of plant phenolic compounds. LWT Food Sci Technol 41(4):652–659.

Raicht RF, Cohen BI and Fazzini EP. 1980. Protective effect of plant sterols against chemically induced colon tumors in rats. Cancer Res 40:403–405.

Ranalli A, Contento S, Lucera L, Pavone G, Di Giacomo G, Aloisio L, Di Gregorio C, Mucci A and Kourtikakis I. 2004. Characterization of carrot root oil arising from supercritical fluid carbon dioxide extraction. J Agric Food Chem 52(15):4795–4801.

Rao AV and Agarwal S. 2000. Role of antioxidant lycopene in cancer and heart disease. J Am Coll Nutr 19(5):563–569.

Rao AV and Shen H. 2002. Effect of low dose lycopene intake on lycopene bioavailability and oxidative stress. Nutr Res 22(10):1125–1131.

Rates SMK, Schapoval EES, Souza IA and Henriques AT. 1993. Chemical constituents and pharmacological activities of *Peschiera australis*. Int J Pharmacogn 31(4):288–294.

Reid RC, Prausnitz JM and Poling BE. 1987. The Properties of Gases and Liquids. New York: McGraw-Hill.

Revilla E and Ryan JM. 2000. Analysis of several phenolic compounds with potential antioxidant properties in grape extracts and wines by high-performance liquid chromatography—photodiode array detection without sample preparation. J Chromatogr 881(1–2):461–469.

Rice-Evans CA, Miller NJ and Paganga G. 1996. Structure–antioxidant activity relationships of flavonoids and phenolic acids. Free Radic Biol Med 20(7):933–956.

Rissanen TH, Voutilainen S, Nyyssonen K, Salonen R, Kaplan GA and Salonen JT. 2003. Serum lycopene concentrations and carotid atherosclerosis: The Kuopio Ischaemic Heart Disease Risk Factor Study. Am J Clin Nutr 77(1):133–138.

Rozzi NL, Singh RK, Vierling RA and Watkins BA. 2002. Supercritical fluid extraction of lycopene from tomato processing byproducts. J Agric Food Chem 50(9):2638–2643.

Ruiz del Castillo ML, Gomez-Prieto MS, Herraiz M and Santa-Maria G. 2003. Lipid composition in tomato skin supercritical fluid extracts with high lycopene content. J Am Oil Chem Soc 80(3):271–274.

Saldaña MDA. 1997. Extraction of caffeine, trigonelline and chlorogenic acid from Brazilian coffee beans using supercritical CO_2 MSc. Thesis, UNICAMP, Campinas, Brazil.

Saldaña MDA, Mazzafera P and Mohamed RS. 1999. Extraction of purine alkaloids from maté (*Ilex paraguariensis*) using supercritical CO_2. J Agric Food Chem 47:3804–3808.

Saldaña MDA, Mohamed RS and Mazzafera P. 2000. Supercritical carbon dioxide extraction of methylxanthines from maté tea leaves. Braz J Chem Eng 17:251–259.

Saldaña MDA, Zetzl C, Mohamed RS and Brunner G. 2002a. Decaffeination of guaraná seeds in a microextraction column using water-saturated CO_2. J Supercrit Fluids 22:119–127.

Saldaña MDA, Zetzl C, Mohamed RS and Brunner G. 2002b. Extraction of methylxanthines from guaraná seeds, maté leaves, and cocoa beans using supercritical carbon dioxide and ethanol. J Agric Food Chem 50:4820–4826.

Saldaña MDA, Mohamed RS and Mazzafera P. 2002c. Extraction of cocoa butter from Brazilian cocoa beans using supercritical CO_2 and ethane. Fluid Phase Equilib 194–197:885–894.

Saldaña MDA, Li S, Guigard SE and Temelli F. 2006. Comparison of the solubility of β-carotene in supercritical CO_2 based on a binary and a multicomponent complex system. J Supercrit Fluids 37(3):342–349.

Saldaña MDA, Tomberli B, Guigard SE, Goldman S, Gray CG and Temelli F. 2007. Determination of vapor pressure and solubility correlation of phenolic compounds in supercritical CO_2. J Supercrit Fluids 40(1):7–19.

Sanders N. 1993. Food legislation and the scope for increased use of near-critical fluid extraction operations in food, flavoring and pharmaceutical industries. In: King MB and Bott TR, editors. Extraction of Natural Products Using Near-Critical Solvents. Glasgow: Blackie Academic, pp. 34–49.

Scientific Committee on Food. 2002. General View of the Scientific Committee on Food on the Long-term Effects of the Intake of Elevated Levels of Phytosterols from Multiple Dietary Sources, with Particular Attention to the Effects on β-carotene. SCF/CS/NF/DOS/20 ADD 1 Final, European Commission, Brussels, Belgium.

Shen Z, Palmer MV, Ting SST and Fairclough RJ. 1996. Pilot scale extraction of rice bran oil with dense carbon dioxide. J Agric Food Chem 44(10):3033–3039.

Shen Z, Palmer MV, Ting SST and Fairclough RJ. 1997. Pilot scale extraction and fractionation of rice bran oil using supercritical carbon dioxide. J Agric Food Chem 45(12):4540–4544.

Shi J. 2001. European Patent Appl. WO2001CA00478 20010406. European Patent WO0179355. Separation of carotenoids from fruits and vegetables.

Shi J and Le Maguer M 2000. Lycopene in tomatoes: chemical and physical properties affected by food processing. Crit Rev Food Sci 40(1):1–42.

Simándi B, Kristo SzT, Kéry Á, Selmeczi LK, Kmecz I and Kemény S. 2002. Supercritical fluid extraction of dandelion leaves. J Supercrit Fluids 23:135–142.

Sotelo A and Alvarez RG. 1991. Chemical composition of wild *Theobroma* species and their comparison to the cacao bean. J Agric Food Chem 39:1940–1943.

Sovova H, Kucera J and Jez J. 1994. Rate of the vegetable oil extraction with supercritical CO_2-II. Extraction of grape oil. Chem Eng Sci 49(3):415–420.

Subra P, Castellani S, Jestin P and Aoufi A, 1998, Extraction of β-carotene with supercritical fluids: experiments and modeling. J Supercrit Fluids 12:261–269.

Sun M and Temelli F. 2006. Supercritical carbon dioxide extraction of carotenoids from carrot using canola oil as a continuous co-solvent. J Supercrit Fluids 37(3):397–408.

Szentmihalyi K, Vinkler P, Lakatos B, Illes V and Then M. 2002. Rose hip (*Rosa canina* L.) oil obtained from waste hip seeds by different extraction methods. Bioresour Technol 82(2):195–201.

Uquiche E, Jeréz M and Ortíz J. 2008. Effect of pretreatment with microwaves on mechanical extraction yield and quality of vegetable oil from Chilean hazelnuts (*Gevuina avellana* Mol). Innov Food Sci Emerg Technol 9(4):495–500.

Van Het Hof KH, de Boer BCJ, Tijburg LBM, Lucius BRHM, Zijp I, West CE, Hautvast JGAJ and Weststrate JA. 2000. Carotenoid bioavailability in humans from tomatoes processed in different ways determined from the carotenoid response in the triglyceride-rich lipoprotein fraction of plasma after a single consumption and in plasma after four days of consumption. J Nutr 130: 1189–1196.

Vasapollo G, Longo L, Rescio L and Ciurlia L 2004. Innovative supercritical CO_2 extraction of lycopene from tomato in the presence of vegetable oil as co-solvent. J Supercrit Fluids 29(1–2): 87–96.

Vatai T, Škerget M and Eljko Knez Z. 2009. Extraction of phenolic compounds from elder berry and different grape marc varieties using organic solvents and/or supercritical carbon dioxide. J Food Eng 90(2):246–254.

Wang J, Xiao X and Li G. 2008. Study of vacuum microwave-assisted extraction of polyphenolic compounds and pigment from Chinese herbs. J Chromatogr A 1198-1199:45–53.

Wang ZY, Huang M, Lou Y, Xie J, Reuhl K, Newmark H, Ho C, Yang C and Conney A. 1994. Inhibitory effects of black tea, green tea, decaffeinated black tea, and decaffeinated green tea on ultraviolet B light-induced skin carcinogenesis in 7,12-dimethylbenz[*a*]anthracene-initiated SKH-1 mice. Cancer Res 54(13):3428–3435.

Zhang C, Brusewitz GH, Maness NO and Gasem KAM. 1995. Feasibility of using supercritical carbon dioxide for extracting oil from whole pecans. Trans ASAE 38(6):1763–1767.

10 Methods of Analysis of Antioxidant Capacity of Phytochemicals

Nuria Grigelmo-Miguel, Mª Alejandra Rojas-Graü, Robert Soliva-Fortuny, and Olga Martín-Belloso

Introduction

Many phytochemicals are regarded as protective food components because of their defensive potential against damaging reactive oxygen species within the human body.

In recent years, many analytical methods have been developed to determine the antioxidant activity in foods and biological systems. These methodologies can be divided into two groups, depending on the desired objective: (1) assays that aim at evaluating the degree of oxidative stress of a biological component, for instance, lipids, proteins, or DNA, and (2) assays through which the ability of a substance to scavenge radical species is measured (Sánchez-Moreno 2002). This chapter describes most of these assays as well as their application in measuring antioxidant activity of fruits and vegetables (Table 10.1).

Methodologies Used to Evaluate Oxidative Stress

Although cells possess an array of defensive mechanisms and repairing systems against reactive oxygen species (ROS), these are not always sufficient, leading to oxidative stress in which the production of ROS overwhelms the antioxidant defenses of the organism. This results in damage to biological components such as lipids, proteins, or DNA.

Methods Based on Lipid Oxidation

Antioxidants may intervene at any of the three major steps of the oxidative process: initiation (oxygen consumption), propagation (conjugated dienes and peroxides formation), or termination (lipid peroxidation products).

Oxygen Absorption Methods

The duration of the initiation phase and its extension in the presence of antioxidant agents can be measured by assessing the oxygen consumption patterns. The measurement methods can be manometric (Drozdowski and Szukalska 1987), gravimetric, or polarographic (Roginsky and Barsukova 2001). Azuma and others (1999) used this method to measure the antioxidative activities of several vegetable extracts. In this same way, Roginsky and Barsukova (2001) determined the antioxidant capacity of red and white wines, green and black teas, beer, and soluble coffee.

Oxygen absorption methods have limited sensitivity and require high levels of oxidation as the endpoint for induction periods (Frankel 1993). In foods, antioxidant

Table 10.1. Methods for assessing antioxidant activity

Classification Principle	Marker	Method
Oxidative stress	Lipid oxidation	*Oxygen absorption*: Manometric, polarographic *Diene conjugation*: HPLC, spectrophotometry (234 nm) *Lipid hydroperoxides*: HPLC, GC-MS, chemiluminescence, spectrophotometry *Iodine liberation*: Titration *Thiocyanate*: Spectrophotometry (500 nm) *Hydrocarbons*: GC *Cytotoxic aldehydes*: LPO-586, HPLC, GC, GC-MS *Hexanal and related end products*: Sensory, physicochemical, Cu(II) induction method, GC *TBARS*: Spectrophotometry (532–535 nm), HPLC *Rancimat*: Conductivity *F_2-iP*: GC/MS, HPLC/MS, immunoassays
	Protein peroxidation	*Modified tyrosines*: GC/MS, HPLC, immunoassays *Protein carbonyls*: Atomic absorption spectroscopy, fluorescence spectroscopy, HPLC
	DNA damage	*8-Hydroxydeoxyguanosine*: HPLC, ELISA
Scavenging radical species	Superoxide radical	Spectrophotometry (560 nm) Gas chromatography Electron spin resonance spectrometry Photochemiluminiscence
	Hydrogen peroxide	Spectrophotometry (230 nm) Chemiluminescence Briggs-Rauscher oscillation
	Hydroxyl radical	Spectrophotometry (532 nm) Electron spin resonance spectrometry
	Hypochloric acid	Spectrophotometry (410 nm)
	Singlet oxygen	Spectrophotometry (440 nm) Chemiluminescence
	Peroxynitrite	Spectrophotometry (275–500 nm) HPLC
Scavenging radical species	Peroxyl radical	*ORAC*: Fluorescence *TRAP*: Chemiluminescence, fluorescence *DCFH-DA*: Fluorescence, spectrophotometry (480 nm) *TOSC*: Gas chromatography *β-Carotene bleaching*: Spectrophotometry (470 nm) *Crocin*: Spectrophotometry (443 nm)
	ABTS radical cation	Spectrophotometry (660–820 nm)
	DMPD radical cation	Spectrophotometry (505 nm)
	DPPH radical	Electron spin resonance spectrometry Spectrophotometry (515–528 nm) Spectrophotometry/HPLC
Other procedures	Folin–Ciocalteu	Spectrophotometry (750 nm)
	FRAP	Spectrophotometry (593 nm)
	CUPRAC	Spectrophotometry (450 nm)
	Cyclic voltammetry	Voltammetry

HPLC: high-performance liquid chromatography, GC-MS: gas chromatography–mass spectrometry, GC: gas chromatography, ORAC: oxygen-radical absorbance capacity, TRAP: total radical-trapping antioxidant parameter, DCFH-DA: dichlorofluorescein–diacetate, TOSC: total oxyradical scavenging capacity, ABTS: 2,2-azinobis(3-ethylbenzothiazoline-6-sulfonate), DMPD: *N,N*-dimethyl-*p*-phenylenediamine, DPPH: 2,2-diphenyl-1-picrylhydrazyl, FRAP: ferric reducing antioxidant power, CUPRAC: cupric reducing antioxidant capacity.

concentrations are usually lower than the levels needed to perform this assay; therefore the sensitivity of these kind of methods may not be sufficient.

Determination of Conjugated Dienes

Measurement of diene conjugation by absorption at 234 nm is commonly used for determining the oxidative stability of a sample. The usual substrate for the determination of conjugated dienes includes any substance containing polyunsaturated fatty acids, with oxidation being initiated by the addition of copper ions, iron ions, AAPH (2,2A-azobis(2-amidinopropane)hydrochloride), DDPH (2,2-diphenyl-1-picrylhydrazyl), or the application of heat (Antolovich and others 2002). This methodology has been used to measure the antioxidant activities of 44 berry fruit wines and liquors (Heinonen and others 1998a) and a total of 92 phenolic extracts from edible and nonedible plant materials (berries, fruits, vegetables, herbs, cereals, tree materials, plant sprouts, and seeds) (Kähkönen and others 1999).

This method is also used to measure *ex vivo* low-density lipoprotein (LDL) oxidation. LDL is isolated fresh from blood samples, oxidation is initiated by Cu(II) or AAPH, and peroxidation of the lipid components is followed at 234 nm for conjugated dienes (Prior and others 2005). In this specific case the procedure can be used to assess the interaction of certain antioxidant compounds, such as vitamin E, carotenoids, and retinyl stearate, exerting a protective effect on LDL (Esterbauer and others 1989). Hence, Viana and others (1996) studied the *in vitro* antioxidative effects of an extract rich in flavonoids. Similarly, Pearson and others (1999) assessed the ability of compounds in apple juices and extracts from fresh apple to protect LDL. Wang and Goodman (1999) examined the antioxidant properties of 26 common dietary phenolic agents in an *ex vivo* LDL oxidation model. Salleh and others (2002) screened 12 edible plant extracts rich in polyphenols for their potential to inhibit oxidation of LDL *in vitro*. Gonçalves and others (2004) observed that phenolic extracts from cherry inhibited LDL oxidation *in vitro* in a dose-dependent manner. Yildirin and others (2007) demonstrated that grapes inhibited oxidation of human LDL at a level comparable to wine. Coinu and others (2007) studied the antioxidant properties of extracts obtained from artichoke leaves and outer bracts measured on human oxidized LDL. Milde and others (2007) showed that many phenolics, as well as carotenoids, enhance resistance to LDL oxidation.

Diene conjugation measurements often cannot be performed directly on tissues and body fluids because many other interfering substances are present, such as heme proteins, chlorophylls, purines, and pyrimidines, that strongly absorb in the UV region. Extraction of lipids into organic solvents before analysis is a common approach to this problem (Antolovich and others 2002). Following this methodology, Bobek and others (1998) and Bobek (1999) established that rat diets containing tomato, grape, and apple pomace reduced plasma levels of conjugated dienes. Sudheesh and Vijayalakshmi (2005) demonstrated that fractions rich in flavonoids obtained from an extract of pomegranate, orally administered to rats at dose of 10 mg kg^{-1} day^{-1}, showed potential antiperoxidative effect. In the same way, Graversen and others (2008) found that black chokeberry juice, black-currant juice, and α-tocopherol protect phosphatidylcholine against oxidation in a peroxidating liposome system.

The measurement of the formation of conjugated dienes has the advantage that it measures an early stage of the oxidation process. However, even in simple lipid systems, diene conjugation by UV spectroscopy is a generic measurement, providing little information on the structure of the compounds. Selectivity can be enhanced by separation of different conjugated dienes using HPLC or by matrix subtraction using second derivative spectroscopy (Antolovich and others 2002).

Determination of Lipid Hydroperoxides

Specific primary peroxidation products of lipids (ROOH) can be detected in plasma using HPLC (Yamamoto and others 1990) or GC-MS (Hughes and others 1986). A proposed method employs xylenol orange (a dye specific for Fe^{3+}) and is based on the well-known ability of transition metals in their reduced form to catalyze the reduction of peroxides to hydroxyl compounds while the metal is oxidized (Nourooz-Zadeh 1999). This method can be applied to lipid suspensions (liposomes, lipoproteins) as well as to whole plasma. Another interesting method is based on the decomposition of lipid peroxides by hemin, which, in the presence of luminol, results in a transient peak of chemiluminescence; this procedure is highly sensitive and suitable for the detection of the content of lipid peroxides in plasma. Other methods are based on the determination of the activity of enzymes that are used to attack lipid peroxides. For instance, in the glutathione peroxidase-based assay (Maiorino and others 1985), either the consumption of NADPH (by UV-spectroscopy) or the formation of glutathione disulfide (GSSG) is determined; in the cyclooxygenase-based method, the decrease of O_2 is usually determined (Warso and Lands 1985).

Commercially available colorimetric assays allow a broader approach, quantifying all lipid hydroperoxides present in a sample. The instability of lipid hydroperoxides makes measurement difficult, as they are readily broken down both *in vivo* and *in vitro* into alkenes and aldehydes (Wood and others 2006). With this methodology, *in vitro* experiments have studied the antioxidant capacity of hot-water vegetable and plant extracts (Maeda and others 1992). On the other hand, several investigations have aimed at quantifiying the antioxidant effect of target compounds through *in vivo* determinations. Wise and others (1996) measured the ability of dehydrated fruit and vegetable extracts to modify oxidative processes by determining lipid peroxides level in human plasma after dietary supplementation. Leontowicz and others (2001) investigated the effect of diets supplemented with sugarbeet pulp and apple pomace on rat plasma lipids and lipid peroxides. Likewise, Krishnan and Vijayalakshmi (2005) evaluated lipid peroxidation in rats fed with fractions of banana rich in flavonoids. Moreover, Arivazhagan and others (2004) evaluated the effects of aqueous extracts of garlic and neem leaf on the extent of lipid peroxidation of induced gastric carcinogenesis in male Wistar rats.

Iodine Liberation

Iodine liberation is one of the oldest and most commonly used methods for assessing lipid substrate oxidation. In this method, hydroperoxides and peroxides oxidize aqueous iodide to iodine, which is then titrated with standard thiosulfate solution and starch as endpoint indicator. The peroxide value is calculated as milliequivalents of peroxide oxygen per kilogram of sample.

This method is relatively easy to use for anhydrous systems, but not for emulsions, foods, or biological media where the presence of water is detrimental. The method has been applied for determining antioxidant activity of both individual natural polyphenols and vegetable extracts. Methods based on this one have been used to evaluate the peroxide value in walnuts (Wilson-Kakashita and others 1995), mango seeds (Joseph 1995), almonds (Uthman and others 1998), and coconut cream powder (Yusof and others 2007).

Thiocyanate Method

The thiocyanate method involves measurement of the peroxide value using linoleic acid as substrate and has also been widely used to measure the antioxidant activity in plant-based foods such as ginger extracts (Kikuzaki and Nakatani 1993), fruit peels (Larrauri and others 1996 1997), extracts from vegetable by-products (Larrosa and others 2002; Llorach and others 2003; Abas and others 2006; Peschel and others 2006), blueberry juice, wines, and vinegars (Su and Chien 2007).

Measurement of Lipid Peroxidation Products

Oxidative damage to membrane polyunsaturated fatty acids leads to the formation of numerous lipid peroxidation products, some of which can be measured as index of oxidative stress, including hydrocarbons, aldehydes, alcohols, ketones, and short carboxylic acids.

Hydrocarbons

Pentane and ethane (end products of n-6 and n-3 polyunsaturated fatty acid peroxidation, respectively) in expired air are useful markers of *in vivo* lipid peroxidation. Nevertheless, when gas chromatography is used to measure hydrocarbons, some technical difficulties may be experienced because chromatographic resolution of pentane from isoprene and isopentane is extremely difficult to achieve. Another possible problem could be the presence of these gases as contaminants in atmosphere. Furthermore, the production of hydrocarbon gases depends on the presence of metal ions to decompose lipid peroxides. If such ions are only available in limited amounts, this index may be inaccurate.

Miller and others (1998) used ethane analysis to study the effect of a diet rich in fruits and vegetables on lipid peroxidation. On the other hand, Stewart and others (2002) used the formation of pentane to evaluate the antioxidant status of healthy young children in response to a commercially available fruit- and vegetable-based antioxidant supplement.

Cytotoxic Aldehydes

The products formed during lipid peroxidation include unsaturated aldehydes, such as 4-hydroxynonenal. Their quantification is of great interest because of their extremely reactive and cytotoxic properties. This extreme reactivity and metabolic conversion, however, may make them unsuitable as test analytes for *in vivo* antioxidant activity studies except at high levels of oxidative stress. Furthermore, simple chemical tests such as the TBARS (thiobarbituric acid reactive substances) and LPO-586 (colorimetric assay for lipid peroxidation markers) tests are not specific for 4-hydroxynonenal. More selective tests based on derivatization and HPLC, GC, or GC-MS are more suitable.

Ranaivo and others (2004) used the 4-hydroxynonenal index to demonstrate that wine polyphenols decrease oxidative stress, whereas Tomaino and others (2006) studied the *in vitro* protective effect of a wine extract on UVB-induced skin damage.

Hexanal and Related End Products

Decomposition of the primary products of lipid oxidation generates a complex mixture including saturated and unsaturated aldehydes such as hexanal. Hexanal is the most commonly measured end product of lipid oxidation, and both sensory and physicochemical methods are used for its determination. Where other antioxidant activity tests may be nonspecific, physicochemical measurement of hexanal offers the advantage of analyzing a single, well-defined end product.

Meyer and others (1997) and Frankel and Meyer (2000) used an induction method in which Cu(II) is applied to initiate oxidation of LDL. The oxidation progress is subsequently monitored through determination by gas chromatography of the amounts of hexanal in the headspace of the reaction vessels.

Jo and others (2006) applied this assay to determine the antioxidant properties of methanolic extracts from Japanese apricot in chicken breast meat. Likewise, Pearson and others (1998) assessed two types of Japanese green tea from Japan and two of their active compounds, catechin and epicatechin, for their relative abilities to inhibit the oxidation of LDL. Also, Pearson and others (1999) assessed the ability of compounds in apple juices and extracts from fresh apple to protect LDL. Heinonen and others (1998b) observed that berry phenolics inhibited hexanal formation in oxidized human LDL.

Thiobarbituric Acid–Reactive Substance (TBARS) Assay

This test is used for both *in vitro* and *in vivo* determinations. It involves reacting thiobarbituric acid (TBA) with malondialdehyde (MDA), produced by lipid hydroperoxide decomposition, to form a red chromophore with peak absorbance at 532 nm (Fig. 10.1). The TBARS reaction is not specific. Many other substances, including other alkanals, proteins, sucrose, or urea, may react with TBA to form colored species that can interfere with this assay.

Further enhancement in specificity has been achieved by the HPLC separation of the complex prior to measurement (Shih and Hu 1999). Other possible approaches are the extraction of malondialdehyde prior to the formation of the chromogen and/or derivative spectrophotometry.

This assay has been used by some authors to evaluate the *in vitro* effects of antioxidant extracts on LDL oxidation (Viana and others 1996; Cirico and Omaye 2006; Kedage and others 2007; Vayalil 2002; García-Alonso and others 2004; Tarwadi and Agte 2005). Oboh and others (2007) confirmed that hot pepper prevents *in vitro* lipid peroxidation in brain tissues. Indeed, Bub and others (2000) demonstrated that a moderate intervention with vegetable products rich in carotenoids reduces lipid peroxidation in men. Nicolle and others (2003) evaluated the effect of carrot intake on antioxidant status in cholesterol-fed rats. Later on, they showed that lettuce consumption improves cholesterol metabolism and antioxidant status in rats (Nicole and others 2004).

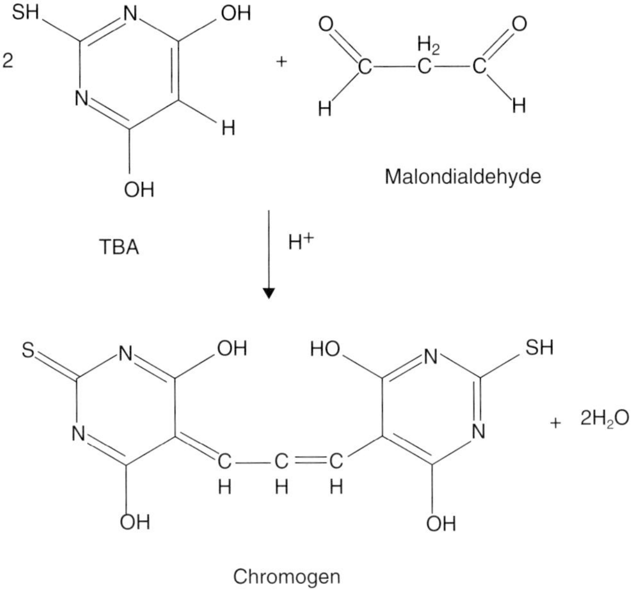

Figure 10.1. Chromophore formed by condensation of thiobarbituric acid (TBA) with malondialdehyde (MDA).

Castenmiller and others (2002) determined the antioxidant properties of differently processed spinach products using TBARS methodology.

Rancimat Assay

Rancimat is an accelerated method to assess oxidative stability of fats and oils. In this test, the sample is subjected to an accelerated oxidative process (by heat in presence of oxygen), where short-chain volatile acids are produced. The acids formed are measured by conductivity.

Mariassyova (2006) studied the antioxidant potential of natural antioxidant concentrates with high contents of flavonoids, carotenoids, and phenolic acids using this method. The assay has been used to evaluate the antioxidant activity of several food products, including herbs and spices (Beddows and others 2000), several Mediterranean and tropical fruits (Murcia and others 2001), truffles and mushrooms subjected to freezing and canning (Murcia and others 2002), apple, goldenrod, and artichoke by-products (Peschel and others 2006), and Egyptian beverages (Allam 2007).

Determination of F_2-isoprostanes (F_2-iP)

The determination of F_2-isoprostanes, oxidation products of arachidonic acid, has been proposed as a more reliable index of oxidative stress *in vivo*, overcoming many of the methodological problems associated with other markers. The isoprostanes have emerged as a most effective method of quantifying the potential of antioxidants to inhibit lipid peroxidation. However, one drawback of this method is that quantification of F_2-iP requires sophisticated techniques, in particular GC/MS and HPLC/MS methods, using negative chemical ionization (Wood and others 2006). Enzyme immunoassays (EIAs) also have been developed for quantification of isoprostanes, being relatively quick and simple. However, it appears that they do not give sufficiently reliable and specific results and should hence be regarded as indices of F_2-iP III-like immunoreactivity.

Thompson and others (1999) measured urinary F_2-iP as a marker of the reduction of oxidative cellular damage by consumption of fruits and vegetables. Stewart and others (2002) studied isoprostanes as indicators in healthy young children in response to a commercially available fruit- and vegetable-based antioxidant supplement. Sánchez-Moreno and others (2004a) assessed the effect of pulsed electric field–processed orange juice intake on biomarkers of antioxidant status in healthy humans, observing a decrease of F_2-iP in plasma. Similarly, Sánchez-Moreno and others (2004b) demonstrated that consumption of a pressurized vegetable soup decreases oxidative stress in healthy humans. Thompson and others (2005) found that 8-isoprostane $F_{2\alpha}$ excretion is reduced in women by increased vegetable and fruit intake. Furthermore, Fowke and others (2006) found that *Brassica* vegetable consumption reduces urinary F_2-isoprostane levels.

Methods Based on Protein Peroxidation

Modified Tyrosines

Protein tyrosine residues constitute key targets for peroxynitrite-mediated nitrations. Attack of various free radicals (ONOO−, $NO_2^{\bullet}$) upon tyrosine generates 3-nitrotyrosine, which can be measured immunologically or by GC/MS or HPLC techniques. The detection of 3-nitrotyrosine was considered a biomarker of peroxynitrite action *in vivo*. Similarly, attack of HOCl and HOBr on tyrosine generates chlorotyrosine and bromotyrosine, respectively, both of which are measured most accurately by GC-MS.

Arteel and Sies (1999) examined procyanidin oligomers of different size, isolated from the seeds of *Theobroma cacao*, for their ability to protect against nitration of tyrosine. Serraino and others (2003) investigated antioxidant activity of the blackberry juice and cyanidin-3-*O*-glucoside on endothelial dysfunction in cells and in vascular rings exposed to peroxynitrite. However, more work is needed in this area, and the confounding effects of oxidized protein/amino acids in the diet need to be elucidated.

Protein Carbonyls

ROS can modify amino acid side chains, with histidine, tryptophan, cysteine, proline, arginine, and lysine among those most susceptible to attack (Brown and Kelly 1994). As a result, carbonyl groups are generated, and these carbonyl concentrations can be measured directly in plasma by using atomic absorption spectroscopy, fluorescence spectroscopy, or HPLC following reaction with 2,4-dinitrophenylhydrazine.

Žitňanová and others (2006) examined the inhibitory effect of extracts from different kinds of fruits and vegetables on the oxidative damage to proteins *in vitro*. Dragsted and others (2004) investigated the relative influence of nutritive and nonnutritive factors in fruit and vegetables on oxidative damage and enzymatic defense. Jacob and others (2003) determined the effect of a moderate intake of antioxidants on biomarkers of oxidative damage assessing *in vivo* protein oxidation in healthy men whose typical diet contained few fruits and vegetables. Consistently, Bloomer at al. (2006) found that supplementation with a fruit and vegetable juice powder concentrate for 2 weeks can attenuate the rise in protein carbonyls after 30 min of aerobic exercise.

Methods Based on DNA Damage

DNA is very susceptible to radical-induced damage. The structural changes in DNA can cause cytotoxicity affecting the replication of DNA *in vivo*.

The most commonly measured product of DNA oxidation is 8-hydroxy-deoxyguanosine (8-OHdG). This product is usually measured by HPLC methods; however, the recent development of commercial ELISA kits is expanding research in this area.

Deng and others (1998) studied the effects of Brussels sprouts on spontaneous and induced oxidative DNA damage in terms of 8-oxo-7,8-dihydro-2′-deoxyguanosine (8-oxodG) in rats. Kasai and others (2000) demonstrated that various plant extracts, such as carrot, apricot, and prune, showed inhibitory effects in an *in vitro* assay of lipid peroxide-induced 8-OHdG formation. Zhu and others (2000) investigated the effect of an aqueous extract of cooked Brussels sprouts on formation of 8-oxodG in calf thymus DNA *in vitro*.

Procedures for Testing Radical Scavenging Species

Free radicals are unstable, highly reactive, and energized molecules having unpaired electrons. Those that regularly interact and damage the major macromolecules in physiological and food-related systems are superoxide ($O_2^{\bullet-}$), hydrogen peroxide (H_2O_2), hydroxyl ($OH^\bullet$), hypochloric acid (HOCl), singlet oxygen ($^1O_2$), peroxynitrite (ONOO−), and peroxyl ($ROO^\bullet$).

Superoxide Radical ($O_2^{\bullet-}$)

Xanthine oxidase is one of the main enzymatic sources of ROS *in vivo*. In normal tissue, xanthine oxidase is a dehydrogenase enzyme that transfers electrons to nicotinamide adenine dinucleotide (NAD), as it oxidizes xanthine or hypoxanthine to uric acid. Under certain stress conditions, the dehydrogenase is converted to an oxidase enzyme and produces superoxide anion ($O_2^{\bullet-}$) and hydrogen peroxide. Thus, xanthine oxidase plus xanthine or hypoxanthine at pH 7.4 can be used to generate the $O_2^{\bullet-}$ anion (MacDonald-Wicks and others 2006), that reduces nitro-blue tetrazolium (NBT) into formazan, which in turn can be spectrophotometrically measured at 560 nm. The reduction of absorbance is estimated as $O_2^{\bullet-}$ scavenging activity compared to the value obtained with no test added sample (Sánchez-Moreno 2002). The foregoing assay has been applied to several vegetable juices (Nishibori and Namiki 1998), some grape extracts (Wood and others 2002), and several vegetables such as spinach, Chinese cabbage, white cabbage, onion, potato, rhubarb, broccoli, green bean, carrot, and tomato (Chu and others 2002; Zhou and Yu 2006). Schmeda-Hirschmann and others (2005) evaluated the free-radical scavenging effect of native food plants gathered in the Argentinian Yungas. Yu and others (2005) used the $O_2^{\bullet-}$ radical scavenging method to measure the antioxidant activity of 11 citrus bioactive compounds. Yang and others

(2006) evaluated 150 edible plants for $O_2^{\bullet -}$ scavenging activity. Wang and Jiao (2000) used hydroxylammonium chloride as oxidant agent instead of NBT to measuring the scavenging activity of some berries.

Another possibility is the determination of the scavenging ability of $O_2^{\bullet -}$ by measuring the reaction of the anion with α-ketomethiolbutyric acid (KMB) to produce ethane, which is measured by gas chromatography. Fresh and air-dried tomato products were evaluated by this methodology (Lavelli and others 1999 2000).

The scavenging ability toward $O_2^{\bullet -}$ can also be measured by using electron spin resonance (ESR) spectrometry. The $O_2^{\bullet -}$ anion is trapped with 5,5-dimethyl-1-pyrroline *N*-oxide (DMPO), and the resultant DMPO–OH adduct is detected by ESR using manganese oxide as internal standard. Noda and others (1997) used this technique to evaluate antioxidant activities of pomegranate fruit extract and its anthocyanidins (delphinidin, cyanidin, and pelargonidin).

The photochemiluminiscence (PCL) assay was initially used by Popov and others (1987). Popov and Lewin (1994 1996) have extensively studied this technique to determine water-soluble and lipid-soluble antioxidants. The PCL assay measures the antioxidant capacity, toward the $O_2^{\bullet -}$ radical, in lipidic and water phase. This method allows the quantification of both the antioxidant capacity of hydrophilic and/or lipophilic substances, either as pure compounds or complex matrices from different origin: synthetic, vegetable, animal, human, etc. The PCL method is based on an approximately 1,000-fold acceleration of the oxidative reactions *in vitro* by the presence of an appropriate photosensitizer. The PCL is a very quick and sensitive method. Chua and others (2008) used this assay to determine the antioxidant potential of *Cinnamomum osmophloeum*, whereas Kaneh and Wang and others (2006) determined the antioxidant capacity of marigold flowers. The antioxidant activity of tree nut oil extracts was also assessed by this method (Miraliakbari and Shahidi 2008).

Hydrogen Peroxide (H_2O_2)

Hydrogen peroxide (H_2O_2) is generated *in vivo* by several oxidase enzymes. Hydrogen peroxide–scavenging activity is easily measured by using peroxidase-based assay systems. This is determined spectrophotometrically from absorption at 230 nm. The most common assay uses horseradish peroxidase and H_2O_2 to oxidize scopoletin into a nonfluorescent product, which can be measured in the presence of a scavenging molecule by inhibiting the oxidation of scopoletin (Sánchez-Moreno 2002). This assay was used to evaluate the antioxidant activity of broccoli amino acids (Martínez-Tomé and others 2001).

Other assays are based on the measure of chemiluminescence. Yoshiki and others (2001) used this technique to evaluate the scavenging activity of some teas and anthocyanin powder of berries. Mansouri and others (2005) tested a series of phenolic acids for their ability to scavenge H_2O_2 using a highly sensitive peroxyoxalate chemiluminescence assay. Almeida and others (2008) employed chemiluminescence to evaluate the *in vitro* scavenging effects on H_2O_2 of an ethanol:water (4:6) extract from *Juglans regia* leaves.

Another method is based on the inhibitory effects of free-radical scavengers toward the oscillations of the Briggs–Rauscher (BR) reaction (Cervellati and others 2001). The

BR oscillating system consists of the iodination and oxidation of an organic substrate (in general, malonic acid or its derivatives) by acidic iodate in the presence of hydrogen peroxide and with the Mn^{2+} ion as catalyst. The antioxidant leads to an immediate cessation of oscillation, and after the so-called inhibition time, the oscillatory behavior is regenerated. This assay is used to evaluate the antioxidative properties of some active principles contained in wines (Kljusurić and others 2005; Höner and others 2002); in fruits, vegetables, and aromatic plants (Höner and Cervellati 2002, 2003; Cervellati and others 2002); and in pasteurized and sterilized commercial red orange juice (Fiore and others 2005).

Other assays have been used to evaluate the antioxidant activity against H_2O_2 of several plant-based products, namely, fruit juices from different cultivars of berries (Wang and Jiao 2000), fractions rich in phenolics isolated from the aqueous by-products obtained during the milling of oil palm fruits (Balasundram and others 2005), cherry laurel fruit and its concentrated juice (Liyana-Pathirana and others 2006), and strawberries and blackberries treated with methyl jasmonate, allyl isothiocyanate, essential oil of *Melaleuca alternifolia*, and ethanol (Chanjirakul and others 2007).

Hydroxyl Radical (OH•)

The hydroxyl radical (OH•) is formed by the combination of Fe(II) and hydrogen peroxide in the so-called Fenton reaction. Usually, the OH• scavenging is measured using 2-deoxy-D-ribose. The OH• radical is formed by combining ascorbic acid, hydrogen peroxide, ferric chloride ($FeCl_3$), and ethylenediaminetetraacetic acid (EDTA) in aqueous solution. The radical reacts with deoxyribose and degrades it into a series of fragments that combine with thiobarbituric acid (TBA) to form a pink chromogen when heated (absorbance = 532 nm). Scavengers of the hydroxyl radical prevent the formation of this chromogen by competing for the radical. This method has been used by some authors to study the antioxidant activity in several vegetables (Chu and others 2002), fruit and vegetable juices (Lichtenthaler and Marx 2005), and berries (Pantelidis and others 2007), as well as in walnut leaf extracts (Almeida and others 2008).

OH• radical has the ability to form nitroxide adducts from the spin trap 5,5-dimethyl-1-pyrroline *N*-oxide (DMPO). This spin trap reacts with OH• radicals, adding a hydroxyl radical scavenger. The adduct DMPO-OH radical exhibits a characteristic ESR response. Furthermore, these adducts have relatively long lives, whereas the duration of OH• is very short. By this method, the scavenging effect of flavonoids from onion skin toward OH• radicals has been evaluated (Suh and others 1999). Certain plant extracts of different polarities (Calliste and others 2001), extracts of pomegranate and its three major anthocyanidins (delphinidin, cyanidin, and pelargonidin), and blood orange juice (Lo Scalzo and others 2004) have also been characterized in terms of their antioxidant potential through this methodology (Noda and others 2002).

Hypochloric Acid (HOCl)

Another source of strong oxidants *in vivo* is neutrophil myeloperoxidase (MPO), which catalyzes oxidation of chloride ions by hydrogen peroxide, resulting in hypochloric acid (HOCl) production. The cytotoxicity of this reaction contributes to the killing of bacteria in the host defense system. However, HOCl generated by MPO might also inactivate α_1-antiproteinase and contribute to proteolytic damage of healthy human tissues in inflammatory disease (Lavelli and others 2000). The ability of a compound to protect α_1-antiproteinase against inactivation by HOCl has been adopted as a method for assessing the antioxidant action of such compounds toward HOCl (Aruoma and others 1993).

In this method, α_1-antiproteinase inhibits the hydrolytic enzyme elastase, and the remaining elastase activity is measured by monitoring increases in absorbance at 410 nm. Martínez-Tomé and others (2001) used a method based on this reaction to measure the antioxidant activity of broccoli amino acids and of Mediterranean spices.

Visioli and others (1998) tested the activity of olive oil phenolics vs. HOCl by evaluating their protective effect in a model solution containing catalase, whose activity was inhibited by hypochlorous acid. Ferreres and others (2006) and Vrchovská and others (2006) investigated the extracts of tronchuda cabbage internal and external leaves for its capacity to act as a scavenger of hypochlorous acid using the HOCl-induced oxidation of 5-thio-2-nitrobenzoic acid (TNB) to 5,5′-dithiobis(2-nitrobenzoic acid (DTNB). Similarly, the antioxidant capacities of tronchuda cabbage seeds (Ferreres and others 2007a) and passion-fruit leaf extracts (Ferreres and others 2007b) have also been studied using HOCl.

Singlet Oxygen ($^1O_2$)

There are various ways to produce singlet oxygen ($^1O_2$) in biological systems. Some biological compounds absorb light and generate $^1O_2$ by photosensitization. The formation of $^1O_2$ *in vivo* without the presence of light is thought to be the result of the spontaneous dismutation of the superoxide anion. Singlet oxygen is also generated from oxidative endogenous processes and activated polymorphonuclear leukocytes (MacDonalds-Wicks and others 2006; Oh and others 2006).

The production of $^1O_2$ by sodium hypochlorite and H_2O_2 can be determined by spectrophotometry in which *N,N*-dimethyl-*p*-nitrosoaniline (NDMA) is used as a selective scavenger of $^1O_2$ and histidine as a selective acceptor of $^1O_2$. The bleaching of NDMA is monitored spectrophotometrically at 440 nm. The extent of $^1O_2$ production can be determined by measuring the decrease in absorbance of NDMA at 440 nm. This method has been used to evaluate the antioxidant activity against $^1O_2$ of fruit juices from different cultivars of berries (Wang and Jiao 2000; Wang and Zheng 2001; Wang and Lin 2003; Chanjirakul and others 2007; Wang and Ballington 2007).

Olas and Wachowicz (2002) investigated the effects of *trans*-resveratrol and vitamin C on oxidative stress in blood platelets. The level of $^1O_2$ in control blood platelets and platelets incubated with resveratrol or vitamin C was recorded using a chemiluminescence method. On the other hand, Oh and others (2006) reported the $^1O_2$ quenching activities of various freshly squeezed fruit and vegetable juices by measuring their inhibitory effects on the rubrene oxidation induced by $^1O_2$ from disproportionation of hydrogen peroxide by sodium molybdate in a microemulsion system.

Peroxynitrite ($ONOO^-$)

Peroxynitrite ($ONOO^-$) is a cytotoxic reactive species that is formed by the reaction of nitric oxide and superoxide. Methods for measuring the scavenging capacity of peroxynitrite usually depend on either the inhibition of tyrosine nitration or the inhibition of dihydrorhodamine 123 (DHR) oxidation to rhodamine 123 (MacDonalds-Wicks and others 2006). Pannala and others (1998) measured the antioxidant capacity of catechin and other polyphenolics by determining the changes in the concentration of tyrosine at 275 nm and of 3-nitrotyrosine (the main product of nitration of tyrosine) at 430 nm in the presence of different concentrations of the antioxidant. This was then confirmed by HPLC separation and quantification of nitrotyrosine. This method was also used by Arteel and Sies (1999) to examine the ability of procyanidin oligomers isolated from the seeds of *Theobroma cacao* to protect against peroxynitrite oxidation.

In the method used by Kooy *and others* (1994), fluorescent intensity was measured by using a spectrofluorometer with the excitation wavelength set at 500 nm and an emission wavelength of 536 nm. The preceding assay has been applied to various products including wines (Paquay and others 1997), flavonoids (Haenen and others 1997; Santos and Mira 2004), procyanidin oligomers from the seeds of *Theobroma cacao* (Arteel and Sies 1999), cyanidin-3-*O*-glucoside in blackberry juice (Serraino and others 2003), and walnut leaf extracts (Almeida and others 2008).

The 3-morpholinosydnonimine system generates superoxide radicals and nitrogen monoxide, forming peroxynitrite, which releases ethane from KMB (α-keto-γ-methiolbutyric acid). This method was used by Lavelli and others (1999) to investigate the radical-scavenging activity of fresh and air-dried tomatoes.

Peroxyl Radical (ROO•)

The peroxyl radical has been the most frequently used compound in assays to measure antioxidant capacity because it is a key component of autoxidation and can be easily produced by the decomposition of azo compounds.

ORAC Assay (Oxygen-Radical Absorbance Capacity)

This assay measures the ability of antioxidant components in test materials to inhibit the decline in β-phycoerythrin (β-PE) fluorescence that is induced by 2,2′ azobis(2-amidinopropane) dihydrochloride (AAPH) as peroxyl radical generator ($ORAC_{ROO\bullet}$), H_2O_2-Cu^{2+} as hydroxyl radical generator ($ORAC_{HO\bullet}$), and Cu^{2+} as a transition metal oxidant ($ORAC_{Cu}$).

In 1993, Cao and others used β-PE as an indicator protein, AAPH as a peroxyl radical generator, and 6-hydroxy-2,5,7,8-tetramethylchroman-2-carboxylic acid (Trolox, a water-soluble vitamin E analog) as a control standard. The ORAC assay combines both inhibition time and inhibition percentage of free-radical action by antioxidants using an area-under-curve technique for quantification. Results are expressed as ORAC units, where 1 ORAC unit equals the net protection produced by 1 μM Trolox. The uniqueness of this assay is that total antioxidant capacity of a sample is estimated by taking the oxidation reaction to completion. This assay has been used in studies on fruit juices (Wang and Lin 2000; Wang and Zheng 2001; Zheng and Wang, 2003), fresh and frozen vegetables (Ninfali and Bacchiocca 2003), and aromatic herbs and spices (Ninfali and others 2005). Furthermore, Mazza and others (2002) applied ORAC to evaluate the overall antioxidant potential associated with the intake of a freeze-dried blueberry powder in human serum after the consumption of a high-fat meal (Mazza and others 2002). This methodology has also been used in determining the bioavailability of vitamin C, carotenoids, anthocyanins, and other phenolic compounds from berries, fruits, vegetables, and wine (Cao and others 1998; Ehlenfeldt and Prior 2001) in humans.

An improved method has been developed and validated using fluorescein as the fluorescent probe (Ou and others 2001). This modification provides a direct measure of hydrophilic chain-breaking antioxidant capacity against peroxyl radical. This method has been applied in vegetables of many kinds (Ou and others 2002; Cho and others 2007) and in tropical fruits (Talcott and others 2003; Mahattanatawee and others 2006).

The foregoing method has been adapted by Dávalos and others (2004) using a conventional fluorescence microplate reader and applied to pure compounds (benzoic and cinnamic acids and aldehydes, flavonoids, and butylated hydroxyanisole) and to wines, as well as to commercial dietary antioxidant supplements. Eberhardt and others (2005) have also proposed a similar method for the determination of the antioxidant activity in broccoli.

The ORAC assay proposed by Ou and others (2001) is limited to hydrophilic antioxidants because of the aqueous environment of the assay. However, lipophilic antioxidants play a critical role in biological defense systems. Huang and others (2002) expanded the assay to the lipidic fraction by introducing a randomly methylated β-cyclodextrin (RMCD) as a water-solubility enhancer for lipophilic antioxidants. Various kinds of foods, including fruit juices and drinks, fruits, vegetables, nuts, and dried fruits, have been evaluated with this method (Zhou and Yu 2006; Wu and others 2004; Kevers and others 2007; Wang and Ballington 2007; Almeida and others 2008; Mullen and others 2007).

A relatively simple but highly sensitive and reliable method for quantifying the ORAC of antioxidants in biological tissues has been automated for use with the COBAS FARA II centrifugal analyzer with a fluorescence-measuring attachment. This method can be used not only for serum but also for biological tissues and food samples, and it can be performed under a wide spectrum of conditions (Cao and others 1995). Thus, this version of the ORAC assay is used to measure antioxidant capacity of vegetables, fruits and their juices (Cao and others 1996; Wang and others 1996; Rababah and others 2005; Prottegente and others 2002), pure compounds, wines, and dietary antioxidant supplements (Dávalos and others 2004) and tea (Cao and others 1996). Also, the antioxidant effect on human plasma of controlled intakes rich in fruits and vegetables has been evaluated (Cao and others 1998).

In 2003, Prior and others described methods for the extraction and analysis of hydrophilic and lipophilic antioxidants, using modifications of the ORAC procedure. These methods provide, for the first time, the ability to obtain a measure of "total antioxidant capacity" in the protein free plasma, using the same peroxyl radical generator for both lipophilic and hydrophilic antioxidants. This assay was also used to measure the total antioxidant capacity of guava fruit extracts (Thaipong and others 2006).

A cellular antioxidant activity (CAA) assay for quantifying the antioxidant activity of phytochemicals, food extracts, and dietary supplements has been developed by Wolfe and Liu (2007). The method measures the ability of compounds to prevent the formation of dichlorofluorescein (DCF) by ABAP-generated peroxyl radicals in human hepatocarcinoma HepG2 cells. The decrease in cellular fluorescence when compared to the control cells indicates the antioxidant capacity of the compounds. The method has been applied to measure the antioxidant activities of selected phytochemicals and fruit extracts.

TRAP Assay (Total Radical-Trapping Antioxidant Parameter)

The original method for measuring the total radical-trapping capacity of biological fluids was developed by Wayner and others (1985). It is based on the evaluation of serum samples by determining their total radical-trapping antioxidant parameter. The thermal decomposition of the water-soluble azo compound ABAP generates the peroxyl radicals at a known and constant rate. After the addition of serum, the peroxyl radicals are scavenged by serum antioxidants. The so-called induction period (the time during which the production of peroxyl radicals is suppressed) determines the TRAP value, which is defined as the number of moles of peroxyl radicals trapped per liter of serum. A chemiluminescence test based on this method was improved by Uotila and others (1994) and applied in citrus fruits by Gorinstein and others (2001).

In 1995, Ghiselli and others introduced a modification of the method using a fluorescent protein. In this method, the rate of peroxidation induced by ABAP was monitored through the loss of fluorescence of the protein R-phycoerythrin (R-PE). The lag phase induced by plasma was compared to that induced by Trolox. Proteins (but not their sulfhydryl groups) interfere with the analysis, partially protecting R-PE when all plasma antioxidants are exhausted. A Trolox-induced lag phase must therefore be measured on each plasma sample. In addition to measuring the antioxidant capacity of plasma, this assay has been used to determine the antioxidant activity of several beverages such as teas, fruit nectars, and juices (Schlesier and others 2002) as well as of vegetables, fruits, and vegetable oils (Pellegrini and others 2003). The method was also used to evaluate the effects of three different peach cultivars on plasma antioxidant activity (Dalla Valle and others 2007).

DCFH-DA Based Assay (Dichlorofluorescein–Diacetate)

DCFH-DA assay was first developed by Valkonen and Kuusi (1997) as an alternative to measure the total peroxyl radical-trapping antioxidant potential of plasma. This assay uses AAPH to generate peroxyl radicals and DCFH-DA as the oxidizable substrate for the generated radicals. The oxidation of DCFH-DA by peroxyl radicals converts DCFH-DA to dichlorofluorescein. DCF is highly fluorescent (excitation 480 nm, emission 526 nm) and also shows absorbance at 504 nm. Therefore, the produced DCF can be monitored either fluorometrically or spectrophotometrically.

The DCFH-DA assay has been used to determine the antioxidant properties of a green tea polyphenol (epigallocatechin 3-gallate) (Tobi and others 2002), green tea polyphenol (epigallocatechin 3-gallate) (Tobi and others 2002), different plant flavonoids (Takamatsu and others 2003), procyanidins from grape seeds (Lu and others 2004), and grape juice (Rozenberg and others 2006).

TOSC Assay (Total Oxyradical Scavenging Capacity)

This method was first reported by Winston and others (1998) and it is based on the oxidation of alpha-keto-γ-methiolbutyric acid (KMBA) to ethylene by peroxyl radicals produced from AAPH. The ethylene formation, which is partially inhibited in the presence of antioxidants, is monitored by gas chromatographic analysis of the reaction vessel headspace.

This assay has been used to determine the antioxidant activity of common vegetables, fruits including apples and strawberries (Eberhardt and others 2000; Chu and others 2002; Sun and others 2002; MacLean and others 2003; Meyers and others 2003), and fruit juices (Lichtenthäler and Marx 2005). No studies have yet reported the application of TOSC assay for the assessment of *in vivo* antioxidant status.

β-Carotene Bleaching

This assay, developed by Taga and others (1984), is based on the coupled oxidation of β-carotene and linoleic acid. The method estimates the relative ability of antioxidant compounds to scavenge the radical of linoleic acid peroxide ($LOO^{\bullet}$) that oxidizes β-carotene in the emulsion phase.

Kanner and others (1994) used the β-carotene bleaching assay to evaluate the antioxidant activity of grapes and wines. Further *in vitro* applications include fruits, vegetables and their juices (Velioglu and others 1998; Makris and Rossiter 2001; Kaur and Kapoor 2002; Ismail and others 2004; Cheung and others 2003); Gorinstein and others 2004; Hassimotto and others 2005; Barros and others 2007; Sun and others 2007), anthocyanins from wild mulberry (Hassimotto and others 2007), chestnut (Barreira and others 2008), and citrus limonoids, flavonoids, and coumarins (Yu and others 2005). The method has been also used to study the effects of dietary supplementation of rats with apple, pear, and grapefruit peels and pulps (Leontowicz and others 2003; Gorinstein and others 2005).

Crocin-based Assays

These assays measure the level of protection provided to the naturally occurring carotenoid derivative crocin from bleaching by the radical generator AAPH. The assay was originally suggested by Bors and others (1984) and modified by Tubaro and others (1998), who used it to show that plasma antioxidant capacity is deeply influenced by the consumption of wine. The addition of a sample containing chain-breaking antioxidants results in the decrease in the rate of crocin decay. The sample is monitored for 10 min at 443 nm.

Later on, a modified microplate-based version suitable for routine determinations was suggested by Lussignoli and others (1999). The assay was originally designed for testing blood plasma; however, nothing prevents the crocin assay from being used for food samples. It has been applied to the study of antioxidant capacity in olive oil phenols (Bonanome and others 2000), certain flavonoids (Finotti and Di Majo 2003; Di Majo and others 2005; Perjési and others 2007), phenolic compounds (Ordoudi and Tsimidou 2006), and capsaicinoids (Perjési and others 2007).

Radical Cation 2,2-Azinobis (3-ethylbenzothiazoline-6-sulfonate) ($ABTS^{\bullet+}$) TEAC Assay (Trolox Equivalent Antioxidant Capacity) or ABTS Assay

The Trolox equivalent antioxidant capacity (TEAC) assay was reported first by Miller and others (1993) and Rice-Evans and Miller (1994). They used the peroxidase activity of metmyoglobin to oxidize 2,2′-azinobis(3-ethylbenzothiazoline-6-sulfonic acid) (ABTS) in the presence of hydrogen peroxide. The TEAC assay is based on the

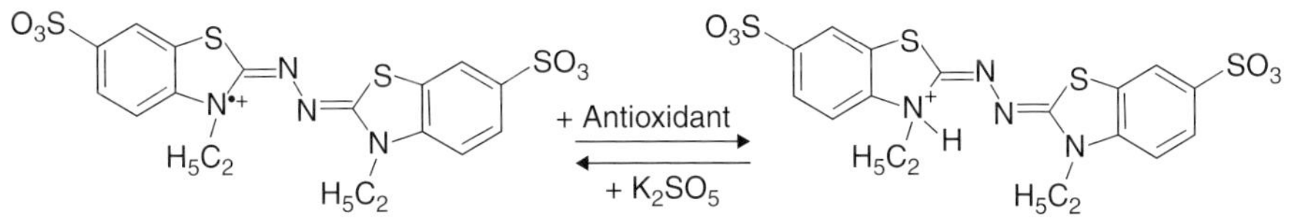

+ Antioxidant

+ K_2SO_5

$ABTS^{\bullet+}$ (λmax = 734 nm) $ABTS^{2-}$ (Colorless)

Figure 10.2. Persulfate oxidation of 2,2′-azinobis(3-ethylbenzothiazoline-6-sulfonic acid)($ABTS^{2-}$).

inhibition by antioxidants of the absorbance of the radical cation $ABTS^{\bullet+}$, which has a characteristic long-wavelength absorption spectrum showing maxima at 660, 734, and 820 nm. Trolox, a water-soluble analog of vitamin E, is used as the reference standard.

Some years later, Miller and others (1996) described a modified TEAC assay that is able to determine the antioxidant activity of carotenoids. In the improved version, $ABTS^{\bullet+}$, the oxidant, is generated by oxidation of 2,2′-azinobis(3-ethylbenzothiazoline-6-sulfonic acid)($ABTS^{2-}$) with manganese dioxide. A similar approach was described by Re and others (1999) in which ABTS was oxidized with potassium persulfate (Fig. 10.2), this version of the TEAC assay is applicable to both water soluble and lipophilic antioxidants (Re and others 1999; Pellegrini and others 1999).

Arnao and others (1996 1999 2001a,b) and Cano and others (1998) introduced some modifications in this assay: the $ABTS^{\bullet+}$ radical cation is generated enzymatically using the system formed by H_2O_2 and horseradish peroxidase (ABTS/H_2O_2/HRP). Several modifications of this protocol have been suggested, depending on the method of $ABTS^{\bullet+}$ generation and a change in the nature of the reference antioxidant: Campos and Lissi (1996) proposed pregeneration of the $ABTS^{\bullet+}$ by heating the radical cation ABTS with a thermolabile azo compound ABAP; Alonso and others (2002) reported the electrochemical generation of $ABTS^{\bullet+}$; Schlesier and others (2002) and De Beer and others (2003) suggested using $K_2S_2O_8$ to generate $ABTS^{\bullet+}$, whereas Kim and others (2002) suggested the use of ascorbic acid as a reference antioxidant instead of Trolox.

The assay has been widely used in many studies related to detection of *in vitro* antioxidant activity of fruits and vegetables because it has the advantage of being quick and simple to perform, making the technique widely accessible. Among other products, ABTS-based assays are applied in a large variety of fruits, vegetables, and their juices, including orange, apple, black currant, grapes, plums, onion, tomato, eggplant, artichoke, spinach, lettuce, asparagus, and broccoli (Miller and Rice-Evans 1997; Alcolea and others 2002; Leong and Shui 2002; Paganga and others 1999; Proteggente and others 2002; Luximon-Ramma and others 2003; Pellegrini and others 2003; Bahorun and others 2004; García-Alonso and others 2004; Chun and others 2005; Nilsson and others 2005; Scalzo and others 2005; Bor and others 2006; Ozgen and others 2006; Zhou and Yu 2006; Lako and others 2007; Stratil and others 2007; Chun and others 2003; Jiménez-Escrig and others 2003; Kim and others 2003; Cano and Arnao 2005; Fiore and others 2005; Oh and others 2006; Pellegrini and others 2007; Sun and others 2007; Santas and others 2008), and in beverages (tea and juices) (Schlesier and others 2002). Žitňanová and others (2006) determined the total antioxidant capacity of extracts from different kinds of fruits and vegetables and examined their *in vitro* inhibitory effect against protein oxidative damage.

The assay has been criticized, because ABTS is not a physiological radical source and thus may not accurately represent *in vivo* effects. But, despite these concerns, the TEAC assay has shown clinical relevance, being sensitive to oxidative insult. Plumb and others (1996) studied the total antioxidant capacity of whole extracts and purified glucosinolates from cruciferous vegetables.Gorinstein and others (2005) compared the influence of naringin vs. red grapefruit juice on plasma lipid levels and plasma antioxidant activity in rats fed cholesterol-containing and cholesterol-free diets. Haruenkit and others (2007) compared *in vitro* and *in vivo* antioxidant capacities of durian, mangosteen, and snake fruit. Indeed, Leontowicz and others (2007) studied the plasma antioxidant activity in rats fed cholesterol-containing diets supplemented with durian at different stages of ripening.

N,N-*Dimethyl*-p-*phenylenediamine Radical Cation (DMPD$^{\bullet+}$): DMPD Assay*

Another assay that is very similar to the ABTS assay is the *N,N*-dimethyl-*p*-phenylenediamine (DMPD assay). In the presence of a suitable oxidant solution at an acidic pH, DMPD is converted to a stable and colored DMPD radical cation (DMPD$^{\bullet+}$). Antioxidants capable of transferring a hydrogen atom to the radical cause the decolorization of the solution, which is spectrophotometrically measured at 505 nm. The reaction is stable, and the endpoint is taken to be the measure of antioxidant efficiency. Antioxidant ability is expressed as Trolox equivalents using a calibration curve plotted with different amounts of Trolox (Fogliano and others 1999). This method is used to measure hydrophilic compounds. The presence of organic acids, especially citric acid, in some extracts may interfere with the DMPD assay, and so this assay should be used with caution in those extracts rich in organic acids (Gil and others 2000).

This assay has been used for the antioxidant capacity evaluation of fresh tomatoes (Leonardi and others 2000; Scalfi and others 2000) and some beverages (red wines, green tea infusion, and pomegranate and orange juices) (Gil and others 2000; Schlesier and others 2002; Fiore and others 2005).

2,2-Diphenyl-1-picrylhydrazyl radical (DPPH$^{\bullet}$) Assay

This method is based on the radical scavenging activity of antioxidants toward the 2,2-diphenyl-1-picrylhydrazyl radical (DPPH$^{\bullet}$). The free radical DPPH$^{\bullet}$ (Fig. 10.3) is reduced to the corresponding hydrazine when it reacts with hydrogen donors (Sánchez-Moreno 2002). The ability can be evaluated by ESR on the basis that the DPPH$^{\bullet}$ signal intensity is inversely related to the antioxidant concentration and the reaction time. More recently, this reaction has been evaluated by the decoloration assay where the decrease in absorbance at 515–528 nm produced by the addition of the antioxidant to DPPH$^{\bullet}$ in methanol or ethanol is measured.

Sánchez-Moreno and others (1998 1999a,b) proposed a new methodology for the evaluation of the antiradical efficiency toward DPPH$^{\bullet}$. Their procedure takes into account not only the concentration of the antioxidant but also the reaction time to reach the plateau of the scavenging reaction, a modification that could be an advantage

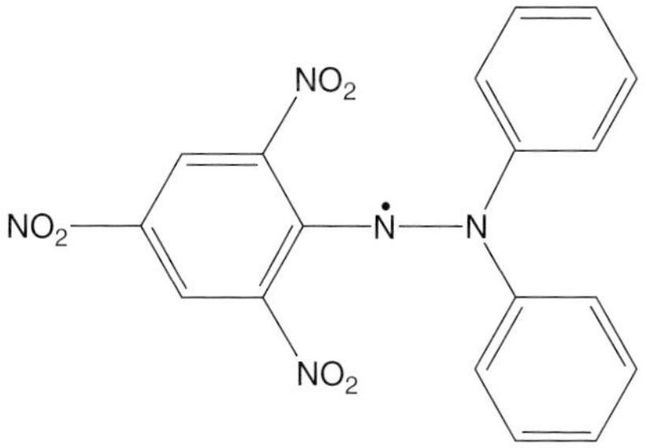

Figure 10.3. Structure of 2,2-diphenyl-1-picrylhydrazyl (DPPH•).

over other methods, which only consider antioxidant concentration. This modified procedure has been applied to determine the scavenging activity of raspberries and cranberries (De Ancos and others 2000) and carotenoids (Jiménez-Escrig and others 2000).

Some posterior modifications for this assay have been suggested. For example, Bandoniene and Murkovic (2002) proposed a version of the DPPH test in which they applied an HLPC determination to quantify the antioxidant activity of apples. Solerrivias and others (2000) suggested a fast and rather simple version of the DPPH test based on thin-layer chromatography with chromametric recording of DPPH quenching. More recently, Polásek and others (2004) suggested a new rapid automated version of the DPPH test based on sequential injection analysis.

Sánchez-Moreno (2002) considered that this assay is an easy and accurate method for determining antioxidant capacity in fruit and vegetable samples. The DPPH assay has been used to determine the antioxidant activity of polyphenols (Sánchez-Moreno and others 1998; Bao and others 2004); flavonols (Jiménez and others 1998 1999; Choi and others 2002); anthocyanin-based natural colorants from berries (Espín and others 2000); ellagic acid derivatives and flavonoids of red raspberry jams (Zafrilla and others 2001); flavonoids and cinnamic acid derivatives (Aaby and others 2004); phenolics, flavonoids, and capsaicinoids from hot pepper fruit (Materska and Perucka 2005); citrus limonoids, flavonoids, and coumarins (Yu and others 2005); several fruits including peels and pulps (guava, apple, pear, strawberries, carob, quince, Jaffa sweeties, berries, persimmons, kiwifruit, grapefruits, orange, lychee, tropical fruits, durian) (Chu and others 2002; Du Toit and others 2001; Jiménez-Escrig and others 2001; Thaipong and others 2006; Leong and Shui 2002; Kondo and others 2002; Orhan and others 2003; Leontowicz and others 2003; Skupień and Oszmiański 2004; Papagiannopoulos and others 2004; Silva and others 2004; Chinnici and others 2004; Gorinstein and others 2004; Nakajima and others 2004; Hamauzu and others 2005; Schmeda-Hirschmann and others 2005; Kondo and others 2005; Jung and others 2005a; Park and others 2006a; Jung and others 2005b; Gorinstein and others 2005; Anagnostopoulou and others 2006; Hukkanen and others 2006; Ozgen and others 2006; Zhao and others 2006; Mahattanatawee and others 2006; Netzel and others 2006; Gorinstein and others 2006; Bae and Suh 2007; Stratil and others 2007; Lim and others 2007; Tsai and others 2007; Li and Jiang 2007; Chanjirakul and others 2007; Haruenkit and others 2007; Xu and others 2008); edible plants; vegetables including tomato, lettuce, mushrooms, cauliflower, corn, potato, sweetpotato, artichoke, broccoli, cabbage, garlic, Italian *Brassicaceae*, peppers, olives, salads, asparagus, and carrot (Martínez-Valverde and others 2002; Muñoz Acuña and others 2002; Pieroni and others 2002; Kang and Saltveit 2002; Cheung and others 2003; Llorach and others 2003; Cevallos-Casals and Cisneros-Zevallos 2003; Jiménez-Escrig and others 2003; Singh and Rajini 2003; Thu and others 2004; Liu and others 2004; Lin and Chan 2005; Borowski and others 2008; Triantis and others 2005; Turkmen and others 2005; Bajpai and others 2005; Ferreres and others 2006, 2007a; Vrchovská and others 2006; Stratil and others 2006; Huang and others 2006; Heimler and others 2006; Deepa and others 2006; Zhou and Yu 2006; Jiang and others 2006; Guil-Guerrero and others 2006; Pereira and others 2006; Maisuthisakul and others 2007; Dasgupta and De 2007; Heimler and others 2007; Roy and others 2007; Barros and others 2007; Sun and others 2007; Jastrzebski and others 2007; Nencini and others 2007; Matsufuji and others 2007; Odriozola-Serrano and others 2008; Yen and others 2008); and beverages: fruit juices, teas, red and white wines, and gazpacho (Gil and others 2000; Du Toit and others 2001; Schlesier and others 2002; Piga and others 2002; De Beer and others 2003; Polovka and others 2003; Kefalas and others 2003; Sousa and others 2004; Fiore and others 2005; Elez-Martínez and Martín-Belloso 2007; Villaño and others 2007; Klimczak and others 2007; Lo Scalzo and others 2004; Jayaprakasha and Patil 2007).

However, the DPPH assay is not suitable for measuring the antioxidant capacity of plasma because protein is precipitated in the presence of the ethanol/methanol solvent used.

Other Procedures

Total Phenols Assay (Folin–Ciocalteu Reagent)

The Folin–Ciocalteu (FC) assay is one of the oldest methods designed to determine the total content of phenolic compounds. This assay was originally intended to analyze protein, taking advantage of the phenol group in tyrosine. Later, Singleton and others (1999) extended this assay to measure total phenols in wine. In this assay, oxidation of phenols by a molybdotungstate reagent yields a colored product with absorbance near 750 nm. General phenolics determined by the FC test are most frequently expressed in gallic acid equivalents.

However, this method is not desirable for products with high vitamin C content, because the FC reagent is nonspecific to phenolic compounds and it can be reduced by many nonphenolic compounds such as L-ascorbic acid, reducing sugars, and soluble proteins (Prior and others 2005). Despite the undefined chemical nature of FC, this method is simple, reproducible, and convenient and therefore has been widely used when studying phenolic antioxidants. Some applications are vegetables, fruits, nuts, and dried fruits (Vinson and others 1998 2001; Velioglu and others 1998; Kähkönen and others 1999; Karakaya and others 2001; Kaur and Kapoor 2002; Imeh and Khokhar 2002; Proteggente and others 2002; Wu and others 2004; Young and others 2005; Stratil and others 2006; Brat and others 2006; Netzel and others 2006; Stratil and others 2007; Roy and others 2007; Heimler and others 2007; Kusznierewicz and others 2008); wines, beers, teas, and soft drinks (López-Vélez and others 2003; Magalhães and others 2006); beverages, juices, and fruit purees (Georgé and others 2005;

Saura-Calixto and Goñi 2006; Abdullakasim and others 2007; Mullen and others 2007), medicinal plants (Surveswaran and others 2007); and commercial vinegars (Sakanaka and Ishihara 2008).

Ferric Reducing Antioxidant Power (FRAP) Assay

This assay was first described in 1996 by Benzie and Strain. The FRAP method is based on the ability of phenolics to reduce a ferroin analog, the Fe^{3+} complex of tripyridyltriazine $Fe(TPTZ)^{3+}$, to the intensely blue colored Fe^{2+} complex $Fe(TPTZ)^{2+}$ in acidic (pH 3.6) conditions. In contrast to many other test systems, it does not use any radical.

The FRAP method is a simple, inexpensive, and robust spectrophotometric technique. However, the relevance of this assay is uncertain, because the assay reaction occurs by electron transfer, which does not mimic physiological situations.

The FRAP assay was originally applied to plasma but has been used for evaluating the antioxidant activity of different plant extracts: fruit juices (Gardner and others 2000; Fiore and others 2005; Mullen and others 2007; Xu and others 2008); tea (Schlesier and others 2002); oils (Pellegrini and others 2003); and relevant polyphenolic compounds from fruits and vegetables (Pulido and others 2000; Proteggente and others 2002; Imeh and Khokhar 2002; Gil and others 2002; Ou and others 2002; Luximon-Ramma and others 2003; Bahorun and others 2004; Guo and others 2003; Nilsson and others 2005; Deepa and others 2006; Jiang and others 2006; Žitňanová and others 2006; Stratil and others 2006 2007; Yang and others 2006; Ozgen and others 2006; Netzel and others 2006; Pantelidis and others 2007; Nencini and others 2007; Santas and others 2008). On the other hand, Nicolle and others (2004) investigated the mean term and postprandial effects of lettuce ingestion on antioxidant protection in rats. Duthie and others (2006) evaluated the effects of cranberry juice consumption on antioxidant status in healthy human volunteers. DeGraft-Johnson and others (2007) investigated *in vitro* the FRAP of 17 plant polyphenols and their metabolites in human plasma.

Cupric Reducing Antioxidant Capacity (CUPRAC)

This method is a variation of the FRAP assay using Cu instead of Fe. It is based on reduction of Cu(II) to Cu(I) by reductants (antioxidants) present in a sample. The CUPRAC assay uses neocuproine (2,9-dimethyl-1,10-phenanthroline), the Cu(I) complex of which absorbs at 450 nm. A dilution curve generated by uric acid standards is used to convert sample absorbance to uric acid equivalents (Apak and others 2004). The method has been applied to measure the antioxidant activity of dietary polyphenols, vitamin C, and vitamin E, but also the antioxidant potential of food matrices such as grape and orange juices, green tea, and blackberry tea (Apak and others 2004); kiwifruit (Park and others 2006b); and roots and stems of rhubarb (Öztürk and others 2007).

Cyclic Voltammetry

The cyclic voltammetry procedure reported by Kohen and others (2000) evaluates the overall reducing power of low-molecular-weight antioxidants in a biological fluid or tissue homogenate. The electrochemical oxidation of a certain compound on an inert carbon glassy electrode is accompanied by the appearance of the current at a certain potential. While the potential at which a cyclic voltammetry peak appears is determined by the redox properties of the tested compound, the value of the current shows the quantity of this compound.

The cyclic voltammetry method has been used for testing fruit and vegetable samples including *Iryanthera juruensis* fruits (Silva and others 2001), prickly pear (Butera and others 2002), orange juice (Sousa and others 2004), and wine (Roginsky and others 2006).

Final Remarks

Many methods are available for determining food antioxidant capacity, which is an important topic in food and nutrition research. However, there is a great need to standardize these methods because the frequent lack of an actual substrate in the procedure, the system composition, and the method of inducing oxidation could limit their accuracy. In fact, antioxidant activities in complex systems cannot be evaluated satisfactorily using a single test, and several test procedures may be required. The search for more specific assays that can be more directly related to oxidative deterioration of foods and biological systems should be the objective of future investigations.

References

Aaby K, Hvattum E and Skrede G. 2004. Analysis of flavonoids and other phenolic compounds using high-performance liquid chromatography with coulometric array detection: relationship to antioxidant activity. J Agric Food Chem 52(15):4595–4603.

Abas F, Lajis NH, Israf DA, Khozirah S and Kalsom YU. 2006. Antioxidant and nitric oxide inhibition activities of selected Malay traditional vegetables. Food Chem 95(4):566–573.

Abdullakasim P, Songchitsomboon S, Techagumpuch M, Balee N, Swatsitang P and Sungpuag P. 2007. Antioxidant capacity, total phenolics and sugar content of selected Thai health beverages. Int J Food Sci Nutr 58(1):77–85.

Alcolea JF, Cano A, Acosta M and Arnao MB. 2002. Hydrophilic and lipophilic antioxidant activities of grapes. Nahrung/Food 46(5):353–356.

Allam SSM. 2007. Antioxidative efficiency of some common traditional Egyptian beverages. Riv Ital Sostanze Grasse 84(2):94–103.

Almeida IF, Fernandes E, Lima JLFC, Costa PC and Bahia, MF. 2008. Walnut (*Juglans regia*) leaf extracts are strong scavengers of pro-oxidant reactive species. Food Chem 106(3):1014–1020.

Alonso ÁM, Domínguez C, Guillén DA and Barroso, CG. 2002. Determination of antioxidant power of red and white wines by a new electrochemical method and its correlation with polyphenolic content. J Agric Food Chem 50(11):3112–3115.

Anagnostopoulou MA, Kefalas P, Papageorgiou VP, Assimopoulou AN and Boskou D. 2006. Radical scavenging activity of various extracts and fractions of sweet orange peel (*Citrus sinensis*). Food Chem 94: 19–25.

Antolovich M, Prenzler PD, Patsalides E, McDonald S and Robards K. 2002. Methods for testing antioxidant activity. Analyst 127(1):183–198.

Apak R, Güçlü K, Mustafa Ö and Karademir SE. 2004. Novel total antioxidant capacity index for dietary polyphenols and vitamins C and E, using their cupric ion reducing capability in the presence of neocuproine: CUPRAC Method. J Agric Food Chem 52(26):7970–7981.

Arivazhagan S, Velmurugan B, Bhuvaneswari V and Nagini S. 2004. Effects of aqueous extracts of garlic (*Allium sativum*) and neem (*Azadirachta indica*) leaf on hepatic and blood oxidant-antioxidant status during experimental gastric carcinogenesis. J Med Food 7(3):334–339.

Arnao MB, Cano A and Acosta M. 1999. Methods to measure the antioxidant activity in plant material. A comparative discussion. Free Radic Res 31(SUPPL.):S89-S96.

Arnao MB, Cano A and Acosta, M. 2001a. The hydrophilic and lipophilic contribution to total antioxidant activity. Food Chem 73(2):239–244.

Arnao MB, Cano A, Alcolea JF and Acosta M. 2001b. Identification of hydrolysable tannins in the reaction zone of *Eucalyptus nitens* wood by high performance liquid chromatography–electrospray ionisation mass spectrometry. Phytochem Anal 12(2):120–127.

Arnao MB, Cano A, Hernández-Ruiz J, García-Cánovas F and Acosta M. 1996. Inhibition by L-ascorbic acid and other antioxidants of the 2,2′-azino-bis(3-ethylbenzthiazoline-6-sulfonic acid) oxidation catalyzed by peroxidase: a new approach for determining total antioxidant status of foods. Anal Biochem 236(2):255–261.

Arteel GE and Sies H. 1999. Protection against peroxynitrite by cocoa polyphenol oligomers. Fed Eur Biochem Soc Lett 462(1-2):167–170.

Aruoma OI, Murcia A, Butler J and Halliwell B. 1993. Evaluation of the antioxidant and prooxidant actions of gallic acid and its derivatives. J Agric Food Chem 41(11):1880–1885.

Azuma K, Ippoushi K, Ito H, Higashio H and Terao J. 1999. Evaluation of antioxidative activity of vegetable extracts in linoleic acid emulsion and phospholipid bilayers. J Sci Food Agric 79(14):2010–2016.

Bae SH and Suh HJ. 2007. Antioxidant activities of five different mulberry cultivars in Korea. LWT Food Sci Technol 40(6):955–962.

Bahorun T, Luximon-Ramma A, Crozier A and Aruoma OI. 2004. Total phenol, flavonoid, proanthocyanidin and vitamin C levels and antioxidant activities of Mauritian vegetables. J Sci Food Agric 84(12):1553–1561.

Bajpai M, Mishra A and Prakash D. 2005. Antioxidant and free radical scavenging activities of some leafy vegetables. Int J Food Sci Nutr 56(7):473–481.

Balasundram N, Ai TY, Sambanthamurthi R, Sundram K and Samman S. 2005. Antioxidant properties of palm fruit extracts. Asia Pac J Clin Nutr 4(4):319–324.

Bandoniene D and Murkovic M. 2002. On-line HPLC-DPPH screening method for evaluation of radical scavenging phenols extracted from apples (*Malus domestica* L.). J Agric Food Chem 50(9):2482–2487.

Bao H, Ren H, Endo H, Takagi Y and Hayashi T. 2004. Effects of heating and the addition of seasonings on the anti-mutagenic and anti-oxidative activities of polyphenols. Food Chem 86(4):517–524.

Barreira JCM, Ferreira ICFR, Oliveira MBPP and Pereira JA. 2008. Antioxidant activities of the extracts from chestnut flower, leaf, skins and fruit. Food Chem 107(3):1106–1113.

Barros L, Baptista P, Correia DM, Morais JS and Ferreira ICFR. 2007. Effects of conservation treatment and cooking on the chemical composition and antioxidant activity of Portuguese wild edible mushrooms. J Agric Food Chem 55(12):4781–4788.

Beddows CG, Jagait C and Kelly MJ. 2000. Preservation of alpha-tocopherol in sunflower oil by herbs and spices. Int J Food Sci Nutr 51(5):327–339.

Benzie IFF and Strain JJ. 1996. The ferric reducing ability of plasma (FRAP) as a measure of "antioxidant power": the FRAP assay. Anal Biochem 239(1):70–76.

Bloomer RJ, Goldfarb AH and McKenzie MJ. 2006. Oxidative stress response to aerobic exercise: comparison of antioxidant supplements. Med Sci Sports Exercise 38(6):1098–1105.

Bobek P. 1999. Dietary tomato and grape pomace in rats: effect on lipids in serum and liver, and on antioxidant status. Br J Biomed Sci 56(2):109–113.

Bobek P, Ozdín L and Hromadová M. 1998. The effect of dried tomato, grape and apple pomace on the cholesterol metabolism and antioxidative enzymatic system in rats with hypercholesterolemia. Nahrung 42(5):317– 320.

Bonanome A, Pagnan A, Caruso D, Toia A, Xamin A, Fedeli E, Berra B, Zamburlini A, Ursini F and Galli G. 2000. Evidence of postprandial absorption of olive oil phenols in humans. Nutr Metab Cardiovasc Dis 10(3):111–120.

Bor JY, Chen HY and Yen GC. 2006. Evaluation of antioxidant activity and inhibitory effect on nitric oxide production of some common vegetables. J Agric Food Chem 54(5):1680–1686.

Borowski J, Szajdek A, Borowska EJ, Ciska E and Zieliński H. 2008. Content of selected bioactive components and antioxidant properties of broccoli (*Brassica oleracea* L.). Eur Food Res Technol 226(3):459–465.

Bors W, Michel C and Saran M. 1984. Inhibition of the bleaching of the carotenoid crocin. A rapid test for quantifying antioxidant activity. Biochim Biophys Acta Lipids Lipid Metab 796(3):312–319.

Brat P, Georgé S, Bellamy A, Du Chaffaut L, Scalbert A, Mennen L, Arnault N and Amiot MJ. 2006. Daily polyphenol intake in France from fruit and vegetables. J Nutr 136(9):2368–2373.

Brown, RK and Kelly, FJ. 1994. Role of free radicals in the pathogenesis of cystic fibrosis. Thorax 49: 738–742.

Bub A, Watzl B, Abrahamse L, Delincée H, Adam S, Wever J, Müller H and Rechkemmer G. 2000. Moderate intervention with carotenoid-rich vegetable products reduces lipid peroxidation in men. J Nutr 130(9):2200–2206.

Butera D, Tesoriere L, Di Gaudio F, Bongiorno A, Allegra M, Pintaudi AM, Kohen R and Livrea MA. 2002. Antioxidant activities of sicilian prickly pear (*Opuntia ficus indica*) fruit extracts and reducing properties of its betalains: betanin and indicaxanthin. J Agric Food Chem 50(23):6895–6901.

Calliste CA, Trouillas P, Allais DP, Simon A and Duroux JL. 2001. Free radical scavenging activities measured by electron spin resonance spectroscopy and B16 cell antiproliferative behaviors of seven plants. J Agric Food Chem 49(7):3321–3327.

Campos AM and Lissi EA. 1996. Total antioxidant potential of Chilean wines. Nutr Res 16(3):385–389.

Cano A and Arnao MB. 2005. Hydrophilic and lipophilic antioxidant activity in different leaves of three lettuce varieties. In J Food Prop 8(3):521–528.

Cano A, Hernández-Ruíz J, García-Cánovas F, Acosta M and Arnao MB. 1998. An end-point method for estimation of the total antioxidant activity in plant material. Phytochem Anal 9(4):196–202.

Cao G, Alessio HM and Cutler RG. 1993. Oxygen-radical absorbance capacity assay for antioxidants. Free Radic Biol Med 14(3):303–311.

Cao G, Booth, SL, Sadowski JA and Prior RL. 1998. Increases in human plasma antioxidant capacity after consumption of controlled diets high in fruit and vegetables. Am J Clin Nutr 68(5):1081–1087.

Cao G, Sofic E and Prior RL. 1996. Antioxidant capacity of tea and common vegetables. J Agric Food Chem 44(11):3426–3431.

Cao G, Verdon CP, Wu AHB, Wang H and Prior RL. 1995. Automated assay of oxygen radical absorbance capacity with the COBAS FARA II. Clin Chem 41(12):1738–1744.

Castenmiller JJM, Linssen JPH, Heinonen IM, Hopia AI, Schwarz K, Hollmann PCH and West CE. 2002. Antioxidant properties of differently processed spinach products. Nahrung/Food 46(4):290–293.

Cervellati R, Höner K, Furrow SD, Neddens C and Costa S. 2001. The Briggs-Rauscher reaction as a test to measure the activity of antioxidants. Helv Chim Acta 84(12):3533–3547.

Cervellati R, Renzulli C, Guerra MC and Speroni E. 2002. Evaluation of antioxidant activity of some natural polyphenolic compounds using the Briggs-Rauscher reaction method. J Agric Food Chem 50(26):7504–7509.

Cevallos-Casals BA and Cisneros-Zevallos L. 2003. Stoichiometric and kinetic studies of phenolic antioxidants from Andean purple corn and red-fleshed sweetpotato. J Agric Food Chem 51(11):3313–3319.

Chanjirakul K, Wang SY, Wang CY and Siriphanich J. 2007. Natural volatile treatments increase free-radical scavenging capacity of strawberries and blackberries. J Sci Food Agric 87(8):1463–1472.

Cheung LM, Cheung PCK and Ooi VEC. 2003. Antioxidant activity and total phenolics of edible mushroom extracts. Food Chem 81(2):249–255.

Chinnici F, Bendini A, Gaiani A and Riponi C. 2004. Radical scavenging activities of peels and pulps from cv. golden delicious apples as related to their phenolic composition. J Agric Food Chem 52(15):4684–4689.

Cho YS, Yeum KJ, Chen CY, Beretta G, Tang G, Krinsky NI, Yoon S, Lee-Kim YC, Blumberg JB and Russell RM. 2007. Phytonutrients affecting hydrophilic and lipophilic antioxidant activities in fruits, vegetables and legumes. J Sci Food Agric 87(6):1096–1107.

Choi CW, Kim SC, Hwang SS, Choi BK, Ahn HJ, Lee MY, Park SH and Kim SK. 2002. Antioxidant activity and free radical scavenging capacity between Korean medicinal plants and flavonoids by assay-guided comparison. Plant Sci 163(6):1161–1168.

Chu YF, Sun J, Wu X and Liu RH. 2002. Antioxidant and antiproliferative activities of common vegetables. J Agric Food Chem 50(23):6910–6916.

Chua M-T, Tung Y-T and Chang S-T. 2008. Antioxidant activities of ethanolic extracts from the twigs of *Cinnamomum osmophloeum*. Bioresource Technol 99(6):1918–1925.

Chun OK, Kim DO, Moon HY, Kang HG and Lee CY. 2003. Contribution of individual polyphenolics to total antioxidant capacity of plums. J Agric Food Chem 51(25):7240–7245.

Chun OK, Kim DO, Smith N, Schroeder D, Han JT and Lee CY. 2005. Daily consumption of phenolics and total antioxidant capacity from fruit and vegetables in the American diet. J Sci Food Agric 85(10):1715–1724.

Cirico TL and Omaye ST. 2006. Additive or synergetic effects of phenolic compounds on human low density lipoprotein oxidation. Food Chem Toxicol 44(4):510–516.

Coinu R, Carta S, Urgeghe, PP, Mulinacci N, Pinelli P, Franconi F and Romani A. 2007. Dose-effect study on the antioxidant properties of leaves and outer bracts of extracts obtained from Violetto di Toscana artichoke. Food Chem 101(2):524–531.

Dalla Valle AZ, Mignani I, Spinardi A, Galvano F and Ciappellano S. 2007. The antioxidant profile of three different peaches cultivars (*Prunus persica*) and their short-term effect on antioxidant status in human. Eur Food Res Technol 225(2):167–172.

Dasgupta N and De B. 2007. Antioxidant activity of some leafy vegetables of India: a comparative study. Food Chem 101(2):471–474.

Dávalos A, Gómez-Cordovés C and Bartolomé B. 2004. Extending applicability of the oxygen radical absorbance capacity (ORAC-Fluorescein) assay. J Agric Food Chem 52(1):48–54.

De Ancos B, Gonzalez EM and Cano MP. 2000. Ellagic acid, vitamin C, and total phenolic contents and radical scavenging capacity affected by freezing and frozen storage in raspberry fruit. J Agric Food Chem 48(10):4565–4570.

De Beer D, Joubert E, Gelderblom WCA and Manley M. 2003. Antioxidant activity of South African red and white cultivar wines: free radical scavenging. J Agric Food Chem 51(4):902–909.

Deepa N, Kaura C, Singh B and Kapoor, HC. 2006. Antioxidant activity in some red sweet pepper cultivars. J Food Compos Anal 19(6-7):572–578.

DeGraft-Johnson J, Kolodziejczyk K, Krol M, Nowak P, Krol B and Nowak D. 2007. Ferric-reducing ability power of selected plant polyphenols and their metabolites: implications for clinical studies on the antioxidant effects of fruits and vegetable consumption. Basic Clin Pharmacol Toxicol 100(5):345–352.

Deng XS, Tuo J, Poulsen HE and Loft S. 1998. Prevention of oxidative DNA damage in rats by brussels sprouts. Free Radic Res 28(3):323–333.

Di Majo D, Giammanco M, La Guardia M, Tripoli E, Giammanco S and Finotti E. 2005. Flavanones in citrus fruit: structure–antioxidant activity relationships. Food Res Int 38(10):1161–1166.

Dragsted LO, Pedersen A, Hermetter A, Basu S, Hansen M, Haren GR, Kall M, Breinholt V, Castenmiller JJM, Stagsted J, Jakobsen J, Skibsted L, Rasmussen SE, Loft S and Sandström B. 2004. The 6-a-day study: effects of fruit and vegetables on markers of oxidative stress and antioxidative defense in healthy nonsmokers. Am J Clin Nutr 79(6):1060–1072.

Drozdowski B and Szukalska, E. 1987. A rapid instrumental method for the evaluation of the stability of fats. J Am Oil Chem Soc 64(7):1008–1011.

Du Toit R, Volsteedt Y and Apostolides Z. 2001. Comparison of the antioxidant content of fruits, vegetables and teas measured as vitamin C equivalents. Toxicology 166(1-2):63–69.

Duthie SJ, Jenkinson AMCE, Crozier A, Mullen W, Pirie L, Kyle J, Yap LS, Christen P and Duthie GG. 2006. The effects of cranberry juice consumption on antioxidant status and biomarkers relating to heart disease and cancer in healthy human volunteers. Eur J Nutr 45(2):113–122.

Eberhardt MV, Kobira K, Keck AS, Juvik JA and Jeffery EH. 2005. Correlation analyses of phytochemical composition, chemical, and cellular measures of antioxidant activity of broccoli (*Brassica oleracea* L. Var. italica). J Agric Food Chem 53(19):7421–7431.

Eberhardt MV, Lee CY and Liu RH. 2000. Antioxidant activity of fresh apples. Nature 405(6789):903–904.

Ehlenfeldt MK and Prior RL. 2001. Oxygen radical absorbance capacity (ORAC) and phenolic and anthocyanin concentrations in fruit and leaf tissues of highbush blueberry. J Agric Food Chem 49(5):2222–2227.

Elez-Martínez P and Martín-Belloso O. 2007. Effects of high intensity pulsed electric field processing conditions on vitamin C and antioxidant capacity of orange juice and gazpacho, a cold vegetable soup. Food Chem 102(1):201–209.

Espín JC, Soler-Rivas C, Wichers HJ and García-Viguera C. 2000. Anthocyanin-based natural colorants: a new source of antiradical activity for foodstuff. J Agric Food Chem 48(5):1588–1592.

Esterbauer H, Striegl G, Puhl H, Oberreither S, Rotheneder M, El-Saadani M and Jurgens G. 1989. The role of vitamin E and carotenoids in preventing oxidation of low density lipoproteins. Ann N Y Acad Sci 570: 254–267.

Ferreres F, Sousa C, Valentão P, Andrade PB, Seabra RM and Gil-Izquierdo Á. 2007b. New C-deoxyhexosyl flavones and antioxidant properties of Passiflora edulis leaf extract. J Agric Food Chem 55(25):10187–10193.

Ferreres F, Sousa C, Valentão P, Seabra RM, Pereira JA and Andrade PB. 2007a. Tronchuda cabbage (*Brassica oleracea* L. var. costata DC) seeds: phytochemical characterization and antioxidant potential. Food Chem 101(2):549–558.

Ferreres F, Sousa C, Vrchovská V, Valentão P, Pereira JA, Seabra, RM and Andrade PB. 2006. Chemical composition and antioxidant activity of Tronchuda cabbage internal leaves. Eur Food Res Technol 222: 88–98.

Finotti E and Di Majo D. 2003. Influence of solvents on the antioxidant property of flavonoids. Nahrung/Food 47(3):186–187.

Fiore A, La Fauci L, Cervellati R, Guerra MC, Speroni E, Costa S, Galvano G, De Lorenzo A, Bacchelli V, Fogliano V and Galvano F. 2005. Antioxidant activity of pasteurized and sterilized commercial red orange juices. Mol Nutr Food Res 49(12):1129–1135.

Fogliano V, Verde V, Randazzo G and Ritieni A. 1999. Method for measuring antioxidant activity and its application to monitoring the antioxidant capacity of wines. J Agric Food Chem 47(3):1035–1040.

Fowke JH, Morrow JD, Motley S, Bostick RM and Ness RM. 2006. Brassica vegetable consumption reduces urinary F2-isoprostane levels independent of micronutrient intake. Carcinogenesis 27(10):2096–2102.

Frankel EN. 1993. In search of better methods to evaluate natural antioxidants and oxidative stability in food lipids. Trends Food Sci Technol 4(7):220–225.

Frankel EN and Meyer AS. 2000. The problems of using onedimensional methods to evaluate multifunctional food and biological antioxidants. J Sci Food Agric 80(13):1925–1941.

García-Alonso M, de Pascual-Teresa S, Santos-Buelga C and Rivas-Gonzalo JC. 2004. Evaluation of the antioxidant properties of fruits. Food Chem 84(1):13–18.

Gardner PT, White TAC, McPhail DB and Duthie GG. 2000. The relative contributions of vitamin C, carotenoids and phenolics to the antioxidant potential of fruit juices. Food Chem 68(4):471–474.

Georgé S, Brat P, Alter P and Amiot MJ. 2005. Rapid determination of polyphenols and vitamin C in plant-derived products. J Agric Food Chem 53(5):1370–1373.

Ghiselli A, Serafini M, Maiani G, Azzini E and Ferro-Luzzi A. 1995. A fluorescence-based method for measuring total plasma antioxidant capability. Free Radic Biol Med 18(1):29–36.

Gil MI, Tomás-Barberán FA, Hess-Pierce B, Holcroft DM and Kader AA. 2000. Antioxidant activity of pomegranate juice and its relationship with phenolic composition and processing. J Agric Food Chem 48(10):4581–4589.

Gil MI, Tomás-Barberán FA, Hess-Pierce B and Kader AA. 2002. Antioxidant capacities, phenolic compounds, carotenoids, and vitamin C contents of nectarine, peach, and plum cultivars from California. J Agric Food Chem 50(17):4976–4982.

Gonçalves B, Landbo AK, Let M, Silva AP, Rosa E and Meyer AS. 2004. Storage affects the phenolic profiles and antioxidant activities of cherries (*Prunus avium* L.) on human low-density lipoproteins. J Sci Food Agric 84(9):1013–1020.

Gorinstein S, Cvikrová M, Machackova I, Haruenkit R, Park YS, Jung ST, Yamamoto K, Martínez Ayala AL, Katrich E and Trakhtenberg S. 2004. Characterization of antioxidant compounds in Jaffa sweeties and white grapefruits. Food Chem 84(4):503–510.

Gorinstein S, Huang D, Leontowicz H, Leontowicz M, Yamamoto K, Soliva-Fortuny R, Martín Belloso O, Martinez Ayala AL and Trakhtenberg S. 2006. Determination of naringin and hesperidin in citrus fruit by high-performance liquid chromatography. The antioxidant potential of citrus fruit. Acta Chromatogr (17):108–124.

Gorinstein S, Leontowicz H, Leontowicz M, Krzeminski R, Gralak M, Delgado-Licon E, Martinez Ayala AL, Katrich E and Trakhtenberg S. 2005. Changes in plasma lipid and antioxidant activity in rats as a result of naringin and red grapefruit supplementation. J Agric Food Chem 53(8):3223–3228.

Gorinstein S, Martín-Belloso O, Park YS, Haruenkit R, Lojek A, Ĉíž M, Caspi A, Libman I and Trakhtenberg S. 2001. Comparison of some biochemical characteristics of different citrus fruits. Food Chem 74(3):309–315.

Graversen HB, Becker EM, Skibsted LH and Andersen ML. 2008. Antioxidant synergism between fruit juice and α-tocopherol. A comparison between high phenolic black chokeberry (*Aronia melanocarpa*) and high ascorbic blackcurrant (*Ribes nigrum*). Eur Food Res Technol 226(4):737–743.

Guil-Guerrero JL, Martínez-Guirado C, Rebolloso-Fuentes M and Carrique-Pérez A. 2006. Nutrient composition and antioxidant activity of 10 pepper (*Capsicum annuum*) varieties. Eur Food Res Technol 224(1):1–9.

Guo C, Yang J, Wei J, Li Y, Xu J and Jiang Y. 2003. Antioxidant activities of peel, pulp and seed fractions of common fruits as determined by FRAP assay. Nutr Res 23(12):1719–1726.

Haenen GRMM, Paquay JBG, Korthouwer REM and Bast A. 1997. Peroxynitrite scavenging by flavonoids. Biochem Biophys Res Commun 236(3):591–593.

Hamauzu Y, Yasui H, Inno T, Kume C and Omanyuda M. 2005. Phenolic profile, antioxidant property, and anti-influenza viral activity of Chinese quince (*Pseudocydonia sinensis* Schneid.), quince (*Cydonia oblonga* Mill.), and apple (*Malus domestica* Mill.) fruits. J Agric Food Chem 53(4):928–934.

Haruenkit R, Poovarodom S, Leontowicz H, Leontowicz M, Sajewicz M, Kowalska T, Delgado-Licon E, Rocha-Guzmán NE, Gallegos-Infante JA, Trakhtenberg S and Gorinstein S. 2007. Comparative study of health properties and nutritional value of durian, mangosteen, and snake fruit: experiments *in vitro* and *in vivo*. J Agric Food Chem 55(14):5842–5849.

Hassimotto NMA, Genovese MI and Lajolo FM. 2007. Identification and characterisation of anthocyanins from wild mulberry (*Morus nigra* L.) growing in Brazil. Food Sci Technol Int 13(1):17–25.

Hassimotto NMA, Genovese MI and Lajolo FM. 2005. Antioxidant activity of dietary fruits, vegetables, and commercial frozen fruit pulps. J Agric Food Chem 53(8):2928–2935.

Heimler D, Isolani L, Vignolini P, Tombelli S and Romani A. 2007. Polyphenol content and antioxidative activity in some species of freshly consumed salads. J Agric Food Chem 55(5):1724–1729.

Heimler D, Vignolini P, Dini MG, Vincieri FF and Romani A. 2006. Antiradical activity and polyphenol composition of local Brassicaceae edible varieties. Food Chem 99(3):464–469.

Heinonen IM, Lehtonen PJ and Hopia AI. 1998a. Antioxidant activity of berry and fruit wines and liquors. J Agric Food Chem 46(1):25–31.

Heinonen IM, Meyer AS and Frankel EN. 1998b. Antioxidant activity of berry phenolics on human low-density lipoprotein and liposome oxidation. J Agric Food Chem 46(10):4107–4112.

Höner K and Cervellati R. 2002b. Measurements of the antioxidant capacity of fruits and vegetables using the BR reaction method. Eur Food Res Technol 215(5):437–442.

Höner K and Cervellati R. 2003. Determination of the antioxidative capacity of apples using the Briggs-Rauscher method. Ernahrungs-Umschau 50(1):13.

Höner K, Cervellati R and Neddens C. 2002. Measurements of the in vitro antioxidant activity of German white wines using a novel method. Eur Food Res Technol 214(4):356–360.

Huang D, Ou B, Hampsch-Woodill M, Flanagan JA and Deemer EK. 2002. Development and validation of oxygen radical absorbance capacity assay for lipophilic antioxidants using randomly methylated β-cyclodextrin as the solubility enhancer. J Agric Food Chem 50(7):1815–1821.

Huang YC, Chang YH and Shao YY. 2006. Effects of genotype and treatment on the antioxidant activity of sweet potato in Taiwan. Food Chem 98(3):529–538.

Hughes H, Smith CV, Tsokos-Kuhn JO and Mitchell JR. 1986. Quantitation of lipid peroxidation products by gas chromatography-mass spectrometry. Anal Biochem 152(1):107–112.

Hukkanen AT, Pölönen SS, Kärenlampi SO and Kokko HI. 2006. Antioxidant capacity and phenolic content of sweet rowanberries. J Agric Food Chem 54(1):112–119.

Imeh U and Khokhar S. 2002. Distribution of conjugated and free phenols in fruits: antioxidant activity and cultivar variations. J Agric Food Chem 50(22):6301–6306.

Ismail A, Marjan ZM and Foong CW. 2004. Total antioxidant activity and phenolic content in selected vegetables. Food Chem 87(4):581–586.

Jacob RA, Aiello GM, Stephensen CB, Blumberg JB, Milbury PE, Wallock LM and Ames BN. 2003. Moderate antioxidant supplementation has no effect on biomarkers of oxidant damage in healthy men with low fruit and vegetables intakes. J Nutr 133(3):740–743.

Jastrzebski Z, Leontowicz H, Leontowicz M, Namiesnik J, Zachwieja Z, Barton H, Pawelzik E, Arancibia-Avila P, Toledo F and Gorinstein S. 2007. The bioactivity of processed garlic (*Allium sativum* L.) as shown in *in vitro* and *in vivo* studies on rats. Food Chem Toxicol 45(9):1626–1633.

Jayaprakasha GK, Patil BS and Bhimanagouda S. 2007. *In vitro* evaluation of the antioxidant activities in fruit extracts from citron and blood orange. Food Chem 101: 410–418.

Jiang H, Ji B, Liang J, Zhou F, Yang Z and Zhang H. 2006. Comparison on the antioxidant capacity of selected fruits and vegetables and their separations. Chem Nat Compd 42(4):410–414.

Jiménez M, Escribano-Cebrián J and García-Carmona, F. 1998. Oxidation of the flavonol fisetin by polyphenol oxidase. Biochim Biophys Acta Gen Subj 1425(3):534–542.

Jiménez M and García-Carmona F. 1999. Myricetin, an antioxidant flavonol, is a substrate of polyphenol oxidase. J Sci Food Agric 79(14):1993–2000.

Jiménez-Escrig A, Dragsted LO, Daneshvar B, Pulido R and Saura-Calixto F. 2003. *In vitro* antioxidant activities of edible artichoke (*Cynara scolymus* L.) and effect on biomarkers of antioxidants in rats. J Agric Food Chem 51(18):5540–5545.

Jiménez-Escrig A, Jiménez-Jiménez I, Sánchez-Moreno C and Saura-Calixto F. 2000. Evaluation of free radical scavenging of dietary carotenoids by the stable radical 2,2-diphenyl-1-picrylhydrazyl. J Sci Food Agric 80(11):1686–1690.

Jiménez-Escrig A, Rincón M, Pulido R and Saura-Calixto F. 2001. Guava fruit (*Psidium guajava* L.) as a new source of antioxidant dietary fiber. J Agric Food Chem 49(11):5489–5493.

Jo SC, Nam KC, Min BR, Ahn DU, Cho SH, Park WP and Lee SC. 2006. Antioxidant activity of *Prunus mume* extract in cooked chicken breast meat. Int J Food Sci Technol 41(1):15–19.

Joseph JK. 1995. Physico-chemical attributes of wild mango (*Irvingia gabonensis*) seeds. Bioresource Technol 53(2):179–181.

Jung KA, Song TC, Han D, Kim IH, Kim YE and Lee CH. 2005b. Cardiovascular protective properties of kiwifruit extracts *in vitro*. Biol Pharm Bull 28(9):1782–1785.

Jung ST, Park YS, Zachwieja Z, Folta M, Barton H, Piotrowicz J, Katrich E, Trakhtenberg S and Gorinstei, S. 2005a. Some essential phytochemicals and the antioxidant potential in fresh and dried persimmon. Int J Food Sci Nutr 56(2):105–113.

Kähkönen MP, Hopia AI, Vuorela HJ, Rauha JP, Pihlaja K, Kujala, TS and Heinonen M. 1999. Antioxidant activity of plant extracts containing phenolic compounds. J Agric Food Chem 47(10):3954–3962.

Kang HM and Saltveit ME. 2002. Antioxidant capacity of lettuce leaf tissue increases after wounding. J Agric Food Chem 50(26):7536–7541.

Kanner J, Frankel E, Granit R, German B and Kinsella JE. 1994. Natural antioxidants in grapes and wines. J Agric Food Chem 42(1):64–69.

Karakaya S, El SN and Ta AA. 2001. Antioxidant activity of some foods containing phenolic compounds. Int J Food Sci Nutr 52(6):501–508.

Kasai H, Fukada S, Yamaizumi Z, Sugie S and Mori H. 2000. Action of chlorogenic acid in vegetables and fruits as an inhibitor of 8-hydroxydeoxyguanosine formation *in vitro* and in a rat carcinogenesis model. Food Chem Toxicol 38(5):467–471.

Kaur C and Kapoor HC. 2002. Anti-oxidant activity and total phenolic content of some Asian vegetables. Int J Food Sci Technol 37(2):153–161.

Kedage VV, Tilak JC, Dixit GB, Devasagayam TPA and Mhatre M. 2007. A study of antioxidant properties of some varieties of grapes *Vitis vinifera* L. Crit Rev Food Sci Nutr 47(2):175–185.

Kefalas P, Kallithraka S, Parejo I and Makris DP. 2003. A comparative study on the *in vitro* antiradical activity and hydroxyl free radical scavenging activity in aged red wines. Food Sci Technol Int 9(6):383–385.

Kevers C, Falkowski M, Tabart J, Defraigne JO, Dommes J and Pincemail J. 2007. Evolution of antioxidant capacity during storage of selected fruits and vegetables. J Agric Food Chem 55(21):8596–8603.

Kikuzaki H and Nakatani N. 1993. Antioxidant effects of some ginger constituents. J Food Sci 58(6):1407–1410.

Kim DO, Jeong SW and Lee CY. 2003. Antioxidant capacity of phenolic phytochemicals from various cultivars of plums. Food Chem 81(3):321–326.

Kim D-O, Lee KW, Lee HJ and Lee CY. 2002. Vitamin C equivalent antioxidant capacity (VCEAC) of phenolic phytochemicals. J Agric Food Chem 50(13):3713–3717.

Klimczak I, Małecka M, Szlachta M and Gliszczyńska-Świgło A. 2007. Effect of storage on the content of polyphenols, vitamin C and the antioxidant activity of orange juices. J Food Compos Anal 20(3–4):313–322.

Kljusurić GJ, Djanovic S, Kruhak I, Kovacevic Ganic K, Komes D and Kurtanjek Ž. 2005. Application of Briggs-Rauscher reaction for measurement of antioxidant capacity of Croatian wines. Acta Aliment 34(4):483–492.

Kohen R, Vellaichamy E, Hrbac J, Gati I and Tirosh O. 2000. Quantification of the overall reactive oxygen species scavenging capacity of biological fluids and tissues. Free Radic Biol Med 28(6):871–879.

Kondo S, Kittikorn M and Kanlayanarat S. 2005. Preharvest antioxidant activities of tropical fruit and the effect of low temperature storage on antioxidants and jasmonates. Potharv Biol Technol 36(3):309–318.

Kondo S, Tsuda K, Muto N and Ueda J. 2002. Antioxidative activity of apple skin or fresh extracts associated with fruit development on selected apple cultivars. Sci Hort 96(1-4):177–185.

Kooy NW, Royall JA, Ischiropoulos H and Beckma JS. 1994. Peroxynitrite-mediated oxidation of dihydrorhodamine 123. Free Radic Biol Med 16(2):149–156.

Krishnan K and Vijayalakshmi NR. 2005. Alterations in lipids & lipid peroxidation in rats fed with flavonoid rich fraction of banana (*Musa paradisiaca*) from high background radiation area. Ind J Med Res 122(6):540–546.

Kusznierewicz B, Bartoszek A, Wolska L, Drzewiecki J, Gorinstein S and Namieśnik J. 2008. Partial characterization of white cabbages (*Brassica oleracea* var. capitata f. alba) from different regions by glucosinolates, bioactive compounds, total antioxidant activities and proteins. LWT Food Sci Technol 41(1):1–9.

Lako J, Trenerry, VC, Wahlqvist M, Wattanapenpaiboon N, Sotheeswaran S and Premier R. 2007. Phytochemical flavonols, carotenoids and the antioxidant properties of a wide selection of Fijian fruit, vegetables and other readily available foods. Food Chem 101(4):1727–1741.

Larrauri JA, Rupérez P, Bravo L and Saura-Calixto F. 1996. High dietary fibre peels: associated powders from orange and lime polyphenols and antioxidant capacity. Food Res Int 29(8):751–762.

Larrauri JA, Rupérez P and Saura-Calixto F. 1997. Effect of drying temperature on the stability of polyphenols and antioxidant activity of red grape pomace peels. J Agric Food Chem 45(4):1390–1393.

Larrosa M, Llorach R, Espín JC and Tomás-Barberán FA. 2002. Increase of antioxidant activity of tomato juice upon functionalisation with vegetable byproduct extracts. Lebensm-Wiss Technol 35(6):532–542.

Lavelli V, Hippeli S, Peri C and Elstner EF. 1999. Evaluation of radical scavenging activity of fresh and air-dried tomatoes by three model reactions. J Agric Food Chem 47(9):3826–3831.

Lavelli V, Peri C and Rizzolo A. 2000. Antioxidant activity of tomato products as studied by model reactions using xanthine oxidase, myeloperoxidase, and copper-induced lipid peroxidation. J Agric Food Chem 48(5):1442–1448.

Leonardi C, Ambrosino P, Esposito F and Fogliano V. 2000. Antioxidative activity and carotenoid and tomatine contents in different typologies of fresh consumption tomatoes. J Agric Food Chem 48(10):4723–4727.

Leong LP and Shui G. 2002. An investigation of antioxidant capacity of fruits in Singapore markets. Food Chem 76(1):69–75.

Leontowicz M, Gorinstein S, Bartnikowska E, Leontowicz H, Kulasek G and Trakhtenberg S. 2001. Sugar beet pulp and apple pomace dietary fibers improve lipid metabolism in rats fed cholesterol. Food Chem 72(1):73–78.

Leontowicz M, Gorinstein S, Leontowicz H, Krzeminski R, Lojek A, Katrich E, Číž M, Martín-Belloso O, Soliva-Fortuny R, Haruenkit R and Trakhtenberg S. 2003. Apple and pear peel and pulp and their influence on plasma lipids and antioxidant potentials in rats fed cholesterol-containing diets. J Agric Food Chem 51(19):5780–5785.

Leontowicz M, Leontowicz H, Jastrzebski Z, Jesion I, Haruenkit R, Poovarodom S, Katrich E, Tashma Z, Drzewiecki J, Trakhtenberg S and Gorinstein S. 2007. The nutritional and metabolic indices in rats fed cholesterol-containing diets supplemented with durian at different stages of ripening. BioFactors 29(2-3):123–136.

Li J and Jiang Y. 2007. Litchi flavonoids: isolation, identification and biological activity. Molecules 12(4):745–758.

Lichtenthäler R and Marx F. 2005. Total oxidant scavenging capacities of common European fruit and vegetable juices. J Agric Food Chem 53(1):103–110.

Lim YY, Lim TT and Tee JJ. 2007. Antioxidant properties of several tropical fruits: a comparative study. Food Chem 103(3):1003–1008.

Lin CH and Chan CY. 2005. Textural change and antioxidant properties of broccoli under different cooking treatments. Food Chem 90(1-2):9–15.

Liu JK, Hu L, Dong ZJ and Hu Q. 2004. DPPH radical scavenging activity of ten natural *p*-terphenyl derivatives obtained from three edible mushrooms indigenous to China. Chem Biodiv 1(4):601–605.

Liyana-Pathirana CM, Shahidi F and Alasalvar C. 2006. Antioxidant activity of cherry laurel fruit (*Laurocerasus officinalis* Roem.) and its concentrated juice. Food Chem 99(1):121–128.

Llorach R, Espín JC, Tomás-Barberán FA and Ferreres F. 2003. Valorization of cauliflower (*Brassica oleracea* L. var. botrytis) by-products as a source of antioxidant phenolics. J Agric Food Chem 51(8):2181–2187.

Lo Scalzo R, Iannoccari T, Summa C, Morelli R and Rapisarda P. 2004. Effect of thermal treatments on antioxidant and antiradical activity of blood orange juice. Food Chem 85(1):41–47.

López-Vélez M, Martínez-Martínez F and Del Valle-Ribes C. 2003. The study of phenolic compounds as natural antioxidants in wine. Crit Rev Food Sci Nutr 43(3):233–244.
Lu Y, Zhou WZ, Chang Z, Chen W-X and Li L. 2004. Procyanidins from grape seeds protect against phorbol ester-induced oxidative cellular and genotoxic damage. Acta Pharmacol Sin 25(8):1083–1089.
Lussignoli S, Fraccaroli M, Andrioli G, Brocco G and Bellavite P. 1999. A microplate-based colorimetric assay of the total peroxyl radical trapping capability of human plasma. Anal Biochem 269(7):38–44.
Luximon-Ramma A, Bahorun T and Crozier A. 2003. Antioxidant actions and phenolic and vitamin C contents of common Mauritian exotic fruits. J Sci Food Agric 83(5):496–502.
MacDonald-Wicks LK, Wood LG and Garg ML. 2006. Methodology for the determination of biological antioxidant capacity in vitro: A review. J Sci Food Agric 86(13):2046–2056.
MacLean DD, Murr DP and DeEll JR. 2003. A modified total oxyradical scavenging capacity assay for antioxidants in plant tissues. Potharv Biol Technol 29(2):183–194.
Maeda H, Katsuki T, Akaike T and Yasutake R. 1992. High correlation between lipid peroxide radical and tumor-promoter effect—suppression of tumor promotion in the Epstein-Barr-virus lymphocyte-B system and scavenging of alkyl peroxide radicals by various vegetable extracts. Jpn J Cancer Res 83(9):923–928.
Magalhães LM, Segundo MA, Reis S, Lima JLFC and Rangel AOSS. 2006. Automatic method for the determination of Folin-Ciocalteu reducing capacity in food products. J Agric Food Chem 54(15):5241–5246.
Mahattanatawee K, Manthey JA, Luzio G, Talcott ST, Goodner K and Baldwin EA. 2006. Total antioxidant activity and fiber content of select Florida-grown tropical fruits. J Agric Food Chem 54(19):7355–7363.
Maiorino M, Roveri A, Ursini F and Gregolin C. 1985. Enzymatic determination of membrane lipid peroxidation. J Free Radic Biol Med 1(3):203–207.
Maisuthisakul P, Suttajit M and Pongsawatmanit R. 2007. Assessment of phenolic content and free radical-scavenging capacity of some Thai indigenous plants. Food Chem 100(4):1409–1418.
Makris DP and Rossiter JT. 2001. Domestic processing of onion bulbs (*Allium cepa*) and asparagus spears (*Asparagus officinalis*): effect on flavonol content and antioxidant status. J Agric Food Chem 49(7):3216–3222.
Mansouri A, Makris DP and Kefalas P. 2005. Determination of hydrogen peroxide scavenging activity of cinnamic and benzoic acids employing a highly sensitive peroxyoxalate chemiluminescence-based assay: structure–activity relationships. J Pharm Biomed Anal 39(1-2):22–26.
Mariassyova M. 2006. Antioxidant activity of some herbal extracts in rapeseed and sunflower oils. J Food Nutr Res 45(3):104–109.
Martínez-Tomé M, García-Carmona F and Murcia MA. 2001. Comparison of the antioxidant and pro-oxidant activities of broccoli amino acids with those of common food additives. J Sci Food Agric 81(10):1019–1026.
Martínez-Valverde I, Periago MJ, Provan G and Chesson A. 2002. Phenolic compunds, lycopene and antioxidant activity in commercial varieties of tomato (*Lycopersicum esculentum*). J Sci Food Agric 82(3):323–330.
Materska M and Perucka I. 2005. Antioxidant activity of the main phenolic compounds isolated from hot pepper fruit (*Capsicum annuum* L.). J Agric Food Chem 53(5):1750–1756.
Matsufuji H, Ishikawa K, Nunomura, O, Chino M and Takeda M. 2007. Anti-oxidant content of different colored sweet peppers, white, green, yellow, orange and red (*Capsicum annuum* L.). Int J Food Sci Technol 42(12):1482–1488.
Mazza G, Kay CD, Cottrell T and Holub BJ. 2002. Absorption of anthocyanins from blueberries and serum antioxidant status in human subjects. J Agric Food Chem 50(26):7731–7737.
Meyer AS, Yi OS, Pearson DA, Waterhouse AL and Frankel EN. 1997. Inhibition of human low-density lipoprotein oxidation in relation to composition of phenolic antioxidants in grape (*Vitis vinifera*). J Agric Food Chem 45(5):1638–1643.
Meyers KJ, Watkins CB, Pritts MP and Liu RH. 2003. Antioxidant and antiproliferative activities of strawberries. J Agric Food Chem 51(23):6887–6892.
Milde J, Elstner EF and GraBmann J. 2007. Synergistic effects of phenolics and carotenoids on human low-density lipoprotein oxidation. Mol Nutr Food Res 51(8):956–961.
Miller ER, Appel LJ and Risby TH. 1998. Effect of dietary patterns on measures of lipid peroxidation: results from a randomized clinical trial. Circulation. J Am Heart Assoc 98(22):2390–2395.
Miller NJ, Rice-Evans C and Davies MJ. 1993. A new method for measuring antioxidant activity. Biochem Soc Trans 21(2):95S

Miller NJ and Rice-Evans CA. 1997. The relative contributions of ascorbic acid and phenolic antioxidants to the total antioxidant activity of orange and apple fruit juices and blackcurrant drink. Food Chem 60(3):331–337.

Miller NJ, Sampson J, Candeias LP, Bramley PM and Rice-Evans CA. 1996. Antioxidant activities of carotenes and xanthophylls. FEBS Lett 384(3):240–242.

Miraliakbari H and Shahidi F. 2008. Antioxidant activity of minor components of tree nut oils. Food Chem 111(2):421–427.

Mullen W, Marks SC and Crozier A. 2007. Evaluation of phenolic compounds in commercial fruit juices and fruit drinks. J Agric Food Chem 55(8):3148–3157.

Muñoz Acuña U, Atha DE, Ma J, Nee MH and Kennelly EJ. 2002. Antioxidant capacities of ten edible North American plants. Phytother Res 16(1):63–65.

Murcia MA, Jiménez AM and Martínez-Tomé M. 2001. Evaluation of the antioxidant properties of Mediterranean and tropical fruits compared with common food additives. J Food Protect 64(12):2037–2046.

Murcia MA, Martínez-Tomé M, Jiménez AM, Vera AM, Honrubia M and Parras P. 2002. Antioxidant activity of edible fungi (truffles and mushrooms): losses during industrial processing. J Food Protect 65(10):1614–1622.

Nakajima J, Tanaka I, Seo S, Yamazaki M and Saito K. 2004. LC/PDA/ESI-MS profiling and radical scavenging activity of anthocyanins in various berries. J Biomed Biotechnol 2004 (5):241–247.

Nencini C, Cavallo F, Capasso A, Franchi GG, Giorgio G and Micheli L. 2007. Evaluation of antioxidative properties of *Allium* species growing wild in Italy. Phytother Res 21(9):874–878.

Netzel M, Netzel G, Tian Q, Schwartz S and Konczak I. 2006. Sources of antioxidant activity in Australian native fruits. Identification and quantification of anthocyanins. J Agric Food Chem 54(26):9820–9826.

Nicolle C, Cardinault N, Aprikian O, Busserolles J, Grolier P, Rock E, Demigné C, Mazur A, Scalbert A, Amouroux P and Rémésy C. 2003. Effect of carrot intake on cholesterol metabolism and on antioxidant status in cholesterol-fed rat. Eur J Nutr 42(5):254–261.

Nicolle C, Cardinault N, Gueux E, Jaffrelo L, Rock E, Mazur A, Amouroux P and Rémésy C. 2004. Health effect of vegetable-based diet: lettuce consumption improves cholesterol metabolism and antioxidant status in the rat. Clin Nutr 23(4):605–614.

Nilsson J, Pillai D, Önning G, Persson C, Nilsson Å and Åkesson B. 2005. Comparison of the 2,2′-azinobis-3-ethylbenzothiazoline-6-sulfonic acid (ABTS) and ferric reducing antioxidant power (FRAP) methods to asses the total antioxidant capacity in extracts of fruit and vegetables. Mol Nutr Food Res 49(3):239–246.

Ninfali P and Bacchiocca M. 2003. Polyphenols and antioxidant capacity of vegetables under fresh and frozen conditions. J Agric Food Chem 51(8):2222–2226.

Ninfali P, Mea G, Giorgini S, Rocchi M and Bacchiocca M. 2005. Antioxidant capacity of vegetables, spices and dressings relevant to nutrition. Br J Nutr 93(2):257–266.

Nishibori S and Namiki K. 1998. Superoxide anion radical-scavenging ability of fresh and heated vegetable juices. J Jpn Soc Food Sci 45(2):144–148.

Noda Y, Anzai K, Mori A, Kohno M, Shinmei M and Packer L. 1997. Hydroxyl, end superoxide anion radical scavenging activities of natural source antioxidants using the computerized JES-FR30 ESR spectrometer system. Biochem Mol Biol Int 42(1):35–44.

Noda Y, Kaneyuki T, Mori A and Packer L. 2002. Antioxidant activities of pomegranate fruit extract and its anthocyanidins: delphinidin, cyanidin, and pelargonidin. J Agric Food Chem 50(1):166–171.

Nourooz-Zadeh J. 1999. Ferrous ion oxidation in presence of xylenol orange for detection of lipid hydroperoxides in plasma. Methods Enzymol 300: 58–62.

Oboh G, Puntel RL and Rocha JBT. 2007. Hot pepper (*Capsicum annuum*, Tepin and *Capsicum chinese*, Habanero) prevents Fe^{2+}-induced lipid peroxidation in brain—*in vitro*. Food Chem 102(1):178–185.

Odriozola-Serrano I, Soliva-Fortuny R and Martín-Belloso O. 2008. Effect of minimal processing on bioactive compounds and color attributes of fresh-cut tomatoes. LWT Food Sci Technol 41(2):217–226.

Oh YS, Jang ES, Bock JY, Yoon SH and Jung AY. 2006. Singlet oxygen quenching activities of various fruit and vegetable juices and protective effects of apple and pear juices against hematolysis and protein oxidation induced by methylene blue photosensitization. J Food Sci 71(4):260–268.

Olas B and Wachowicz B. 2002. Resveratrol and vitamin C as antioxidants in blood platelets. Thromb Res 106(2):143–148.

Ordoudi SA and Tsimidou MZ. 2006. Crocin bleaching assay (CBA) in structure-radical scavenging activity studies of selected phenolic compounds. J Agric Food Chem 54(25):9347–9356.

Orhan I, Aydin A, Çölkesen A, Sener B and Isimer AI. 2003. Free radical scavenging activities of some edible fruit seeds. Pharm Biol 41(3):163–165.

Ou B, Hampsch-Woodill M and Prior RL. 2001. Development and validation of an improved oxygen radical absorbance capacity assay using fluorescein as the fluorescent probe. J Agric Food Chem 49(10):4619–4626.

Ou B, Huang D, Hampsch-Woodill M, Flanagan, JA and Deemer EK. 2002. Analysis of antioxidant activities of common vegetables employing oxygen radical absorbance capacity (ORAC) and ferric reducing antioxidant power (FRAP) assays: a comparative study. J Agric Food Chem 50(11):3122–3128.

Ozgen M, Reese RN, Tulio AZ Jr., Scheerens JC and Miller AR. 2006. Modified 2,2-azino-bis-3-ethylbenzothiazoline-6-sulfonic acid (ABTS) method to measure antioxidant capacity of selected small fruits and comparison to ferric reducing antioxidant power (FRAP) and 2,2′-diphenyl-1-picrylhydrazyl (DPPH) methods. J Agric Food Chem 54(4):1151–1157.

Öztürk M, Aydoğmuş-Öztürk F, Duru ME and Topçu G. 2007. Antioxidant activity of stem and root extracts of rhubarb (*Rheum ribes*): an edible medicinal plant. Food Chem 103(2):623–630.

Paganga G, Miller N and Rice-Evans CA. 1999. The polyphenolic content of fruit and vegetables and their antioxidant activities. What does a serving constitute? Free Radic Res 30(2):153–162.

Pannala A, Razaq R, Halliwell B, Singh S and Rice-Evans CA. 1998. Inhibition of peroxynitrite dependent tyrosine nitration by hydroxycinnamates: nitration or electron donation? Free Radic Biol Med 24(4):594–606.

Pantelidis GE, Vasilakakis M, Manganaris GA and Diamantidis GR. 2007. Antioxidant capacity, phenol, anthocyanin and ascorbic acid contents in raspberries, blackberries, red currants, gooseberries and Cornelian cherries. Food Chem 102(3):777–783.

Papagiannopoulos M, Wollseifen HR, Mellenthin A, Haber B and Galensa R. 2004. Identification and quantification of polyphenols in carob fruits (*Ceratonia siliqua* L.) and derived products by HPLC-UV-ESI/MS^{n}. J Agric Food Chem 52(12):3784–3791.

Paquay JBG, Haenen GRMM, Korthouwer REM and Bast A. 1997. Peroxynitrite scavenging by wines. J Agric Food Chem 45(9):3357–3358.

Park YS, Jung ST, Kang SG, Delgado-Licon E, Katrich E, Tashma Z, Trakhtenberg S and Gorinstein S. 2006b. Effect of ethylene treatment on kiwifruit bioactivity. Plant Foods Hum Nutr 61(3):151–156.

Park YS, Jung ST, Kang SG, Delgado-Licon E, Martinez Ayala AL, Tapia MS, Martín-Belloso O, Trakhtenberg S and Gorinstein S. 2006a. Drying of persimmons (*Diospyros kaki* L.) and the following changes in the studied bioactive compounds and the total radical scavenging activities. LWT Food Sci Technol 39(7):748–755.

Pearson DA, Frankel EN, Aeschbach R and German JB. 1998. Inhibition of endothelial cell mediated low-density lipoprotein oxidation by green tea extracts. J Agric Food Chem 46(4):1445–1449.

Pearson DA, Tan CH, German JB, Davis PA and Gershwin ME. 1999. Apple juice inhibits human low density lipoprotein oxidation. Life Sci 64(21):1913–1920.

Pellegrini N, Colombi B, Salvatore S, Brenna, OV, Galaverna G, Del Rio D, Bianchi M, Bennett RN and Brighenti F. 2007. Evaluation of antioxidant capacity of some fruit and vegetable foods: efficiency of extraction of a sequence of solvents. J Sci Food Agric 87(1):103–111.

Pellegrini N, Re R, Yang M and Rice-Evans C. 1999. Screening of dietary carotenoids and carotenoid-rich fruit extracts for antioxidant activities applying 2,2′-azinobis(3-ethylenebenzothiazoline-6-sulfonic acid) radical cation decolorization assay. Methods Enzymol 299: 379–389.

Pellegrini N, Serafini M, Colombi B, Del Rio D, Salvatore S, Bianchi M and Brighenti F. 2003. Total antioxidant capacity of plant foods, beverages and oils consumed in Italy assessed by three different *in vitro* assays. J Nutr 133(9):2812–2819.

Pereira JA, Pereira APG, Ferreira ICFR, Valentão P, Andrade PB, Seabra R, Estevinho L and Bento A. 2006. Table olives from Portugal: phenolic compounds, antioxidant potential, and antimicrobial activity. J Agric Food Chem 54(22):8425–8431.

Perjési P, Kuzmaa M, Fodora K and Rozmera, Z. 2007. Application of crocin bleaching and deoxyribose degradation tests to assess antioxidant capacity of capsaicinoids and some selected flavonoids. Eur J Pharm Sci 32: 39.

Peschel W, Sánchez-Rabaneda F, Diekmann W, Plescher A, Gartzía I, Jiménez D, Lamuela-Raventós R, Buxaderas S and Codina, C. 2006. An industrial approach in the search of natural antioxidants from vegetable and fruit wastes. Food Chem 97(1):137–150.

Pieroni A, Janiak V, Dürr, C. M, Lüdeke S, Trachsel E and Heinrich, M. 2002. *In vitro* antioxidant activity of non-cultivated vegetables of ethnic Albanians in southern Italy. Phytother Res 16(5):467–473.

Piga A, Agabbio M, Gambella F and Nicoli, MC. 2002. Retention of antioxidant activity in minimally processed mandarin and Satsuma fruits. Lebensm-Wiss Technol 35(4):344–347.

Plumb, GW, Lambert N, Chambers, SJ, Wanigatunga S, Heaney, RK, Plumb, JA, Aruoma, OI, Halliwell B, Miller, NJ and Williamson, G. 1996. Are whole extracts and purified glucosinolates from cruciferous vegetables antioxidants?. Free Radic Res 25(1):75–86.

Polásek M, Skála P, Opletal L and Jahodár, L. 2004. Rapid automated assay of anti-oxidation/radical-scavenging activity of natural substances by sequential injection technique (SIA) using spectrophotometric detection. Anal Bioanal Chem 379(5–6):754–758.

Polovka M, Brezová V and Staško, A. 2003. Antioxidant properties of tea investigated by EPR spectroscopy. Biophys Chem 106(1):39–56.

Popov IN and Lewin G. 1994. Photochemiluminescent detection of antiradical activity. II. Testing of nonenzymic water-soluble antioxidants. Free Radic Biol Med 17(3):267–271.

Popov IN and Lewin G. 1996. Photochemiluminescent detection of antiradical activity. IV. Testing of lipid-soluble antioxidants. J Biochem Biophys Methods 31(1–2):1–8.

Popov IN, Lewin G and von Baehr R. 1987. Photochemiluminescent detection of antiradical activity. I. Assay of superoxide dismutase. Biomed Biochim Acta 46(11):775–779.

Prior RL, Hoang H, Gu L, Wu X, Bacchiocca M, Howard L, Hampsch-Woodill M, Huang D, Ou B and Jacob R. 2003. Assays for hydrophilic and lipophilic antioxidant capacity (oxygen radical absorbance capacity ($ORAC_{FL}$)) of plasma and other biological and food samples. J Agric Food Chem 51(11):3273–3279.

Prior RL, Wu X and Schaich K. 2005. Standardized methods for the determination of antioxidant capacity and phenolics in foods and dietary supplements. J Agric Food Chem 53(10):4290–4302.

Proteggente AR, Pannala AS, Paganga G, van Buren L, Wagner E, Wiseman S, van de Put F, Dacombe C and Rice-Evans CA. 2002. The antioxidant activity of regularly consumed fruit and vegetables reflects their phenolic and vitamin C composition. Free Radic Res 36(2):217–233.

Pulido R, Bravo L and Saura-Calixto, F. 2000. Antioxidant activity of dietary polyphenols as determined by a modified ferric reducing/antioxidant power assay. J Agric Food Chem 48(8):3396–3402.

Rababah, TM, Ereifej, KI and Howard, L. 2005. Effect of ascorbic acid and dehydration on concentrations of total phenolics, antioxidant capacity, anthocyanins, and color in fruits. J Agric Food Chem 53(11):4444–4447.

Ranaivo, HR, Diebolt M and Andriantsitohaina, R. 2004. Wine polyphenols induce hypotension, and decrease cardiac reactivity and infarct size in rats: involvement of nitric oxide. Br J Pharmacol 142(4):671–678.

Re R, Pellegrini N, Proteggente A, Pannala A, Yang M and Rice-Evans, C. 1999. Antioxidant activity applying an improved ABTS radical cation decolorization assay. Free Radic Biol Med 26(9-10):1231–1237.

Rice-Evans C and Miller, NJ. 1994. Total antioxidant status in plasma and body fluids. Methods Enzymol 234: 279–293.

Roginsky V and Barsukova, T. 2001. Superoxide dismutase inhibits lipid peroxidation in micelles. Chem Phys Lipids 111(1):87–91.

Roginsky V, de Beer D, Harbertson, JF, Kilmartin, PA, Barsukova T and Adams, DO. 2006. The antioxidant activity of Californian red wines does not correlate with wine age. J Sci Food Agric 86(5):834–840.

Roy MK, Takenaka M, Isobe S and Tsushida T. 2007. Antioxidant potential, anti-proliferative activities, and phenolic content in water-soluble fractions of some commonly consumed vegetables: effects of thermal treatment. Food Chem 103(1):106–114.

Rozenberg O, Howell A and Aviram M. 2006. Pomegranate juice sugar fraction reduces macrophage oxidative state, whereas white grape juice sugar fraction increases it. Atherosclerosis 188(1):68–76.

Sakanaka S and Ishihara Y. 2008. Comparison of antioxidant properties of persimmon vinegar and some other commercial vinegars in radical-scavenging assays and on lipid oxidation in tuna homogenates. Food Chem 107(2):739–744.

Salleh MN, Runnie I, Roach, PD, Mohamed S and Abeywardena MY. 2002. Inhibition of low-density lipoprotein oxidation and up-regulation of low-density lipoprotein receptor in HepG2 cells by tropical plant extracts. J Agric Food Chem 50(13):3693–3697.

Sánchez-Moreno C. 2002. Review: Methods used to evaluate the free radical scavenging activity in foods and biological systems. Food Sci Technol Int 8(3):121–137.

Sánchez-Moreno C, Cano MP, De Ancos B, Plaza L, Olmedilla B, Granado F, Elez-Martínez P, Martín-Belloso O and Martín A. 2004a. Pulsed electric fields–processed orange juice consumption increases plasma vitamin C and decreases F2-isoprostanes in healthy humans. J Nutr Biochem 15(10):601–607.
Sánchez-Moreno C, Cano MP, De Ancos B, Plaza L, Olmedilla B, Granado F and Martín A. 2004b. Consumption of high-pressurized vegetable soup increases plasma vitamin C and decreases oxidative stress and inflammatory biomarkers in healthy humans. J Nutr 134(11):3021–3025.
Sánchez-Moreno C, Larrauri JA and Saura-Calixto F. 1999b. Free radical scavenging capacity and inhibition of lipid oxidation of wines, grape juices and related polyphenolic constituents. Food Res Int 32(6):407–412.
Sánchez-Moreno C, Larrauri, JA and Saura-Calixto, F. 1998. A procedure to measure the antiradical efficiency of polyphenols. J Sci Food Agric 76(2):270–276.
Sánchez-Moreno C, Larrauri, JA and Saura-Calixto F. 1999a. Free radical scavenging capacity of selected red, rose and white wines. J Sci Food Agric 79(10):1301–1304.
Santas J, Carbó R, Gordon MH and Almajano MP. 2008. Comparison of the antioxidant activity of two Spanish onion varieties. Food Chem 107(3):1210–1216.
Santos MR and Mira L. 2004. Protection by flavonoids against the peroxynitrite-mediated oxidation of dihydrorhodamine. Free Radic Res 38(9):1011–1018.
Saura-Calixto F and Goñi I. 2006. Antioxidant capacity of the Spanish Mediterranean diet. Food Chem 94(3):442–447.
Scalfi L, Fogliano V, Pentangelo A, Graziani G, Giordano I and Ritieni A. 2000. Antioxidant activity and general fruit characteristics in different ecotypes of Corbarini small tomatoes. J Agric Food Chem 48(4):1363–1366.
Scalzo J, Politi A, Pellegrini N, Mezzetti B and Battino M. 2005. Plant genotype affects total antioxidant capacity and phenolic contents in fruit. Nutrition 21(2):207–213.
Schlesier K, Harwat M, Böhm V and Bitsch R. 2002. Assessment of antioxidant activity by using different *in vitro* methods. Free Radic Res 36(2):177–187.
Schmeda-Hirschmann G, Feresin G, Tapia A, Hilgert N and Theoduloz C. 2005. Proximate composition and free radical scavenging activity of edible fruits from the Argentinian Yungas. J Sci Food Agric 85(8):1357–1364.
Serraino I, Dugo L, Dugo P, Mondello L, Mazzon E, Dugo G, Caputi AP and Cuzzocre S. 2003. Protective effects of cyanidin-3-*O*-glucoside from blackberry extract against peroxynitrite-induced endothelial dysfunction and vascular failure. Life Sci 73(9):1097–1114.
Shih MK and Hu ML. 1999. Relative roles of metal ions and singlet oxygen in UVA-induced liposomal lipid peroxidation. J Inorg Biochem 77(3-4):225–230.
Silva BM, Andrade PB, Valentão P, Ferreres F, Seabra RM and Ferreira MA. 2004. Quince (*Cydonia oblonga* Miller) fruit (pulp, peel, and seed) and jam: antioxidant activity. J Agric Food Chem 52(15):4705–4712.
Silva DHS, Pereira FC, Zanoni MVB and Yoshida M. 2001. Lipophilic antioxidants from *Iryanthera juruensis* fruits. Phytochemistry 57(3):437–442.
Singh N and Rajini PS. 2003. Free radical scavenging activity of an aqueous extract of potato peel. Food Chem 85: 611–616.
Singleton VL, Orthofer R and Lamuela-Raventos RM. 1999. Analysis of total phenols and other oxidation substrates and antioxidants by means of Folin-Ciocalteu reagent. Methods Enzymol 299: 152–178.
Skupień K and Oszmiański J. 2004. Comparison of six cultivars of strawberries (*Fragaria* x *ananassa* Duch.) grown in northwest Poland. Eur Food Res Technol 219(1):66–70.
Solerrivias C, Espin JC and Wickers HJ. 2000. An easy and fast test to compare total free-radical scavenger capacity of foodstuffs. Phytochem Anal 11: 330–338.
Sousa WR, da Rocha C, Cardoso CL, Silva DHS and Zanoni MVB. 2004. Determination of the relative contribution of phenolic antioxidants in orange juice by voltammetric methods. J Food Compos Anal 17(3):619–633.
Stewart RJ, Askew EW, McDonald CM, Metos J, Jackson WD, Balon TW and Prior RL. 2002. Antioxidant status of young children: response to an antioxidant supplement. J Am Diet Assoc 102(11):1652–1657.
Stratil P, Klejdus B and Kubáň V. 2006. Determination of total content of phenolic compounds and their antioxidant activity in vegetables: evaluation of spectrophotometric methods. J Agric Food Chem 54(3):607–616.

Stratil P, Klejdus B and Kubáň V. 2007. Determination of phenolic compounds and their antioxidant activity in fruits and cereals. Talanta 71(4):1741–1751.

Su MS and Chien PJ. 2007. Antioxidant activity, anthocyanins, and phenolics of rabbiteye blueberry (*Vaccinium ashei*) fluid products as affected by fermentation. Food Chem 104(1):182–187.

Sudheesh S and Vijayalakshmi NR. 2005. Flavonoids from *Punica granatum*—potential antiperoxidative agents. Fitoterapia 76(2):181–186.

Suh HJ, Leeb JM, Chob JS, Kima YS and Chung SH. 1999. Radical scavenging compounds in onion skin. Food Res Int 32(10):659–664.

Sun JS, Chu YF, Wu X and Liu RH. 2002. Antioxidant and antiproliferative activities of common fruits. J Agric Food Chem 50(25):7449–7454.

Sun T, Powers JR and Tang J. 2007. Evaluation of the antioxidant activity of asparagus, broccoli and their juices. Food Chem 105(1):101–106.

Surveswaran S, Cai YZ, Corke H and Sun M. 2007. Systematic evaluation of natural phenolic antioxidants from 133 Indian medicinal plants. Food Chem 102(3):938–953.

Taga MS, Miller EE and Pratt DE. 1984. Chia seeds as a source of natural lipid antioxidants. J Am Oil Chem Soc 61(5):928–931.

Takamatsu S, Galal AM, Ross SA, Ferreira D, ElSohly MA, Ibrahim ARS and El-Feraly FS. 2003. Antioxidant effect of flavonoids on DCF production in HL-60 cells. Phytother Res 17(8):963–966.

Talcott ST, Percival SS, Pittet-Moore J and Celoria C. 2003. Phytochemical composition and antioxidant stability of fortified yellow passion fruit (*Passiflora edulis*). J Agric Food Chem 51(4):935–941.

Tarwadi K and Agte V. 2005. Antioxidant and micronutrient quality of fruit and root vegetables from the Indian subcontinent and their comparative performance with green leafy vegetables and fruits. J Sci Food Agric 85(9):1469–1476.

Thaipong K, Boonprakoba U, Crosby K, Cisneros-Zevallos L and Byrne DH. 2006. Comparison of ABTS, DPPH, FRAP, and ORAC assays for estimating antioxidant activity from guava fruit extracts. J Food Compos Anal 19(6-7):669–675.

Thompson HJ, Heimendinger J, Haegele A, Sedlacek SM, Gillette C, O'Neill C, Wolfe P and Conry C. 1999. Effect of increased vegetable and fruit consumption on markets of oxidative cellular damage. Carcinogenesis 20(12):2261–2266.

Thompson HJ, Heimendinger J, Sedlacek S, Haegele A, Diker A, O'Neill C, Meinecke B, Wolfe P, Zhu Z and Jiang W. 2005. 8-Isoprostane F2α excretion is reduced in women by increased vegetable and fruit intake. Am J Clin Nutr 82(4):768–776.

Thu NN, Sakurai C, Uto H, Van Chuyen N, Lien DTK, Yamamoto S, Ohmori R and Kondo K. 2004. The polyphenol content and antioxidant activities of the main edible vegetables in northern Vietnam. J Nutr Sci Vitaminol 50(3):203–210.

Tobi SE, Gilbert M, Paul N and McMillan TJ. 2002. The green tea polyphenol, epigallocatechin-3-gallate, protects against the oxidative cellular and genotoxic damage of UVA radiation. Int J Cancer 102(5):439–444.

Tomaino A, Cristani M, Cimino F, Speciale A, Trombetta D, Bonina F and Saija A. 2006. *In vitro* protective effect of a Jacquez grapes wine extract on UVB-induced skin damage. Toxicol Vitr 20(8):1395–1402.

Triantis T, Stelakis A, Dimotikali D and Papadopoulos K. 2005. Investigations on the antioxidant activity of fruit and vegetable aqueous extracts on superoxide radical anion using chemiluminescence techniques. Anal Chim Acta 536(1–2):101–105.

Tsai HL, Chang SKC and Chang SJ. 2007. Antioxidant content and free radical scavenging ability of fresh red pummelo [*Citrus grandis* (L.) Osbeck] juice and freeze-dried products. J Agric Food Chem 55(8):2867–2872.

Tubaro F, Ghiselli A, Rapuzzi P, Maiorino M and Ursini F. 1998. Analysis of plasma antioxidant capacity by competition kinetics. Free Radic Biol Med 24(7-8):1228–1234.

Turkmen N, Sari F and Velioglu YS. 2005. The effect of cooking methods on total phenolics and antioxidant activity of selected green vegetables. Food Chem 93(4):713–718.

Uotila JT, Kirkkola A-L, Rorarius M, Tuimala RJ and Metsä-Ketelä T. 1994. The total peroxyl radical-trapping ability of plasma and cerebrospinal fluid in normal and preeclamptic parturients. Free Radic Biol Med 16(5):581–590.

Uthman RS, Toma RB, Garcia R, Medora, Nilufer P and Cunningham S. 1998. Lipid analyses of fumigated vs. irradiated raw and roasted almonds. J Sci Food Agric 78(2):261–266.

Valkonen M and Kuusi T. 1997. Spectrophotometric assay for total peroxyl radical-trapping antioxidant potential in human serum. J Lipid Res 38(4):823–833.
Vayalil PK. 2002. Antioxidant and antimutagenic properties of aqueous extract of date fruit (*Phoenix dactylifera* L. Arecaceae). J Agric Food Chem 50(3):610–617.
Velioglu YS, Mazza G, Gao L and Oomah BD. 1998. Antioxidant activity and total phenolics in selected fruits, vegetables, and grain products. J Agric Food Chem 46(10):4113–4117.
Viana M, Barbas C, Bonet B, Bonet MV, Castro M, Fraile MV and Herrera E. 1996. *In vitro* effects of a flavonoid-rich extract on LDL oxidation. Atherosclerosis 123(1–2):83–91.
Villaño D, Fernández-Pachón MS, Moyá ML, Troncoso AM and García-Parrilla MC. 2007. Radical scavenging ability of polyphenolic compounds towards DPPH free radical. Talanta 71(1):230–235.
Vinson JA, Hao Y, Su X and Zubik L. 1998. Phenol antioxidant quantity and quality in foods: vegetables. J Agric Food Chem 46(9):3630–3634.
Vinson JA, Su X, Zubik L and Bose P. 2001. Phenol antioxidant quantity and quality in foods: fruits. J Agric Food Chem 49(11):5315–5321.
Visioli F, Bellomo G and Galli C. 1998. Free radical-scavenging properties of olive oil polyphenols. Biochem Biophys Res Commun 247(1):60–64.
Vrchovská V, Sousa C, Valentão P, Ferreres F, Pereira JA, Seabra, RM and Andrade PB. 2006. Antioxidative properties of tronchuda cabbage (*Brassica oleracea* L. var. costata DC) external leaves against DPPH, superoxide radical, hydroxyl radical and hypochlorous acid. Food Chem 98(3):416–425.
Wang H, Cao G and Prior RL. 1996. Total antioxidant capacity of fruits. J Agric Food Chem 44(3):701–705.
Wang M, Tsao R, Zhang S, Dong Z, Yang R, Gong J and Pei Y. 2006. Antioxidant activity, mutagenicity/anti-mutagenicity, and clastogenicity/anti-clastogenicity of lutein from marigold flowers. Food Chem Toxicol 44(9):1522–1529.
Wang SY and Ballington JR. 2007. Free radical scavenging capacity and antioxidant enzyme activity in deerberry (*Vaccinium stamineum* L.). LWT Food Sci Technol 40(8):1352–1361.
Wang SY and Jiao H. 2000. Scavenging capacity of berry crops on superoxide radicals, hydrogen peroxide, hydroxyl radicals, and singlet oxygen. J Agric Food Chem 48(11):5677–5684.
Wang SY and Lin HS. 2000. Antioxidant activity in fruits and leaves of blackberry, raspberry, and strawberry varies with cultivar and developmental stage. J Agric Food Chem 48(2):140–146.
Wang SY and Lin HS. 2003. Compost as a soil supplement increases the level of antioxidant compounds and oxygen radical absorbance capacity in strawberries. J Agric Food Chem 51(23):6844–6850.
Wang SY and Zheng W. 2001. Effect of plant growth temperature on antioxidant capacity in strawberry. J Agric Food Chem 49(10):4977–4982.
Wang W and Goodman MT. 1999. Antioxidant properties of dietary phenolic agents in a human LDL-oxidation *ex vivo* model: interaction of protein binding activity. Nutr Res 19(2):191–202.
Warso MA and Lands WEM. 1985. Presence of lipid hydroperoxide in human plasma. J Clin Invest 75(2):667–671.
Wayner DDM, Burton GW, Ingold KU and Locke S. 1985. Quantitative measurement of the total, peroxyl radical–trapping antioxidant capability of human blood plasma by controlled peroxidation. The important contribution made by plasma proteins. FEBS Lett 187(1):33–37.
Wilson-Kakashita G, Gerdes DL and Hall WR. 1995. The effect of gamma irradiation on the quality of English walnuts (*Juglans regia*). Lebensm Wiss Technol 28(1):17–20.
Winston GW, Regoli F, Dugas JR, Fong JH and Blanchard KA. 1998. A rapid gas chromatographic assay for determining oxyradical scavenging capacity of antioxidants and biological fluids. Free Radic Biol Med 24(3):480–493.
Wise JA, Morin RJ, Sanderson R and Blum K. (1996). Changes in plasma carotenoid, alpha-tocopherol, and lipid peroxide levels in response to supplementation with concentrated fruit and vegetable extracts: a pilot study. Curr Ther Res 57(6):445–461.
Wolfe KL and Liu RH. 2007. Cellular antioxidant activity (CAA) assay for assessing antioxidants, foods, and dietary supplements. J Agric Food Chem 55(22):8896–8907.
Wood JE, Senthilmohan ST and Peskin AV. 2002. Antioxidant activity of procyanidin-containing plant extracts at different pHs. Food Chem 77(2):155–161.
Wood LG, Gibson PG and Garg ML. 2006. A review of the methodology for assessing *in vivo* antioxidant capacity. J Sci Food Agric 86(13):2057–2066.

Wu X, Beecher GR, Holden JM, Haytowitz DB, Gebhardt SE and Prior RL. 2004. Lipophilic and hydrophilic antioxidant capacities of common foods in the United States. J Agric Food Chem 52(12):4026–4037.

Xu G, Liu D, Chen J, Ye X, Maa Y and Shi J. 2008. Juice components and antioxidant capacity of citrus varieties cultivated in China. Food Chem 106(2):545–551.

Yamamoto H, Manabe T and Okuyama T. 1990. Apparatus for coupled high-performance liquid chromatography and capillary electrophoresis in the analysis of complex protein mixtures. J Chromatogr 515: 659–666.

Yang RY, Tsou SCS, Lee TC, Wu WJ, Hanson PM, Kuo G, Engle LM and Lai PY. 2006. Distribution of 127 edible plant species for antioxidant activities by two assays. J Sci Food Agric 86(14):2395–2403.

Yen YH, Shih CH and Chang CH. 2008. Effect of adding ascorbic acid and glucose on the antioxidative properties during storage of dried carrot. Food Chem 107(1):265–272.

Yildirin N, Arat N, Dogan MS, Sokmen Y and Ozcan F. 2007. Comparison of traditional risk factors, natural history and angiographic findings between coronary heart disease patients with age <40 and ≥ 40 years old. Anadol Kardiyol Dergisi 7(2):124–127.

Yoshiki Y, Iida T, Akiyama Y, Okubo K, Matsumoto H and Sato M. 2001. Imaging of hydroperoxide and hydrogen peroxide–scavenging substances by photon emission. Luminescence 16(5):327–335.

Young JE, Zhao X, Carey EE, Welti R, Yang SS and Wang W. 2005. Phytochemical phenolics in organically grown vegetables. Mol Nutr Food Res 49(12):1136–1142.

Yu J, Wang L, Walzem RL, Miller EG, Pike LM and Patil BS. 2005. Antioxidant activity of citrus limonoids, flavonoids, and coumarins. J Agric Food Chem 53(6):2009–2014.

Yusof N, Ahmad Ramli RA and Ali F. 2007. Chemical, sensory and microbiological changes of gamma irradiated coconut cream powder. Radiat Phys Chem 76(11–12):1882–1884.

Zafrilla P, Ferreres F and Tomás-Barberán FA. 2001. Effect of processing and storage on the antioxidant ellagic acid derivatives and flavonoids of red raspberry (*Rubus idaeus*) jams. J Agric Food Chem 49(8):3651–3655.

Zhao M, Yang B, Wang J, Li B and Jiang Y. 2006. Identification of the major flavonoids from pericarp tissues of lychee fruit in relation to their antioxidant activities. Food Chem 98(3):539–544.

Zheng W and Wang SY. 2003. Oxygen radical absorbing capacity of phenolics in blueberries, cranberries, chokeberries, and lingonberries. J Agric Food Chem 51(2):502–509.

Zhou K and Yu L. 2006. Total phenolic contents and antioxidant properties of commonly consumed vegetables grown in Colorado. LWT Food Sci Technol 39(10):1155–1162.

Zhu C, Poulsen HE and Loft S. 2000. Inhibition of oxidative DNA damage *in vitro* by extracts of Brussels sprouts. Free Radic Res 33(2):187–196.

Žitňanová I, Ranostajová S, Sobotová H, Demelová D, Pecháň I and Ďuračková Z. 2006. Antioxidative activity of selected fruits and vegetables. Biol Bratislava 61(3):279–284.

11 Phytochemical Changes in the Postharvest and Minimal Processing of Fresh Fruits and Vegetables

Gustavo A. González-Aguilar*, J. Fernando Ayala-Zavala, Laura A. de la Rosa, and Emilio Alvarez-Parrilla

Introduction

Consumption of fruits and vegetables is increasing considerably in the daily diet because they supply high levels of biologically active compounds, called phytochemicals, that impart health benefits beyond basic nutrition (Robles-Sanchez and others 2007). Horticultural produce provide an optimal mixture of phytochemicals. The most thoroughly investigated dietary components in fruits and vegetables acting as antioxidants are phenolic acids, flavonoids, anthocyanins, lycopene, vitamins A, B, C, tocopherols, and sulfides (Robles-Sanchez and others 2007).

It is important to consider that the antioxidant content of fresh fruit and vegetable tissues can be affected by maturity, agricultural practices, postharvest handling, minimal processing, and storage conditions (Sacchetti and others 2008). The types of stresses, handling, and treatments to which fruits and vegetables are exposed, such as high-temperature storage, UV-C irradiation, maturity effectors, modified atmospheres, minimal processing, and antimicrobial and antibrowning agents, among others, also affect their antioxidant content (Ayala-Zavala and others 2005). However, the presence of high amounts of bioactive compounds in fresh tissue does not always ensure their bioavailability once they react against oxidative agents (Robles-Sanchez and others 2007).

As has been explained in previous chapters, the antioxidant capacity of fruits and vegetables is a function of the amounts and types of phytochemicals that are present in the fresh tissues. However, the individual contribution to the total antioxidant capacity varies widely. Various studies have demonstrated that phenols and flavonoids contribute to a higher extent than ascorbic acid, carotenoids, and others to the antioxidant capacity of fruits and vegetables (Robles-Sanchez and others 2007). It has been observed that a given content of vitamin E in fruits contributes significantly more to the antioxidant capacity than the same content of ascorbic acid.

The purpose of this chapter is to collect and explain information regarding changes in bioactive phytochemical compounds, including their content and the antioxidant capacity of fresh fruits and vegetables during various postharvest handling and minimal processing, as well as to disseminate knowledge on the health benefits of bioactive compounds.

* Corresponding author: Coordinación de Tecnología de Alimentos de Origen Vegetal, Centro de Investigación en Alimentación y Desarrollo, AC. Carretera a la Victoria km. 0.6. Apartado Postal 1735, Hermosillo, Sonora, México (83000).

Phytochemical Content and Bioactivity During the Postharvest

As phytochemical content and activity is becoming an increasingly important parameter of fruit and vegetable quality, it is of great interest to evaluate and understand changes in antioxidant status during postharvest storage of fresh fruits and vegetables (Ayala-Zavala and others 2007).

Storage Temperature

Metabolic rates of fruits and vegetables are directly related to storage temperatures within a given range. The higher the rate of respiration, the faster the produce deteriorates. Lower temperatures slow respiration rates, as well as ripening and senescence processes, which prolongs the storage life of fruits and vegetables (González-Aguilar and others 2004). Low temperatures also slow the growth of pathogenic and deteriorative microorganisms that cause spoilage and compromise safety of fruits and vegetables during the storage period (Lamikanra 2002). Therefore, this metabolic effect of temperature should be reflected in the phytochemical status of the stored horticultural produce. Koca and Karadeniz (2008) reported that cold storage (6 months at 0°C) did not affect β-carotene, α-carotene, and total carotenoids of carrots, as well as provitamin A activity, whereas the level of lutein, a minor component in carrot, decreased by 38%. Similarly, Kidmose et al. (2006) reported that carrot carotenoids were rather stable compounds during storage as no differences were seen in the content of α- and β-carotene during 4 months of refrigerated storage (1°C, 98% RH). There are many previous investigations, which reported no or only a minor degradation of carrot carotenoids during cold storage (Kopas-Lane and Warthesen 1995; Howard and others 1999). On the other hand, some authors observed that α- and β-carotene concentration decreased in minimally processed carrots stored at 1 and 2°C (Li and Barth 1998). Booth (1951), however, reported carotenoid chromogenesis for freshly harvested carrots while stored at 6°C in the dark. After 60 days of storage, total carotenoids were 11% higher than initial levels and subsequent decreases were reported. In any case, data showed that degradation did not occur during storage at 4 and 10°C; this is particularly interesting because of the dietary importance of carrot as a source of vitamin A precursors.

Anthocyanins and Phenol Content

The effect of storage temperatures (0, 5, 10°C) on the total anthocyanin content of strawberry fruit has been reported (Ayala-Zavala and others 2004). The content of these constituents was significantly affected by the storage temperature (Fig. 11.1, I). Anthocyanin content decreased in strawberry fruit stored at 0 and 5°C during the first 5 days. Meanwhile, anthocyanin content in fruit stored at 10°C increased gradually during the storage period and reached its highest values near the end of the storage period (13 days) (Ayala-Zavala and others 2004). Anthocyanins occur almost universally, and they are largely responsible for the red color of ripe strawberries. Two anthocyanidin glycosides, pelargonidin 3-glucoside and cyanidin 3-glucoside, contribute primarily to the red color of strawberries (Seeram and others 2004). The antioxidant capacity of anthocyanidins may be one of their most significant biological properties.

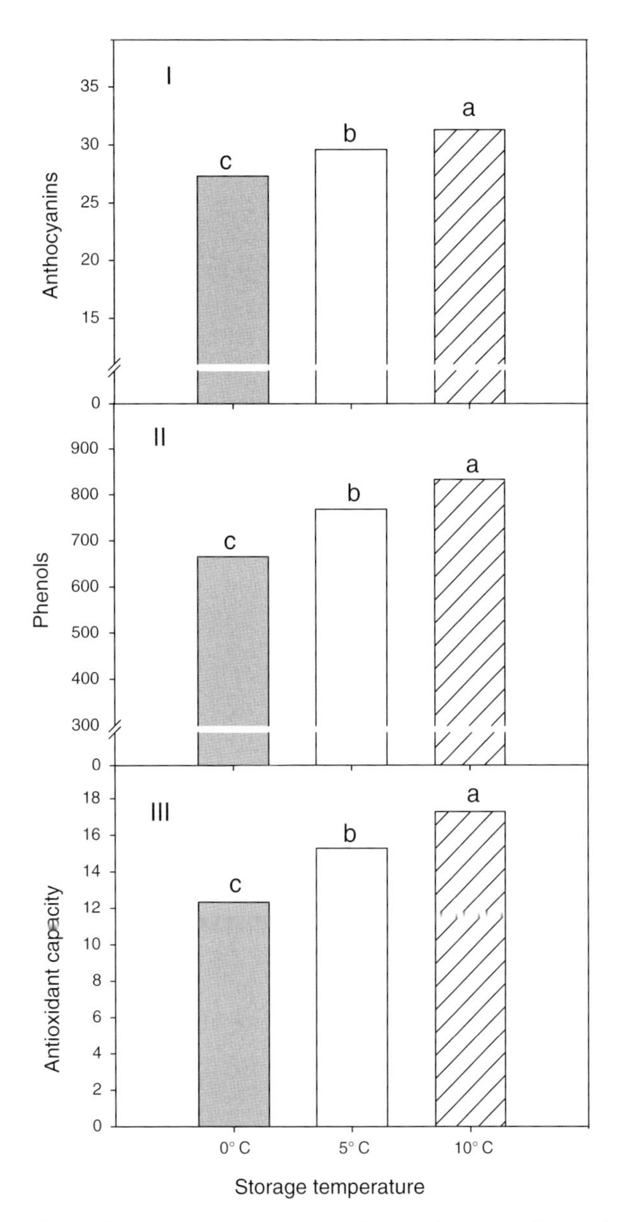

Figure 11.1. Effect of storage temperature (0, 5, 10°C) on (I) total anthocyanins (mg/100 g FW), (II) total phenols (mg/100 g FW), and (III) antioxidant capacity measured as ORAC (μmol TE/g FW) of strawberry fruit (cv. Chandler). Bars show the final values after treatments. Different letters on top of the bars indicate statistical differences among treatments ($p < 0.05$).

The effect of storage temperatures on total phenol compounds on strawberry fruit has been also reported (Ayala-Zavala and others 2004). Total phenol compounds increased in berries stored at 5°C and 10°C (Fig. 11.1, II). However, strawberry fruit stored at 0°C maintained a constant value of total phenol compounds during the storage period. Both temperature and storage time had a significant effect on total phenol compounds of strawberry fruits (Ayala-Zavala and others 2004).

Concentration patterns of phytochemical bioactive compounds in apples during storage are variable. Leja and others (2003) found an increase in total phenols of Jonagold apple from 520 mg/100 g after 120 days of storage at 0°C, to 640 mg/100 g after 120 days at 0°C + 7 days at 16°C; however, the amount of anthocyanins decreased during storage from 158 mg/100 g to 119 and 103 mg/100 g (Leja and others 2003). In the same study, the authors related the total phenol increase during storage with the increase of ethylene in the same period, because this hormone stimulates activity of phenylalanine ammonia lyase (PAL), a key enzyme in the biosynthesis of phenol compounds.

Ascorbic Acid Content

Temperature control is the most important tool to extend postharvest life and quality of fresh fruits and vegetables (Lee and Kader 2000). Delays between harvesting and cooling or processing can result in direct losses due to water loss and decay and indirect losses such as those in flavor and nutritional quality. Conditions favorable to wilting resulted in a more rapid loss of vitamin C in fresh produce (Lee and Kader 2000). The use of film packaging and storage in high-relative-humidity atmospheres prevented water loss and reduce ascorbic acid degradation of fresh fruits and vegetables. In general, the extent of loss in ascorbic acid content in response to elevated temperatures has been studied more in vegetables than in acidic fruits, such as citrus, because ascorbic acid is more stable under acidic conditions (Lee and Kader 2000).

Some horticultural crops such as sweet potatoes, bananas, and pineapples can suffer from chilling injury at low temperatures (Lee and Kader 2000). Chilling injury causes accelerated losses in ascorbic acid content of chilling-sensitive crops. Destruction of ascorbic acid can occur before development of any visible symptoms of chilling injury (Lee and Kader 2000).

Antioxidant Capacity

Storage temperatures significantly affect the antioxidant capacity of several fruits (Ayala-Zavala and others 2004). The antioxidant capacity expressed as oxygen-radical absorbance capacity (ORAC) values in strawberries changed very little during storage at 0°C. However, significant increases of ORAC values were found in strawberries stored at 5 and 10°C (Ayala-Zavala and others 2004). The higher the storage temperature, the greater the increase in ORAC (Fig. 11.1, III). One explanation for this difference could be related to different total phenol and anthocyanin content of fruit under the different storage temperatures. Strawberry storage at 10°C resulted in significantly increased total phenols and anthocyanin content (Ayala-Zavala and others 2004). However, even though antioxidant activity was the highest at 10°C, this elevated temperature may not be optimal for obtaining the best quality of strawberry fruit.

Light Conditions During the Storage Period

Light exposure appears to have little or no effect on either the ascorbic acid or carotenoid content of the edible portion of fruits and vegetables, whereas in some cases, phenol content can be increased by exposure to elevated light or UV irradiation. Pinot Noir grapes on sun-exposed clusters had a tenfold higher quercetin glycoside content compared with grapes from the shaded part of the cluster (Price and others 1995). In the same study, anthocyanin content was not affected by light exposure. Because flavonoids are known to absorb UV irradiation, it is believed that flavonoid production is stimulated to protect plant tissues from UV damage.

In apple fruit, light is required to stimulate localized anthocyanin production during fruit ripening (Lancaster and others 2000). The content of both anthocyanins and quercetin was much greater in sun-exposed portions of both Elstar and Jonagold apple fruit. Not all phenol synthesis is responsive to light (Lancaster and others 2000). Although levels of anthocyanins and quercetin compounds were much higher in sun-exposed apples, the contents of other major apple phenols, including catechins, phloridzin, and chlorogenic acid, were not different between apples from the sun-exposed and shaded portions of the tree (Awad and others 2000).

Modified/Controlled Atmospheres

Several studies have reported that modified/controlled atmosphere packaging delayed senescence and microbial growth in fruits and vegetables (Ayala-Zavala and others 2007). On the other hand, it has been observed that the antioxidant content and bioactivity could vary depending on the kind of treated fruit and treatment (Ayala-Zavala and others 2005).

One illustrative example of the effect of modified atmospheres on the phytochemical content of the stored fruit is treatment with high oxygen atmospheres. The application of high oxygen atmospheres could be an effective way of preserving strawberry quality (Ayala-Zavala and others 2007). Increased oxygen concentration in the internal and external fruit atmosphere could cause an increase in the production of free radicals that could result in oxidative stress in the fruit tissue, triggering responses of the antioxidant system and affecting phytochemical content and activity (Ayala-Zavala and others 2007).

Anthocyanins and Phenol content

Controlled atmospheres caused a decrement in the anthocyanin content of strawberry fruit stored within atmospheres with high oxygen at 5°C (Fig. 11.2, I) (Ayala-Zavala and others 2007). Anthocyanin content decreased for all atmospheres tested; however, high oxygen caused a higher decrement. This effect was attributed to the reaction of anthocyanins to inhibit radical activity at high oxygen concentrations, causing depletion of the antioxidant pigments (Ayala-Zavala and others 2007).

High oxygen atmospheres induced an increment on the total phenol content; these compounds were significantly affected by the high oxygen atmosphere (Ayala-Zavala and others 2007). The total phenol content of those fruits stored under the lower concentrations of oxygen (20–40 kPa) was affected to a lesser extent (Fig. 11.2, II). There was a correlation between high oxygen concentration (>21 kPa) and high levels of phenol compounds. High oxygen atmospheres also increased the phenol content of

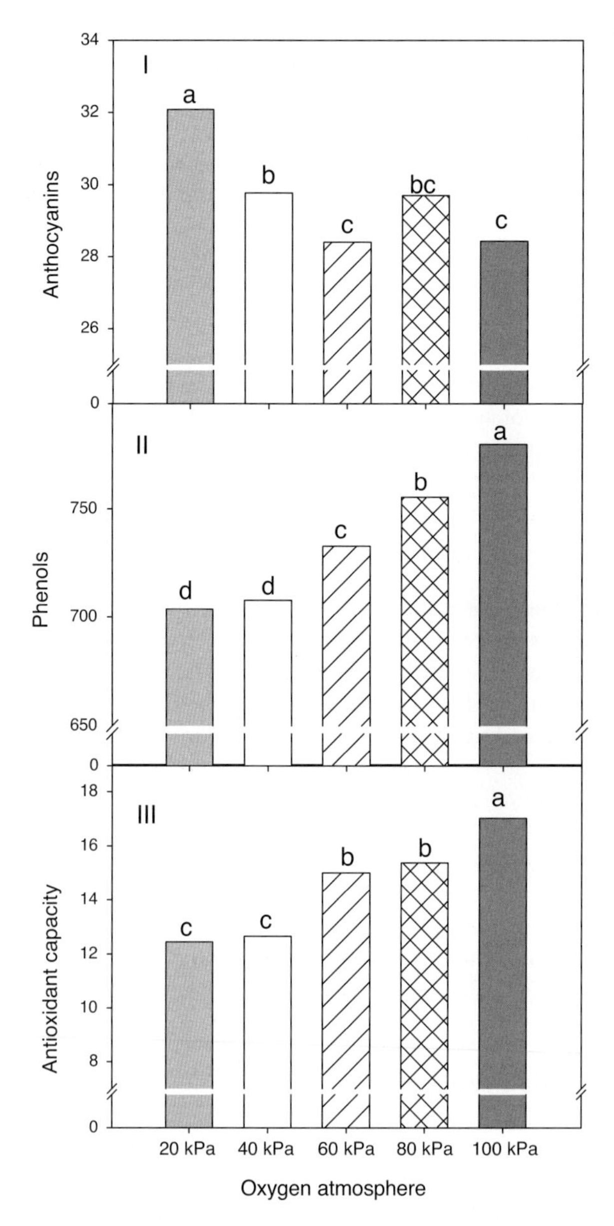

Figure 11.2. Effect of high oxygen atmospheres (20, 40, 60, 80, 100 kPa O_2) on (I) total anthocyanins (mg/100 g FW), (II) total phenols (mg/100 g FW), and (III) antioxidant capacity measured as ORAC (μmol TE/g FW) of strawberry fruit (cv. Allstar) stored at 7.5°C. Bars show the final values after treatments. Different letters on top of the bars indicate statistical differences among treatments ($p < 0.05$).

blueberries during storage (Zheng and others 2008). The increment of the total phenol compounds could be a response to the oxidative stress caused by the high oxygen concentrations (Srilaong and Tatsumi 2003).

It seems that the effect of the induction of phenol-content increment by high oxygen atmospheres is affected also by the exposure time of the product to the given atmosphere (Ayala-Zavala and others 2007). Awad and others found no losses of flavonoids in apples stored under conventional or controlled atmosphere (Awad and others 2000). No changes in the concentration of simple phenols, flavonoids, and anthocyanins were observed for Delicious and Ralls apples held for 4 to 5 months under refrigeration (Ju and others 1996). However, they found a decrease in simple phenols in earlier harvested apples after 3 months of cold storage. However, after 7 days at 20°C storage, phenols and flavonoid content decreased rapidly.

Antioxidant Capacity

The antioxidant capacity of strawberry fruit showed a similar behavior to that of the total phenol compounds (Ayala-Zavala and others 2007). In Figure 11.2, III, it is possible to see that antioxidant capacity of berries increased when the internal oxygen concentration increased from 60 to 100 kPa. Antioxidant capacity (ORAC) continued to increase during the storage period and reached maximum values of 21.5, 19, and 18.2 μmol Trolox equivalents (TE)/g on the tenth day, for those fruits stored under 100, 80, and 60 kPa oxygen, respectively. However, no noticeable changes were observed in those fruits stored under 20 or 40 kPa oxygen, with the exception of a small increase on the fifth day of storage at 5°C (Ayala-Zavala and others 2007). Previous studies indicate a linear correlation between phenol content and antioxidant capacity of several berries (Zheng and others 2008). The high values of antioxidant capacity observed in those berries stored under high oxygen atmospheres could also be attributed to the high level of phenol compounds.

Treatment with Natural Products

Research on the effect of several natural additive compounds has been undertaken, evaluating their mode of action, activity, toxicology, and effect on sensory, biochemical, and physiological properties of the treated fresh horticultural produce (Ayala-Zavala and others 2008b). Use of such additives is intended to meet consumer demands for healthy and safe produce with excellent quality. However, these treatments have been shown to have a significant effect on the phytochemical content and activity of the treated produce. Among the traditional natural treatments used to prevent deterioration of fruits and vegetables, honey has been recently reported to be a good food protector (Viuda-Martos and others 2008). Honey contains a wide range of substances that may act in this way, including ascorbic acid, small peptides, flavonoids, α-tocopherol, and enzymes such as glucose oxidase, catalase, and peroxidase (Jeon and Zhao 2005). Therefore, honey may be a natural alternative to sulfites for controlling enzymatic browning during fruit and vegetable processing.

Anthocyanins and Phenol Content

Ayala-Zavala and others (2005) tested the effects of different natural antimicrobial volatiles (methyl jasmonate, ethanol, and their combination) on the phytochemical content and antioxidant capacity of strawberry fruit. Total anthocyanin content was significantly affected by treatment with natural antimicrobials (Fig. 11.3, I). Anthocyanin content decreased continuously in all treatments. However, anthocyanin content of fruits treated with ethanol vapors decreased at a higher rate. Strawberries treated with the combination methyl jasmonate–ethanol showed the highest values of phenol content (Fig. 11.3, II). A sharp increase in total phenol compounds was observed during the first 5 days at 7.5°C in berries treated with the volatiles used. Afterward, a decrease was seen in these treatments. Untreated berries showed the lowest values during the storage period. Both treatments and storage period showed a significant effect on the total phenol compounds of strawberry fruit (Ayala-Zavala and others 2005).

Antioxidant Capacity

It has been observed that treatment with natural antimicrobial volatiles also affected the antioxidant capacity of fruits (Ayala-Zavala and others 2005). ORAC values of control strawberries changed during storage at 7.5°C (Fig. 11.3, III). However, significant increases in antioxidant capacity values were observed in strawberries treated with methyl jasmonate, methyl jasmonate–ethanol, and ethanol. One explanation for this difference could be associated with differences on total phenol content (Ayala-Zavala and others 2005).

Treatment of strawberry with methyl jasmonate resulted in a significant increase in total phenol content. However, even though antioxidant activity was the highest in those berries treated with methyl jasmonate, the combination methyl jasmonate–ethanol was the most effective in extending the shelf life. It appears that methyl jasmonate and ethanol treatments had an additive effect in maintaining quality of strawberries but not in retaining high antioxidant activity.

Changes in Phytochemical Content and Bioactivity in Fresh-Cut Produce

Fresh-cut fruits and vegetables are products that are partially prepared so that no additional preparation is necessary for their use (Ayala-Zavala and others 2008a). They are prepared for restaurants, fast food outlets, and retail markets. "Fresh-cut" is also defined as any fruit or vegetable or combination that has been trimmed, peeled, washed, and cut into 100% useable product that is then bagged or prepackaged and remains in a fresh state (Lamikanra 2002).

The basic requirements for preparation of fresh-cut fruits or vegetables are as follows: high-quality raw material, strict hygiene and good manufacturing practices, low temperatures during processing, careful cleaning and/or washing before and after peeling, use of mild processing aids in wash water for disinfection or prevention of browning and texture loss, minimization of damage during peeling, cutting, slicing and shredding operations, gentle draining to remove excess moisture, correct packaging materials and methods, and correct temperature during distribution and handling (Lamikanra 2002).

Fresh-cut fruits and vegetables are highly perishable products because of their intrinsic characteristics and the minimal processing (Ayala-Zavala and others 2008a). Microbial growth, decay of sensory attributes, and loss of nutrients are among the

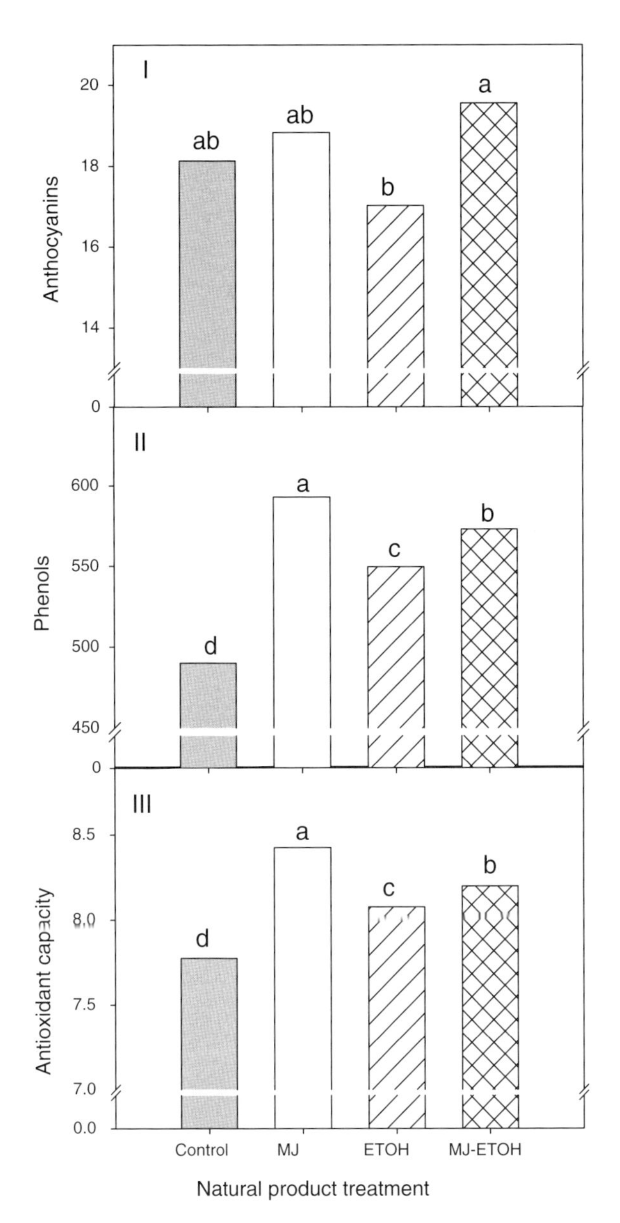

Figure 11.3. Effect of natural product treatment (methyl jasmonate: MJ 22 μg/liter, ethanol: ETOH 400 μL/liter, methyl jasmonate–ethanol (MJ-ETOH) on (I) total anthocyanins (mg/100 g FW), (II) total phenols (mg/100 g FW), and (III) antioxidant capacity measured as ORAC (μmol TE/g FW) of strawberry fruit (cv. Chandler) stored at 5°C. Bars show the final values after treatments. Different letters on top of the bars indicate statistical differences among treatments ($p < 0.05$).

major causes of compromised safety and quality of fresh-cut produce. These problems are caused by the steps involved in the minimal processing, such as peeling and cutting, which promote an increment in the metabolic rate, enzymatic reactions, and released juice (Ayala-Zavala and others 2008b).

Fresh-cut products are wounded tissues, and consequently they deteriorate more rapidly and their physiology differs from that of intact fruit and vegetables (Lamikanra 2002). The processes of peeling, coring, chopping, slicing, dicing, or shredding injure cells, releasing their contents at the sites of wounding. Subcellular compartmentalization is disrupted at the cut surfaces, and the mixing of substrates and enzymes that are normally separated can initiate reactions that normally do not occur (González-Aguilar and others 2005), which could affect the phytochemical content and antioxidant capacity of the produce.

Cutting and Shape Effect

Fresh cut fruits are cut in a wide variety of shapes, and the cutting shapes could influence the degree of damage caused by the wound (Rivera-Lopez and others 2005). Minimal processing includes the unit operation of size reduction, and a given shape has to be given to the product depending on its final use. In the next section the effect of this unit operation on phytochemical constituents and bioactivity is discussed.

Ascorbic Acid Content

The influences of minimal processing on ascorbic acid content of fresh-cut fruits were evaluated in comparison to whole fruits stored for the same duration but prepared on the day of sampling (Gil and others 2006). There was a significant loss of ascorbic acid but an increase in dehydroascorbic acid throughout storage for fresh-cut pineapple, resulting in greater ascorbic acid content of fresh-cut versus whole fruit. Fresh-cut mango cubes showed a decrease in ascorbic acid and an increase in dehydroascorbic acid during storage, maintaining the content of ascorbic acid close to the initial value for both fresh-cut and whole fruit, except at day 9 when the high AA and vitamin C contents of whole and fresh-cut fruit were rather different between them. Vitamin C in whole cantaloupes was better preserved than in fresh-cut cubes during storage, although differences were minimal. There were no significant differences in the content of ascorbic acid between fresh-cut and whole watermelon, strawberry, and kiwifruit (Gil and others 2006).

Ascorbic acid content was not affected by cutting shape for both cubes and slices of fresh-cut papaya (Rivera-Lopez and others 2005). Cubes and slices contained 65.47 mg/100 g of ascorbic acid at day 0. After 6 days of storage, no changes in ascorbic acid content were observed in cubes and slices stored at 5°C, whereas those fruits stored at 10°C and 20°C presented an average of 5% and 63% lower ascorbic acid content, respectively, within the same period. Ascorbic acid content of cubes and slices stored at 5°C decreased by 29.6 and 26.38% after 18 days of storage, respectively. Cubes and slices stored at 10°C decreased by 27.30% and 38.92%, respectively, from their initial ascorbic acid content after 14 days of storage (Rivera-Lopez and others 2005). Even though this study did not observe significant differences among the cutting shapes of papaya fruit, the cutting step per se could affect the ascorbic acid content.

Ascorbic acid retention in some fresh-cut produce is affected by the nature of the slicing method used (Lee and Kader 2000). Higher levels of ascorbic acid were retained in samples that had been prepared by manually tearing the lettuce into strips. Lettuce shredded using a sharp knife initially retained 18% less ascorbic acid than the torn samples. The retention of ascorbic acid in the products sliced by machine was 25–63% lower than that in lettuce shredded by manual tearing. Using a blunt machine blade resulted in 10% lower ascorbic acid levels than when a sharp blade was used (Barry-Ryan and O'Beirne 1999). Losses in ascorbic acid occur when vegetables are severely cut or shredded (Lee and Kader 2000).

Under stress conditions, such as cutting or light exposure, ascorbate oxidase has been described as promoting the transformation of ascorbic acid to dehydroascorbic acid (Wright and Kader 1997b). However, because ascorbic acid can be easily converted into dehydroascorbic acid, it is necessary to measure both ascorbic and dehydroascorbic acids to observe that the content of vitamin C was well preserved in fresh-cut fruit.

Carotenoid Content

The influences of minimal processing and storage on carotenoid content of fresh-cut fruits were evaluated in comparison to whole fruits stored for the same duration but prepared on the day of sampling (Gil and others 2006). During storage at 5°C, the total carotenoid content of cubes from whole pineapple was stable during storage, and no significant changes were observed after 9 days. However, after 3 days, there was a decrease in total carotenoids of fresh-cut samples, resulting in a 25% reduction relative to the whole fruit. The total carotenoid content of whole Ataulfo mangoes was quite stable during storage, and no significant changes were observed when compared with the initial contents. Fresh-cut mango cubes had slightly lower total carotenoid content than did cubes obtained from the whole stored fruit. The effect of slicing and storage was shown only after 9 days of storage, when a 25% reduction in total carotenoid content of fresh-cut mango with respect to the initial values was observed. The total carotenoid content in both fresh-cut and whole San Joaquin Gold cantaloupe cubes decreased after cutting, but it was followed by a steady content over the time of storage. There was no significant change in the total carotenoid content of fresh-cut slices and sliced strawberry fruit over the time of storage with the exception of day 9, when lower and higher contents of total carotenoids were found for fresh-cut slices and sliced fruit, respectively. The carotenoid content of kiwifruits did not change significantly in slices from either fresh-cut or whole stored fruit over the storage period (Gil and others 2006).

The cutting shape of fresh-cut papaya fruit was not shown to have immediate effect on β-carotene content (Rivera-Lopez and others 2005). Fresh-cut papaya cubes and slices stored at 5°C did not present any change in β-carotene content during 10 days of storage. No changes were observed in β-carotene content after 6 days of storage at 10°C; however, after 14 days of storage, there was a depletion of 62% for cubes and 63.4% for papaya slices. Fresh-cut papaya cubes and slices stored at 20°C showed the highest β-carotene losses. At this temperature and after 6 days of storage, β-carotene content of cubes and slices decreased by 57.4% and 60.63%, respectively.

Wright and Kader reported that there were no significant changes in the β-carotene content of sliced peach fruit over the time of storage (Wright and Kader 1997a), with the exception of fruit stored under air + 12% CO_2; the treatment resulted in a lower concentration of β-carotene. Persimmon fruit slices stored under air showed a decrease in β-carotene over the course of the study; however, overall, there was no significant difference among treatments (Wright and Kader 1997a). Enzymatic carotene degradation is mediated by lipoxygenase; β-carotene is lost through secondary reactions or co-oxidation during fatty acid oxidation (Lamikanra 2002).

Antioxidant Capacity

No significant effect of the cutting shape on the antioxidant capacity expressed as ORAC value was observed in fresh-cut papaya fruit (Rivera-Lopez and others 2005); however, storage temperature significantly affected it. Fresh-cut papaya cubes and slices changed slightly during storage at 5°C. However, significant reductions in ORAC values were found in fresh-cut papaya at 10°C and 20°C, primarily during final storage stage for cubes, whereas only at 10°C were ORAC differences significant ($P < 0.05$) for slices (Rivera-Lopez and others 2005).

Minimal processing steps could be expected to induce a rapid enzymatic depletion of several natural antioxidants. Depletion of antioxidant capacity in cubed and sliced papaya fruit could be associated with depletion of both ascorbic acid and β-carotenes, but in general, the highest loss of these important nutrients was reached before the product was unacceptable or spoiled (Rivera-Lopez and others 2005).

Sanitation Effect

Preserving the microbiological safety of processed produce should not have to conflict with nutritional quality. Although some sanitizing compounds are generally perceived as harsh chemicals that oxidize and destroy nutriments, it is important to consider the effect of those treatments and new alternatives on the nutritional quality of produce (Ruiz-Cruz and others 2007). Therefore, minimal processing with low or no food safety risk with optimal nutritional content will lead to increased consumption and meeting of consumer demands. In this context the effect of fruit sanitation on the phytochemical content and bioactivity must be contemplated.

Carotenoid content

Carotene content is a critical factor for several fruits and vegetables such as carrots, because consumers consider this vegetable as a major single source of provitamin A, providing 17% of the total vitamin A consumption (Barry-Ryan and O'Beirne 2000). Recently, the demand for carotenoids, especially β-carotene, has increased because of its health benefits (Ruiz-Cruz and others 2007).

Carotenoid content of fresh-cut carrot sanitized with various compounds decreased continuously during storage to a different extent among treatments (Ruiz-Cruz and others 2007). The carotene content of control carrots (washed with water) was reduced rapidly after 6 days of storage at 5°C and continued decreasing during storage. Carotene content of carrots washed with acidified sodium chlorite (ASC) decreased by 50% after 21 days at 5°C. However, carrots washed with the rest of the sanitizers and control decreased by 65% after 21 days of storage. These results show that carrots washed with ASC retained higher carotene content compared with carrots washed with other sanitizers (Ruiz-Cruz and others 2007).

Minimal processing may induce lipoxygenase-associated free radicals that oxidize β-carotene (Lamikanra 2002). Moreover, processes such as cutting may also increase exposure of carotenoids to oxygen compared to that in whole fruit. The rapid loss of carotene observed in carrots could be due to oxidative changes resulting from exposure to light and oxygen during processing (as cutting, because there is major surface contact) and may be related to losses in nutritive value.

Phenol Content

Phenol content (measured as gallic acid equivalents) increased initially in fresh-cut carrot treated with different sanitizers and later decreased in a different pattern for each treatment (Ruiz-Cruz and others 2007). Washing treatments significantly affected phenol content. Comparing sanitized shredded carrots with controls (unwashed and water washed), a sharp increase with a maximum value at days 3 and 6 (5.6 to 6 mg/100 g) was found, followed by a decline. Final phenol concentration was 0.7 to 1.3 mg/100 g for all treatments at the end of the storage period (Ruiz-Cruz and others 2007).

It is well known that PAL is the key enzyme of the phenylpropanoid pathway, catalyzing the first and committed step in the biosynthesis of these compounds (Lancaster and others 2000). The increase in phenol content could be result of PAL induction by cutting. In general, PAL activity increases as a response of fruit tissue to various stresses such as peeling and cutting (Cisneros-Zevallos and others 1995). The extent of PAL activity induced by peeling and cutting appears to be related more to the cultivar and maturity stage of the produce than to the sanitizer treatment.

Antioxidant Capacity

The antioxidant capacity expressed as ORAC value of shredded carrots washed with the different sanitizers ranged from 9 to 11.5 μmol TE/g (Ruiz-Cruz and others 2007). An increase in antioxidant capacity of washed carrots was observed during the first 3 days of storage. Afterward, a decline in ORAC value was observed. Control carrots showed a rapid decrease from day 3 to 10, and they continued losing antioxidant capacity until the end of the storage period. However, washed carrots remained stable from day 10 to the end of storage at 5°C. Carrots washed with the various sanitizers showed a similar pattern, but to a different extent. Carrots washed with 500 ppm ASC showed the highest values during the storage period at 5°C. The rapid increase in ORAC value after day 3 in control samples may be due to cutting stress that induces PAL activity and phenol accumulation, where a good correlation was found between ORAC and phenol compounds ($r^2 = 0.84$). In the same way, no noticeable changes were observed in washed carrots with the different sanitizers (Ruiz-Cruz and others 2007).

The antioxidant capacity retained in carrots washed with ASC could be attributed to retention of phenol and flavonoid compounds as well as carotene content (Ruiz-Cruz and others 2007). These results suggest that the use of sanitizers such as ASC is helpful in preserving the antioxidant capacity of carrots. Similar ORAC values were observed between different sanitizers and both water conditions, without significant changes. This is a clear example of a postharvest treatment maintaining the nutritional value of produce.

Modified Atmosphere Packaging of Fresh-Cut Produce

Modified atmosphere packaging is a common technique that describes the practice of modifying the composition of the internal atmosphere of a package in order to improve the shelf life (González-Aguilar and others 2004). The modification process often tries to lower the amount of oxygen, in order to slow down the growth of aerobic microorganisms and the speed of oxidation reactions (Lamikanra 2002). The removed oxygen can be replaced with nitrogen (N_2), commonly acknowledged as an inert gas, or carbon dioxide, which can lower the pH or inhibit the growth of bacteria and also affect the phytochemical status of the packed fresh-cut fruits and vegetables.

Ascorbic Acid Content

Modified atmospheres could be used to prevent losses of antioxidant nutrients in horticultural commodities during storage (Lamikanra 2002). Diced, blanched peppers stored in 2 and 4 mL/100 mL of oxygen retained more ascorbic acid than air-stored peppers. Fresh-cut kiwifruit stored under low oxygen was higher in ascorbic acid, whereas those from higher carbon dioxide were lower in ascorbic acid content than air-control slices (Agar and others 1999). Dehydroascorbic acid production during storage of blanched sweet green peppers was also retarded under low oxygen conditions (Agerlin-Peterson and Berends 1993).

Fresh-cut peppers packed under modified atmospheres packaging and vacuum and stored at 5 and 10°C showed different patterns in the phytochemical status (González-Aguilar and others 2004). Peppers stored at 10°C did not present apparent changes in ascorbic acid contents in both kinds of packages. Fresh-cut peppers under modified atmosphere and vacuum packages showed the highest values of ascorbic acid during storage at 5°C (González-Aguilar and others 2004). Howard and Hernandez-Brenes (1998) found that ascorbic acid of fresh-cut peppers was retained with modified atmosphere packaging stored at 4.4°C but decreased after the shelf-life period (3 days at 13°C). Therefore, once more it can be observed that the storage temperature is a detrimental factor that significantly affects ascorbic acid retention (Howard and Hernandez-Brenes 1998).

Carotenoid Content

Various atmospheres applied in packages of fresh-cut peach fruit appear to have no effect on quality attributes, although changes in the flesh color that occur in other fruits may have been masked by the rapid browning of the sliced fruit (Wright and Kader 1997a). There were no significant changes in the carotenoid content of the sliced peach fruit over the time of storage, with the exception of fruit stored under air + 12% CO_2; this treatment resulted in a lower concentration of carotene. The cryptoxanthin content of the fruit tended to increase in the fruit stored under 2% oxygen or 2% oxygen + 12% carbon dioxide, whereas the air + 12% oxygen treatment resulted in lower levels (Wright and Kader 1997a).

Persimmon fruit slices stored under air showed a decrease in carotene content (Wright and Kader 1997a). The air + 12% carbon dioxide treatment resulted in a loss of cryptoxanthin over the first 3 days, followed by a slight recovery. Levels of carotene were highly variable for fruit stored under air. Fruit stored under 2% oxygen + 12% carbon dioxide tended to maintain a concentration higher than the original (Wright and Kader 1997a).

UV-C Irradiation

UV-C technology is widely used as an alternative to chemical sterilization and microorganism reduction in food products (Lamikanra 2002; Fan and others 2008). Ultraviolet light also induces biological stress in plants and defense mechanisms in plant tissues with the consequent production of phytochemical compounds (Lee and Kader 2000). Phytoalexin accumulation could be accompanied by other inducible defenses such as cell-wall modifications, defense enzymes, and antioxidant activity, which have been reported with health benefits (González-Aguilar and others 2007). It is well documented that UV-C irradiation has an effect in secondary metabolism.

Ascorbic Acid Content

UV-C had a negative effect in the maintenance of ascorbic acid during storage of fresh-cut mangoes (González-Aguilar and others 2007). Ascorbic acid decreased significantly in those fresh-cut mangoes that were irradiated (Fig. 11.4, I). The highest level of ascorbic acid was observed in controls (0 min of irradiation time), being significantly different from the rest of the treatments (González-Aguilar and others 2007).

Fresh-cut fruits irradiated for 1, 3, and 5 min did not show significant differences in ascorbic acid content (González-Aguilar and others 2007). However, the lowest values of ascorbic acid were presented in the fruit irradiated for 10 min. It appears that exposure of fresh-cut mangoes to UV-C irradiation caused a significant reduction in ascorbic acid content. This behavior can be attributed to the increased oxidation of ascorbic acid caused by the increment in UV-C exposure time. González-Aguilar and others (2007) reported that irradiation oxidized a portion of total ascorbic acid to dehydroascorbic acid; however, both forms of the vitamin are biologically active, suggesting minimum nutritional impact.

Carotenoid Content

Compared to ascorbic acid, β-carotene content was similarly affected by irradiation and storage time (González-Aguilar and others 2007). The major reduction of β-carotene was observed in fresh-cut mangoes UV-C irradiated for 10 and 5 min, followed by those treated for 3 and 1 min (Fig. 11.4, II). Because β-carotene has a role in antioxidant defense, a further aspect of increased UV-C could, therefore, increase the oxidative stress of β-carotene on mango tissue. Control fruit presented significantly the lowest reduction in β-carotene content. The major losses of β-carotene were observed during the first 3 days at 5°C on UV-C treated fruit. Thereafter, no significant change was observed in exposures for 5 and 10 min, with slight decreases in fruit treated for 1 and 3 min. There was no significant difference ($p > 0.05$) between mangoes irradiated for 5 and 10 min; the lowest values were shown during the storage period, being significantly different from those for fresh-cut fruit irradiated for 1 and 3 min (González-Aguilar and others 2007).

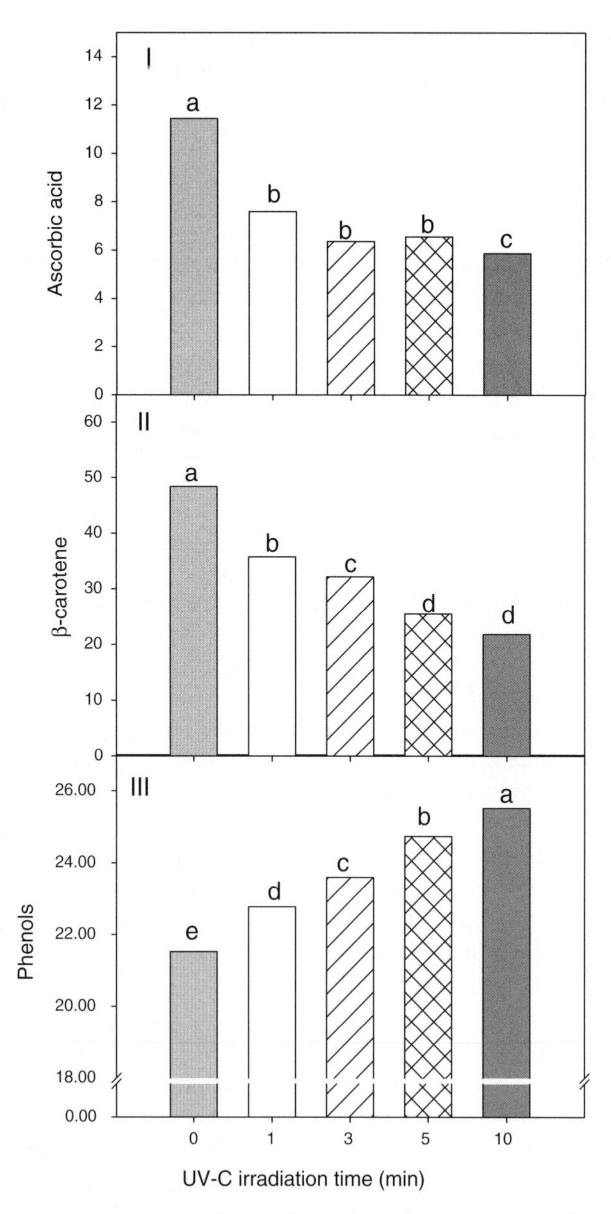

Figure 11.4. Effect of UV-C irradiation time (0, 1, 3, 5, 10 min) on (I) ascorbic acid (mg/100 g FW), (II) β-carotene (mg/100 g FW), and (III) total phenols (mg/100 g FW) of fresh-cut mango (cv. Tommy Atkins) stored at 5°C. Bars show the final values after treatments. Different letters on top of the bars indicate statistical differences among treatments ($p < 0.05$).

Phenol Content

Irradiation and storage time significantly affected the total phenol content of fresh-cut mango after the storage period at 5°C (Fig. 4, III) (González-Aguilar and others 2007). After 3 days of storage a sharp increase in phenols was observed in all UV-C treated fresh-cut mangoes, but to a different extent. Thereafter, no significant changes were observed, except for mangoes treated for 10 min, which presented a continuous increase. Fresh-cut fruit treated for 10 min presented the highest level of total phenol compounds, followed by fruit treated for 5, 3, and 1 min. However, the lowest phenol content was for controls. It has been observed that during the storage period a slight increment in the total phenol content was detected in all the treatments. In this context, UV-C irradiation time and the storage time had a positive effect on total phenol content of fresh-cut mango fruit (González-Aguilar and others 2007).

A similar effect was observed in other fruits and vegetables, where UV-C treated strawberries showed a higher increment of phenols and PAL activity 12 hours after treatment than unirradiated (control)(Pan and others 2004), which could be the reason for the increment in total phenol constituents (Lancaster and others 2000). UV-C and UV-B caused a two- and threefold increase in content of resveratrol (a grape phenol constituent). Thus, mature "Napoleon" grapes that had been irradiated with UV-C light can provide up to 3 mg of resveratrol per serving (Cantos and others 2001). Therefore, UV-C treatments clearly cause a benefit effect, increasing total phenol content, which can be mainly attributed to the increment of PAL activity.

Flavonoid Content

Flavonoid changes in fresh-cut fruit treated with different exposure times of UV-C were evaluated in mango slices stored for 15 days at 5°C (González-Aguilar and others 2007). Irradiation significantly affected the flavonoid content of fresh-cut mango after the storage period, similar to that observed in phenol content (Fig. 11.5, I). The flavonoid content of fresh-cut fruit irradiated for 10 and 5 min increased rapidly after 3 days of storage and showed a continuous increase during storage at 5°C, followed by those fruit treated for 3, 1, and 0 min. However, flavonoid content in controls was constant during the whole storage period (González-Aguilar and others 2007).

Flavonoids act in plants as antioxidants, antimicrobials, photoreceptors, visual attractors, feeding repellents, and light screeners (Schauss and others 2006). Previous studies reported that UV-C light treatment had no significant effect on grape and pomegranate anthocyanin content (Cantos and others 2001; Lopez-Rubira and others 2005). However, there is general agreement that the induction of flavonoids is a specific acclimation response, supporting the hypothesis of the existence of different UV-signaling pathways in plant tissues (Rivera-Pastrana and others 2007).

Antioxidant Capacity

Irradiation of fresh-cut mangoes improved the antioxidant capacity of the product (Fig. 11.5, II), even when a long exposure could reduce the ascorbic acid and β-carotene content (González-Aguilar and others 2007). This was not proportional to the antioxidant capacity, which was influenced more by total phenols and flavonoid content. It seems that the antioxidant capacity improvement could be an additional factor to give added value in the fresh-cut mango industry.

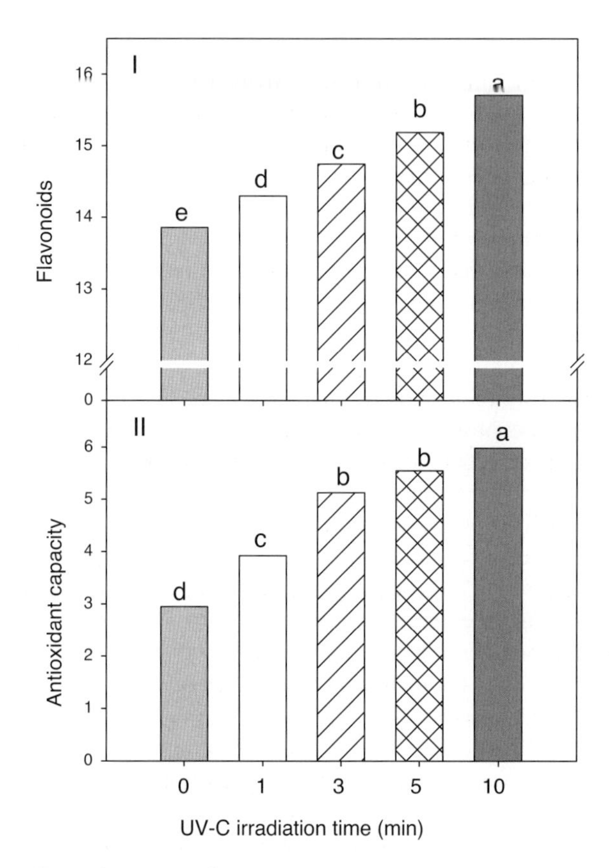

Figure 11.5. Effect of UV-C irradiation time (0, 1, 3, 5, 10 min) on (I) total flavonoids (mg/100 g FW) and (II) antioxidant capacity measured as ORAC (μmol TE/g FW) of fresh-cut mango (cv. Tommy Atkins) stored at 5°C. Bars show the final values after treatments. Different letters on top of the bars indicate statistical differences among treatments ($p < 0.05$).

The antioxidant capacity of fresh-cut mango was increased by the irradiation treatment: the longer the irradiation exposure time, the higher the antioxidant capacity of the mango tissue (González-Aguilar and others 2007). Fruit treated for 10 min presented the highest values of antioxidant capacity, followed by fruits treated for 5, 3, and 1 min, which were statistically different. Nevertheless, the lowest antioxidant capacity was found in controls. UV-C irradiation time had a positive effect on antioxidant capacity, mainly influenced by total phenols and flavonoids of fresh-cut mango, even when ascorbic acid and β-carotene content was decreased by the UV-C light (González-Aguilar and others 2007).

As discussed previously, UV-C irradiation increased the phenol and flavonoid content; such compounds present high radical-quenching activity by themselves (Robles-Sanchez and others 2007). Few studies have been reported specifically about the effect of UV-C irradiation on antioxidant capacity of treated fruit. However, Fan and others

(2003) suggested that ionizing irradiation increased phenol content and antioxidant capacity of endive and of romaine and iceberg lettuce (Fan and others 2003). Increased consumption of diets rich in antioxidants and phenols may contribute to reducing human diseases, and irradiation processing may produce a healthier vegetable product (Robles-Sanchez and others 2007).

Natural Antimicrobial Volatiles

Natural antimicrobial compounds are a re-emerging alternative to fresh-cut produce preservation (Ayala-Zavala and others 2008b). The antimicrobial power of plant and herb extracts has been recognized for centuries and has mainly been used as natural medicine. Plants produce a wide range of volatile compounds, some of which are important flavor quality factors in fruits, vegetables, spices, and herbs (Ayala-Zavala and others 2008b). A number of volatile compounds inhibit the growth of microorganisms (Ayala-Zavala and others 2008c). The ability of plant volatiles to inhibit microorganism growth is one of the reasons why there is increased interest in using them to control postharvest and postprocessing decay of fruits and vegetables (Yao and Tian 2005). Plant volatiles have been widely used as food flavoring agents, and many are generally recognized as safe (GRAS); in addition, they have a given antioxidant status per se.

Lycopene Content

Lycopene is an isoprenoid compound that provides red color to fruits and vegetables (Kubota and Thomson 2006). Treating fresh-cut tomatoes with methyl jasmonate significantly increased their lycopene content from 34 to 73.3 mg/kg (Fig. 11.6, I) (Ayala-Zavala and others 2008c). Methyl jasmonate increased chlorophyll loss and lycopene synthesis as a consequence of its senescence-promoting effects (González-Aguilar and others 2001). This increase in lycopene content of fresh-cut tomato treated with methyl jasmonate might be due to the effect of the treatment on the isoprenoid biosynthesis pathway. Methyl jasmonate has been shown to be involved in wounding stresses and affecting the isoprenoid biosynthesis pathway (Karakurt and Huber 2003). Methyl jasmonate treatments have been shown to accelerate the ripening process and to enhance chlorophyll degradation of tropical fruits (Karakurt and Huber 2003). On the other hand, fresh-cut tomatoes treated with methyl jasmonate–ethanol as well as control fruits retained their initial lycopene for a period of 15 days at 4°C (Ayala-Zavala and others 2008c).

Although methyl jasmonate has been shown to have an important influence on the ripening process of fresh-cut tomato, increasing lycopene synthesis, the mixture of this antimicrobial with ethanol did not have such an effect (Ayala-Zavala and others 2008c). On the contrary, lycopene content in fresh-cut tomatoes treated with ethanol decreased slightly throughout storage. This trend could be related to the suppression of lycopene synthesis or its degradation. It has been demonstrated that ethanol inhibited the ripening process, which involves a chlorophyll loss and a decrease in synthesis of lycopene, pigments that contribute to skin and flesh color of tomatoes (Ayala-Zavala and others 2008c).

Previous publications reported that ethanol treatment retarded ripening of several fruits, including whole tomatoes and tomato slices (Saltveit and Mencarelli 1988). Also, ethanol can slow down tomato ripening by inhibiting synthesis and action of ethylene,

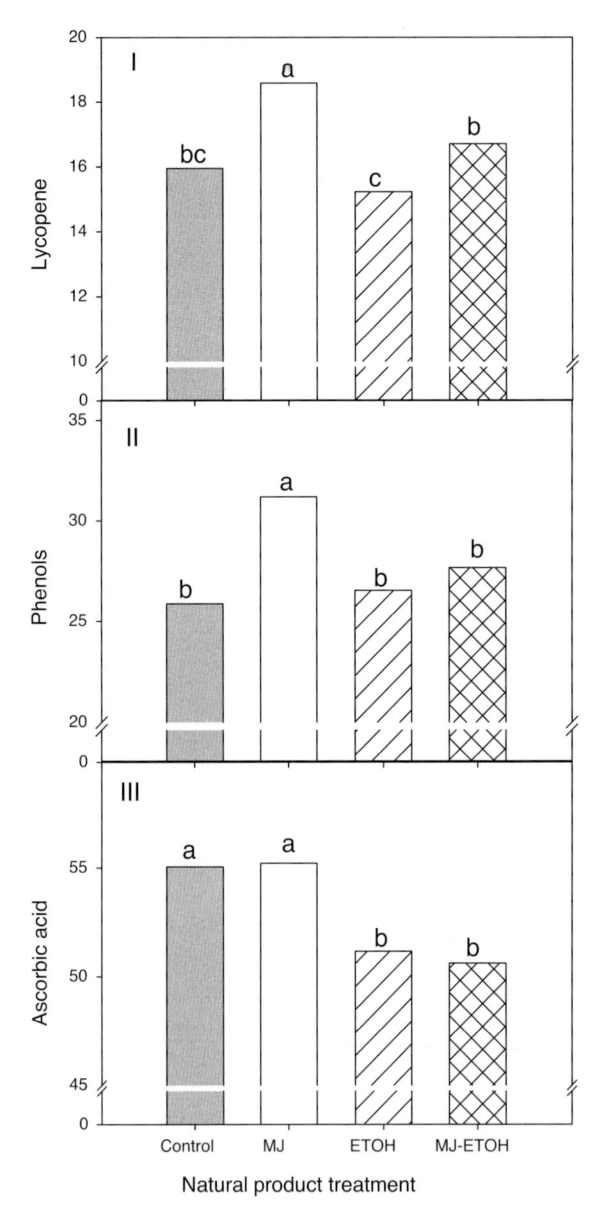

Figure 11.6. Effect of natural product treatment (methyl jasmonate: MJ 22.4 μg/liter, ethanol: ETOH 300 μL/liter, methyl jasmonate–ethanol (MJ-ETOH) on (I) lycopene (mg/100 g FW), (II) total phenols (mg/100 g FW), and (III) ascorbic acid (mg/100 g FW) of fresh-cut tomato stored at 5°C. Bars show the final values after treatments. Different letters on top of the bars indicate statistical differences among treatments ($p < 0.05$).

delaying the loss of green color as a consequence of lycopene synthesis (Yanuriati and others 1999). The changes in lycopene content throughout storage time might have affected color parameters (Ayala-Zavala and others 2008c). A good correlation between lycopene content and a^* value was observed ($r^2 = 0.9253$), which is consistent with the fact that an intense red color should be indicative of a higher lycopene content. Moreover, a negative correlation was found between lycopene concentrations and h^* results ($R^2 = -0.9248$) (Ayala-Zavala and others 2008c).

Phenol Content

The effect of natural antimicrobials on total phenols of fresh-cut tomatoes during storage at 5°C was evaluated previously (Ayala-Zavala and others 2008c). Total phenol content was significantly affected by storage period and treatments. Phenol compounds increased during the storage period after volatile compounds application, whereas control fresh-cut fruits maintained the initial values throughout the storage time. Methyl jasmonate treated-fruit contained much higher levels of phenol compounds compared with those obtained by the other treatments. Fresh-cut tomatoes treated with methyl jasmonate–ethanol, garlic oil, and tea tree oil increased gradually, their phenol contents reaching values of 267.4, 244.1, and 228.3 mg/kg, respectively (Ayala-Zavala and others 2008c). Methyl jasmonate has shown to increase *in vivo* activity of PAL, which is the key enzyme that uses phenylalanine to synthesize phenol compounds (González-Aguilar and others 2006).

Methyl jasmonate is a molecule used in plant tissue as a direct defense against stress, activating genes for PAL production that catalyze secondary metabolites. An increase in PAL activity would promote the synthesis of phenol compounds in those fruits treated with methyl jasmonate, as a defensive response to stress (González-Aguilar and others 2006). Natural antimicrobial treatments, in general, maintained or increased the content of phenol compounds in fresh-cut tomatoes, which in turn affected the antioxidant status of the product. Those living tissues exposed to stress can activate their natural defenses, in the form of an increment in phenol compound synthesis (Ayala-Zavala and others 2008c).

Several secondary metabolites, including methyl jasmonate and exogenous phenol compounds presented in tea tree oil, have become well recognized as vital defensive compounds protecting plants from pathogen attack (Ayala-Zavala and others 2008b). The plant synthesis of such secondary metabolites has been related to the exposure of tissues to various stresses. The shikimic acid and phenylpropanoid pathways, which are the most affected by the signal induced by the methyl jasmonate molecule, have long been studied as the basis of the defensive response (Ayala-Zavala and others 2008c).

Ascorbic Acid Content

The ascorbic acid concentration of fresh-cut tomatoes treated with natural volatile antimicrobial agents increased during the first days of storage at 5°C and tended to stabilize during subsequent days, except for fresh-cut tomatoes treated with ethanol and control fruits (Ayala-Zavala and others 2008c). The highest ascorbic acid content throughout the storage time was found in fresh-cut tomatoes treated with methyl jasmonate, leading to levels of 127.7 mg/kg at 15 days of storage (Fig. 11.6, III). Treatment of plant cells with methyl jasmonate induces many metabolic processes that require ascorbic acid; as a result, methyl jasmonate can enhance ascorbic acid synthesis in order to sustain secondary pathways and to preserve the redox status of plant cells (González-Aguilar and others 2006). On the other hand, fresh-cut tomatoes treated with methyl jasmonate–ethanol did not show significant differences in ascorbic acid content throughout cold storage. A significant increase of ascorbic acid was reported in raspberries treated with methyl jasmonate (Chanjirakul and others 2006). In addition, the increase in ascorbic acid content in tomatoes might be due to ripening. As fruits ripen, the ascorbic acid content in the flesh increases even to the overripe stage (Ayala-Zavala and others 2008c).

In contrast, fresh-cut tomatoes treated with ethanol and control fruits retained their initial ascorbic acid content for a period of 21 days at 4°C (Ayala-Zavala and others 2008c). It is known that the precursors of biosynthesis of ascorbic acid are simple sugars such as galactose and mannose. The higher the content of these sugars in the fruit, the greater the production of ascorbic acid. According to this fact, the lower synthesis of ascorbic acid in fresh-cut tomato treated with ethanol could be related to a low content of galactose and mannose as a result of ethanol effects on ripening inhibition (Ayala-Zavala and others 2008c).

Natural Antioxidant Treatments

It is well known that minimally processed fruits and vegetables are generally more perishable than the original raw materials (Lamikanra 2002). Mechanical stress during processing results in cellular delocalization of enzymes and their substrates, leading to biochemical deterioration such as enzymatic browning (Ayala-Zavala and others 2008a). An important problem in which reactive oxygen species are involved is the limited shelf life of those commodities due to surface browning (González-Aguilar and others 2005). The phenomenon is usually caused by the enzyme polyphenoloxidase (PPO), which, in the presence of oxygen, converts phenol compounds into dark colored pigments (González-Aguilar and others 2005). If phenol compounds are affected by this deteriorative process, the phytochemical status of fresh-cut fruits and vegetables could be compromised.

Nowadays, numerous research efforts pursue the development of new ways to prevent browning in fresh-cut produce. Reducing agents play a relevant role in the prevention of enzymatic browning either by reducing *o*-quinones to colorless diphenols or by reacting irreversibly with *o*-quinones to form stable colorless products (Oms-Oliu and others 2006). Ascorbic acid is extensively used to avoid enzymatic browning of fruit due to the reduction of the *o*-quinones, generated by the action of the PPO enzymes, back to the phenol substrates (González-Aguilar and others 2008; González, 2005). Several thiol-containing compounds such as cysteine, *N*-acetylcysteine, and reduced glutathione have been investigated as inhibitors of enzymatic browning (Oms-Oliu and others 2006; Rojas-Grau and others 2007; Rojas-Grau and others 2006; González-Aguilar and others 2005). They have been applied in combination with organic acids and calcium salts to prevent enzymatic browning of fruits (Pizzocaro and others 1993; Senesi and others 1999; Soliva-Fortuny and others 2001, 2002). These compounds have been shown to delay browning of fresh-cut produce and as a result maintain a higher level of phytochemicals.

However, some researchers have established that ascorbic acid is consumed during antibrowning reactions (Luo and Barbosa-Cánovas 1997; Ozoglu and Bayindirli 2004; Rojas-Grau and others 2006). Sapers (1993) reported that once the ascorbic acid has been completely oxidized to dehydroascorbic acid, quinones can again accumulate and undergo browning. Because the individual effect of ascorbic acid is temporary, other alternatives to control browning should be sought.

Ascorbic Acid Content

Changes in ascorbic acid of pineapple slices treated with different antibrowning agents have been evaluated (González-Aguilar and others 2005). No significant changes in ascorbic acid levels were observed in controls or slices treated with acetylcysteine during the storage period at 10°C (González-Aguilar and others 2005). In contrast, it has been found that dehydrated pineapple and guava pretreated with cysteine hydrochloride had increased ascorbic acid retention and reduced color change during storage (Mohamed and others 1993). As is logical, ascorbic acid content was significantly increased after treatments with 0.05 M of ascorbic acid applied to fresh-cut pineapples (González-Aguilar and others 2005). However, after 3 days, ascorbic acid content of fresh-cut pineapple treated with ascorbic acid decreased continuously from 45 to 18 mg/100 g at the end of the storage period (González-Aguilar and others 2005). With similar results, Gorny and others found that the ascorbic acid content of pear slices treated with exogenous AA (2%), dropped to endogenous control levels after 3 days at 0°C (Gorny and others 2002). It appears that ascorbic acid is most likely converted to dehydroascorbic acid and further degraded to 2,3-diketogluconic acid (González-Aguilar and others 2005).

Cocci and others studied the effect of antioxidant dipping treatment (1% ascorbic acid and 1% citric acid for 3 min) and modified atmosphere in fresh cut apples (Cocci and others 2006). As a result of the antibrowning treatment the ascorbic acid treated samples had about 20-fold higher ascorbic acid content than nontreated samples at the beginning of storage, and content remained higher until the sixth day of refrigeration; total polyphenols were also higher for treated samples compared to those not treated (Cocci and others 2006). Results showed that the treatment used served the antibrowning purpose and in addition compensated the losses in nutritional properties (Cocci and others 2006). Robles-Sánchez and others (2009) reported that dipping treatments affected ($P < 0.05$) total phenols, vitamin C, vitamin E, and gallic and *p*-OH-benzoic acids, whereas storage time affected ($P < 0.05$) all the parameters studied. The initial vitamin C values were of 11 and 44 mg/100 g of fresh weight for control and treated fresh-cut mango cubes, respectively. As expected, a fourfold increase in vitamin C was observed immediately after the treatment with AA + CA + $CaCl_2$. However, in this study vitamin C losses during storage of fresh cut mangoes were minimal in both control and treated tissue. A higher level of vitamin C in treated fresh-cut cubes was observed compared to controls.

Another option to overcome the nutritional losses is to use the preserving treatment in fresh-cut fruits to increase the nutritional content by the use of edible coatings and vacuum impregnation.

Phenol Content

A sharp reduction in phenols was observed in control pineapple slices after cold storage; however, the antibrowning agents significantly reduced the decline in total phenols (Cocci and others 2006). Treatments with acetylcysteine and isoascorbic acid were significantly more effective in preventing phenol reduction during cold storage of slices. However, the effectiveness of antibrowning compounds in reducing phenol content diminished with length of storage (Cocci and others 2006). Recently, it was reported that dipping treatment increased total phenol content of fresh-cut treated mango cubes with respect to control cubes (91.39 and 51.90 mg/100 g of fresh weight, respectively) (Robles-Sánchez and others 2009). No significant losses of total phenols were found at the end of storage, although some fluctuations were observed during the whole period. Control samples exhibited a minimal reduction in total phenol levels, showing losses of about 4% at the end of the storage. With respect to total flavonoid content, no significant changes ($P < 0.05$) were observed between fresh-cut control and treated cubes.

Other Preservation Technologies: Nonthermal Approaches

Reducing microbial growth and spoilage in fresh-cut and other minimally processed fruit and vegetable products has also been achieved through the use of "nonthermal sterilization techniques," which include high-intensity pulsed electric fields (PEF), high hydrostatic pressures (HPP), and ionizing (gamma) radiation (IR), among others. Many papers are available on the modes of action of the different techniques against spoilage and pathogenic microorganisms or on their effectiveness on the microbial load reduction in model and real systems (Corbo and others 2009). However studies assessing the effect of these emerging technologies on degradation of nutrients and bioactive compounds of food products are still scarce. Table 11.1 summarizes a few of the most recent reports on this topic with emphasis on the most important bioactive compounds. Most of these studies have been carried out in minimally processed fruit juices; however, HHP and IR treatment of fresh-cut fruits and vegetables has been described (Fan and others 2008; Wolbang and others 2008; McInerney and others 2007). In general, the nonthermal treatments induce a slight loss of bioactive compounds in vegetal foods and may even enhance their recovery and retention throughout storage. For example, PEF treatment extended the period of retention of ascorbic acid and lycopene in tomato juices stored at 4°C, as compared with thermal pasteurization (Odriozola-Serrano and others 2008b).

Ascorbic acid is probably the most labile bioactive compound in fruit juices and fruit and vegetable pieces, as we described in the first part of this chapter. Retention of this phytochemical after the nonthermal treatments ranged from 47% to 100%, depending on the intensity of the applied treatment and the product. For example, the greatest losses of vitamin C were found in fresh-cut red lettuce and melon treated with IR and HHP (Fan and others 2008; Wolbang and others 2008), respectively. However, the use of gamma radiation in various vegetables retained 100% of their total ascorbic acid content (Fan and others 2008).

Phenolic compounds are far more stable than vitamin C, showing retention/recovery values of 86–140%; of this group, anthocyanins showed the lowest recovery rates. Phenolic compounds are also retained during storage. Odriozola-Serrano and others

Table 11.1. Retention of bioactive compounds in minimally processed fruit and vegetable products treated with nonthermal preservation technologies

Technology	Bioactive	Food product	Retention	Reference
Pulsed electric fields	Total phenols and catechins	Tea infusions	100%	Zhao and others 2009
	Total anthocyanins	Strawberry juice	96.1–100.5%[1]	Odriozola–Serrano and others 2008a
	Ascorbic acid	Strawberry juice	$\geq$ 87%[1]	Odriozola–Serrano and others 2008a
	Antioxidant capacity	Strawberry juice	$\geq$ 81.3%[1]	Odriozola–Serrano and others 2008a
	Lycopene	Tomato juice	107.6%	Odriozola–Serrano and others 2008b
	Ascorbic acid	Tomato juice	86.5%	Odriozola–Serrano and others 2008b
	Total phenolics	Tomato juice	100%	Odriozola–Serrano and others 2008b
	Antioxidant capacity	Tomato juice	100%	Odriozola–Serrano and others 2008b
	Total carotenoids	Orange juice	90.4–93.6%[1]	Cortés and others 2006
	Vitamin A	Orange juice	86.1–106%[1]	Cortés and others 2006
	Total Carotenoids	Orange juice	100%	Sánchez–Moreno and others 2005
	Flavanones	Orange juice	100%	Sánchez–Moreno and others 2005
	Antioxidant capacity	Orange juice	100%	Sánchez–Moreno and others 2005
High hydrostatic pressures	Antioxidant capacity	Strawberry puree	80.6–100%[1]	Patras and others 2009
		Blackberry puree	100–168%[1]	Patras and others 2009
	Total phenolics	Strawberry puree	100–109.8%[1]	Patras and others 2009
		Blackberry puree	100–104.9%[1]	Patras and others 2009
	Anthocyanins	Strawberry puree	86–100%[1]	Patras and others 2009
		Blackberry puree	100%	Patras and others 2009
	Ascorbic acid	Strawberry puree	90.7–94.6%[1]	Patras and others 2009
	Beta-carotene	Melon pieces	123%[2]	Wolbang and others 2008
	Ascorbic acid	Melon pieces	43%[2]	Wolbang and others 2008
	Antioxidant capacity	Melon pieces	60%[2]	Wolbang and others 2008

(Continued)

Table 11.1. *Continued*

Technology	Bioactive	Food product	Retention	Reference
	Total carotenoids	Carrots, green beans, and broccoli	100%	McInerney and others 2007
	Antioxidant capacity	Carrots	79–100%[1]	McInerney and others 2007
		Green beans	125–193%[1]	McInerney and others 2007
		Broccoli	100%	McInerney and others 2007
	Total Carotenoids	Orange juice	154%	Sánchez–Moreno and others 2005
	Provitamin A carotenoids	Orange juice	139%	Sánchez–Moreno and others 2005
	Flavanones: naringenin	Orange juice	120%	Sánchez–Moreno and others 2005
	Flavanones: hesperitin	Orange juice	140%	Sánchez–Moreno and others 2005
	Antioxidant capacity	Orange juice	100%	Sánchez–Moreno and others 2005
Radiation	Ascorbic acid	Red lettuce	47%	Fan and others 2008
		Green lettuce	76%	Fan and others 2008
		Iceberg lettuce, romaine lettuce, spinach, tomato, etc.	100%	Fan and others 2008

[1]Retention depended on treatment variables. [2]Retention varied among cultivars.

(2008b) found no loss on total phenolic compounds of PEF-treated tomato juice during 91 days of storage at 4°C. HHP treatment increased the content of total phenolics in berry puree (Patras and others 2009) and flavanones in orange juice (Sanchez-Moreno and others 2005) and also of carotenoids in orange juice (Sánchez-Moreno and others 2005) and β-carotene in melon pieces (Wolbang and others 2008).

Carotenoids showed good retention rates (86–100%) in minimally processed juices treated with PEF and fresh-cut vegetables treated with HHP.

Antioxidant capacity of fruits and vegetables depends on the total concentrations of phytochemicals, mainly ascorbic acid, phenolic compounds (including flavonoids), and carotenoids. However, as previously stated, the individual contribution of each compound to the total antioxidant capacity varies widely and is difficult to quantify in a whole food product.

Antioxidant capacity can be measured by several techniques, each of which has its own limitations. Antioxidant capacity of fruit juices and purees was evaluated by the DPPH method (which measures the radical-scavenging activity against a nonphysiological free radical), and treatment of these products with PEF or HHP resulted in losses of 20% or less antioxidant activity; in one case antioxidant capacity of blackberry puree was increased by HHP treatment (Patras and others 2009). In this study phenolic compounds were also increased by HHP treatment, although slightly; in contrast, ascorbic acid was decreased, suggesting that the antioxidant capacity of blackberry puree is influenced more by phenolic compounds than by ascorbic acid as was suggested in the past.

Antioxidant activity of HHP-treated fresh-cut fruits and vegetables was evaluated by means of the FRAP method (ferric-reducing antioxidant power). In melon pieces 40% of antioxidant capacity was lost after treatment; a similar effect was observed in ascorbic acid and a contrary effect in β-carotene (its concentration was raised by HHP) (Wolbang and others 2008). Antioxidant capacity was retained 79–100% in HHP-treated carrots and broccoli and increased 25–93% in green beans (McInerney and others 2007). These authors suggested that changes induced by HHP treatment in the tissue matrix of green beans could have resulted in the release of more antioxidant compounds into the extracellular environment and hence an increase in the antioxidant activity of the whole product. In green beans, lutein bioavailability was also increased by pressure treatment, whereas in broccoli β-carotene availability was reduced. High-pressure treatment had no effect on carotenoid bioavailability of carrots (McInerney and others 2007).

It may be concluded that PEF, HHP, and IR are adequate techniques for the retention of bioactive compounds in fruit and vegetable products and may even enhance bioactivity of juices, purees, and fresh-cut produce. A greater degradation of ascorbic acid in comparison with phenolics and carotenoids is usually observed.

Future Directions

Considering the increasing demands of consumers for healthy products, and of producers for emerging and effective technologies to prevent deterioration of fresh and minimally processed fruits and vegetables, it is important to evaluate the effect of new treatments on the phytochemical status of the treated produce, because phytochemical content and bioactivity must be contemplated as one more major quality attribute. It is important to determine the changes of individual contribution of the various phytochemical compounds to the total antioxidant capacity of fruits and vegetables, under different storage conditions and postharvest treatments and processing. Addition of exogenous phytochemicals as natural preservatives for fresh and minimally processed fruits and vegetables, with antioxidant and antimicrobial activities and capable of inducing defense responses of tissues, is another interesting area for future research.

As research on antioxidant phytochemicals increases, inclusion of antioxidant capacity information on the product label should be considered necessary, in order to give more information to consumers on the antioxidant status that contributes to the overall quality of fresh fruits and vegetables. Development of intelligent packaging that indicates the changes in quality attributes of fresh fruits and vegetables will be also useful for consumers to know the antioxidant and microbial status of the packed fruit. With this information, consumers will be aware of the effect of treatments and storage on the bioactive compound content in fruit and vegetables, in order to choose the healthiest products.

References

Agar IT, Massantini R, Hess-Pierce B and Kader AA. 1999. Postharvest CO_2 and ethylene production and quality maintenance of fresh-cut kiwifruit slices. J Food Sci 64(3):433–440.

Agerlin-Peterson M. and Berends H. 1993. Ascorbic acid content of blanched sweet green pepper during chilled storage in modified atmospheres. LWT Food Sci Technol 197:546–549.

Awad MA, de Jager A and van Westing LM. 2000. Flavonoid and chlorogenic acid levels in apple fruit: characterisation of variation. Sci Hort Amsterdam 83(3-4):249–263.

Ayala-Zavala JF, Del-Toro-Sanchez L, Alvarez-Parrilla E and González-Aguilar GA. 2008a. High relative humidity in-package of fresh-cut fruits and vegetables: advantage or disadvantage considering microbiological problems and antimicrobial delivering systems? J Food Sci 73(4):R41–R47.

Ayala-Zavala JF, del-Toro-Sánchez L, Alvarez-Parrilla E, Soto-Valdez H, Martín-Belloso O, Ruiz-Cruz S and González-Aguilar GA. 2008b. Natural antimicrobial agents incorporated in active packaging to preserve the quality of fresh fruits and vegetables. Stewart Postharvest Rev 4:1–9.

Ayala-Zavala JF, Oms-Oliu G, Odriozola-Serrano I, González-Aguilar GA, Alvarez-Parrilla E and Martin-Belloso O. 2008c. Bio-preservation of fresh-cut tomatoes using natural antimicrobials. Eur Food Res Technol 226(5):1047–1055.

Ayala-Zavala JF, Wang SY, Wang CY and González-Aguilar GA. 2004. Effect of storage temperatures on antioxidant capacity and aroma compounds in strawberry fruit. LWT Food Sci Technol 37(7):687–695.

Ayala-Zavala JF, Wang SY, Wang CY and González-Aguilar GA. 2005. Methyl jasmonate in conjunction with ethanol treatment increases antioxidant capacity, volatile compounds and postharvest life of strawberry fruit. Eur Food Res Technol 221(6):731–738.

Ayala-Zavala JF, Wang SY, Wang CY and González-Aguilar GA. 2007. High oxygen treatment increases antioxidant capacity and postharvest life of strawberry fruit. Food Technol Biotechnol 45(2):166–173.

Barry-Ryan C and O'Beirne D. 1999. Ascorbic acid retention in shredded iceberg lettuce as affected by minimal processing. J Food Sci 64(3):498–500.

Barry-Ryan C and O'Beirne D. 2000. Effects of peeling methods on the quality of ready-to-use carrot slices. Int J Food Sci Technol 35(2):243–254.

Booth BH. 1951. Chromogenesis in stored carrots. J Sci Food Agric 2:353–358.

Cantos E, Espin JC and Tomas-Barberan FA. 2001. Postharvest induction modeling method using UV irradiation pulses for obtaining resveratrol-enriched table grapes: A new "functional" fruit? J Agric Food Chem 49(10):5052–5058.

Cisneros-Zevallos L, Saltveit ME and Krochta JM. 1995. Mechanism of surface white discoloration of peeled (minimally processed) carrots during storage. J Food Sci 60(2):320–323, 333.

Cocci E, Rocculi P, Romani S and Dalla Rosa M. 2006. Changes in nutritional properties of minimally processed apples during storage. Postharvest Biol Technol 39:265–271.

Corbo MR, Bevilacqua A, Campaniello D, D'Amato D, Speranza D and Sinigaglia N. 2009. Prolonging microbial shelf life of foods through the use of natural compounds and non-thermal approaches – a review. Int J Food Sci Technol 44:223–241.

Cortés C, Esteve MJ, Rodrigo D, Torregosa F and Frígola A. 2006. Changes of colour and carotenoids contents during high intensity pulsed electric field treatment in orange juices. Food Chem Toxicol 44:1932–1939.

Chanjirakul K, Wang CY, Wang SY and Siriphanich J. 2006. Effect of natural volatile compounds on antioxidant capacity and antioxidant enzymes in raspberries. Postharvest Biol Technol 40:106–115.

Fan X, Niemira BA and Prakash A. 2008. Irradiation of fresh fruits and vegetables. Food Technol-Chicago 62(3):36.

Fan XT, Toivonen PMA, Rajkowski KT and Sokorai KJB. 2003. Warm water treatment in combination with modified atmosphere packaging reduces undesirable effects of irradiation on the quality of fresh-cut iceberg lettuce. J Agric Food Chem 51(5):1231–1236.

Gil MI, Aguayo E and Kader AA. 2006. Quality changes and nutrient retention in fresh-cut versus whole fruits during storage. J Agric Food Chem 54(12):4284–4296.

González-Aguilar GA, Ayala-Zavala JF, Ruiz-Cruz S, Acedo-Felix E and Diaz-Cinco ME. 2004. Effect of temperature and modified atmosphere packaging on overall quality of fresh-cut bell peppers. LWT Food Sci Technol 37(8):817–826.

González-Aguilar GA, Buta JG and Wang CY. 2001. Methyl jasmonate reduces chilling injury symptoms and enhances colour development of 'Kent' mangoes. J Sci Food Agric 81(13):1244–1249.
González-Aguilar GA, Celis J, Sotelo-Mundo RR, de la Rosa LA, Rodrigo-Garcia J and Alvarez-Parrilla E. 2008. Physiological and biochemical changes of different fresh-cut mango cultivars stored at 5 degrees C. Int J Food Sci Technol 43(1):91–101.
González-Aguilar GA, Ruiz-Cruz S, Soto-Valdez H, Vazquez-Ortiz F, Pacheco-Aguilar R and Wang CY. 2005. Biochemical changes of fresh-cut pineapple slices treated with antibrowning agents. Int J Food Sci Technol 40(4):377–383.
González-Aguilar GA, Villegas-Ochoa MA, Martinez-Tellez MA, Gardea AA and Ayala-Zavala JF. 2007. Improving antioxidant capacity of fresh-cut mangoes treated with UV-C. J Food Sci 72(3):S197–S202.
Gorny J, Hess-Pierce B, Cifuentes R and Kader A. 2002. Quality changes in fresh-cut pear slices as affected by controlled atmospheres and chemical preservatives. Postharvest Biol Technol 24:271–278.
Howard LR and Hernandez-Brenes C. 1998. Antioxidant content and market quality of jalapeno pepper rings as affected by minimal processing and modified atmosphere packaging. J Food Quality 21(4):317–327.
Howard LA, Wong AD, Perry AK and Klein. 1999. β-Carotene and ascorbic acid retention in fresh and processed vegetables. J Food Sci 64(5):929–936.
Jeon M and Zhao Y. 2005. Honey in combination with vacuum impregnation to prevent enzymatic browning of fresh-cut apples. Int J Food Sci Nutr 56(3):165–176.
Ju Z, Yuan Y, Liu C, Zhan S and Wang M. 1996. Relationships among simple phenol, flavonoid and anthocyanin in apple fruit peel at harvest and scald susceptibility. Postharvest Biol Technol 8:83–93.
Karakurt Y and Huber DJ. 2003. Activities of several membrane and cell-wall hydrolases, ethylene biosynthetic enzymes, and cell wall polyuronide degradation during low-temperature storage of intact and fresh-cut papaya (*Carica papaya*) fruit. Postharvest Biol Technol 28(2):219–229.
Kidmose U, Hansen SL, Christensen LP, Edelenbos M, Larser E and Nørb R. 2006. Effects of Genotype, Root Size, Storage, and Processing on Bioactive Compounds in Organically Grown Carrots (Daucus carota L.). J. Food Sci. 69(9):S388–S394.
Koca N and Karedeniz F. 2008. Changes of bioactive compounds and anti-oxidant activity during cold storage of carrots. Int J Food Sci Technol. 43(11):2019–2025.
Kopas-Lane LM and Warthesen JJ. 1995. Carotenoid Photostability in Raw Spinach and Carrots During Cold Storage. J. Food Sci. 60(4):773–776.
Kubota C and Thomson CA. 2006. Controlled environments for production of value-added food crops with high phytochemical concentrations: lycopene in tomato as an example. Hortscience 41(3):522–525.
Lamikanra O. 2002. Fresh-cut fruits and vegetables; science, technology, and market. Boca Raton: CRC Press.
Lancaster JE, Reay PF, Norris J and Butler RC. 2000. Induction of flavonoids and phenolic acids in apple by UV-B and temperature. J Hort Sci Biotech 75(2):142–148.
Lee SK and Kader AA. 2000. Preharvest and postharvest factors influencing vitamin C content of horticultural crops. Postharvest Biol Technol 20(3):207–220.
Leja M, Mareczek A and Ben J. 2003. Antioxidant properties of two apple cultivars during long-term storage. Food Chem 80(3):303–307.
Li P and Barth MM. 1998. Impact of edible coating on nutritional and physiological changes in lightly-processed carrots. Postharvest Biol Technol 14:51–60.
Lopez-Rubira V, Conesa A, Allende A and Artes F. 2005. Shelf life and overall quality of minimally processed pomegranate arils modified atmosphere packaged and treated with UV-C. Postharvest Biol Technol 37(2):174–185.
Luo Y and Barbosa-Cánovas GV. 1997. Enzymatic browning and its inhibition in new apple cultivar slices using 4-hexylresorcinol in combination with ascorbic acid. Food Sci Technol Int 3:195–201.
McInerney JK, Seccafien CA, Stewart CM and Bird AR. 2007. Effects of high pressure processing on antioxidant activity, and total carotenoid content and availability, in vegetables. Innov Food Sci Emerg Technol 8:543–548.
Mohamed S, Kyi KMM and Sharif ZM. 1993. Protective effect of cysteine-HCl on vitamin C in dehydrated pickled/candied pineapples and guava. J Sci Food Agric 61:133–136.
Odriozola-Serrano I, Soliva-Fortuny R, Gimeno-Añó V and Martin-Belloso O. 2008a. Kinetic study of anthocyanins, vitamin C, and antioxidant capacity in strawberry juices treated by high-intensity pulsed electric fields. J Agric Food Chem 56:8387–8393.

Odriozola-Serrano I, Soliva-Fortuny R and Martin-Belloso O. 2008b. Changes of health-related compounds throughout cold storage of tomato juice stabilized by thermal or high intensity pulsed electric field treatments. Innov Food Sci Emerg Technol 9:272–279.

Oms-Oliu G, Aguilo-Aguayo I and Martin-Belloso O. 2006. Inhibition of browning on fresh-cut pear wedges by natural compounds. J Food Sci 71(3):S216–S224.

Ozoglu H and Bayindirli A. 2004. Inhibition of enzymatic browning in cloudy apple juice with selected antibrowning agents. Food Control 13:213–221.

Pan J, Vicente AR, Martinez GA, Chaves AR and Civello PM. 2004. Combined use of UV-C irradiation and heat treatment to improve postharvest life of strawberry fruit. J Sci Food Agric 84(14):1831–1838.

Patras A, Brunton NP, DaPieve S and Butler F. 2009. Impact of high pressure processing on total antioxidant activity, phenolic, ascorbic acid, anthocyanin content and colour of strawberry and blackberry purée. Innov Food Sci Emerg Technol doi:10.1016/j.ifset.2008.12.004.

Pizzocaro F Torreggiani D and Gilardi G. 1993. Inhibition of apple polyphenol oxidase (PPO) by ascorbic acid, citric acid and sodium chloride. J Food Proc Preserv 17:21–30.

Price SF, Breen PJ, Valladao M and Watson BT. 1995. Cluster sun exposure and quercetin in Pinot Noir grapes and wine. Am J Enol Vitic 46(2):187–194.

Rivera-Lopez J, Vazquez-Ortiz FA, Ayala-Zavala JF, Sotelo-Mundo RR and González-Aguilar GA. 2005. Cutting shape and storage temperature affect overall quality of fresh-cut papaya cv. ‘Maradol.’ J Food Sci 70(7):S482–S489.

Rivera-Pastrana DM, Bejar AAG, Martinez-Tellez MA, Rivera-Dominguez M and González-Aguilar GA. 2007. Postharvest biochemical effects of UV-C irradiation on fruit and vegetables. Revista Fitotecnia Mexicana 30(4):361–372.

Robles-Sánchez M, Gorinstein S, Martin-Belloso O, Astiazaran-Garcia H, González-Aguilar GA and Cruz-Valenzuela R. 2007. Minimal processing of tropical fruits: antioxidant potential and its impact on human health. Interciencia 32(4):227–232.

Robles-Sánchez RM, Rojas-Graü MA, Odriozola-Serrano I, González-Aguilar GA and Martín-Belloso O. 2009. Effect of minimal processing on bioactive compounds and antioxidant activity of fresh-cut ‘Kent’ mango (*Mangifera indica* L.). Postharvest Biol Technol 51(3):384–390.

Rojas-Grau MA, Grasa-Guillem R and Martin-Belloso O. 2007. Quality changes in fresh-cut Fuji apple as affected by ripeness stage, antibrowning agents, and storage atmosphere. J Food Sci 72(1):S036–S043.

Rojas-Grau MA, Sobrino-Lpez A, Soledad Tapia M and Martin-Belloso O. 2006. Browning inhibition in fresh-cut ‘Fuji’ apple slices by natural antibrowning agents. J Food Sci 71(1):S59–S65.

Ruiz-Cruz S, Islas-Osuna MA, Sotelo-Mundo RR, Vazquez-Ortiz F and González-Aguilar GA. 2007. Sanitation procedure affects biochemical and nutritional changes of shredded carrots. J Food Sci 72(2):S146–S152.

Sacchetti G, Cocci E, Pinnavaia G, Mastrocola D and Rosa MD. 2008. Influence of processing and storage on the antioxidant activity of apple derivatives. Int J Food Sci Technol 43(5):797–804.

Sánchez-Moreno C, Plaza L, Elez-Martínez P, De Ancos B, Martín-Belloso O and Cano MP. 2005. Impact of high pressure and pulsed electric fields on bioactive compounds and antioxidant activity of orange juice in comparison with traditional thermal processing. J Agric Food Chem 53(11):4403–4409.

Saltveit ME and Mencarelli F. 1988. Inhibition of ethylene synthesis and action in ripening tomato fruit by ethanol vapors. J Am Soc Hort Sci 113:572–576.

Sapers GM. 1993. Browning of foods. Control by sulfites, antioxidants and other means. Food Technol 47:75–84.

Schauss AG, Wu XL, Prior RL, Ou BX, Patel D, Huang DJ and Kababick JP. 2006. Phytochemical and nutrient composition of the freeze-dried Amazonian palm berry, *Euterpe oleracea* Mart. (Acai). J Agric Food Chem 54(22):8598–8603.

Seeram NP, Adams LS, Hardy ML and Heber D. 2004. Total cranberry extract versus its phytochemical constituents: antiproliferative and synergistic effects against human tumor cell lines. J Agric Food Chem 52(9):2512–2517.

Senesi E, Galvis A and Fumagalli G. 1999. Quality indexes and internal atmosphere of packaged fresh-cut pears (Abate Fetel and Kaiser varieties). Ital J Food Sci 2:111–120.

Soliva-Fortuny RC, Biosca-Biosca M, Grigelmo-Miguel N and Martín-Belloso O. 2002. Browning, polyphenol oxidase activity and headspace gas composition during storage of minimally processed pears using modified atmosphere packaging. J Agric Food Chem 82:1490–1496.

Soliva-Fortuny RC, Grigelmo-Miguel N, Odriozola-Serrano I, Gorinstein S, and Martín-Belloso O. 2001. Browning evaluation of ready-to-eat apples as affected by modified atmosphere packaging. J Agric Food Chem 49:3685–3690.

Srilaong V and Tatsumi Y. 2003. Changes in respiratory and antioxidative parameters in cucumber fruit (*Cucumis sativus* L.) stored under high and low oxygen concentrations. J Jap Soc Hort Sci 72(6):525–532.

Viuda-Martos M, Ruiz-Navajas Y, Fernández-López J and Pérez-Álvarez JA. 2008. Functional properties of honey, propolis, and royal jelly. J Food Sci 73(9):R117–R124.

Wolbang CM, Fitos JL and Treeby MT. The effect of high pressure processing on nutritional value and quality attributes of *Cucumis melo* L. Innov Food Sci Emerg Technol 9:196–200.

Wright KP and Kader AA. 1997a. Effect of controlled-atmosphere storage on the quality and carotenoid content of sliced persimmons and peaches. Postharvest Biol Technol 10:89–97.

Wright KP and Kader AA. 1997b. Effect of slicing and controlled-atmosphere storage on the ascorbate content and quality of strawberries and persimmons. Postharvest Biol Technol 10:39–48.

Yanuriati A, Savage GP and Rowe RN. 1999. The effects of ethanol treatment on the metabolism, shelf life and quality of stored tomatoes at different maturities and temperatures. J Sci Food Agric 79(7):995–1002.

Yao HJ and Tian SP. 2005. Effects of biocontrol agent and methyl jasmonate on postharvest diseases of peach fruit and the possible mechanisms involved. J Appl Microbiol 98:941–950.

Zhao W, Yang R, Wang M and Lu R. 2009. Effects of pulsed electric fields on bioactive components, colour and flavour of green tea infusions. Int J Food Sci Technol 44:312–321.

Zheng YH, Yang ZF and Chen XH. 2008. Effect of high oxygen atmospheres on fruit decay and quality in Chinese bayberries, strawberries and blueberries. Food Control 19(5):470–474.

12 Quality Loss of Fruits and Vegetables Induced by Microbial Growth

Saul Ruiz-Cruz* and Sofia Arvizu-Medrano

Introduction

Fruits and vegetables form an important component in human nutrition, being rich sources of sugars, vitamins, minerals, carotenoids, polyphenols, dietary fiber, and other phytochemicals that play a significant role in the health. Furthermore, fruits and vegetables are naturally contaminated with microorganisms, and many of these microorganisms possess pectin-degrading enzymes, enabling them to produce colonization by using fruit nutrients. Moreover, tissue damage caused by cutting or wounding causes cell damage, releasing nutrients and favoring growth of most types of microorganisms. They may also cause spoilage and affect the economic value of produce, not only by decreasing the quality (organoleptic and nutritional) and shelf-life of produce, but also (a matter of public health concern) by causing food-borne disease. Therefore, is important to prevent contamination and growth of microorganisms, in order to reduce degradation of nutrients and maintain fruit safety and sensory attributes. Some of these problems can be solved by improving preharvesting practices, and others need to be addressed through appropriate postharvest handling and processing.

Microbiology of Fresh Fruits and Vegetables

Total Microflora

Microorganisms are natural contaminants of fresh fruits and vegetables, and the types of initial total microflora present in the tissue are commonly related to those found in the environment. Moreover, vegetables can be contaminated during their growth and development, harvesting, processing, distribution, retail sale, and final preparation (Harris and others 2003). Also, small wounds or cuts in tissue occurring during harvesting, processing, and transportation provide easy access and growth of spoilage and pathogens microorganism (Spadaro and Gullino 2004; Francés and others 2006). They may cause spoilage and affect the economic value of the produce by decreasing quality and shelf-life. Depending on the pathogen present in the vegetable, it could cause a public health concern by causing food-borne disease (Nguyen-The and Carlin 2000).

Bacteria occur normally in fresh fruit and vegetable tissues. A wide variety of microorganisms have been found on fresh fruits and vegetables, which include mesophilic bacteria, lactic acid bacteria, coliforms, and yeasts and molds (Nguyen-The and

* Corresponding author: Saul Ruiz Cruz. Instituto Tecnológico de Sonora, Departamento de Biotecnología y Ciencias Alimentarias. 5 de febrero 818 Sur, colonia centro, Ciudad Obregón, Sonora (85000), México e-mail: sruiz@itson.mx.

Table 12.1. Microbial loads predominant in some fresh fruits and vegetables

Produce	APC	Total Coliforms	Molds and Yeasts	Reference
Spinach	5.8	1.5	–	Johnston and others 2005
Cilantro	6.1	1.8	–	
Parsley	5.6	2.3	–	
Lettuce	6.4–8.6	4.2–5.3	2.1–5.4	Grass and others 1994; Thunberg and others 2002; Erkan and Vural 2008
Carrots	5.2	4.3	3.1	Ruiz–Cruz and others 2006
Broccoli	6.3	2.2	3–3.5	Thunberg and others 2002; Stringer and others 2007
Cucumber	7.1	4.1	4.6	Koseky and Isobe 2007
Cantaloupe	6.6	3.0		Johnston and others 2005

APC: aerobic plate count.

Carlin, 1994). Numbers of microbial counts reported on fresh fruits and vegetables are within the range 10^1–10^9 cfu/g, depending upon the fruit or vegetable (Table 12.1). They include many psychrotrophs, such as some *Pseudomonas* and *Erwinia* species that are able to multiply below 10°C. They are mostly gram-negative motile rods, representatives of the Pseudomonadaceae and the Enterobacteriaceae (Martínez and others 2000), with *Pseudomonas* representing approximately 80–90% of the total flora (Nguyen-The and Carlin 2000). For example, fluorescent *Pseudomonas* is the predominant group isolated from carrots, spinach, endive, and chicory (Garg and others 1990; Jacques and Morris 1995; Van Outryve and others 2008). On the other hand, native microflora of fruits is mostly composed of molds and yeasts. Fungi such as *Botrytis cinerea* and *Aspergillus niger* and yeasts such as *Candida*, *Cryptococcus*, *Fabospora*, *Kluyveromyces*, *Pichia*, *Saccharomyces*, and *Zygosaccharomyces* are present in most fresh fruits (Chen 2002).

A recent study by Badosa and others (2008) reported that populations of aerobic plate counts varied from 10^1 to 10^9 cfu/g on fresh produce, with the lowest and highest counts recorded for fruits and sprouts, respectively. The highest incidence level of coliforms was found in ready-to-eat vegetables, ranging from 10^5 to 10^9 cfu/g, whereas yeasts and molds showed their highest incidence level between 10^5 and 10^6 cfu/g. Lactic acid bacteria are also present on fresh-cut produce and have been reported to occur in populations ranging from 10^1 to 10^6 log cfu/g in shredded carrots and mixed vegetables (Manzano and others 1995; Sinigaglia and others 1999; Ruiz-Cruz and others 2006).

Fruit and Vegetable Spoilage

Microbial spoilage appears to be one of the major causes of quality loss of fresh fruits and vegetables by formation of off-flavors, fermented aromas, and tissue decay. The shelf-life of many food products may be accurately predicted by quantifying the population of microbes present on the food product (Zhuang and others 2003). The

Table 12.2. Molds and bacteria associated with spoilage of fruits and vegetables

Organisms	Fruit and vegetable affected
Botrytis	Raspberries, strawberries, grapes, kiwifruit, pears, peaches, plums, cherries, carrots, lettuce, peas, and beans[1]
Penicillium	Onions,[1] citrus fruits[6]
Rhizopus	Raspberries, strawberries, apples, and pears[1]
Alternaria	Cabbage, carrots, peppers, tomatoes[2]
Aspergillus	Onions, tomatoes[2]
Fusarium	Asparagus, carrots[2]
Erwinia	Sprouts and many vegetables[3]
Pseudomonas	Celery, potatoes, lettuce, cabbage, carrots[4,5]

[1]Moss 2008.
[2]Bulgarelli and Brakett 1990.
[3]Rasch and others 2005.
[4]Marchetti and others 1992.
[5]Bennick and others 1998.
[6]Kombrink and Somssich 1995.

decay of vegetables may occur due to abiotic and biotic factors that include physical factors, action of their own hydrolytic enzymes, or microbial contaminants.

Abiotic spoilage is produced by different physical and chemical changes such as hydrolytic action of enzymes, oxidation of fats, breakdown of proteins, and a browning reaction between proteins and sugars. However, in this chapter we focus on microbial deterioration and their effects on bioactive compounds.

"Biotic" includes the microbial actions associated with bacteria, molds, and yeasts of fruits and vegetables and the normal processes of senescence. The species of microorganisms causing food spoilage largely depend upon different factors, for example, type and variety of fruits or vegetables and environmental conditions (storage, temperature, relative humidity, etc.). There are two types of microbial spoilage: (1) spoilage caused by plant pathogens that attack various parts of the plant used as foods, and (2) spoilage caused by saprophytes. Table 12.2 shows the most important microorganisms associated with spoilage of fruits and vegetables.

Microbial Spoilage

Of the many types of microorganisms, only a few affect the quality of fruits and vegetables. Microbial spoilage in fruits and vegetables is mainly caused by yeast and molds. Complex carbohydrates, such as cellulose, lignin, and pectin, limit the bacterial activity in these foods. Despite the high water activity of most fruits and vegetables, the low pH, especially of fruits, gives fungi a competitive advantage over the majority of bacteria. Vegetables, in contrast, have pH values closer to neutrality and, as well as fungi, bacteria play a significant role in their spoilage (Moss 2008). Fungal invasion of these tissues requires that the fungi overcome many barriers. Plants have evolved many mechanisms to prevent fungal attack on living tissue. These may involve physical barriers, chemical barriers, or the production of antifungal metabolites in response to microbial attack (Fawe and others 1998; Franklin and others 2009).

Bacterial Spoilage Characteristics

The types of spoilage caused by bacteria in fruits and vegetables are diverse; they include sensory changes, degradation of compounds, and formation of new substances such as acids, volatile compounds, and polymers. For example, the bacteria produce a set of enzymes such as pectinases, cellulases, proteases, and others that causes maceration and softening of tissue. Off-flavor development is common in contaminated tissues, caused by volatile compounds produced by microflora (Jay 1992).

Souring, Changes in Odor and Taste

Growth of most lactic acid bacteria (LAB) is enhanced in tissues with a pH from 6 to 7. However, many lactobacilli and *Oenococcus oeni* tolerate low-pH conditions (Schillinger and Holzapfel 2006). Some species of *Lactobacillus*, *Pediococcus*, and *Enterococcus* have shown the ability to grow in high-acidity products such as tomato juice (Di Cagno and others 2009). In fruit juices, tomato juice, and citrus products, growing of LAB may produce diacetyl from the citrate present, resulting in an unpleasant buttery flavor (Schillinger and Holzapfel 2006). Levels as low as 5 ppm diacetyl in concentrated orange juice spoil the product. Diacetyl and acetone could be good indicators of contamination of tomato juice by LAB in the early stages of growth (Juven and Weisslowicz 1981).

Lactic acid is generally present in spoiled canned, nonfermented vegetables, and it has been suggested that this can serve as a good indicator of microbial spoilage in this type of foods (Ackland and Reeder 1984).

Gas and Bloater Formation

Canned vegetables may be spoiled by LAB forming carbon dioxide from glucose and blowing the cans. In salad dressings, the formation of carbon dioxide caused by *Lactobacillus fructivorans* leads to bubble-filled dressing and/or blowing of the containers (Smittle and Flowers 1982; Meyer and others 1989). Bloater damage in brined cucumbers results from an increase in gas pressure inside the cucumbers during fermentation. The gas pressure is due to the combined effects of nitrogen trapped inside the cucumbers and carbon dioxide production (Fleming and Pharr 1980). Some LAB may either cause malic acid degradation to lactic acid and carbon dioxide, or produce carbon dioxide by other pathways (McFeeters and others 1984). On the other hand, it appears that the bacteria can enter the living plant tissue by different mechanisms and be harmful to the tissue. When the vegetables are brined, the bacteria multiply in the tissue as well as in the brine. For example, lactobacilli penetrate tomatoes primarily through the stem scar and multiply more rapidly in the fruit than in the brine. During fermentation of tomatoes and cucumbers, the Enterobacteriaceae are mostly suppressed by the lactic-acid-forming bacteria. However, if the latter are excluded during fruit disinfection, the enterobacteria continue to multiply, causing internal bloaters, an increase in pH, and fruit deterioration (Samish and others 1963).

Protein Swell

"Protein swell" is an unusual type of LAB spoilage, because it raises the pH of the spoiled product. This type of LAB spoilage was first reported by Meyer in 1956 in canned fish marinades (Schillinger and Holzapfel 2006). He called it "protein swell" and distinguished it from "carbohydrate swell," where the increase in acidity and CO_2 formation results from heterofermentative utilization of glucose. Normally in LAB spoilage, the pH of produce decreases because of lactic acid production. In "protein swell," proteins are decomposed by enzyme action, and then the decarboxylation of amino acids leads to enhanced CO_2 production. The decrease in acidity related to "protein swell" has been attributed to production of ammonia by bacterial deamination of amino acids. This spoilage type has been reported to affect anchovy-stuffed olives (Harmon and others 1987). *Lactobacillus brevis* has been reported as the main spoilage organism in the anchovy-stuffed olives. Since this species was not proteolytic, it was thought that autolytic enzymes persisting because of inadequate processing released amino acids that were further utilized by *Lactobacillus brevis* (Schillinger and Holzapfel 2006).

Soft Rot

Some psychrotrophs are able to grow in vegetable products, such as *Erwinia carotovora, Pseudomonas fluorescens, P. aeruginosa, P. luteola, Bacillus* spp., *Cytophaga johnsonae, Xanthomonas campestris, and Vibrio fluvialis* (Martínez and others 2000). The pectinolytic species cause soft tissue. Pectinolytic *Pseudomonas* are generally considered the primary cause of the spoilage of fresh produce stored at low temperature (4–5°C). These *Pseudomonas* are psychrotrophic and capable of growing and inducing soft rot of fresh produce at 10°C or below. Pectinolytic strains of *P. fluorescens* biovar II are well known for their ability to cause soft rot of fresh produce after harvest (Liao 2006). Pectinolytic *Erwinia*, in particular *E. carotovora* subsp. *carotovora*, is considered the most common organism associated with spoilage of most vegetables (Barras and others 1994; Tournas 2005). These bacteria use insects as dissemination vectors, for example, the fruit fly *Drosophila melanogaster* (Quevillon-Cheruel and others 2009).

Quorum sensing (QS) has been demonstrated in many bacteria. In this phenomenon, the bacteria produce chemical signal molecules to regulate expression of particular phenotypes, such as virulence factors, biofilm formation, or motility processes (Sperandio and others 2002; Annand and Griffiths 2003; Bearson and Bearson 2008; Zhang and others 2009). Proteolytic and pectinolytic activities in some gram-negative bacteria are regulated by quorum sensing, for instance, *P. aeruginosa* (Passador and others 1993), *Serratia liquefaciens* (Eberl and others 1996), and *Vibrio anguillarum* (Croxatto and others 2002).

Fungal Spoilage

Penicillium species are the major fungi causing spoilage of citrus and pome fruits. *Penicillium expansum* has been isolated from tomatoes, strawberries, avocados, mangoes, and grapes, indicating that this pathogen is present in a wide range of other fruits (Morales and others 2008). Apples and pears are commonly spoiled by this fungus, which causes a characteristic brown, spreading rot. *Penicillium solitum* is a less common but also important pathogen of pomaceous fruit. It is resistant to the fungicides used to control growth of *P. expansum*, and so its role in apple spoilage has increased in recent years (Pitt and others 1991). *Penicillium digitatum* produces destructive brown rots on oranges and less frequently on other types of citrus, but there is little evidence of a host defense (Macarisin and others 2007). *Penicillium brevicompactum* is commonly isolated from a wide range of fruit; however, it is not as aggressive as the pathogens mentioned previously. It has been reported to cause spoilage in stored apples and grapes, mushrooms, cassava and potatoes (Patiño and others 2007), and ginger (Overy and Frisvad 2005).

Dried fruits and concentrated fruit juice are very susceptible to spoilage by yeasts, because of their relatively high sugar and moisture contents. Some species of *Zygosaccharomyces*, *Torulaspora*, and *Lachancea* have been associated with spoilage of these products (Dijksterhuis and Samson 2006).

Rhizopus-soft rot is a threat in all postharvest situations including storage, marketing, and transport of crops. It causes soft rot of avocados, cassava, crucifers, pulses, yams, and sweet potatoes. This spoilage type is mainly caused by *Rhizopus stolonifer* and to a lesser extent by *R. oryzae*, *Mucor piriformis*, and *Gilbertella persicaria* (Dijksterhuis and Samson 2006).

Botrytis cinerea is responsible for gray mold disease in more than 200 host plants. This necrotrophic fungus displays the capacity to kill host cells through the production of toxins and reactive oxygen species and the induction of a plant-produced oxidative burst. Thanks to an arsenal of degrading enzymes, *B. cinerea* is then able to feed on various plant tissues (Choquer and others 2007).

Characteristics of Spoilage Organisms

Pseudomonas

These gram-negative rots are typically present in soil and water and widely distributed among foods, especially vegetables. They are the most important group of bacteria that cause spoilage of fresh fruits and vegetables and refrigerated fresh foods, because many species are psychrotrophic (Jay 1992). They are able to grow in simple media and are characterized by their ability to metabolize a wide range of organic compounds.

Pectinolytic fluorescent pseudomonads, mainly *Pseudomonas fluorescens* and *Pseudomonas viridiflava*, are responsible for postharvest rot of fruits and vegetables during cold storage (Brocklehurst and Lund 1981) as well as wholesale and retail markets (Liao and Wells 1987). The ability of these pseudomonads to cause maceration of plant tissues is primarily due to their ability to produce pectinolytic enzymes capable of degrading pectic components of plant cell walls (Liao and others 1988, 1997). Some enzymes include cellulases, xylanases, and glycoside hydrolases and lipoxygenases (Zhuang and others 1994). Sharma (2005) reported that *Pseudomonas putida* PpG7 strain can utilize limonin, a highly oxygenated triterpenoid compound that is the major cause of delayed bitterness in citrus juice, as a sole source of carbon and energy. It was concluded that the microorganism possesses an enzyme that acts on limonin to convert it into nonbitter metabolites.

Erwinia

Erwinia spp. are commonly present on vegetables at harvest. This bacterium has the characteristic of producing extracellular enzymes that degrade plant cell walls. Moreover, many *Erwinia* spp. such as *E. carotovora* are capable of using compounds as energy that are normally not utilized by most common bacteria (Jay 1992). Moreover, these species produce many enzymes (such as pectate lyase, polygacturonase, cellulases, and proteases) that play an important role in plant and tissue maceration (Chen 2002).

Lactic Acid Bacteria

These bacteria are commonly found on crop surfaces. Lactic acid bacteria may appear abundantly in the same vegetable in one field and rarely in others. They are found more frequently in low-growing vegetables than in tree-borne fruits. In cucumbers the bacteria are more often in the tissue close to the periphery and less often in the central core. In tomatoes their frequency is closer to the stem scar and the central core of the fruit, decreasing toward the fruit periphery. The characteristic of these bacteria is the production of lactic acid from sugar fermentation. They are not able to degrade many compounds, and they use the sugars produced by other microorganisms, resulting in the production of alcohol, organic acid, and carbon dioxide (Chen 2002).

Molds and Yeasts

The molds commonly associated with the spoilage of fruits and vegetables include *Botrytis cinerea* and *Aspergillus niger*, among others. The role of molds as deteriorative organisms is well known. They produce a cocktail of enzymes such as hydrolases that are able to break down cell walls of produce, causing spoilage (Walton 1994). The blue mold rot caused by fungi includes various species of *Penicillium*, *Botrytis cinerea*, and *Monilinia laxa*, as well as other fungi. Some of these fungi produce mycotoxins that are of human health concern (Spadaro and Gullino 2004). Under favorable conditions, yeasts are very effective at fermenting single sugars to produce alcohol and other volatiles that affect fruit quality (Barnett and others 2000). Some yeasts found in fresh fruits and vegetables include *Candida*, *Cryptococcus*, *Fabospora*, *Kluyveromyces*, *Pichia*, *Saccharomyces*, and *Zygosaccharomyces* (Chen 2002).

Degradation of Nutrients by Microorganisms

Many microorganisms possess characteristics enabling colonization of the produce. For example, they produce pectin-degrading enzymes, which enhance tissue softening and breakdown (Watada and others 1996). Moreover, tissue damage caused by cutting and wounding breaks cells, favoring release of nutrients that are used by most types of microflora. For example, microbial load in fresh-cut produce is greater than that in intact produce (Toivonen and DeEll 2002).

Fruits and vegetables contain nutrients such as simple sugars and amino acids and energy reserves such as starches and carbon that are readily used by microorganisms for their growth (Chen 2002). Fresh produce also contains others biopolymers such as cuticle (protective layer of fruits and vegetables), lipids, proteins, vitamins, and phenolic compounds. Microorganisms can secrete substances (such as enzymes, toxins, and polysaccharides) to convert these polymers into soluble products that can be transported into microbial cell for use (Agrios 1997; Chen 2002). These substances secreted by the microorganisms play a significant role in the quality (rot or spoilage) and safety of fresh fruits and vegetables (Liao and others 1999).

All bacteria utilize nutrients as sources of the energy required for all of the biosynthetic processes that bacteria use for their maintenance and reproduction. Bacteria produce enzymes that allow them to oxidize different energy sources; however, the energy sources that different bacteria use depend on the specific enzymes that each bacterium produces. For example, many bacteria often use carbohydrates (such as glucose, a monosaccharide or simple sugar) as energy sources, because many bacteria possess the enzymes required for the degradation and oxidation of these sugars. Few bacteria are able to use complex carbohydrates such as disaccharides (sucrose) or polysaccharides (starch). Disaccharides and polysaccharides are simple sugars that are linked by glycosidic bonds; bacteria must produce enzymes to cleave these bonds and produce simple sugars that can be transported into the cell. If bacteria are not able to produce these enzymes, then the complex carbohydrates are not used. For example, starch is a large polysaccharide consisting of long chains of monomeric glucose (α-amylose and amylopectin) linked by glycosidic bonds. Bacteria that use starch must first produce several types of enzymes called amylases (α-amylases, glucoamylases, β-amylases, and others such as exo-α1,4-gluconases) that break glycosidic bonds and thus free monomeric glucose is produced and used directly by the microorganisms (Nigan and Singh 1995; Chen 2002).

Landete and others (2009) reported that *Lactobacillus plantarum* have the ability to metabolize phenolic compounds found in olive products (such as oleuropein, hydroxytyrosol, and tyrosol, as well as vanillic, *p*-hydroxybenzoic, sinapic, syringic, protocatechuic, and cinnamic acids). For example, oleuropein was metabolized mainly to hydroxytyrosol, whereas protocatechuic acid was decarboxylated to catechol by the enzymatic actions.

Control of Fruit and Vegetable Spoilage

Fruits and vegetables are exposed during their growth in the field and processing steps to microorganisms with spoilage potential. Food preservation implies designing an environment that is hostile to microorganisms with the capacity to change sensory characteristics, and then:

- Inhibit their growth
- Shorten their survival, or
- Cause their inactivation

The most important methods commonly used in produce preservation are low temperature, a_w reduction, low acidity, and preservatives (e.g. organic acids and sulfite). The microbial stability and the sensory quality of most foods are based on passing over a combination of hurdles.

Modified-atmosphere packaging and subsequent storage at low temperature (5–7 days at 1–8°C) have been developed as adequate techniques to prolong shelf-life of raw vegetables. However, in recent decades other methods have been developed to maintain the quality of foods.

Physical Processes

Several decontamination methods exist, but the most versatile treatment among them is processing with ionizing radiation. Decontamination of food by ionizing radiation is a safe, efficient, environmentally clean, and energy-efficient process. Shelf-life improvement of mushrooms and insect disinfestations in dried fruits, nuts, and certain fresh fruits have been described (Thomas 1988). Radiation treatment at doses of 2–7 kGy can effectively eliminate non-spore-forming bacteria (Farkas 1998). Radiation doses of 25 krad for strawberries and 75 krad for carrots were observed as optimum doses that did not cause significant changes in organoleptic quality (Ismail and Afifi 1976). Electron beam also has been applied to reduce yeasts and molds in dried fruits and nuts. Ic and others (2007) reported that 1.09–1.59 kGy were necessary to reach a decimal inactivation of yeasts and molds naturally present in these products.

Intense light pulses (ILP) is a new technology intended for decontamination of food surfaces by killing microorganisms using brief, high-frequency pulses of an intense broad-spectrum light that is rich in UV-C. This treatment alone does not increase vegetable shelf-life, in spite of the reduction in the initial microbial load (Gómez-López and others 2005).

Vacuum-steam-vacuum (VSV) treatment resulted in a 1.0-log reduction of aerobic mesophilic bacteria, a 2.0-log reduction of yeasts and molds, and a 1.5-log reduction of *Pseudomonas* spp. on cantaloupe surfaces. VSV treatment significantly reduced transfer of yeasts and molds and *Pseudomonas* spp. from whole cantaloupe surface to fresh-cut pieces during preparation ($P < 0.05$). Texture and color of the fresh-cut pieces prepared from the VSV-treated whole melons were similar to the controls (Ukuku and others 2006).

Chemical

In the past few years, a wide number of studies have been performed in order to investigate the potential use of plants and plant extracts as sources of antimicrobial compounds. Essential oils are naturally occurring volatile components constituted by variable mixtures of principally terpenoids, especially thymol, *p*-cymene, carvacrol, α-terpinyl acetate, *cis*-myrtanol, menthol, menthyl acetate, carvone, and menthone (McKay and Blumberg 2006; Soković and others 2009). Some essential oils possess great antifungal potential and could be used as a natural preservatives and fungicides, such as oils of *Thymus* and *Mentha* species. Treatment with thymol significantly reduced the severity of decay in strawberries stored at 10°C. Treatments with menthol or eugenol also suppressed fungal growth of strawberries, but to a lesser extent. Strawberries treated with this oil also maintained better fruit quality with higher levels of sugars, organic acids, phenolic compounds, anthocyanins, flavonoids, and oxygen radical absorbance capacity than the untreated fruits (Wang and others 2007).

Novel monosubstituted carbohydrate fatty acid (CFA) esters and ethers have shown a bacteriostatic effect. Among the carbohydrate derivatives synthesized, lauric ether of methyl alpha-D-glucopyranoside and lauric ester of methyl α-D-mannopyranoside showed MIC values of 0.04 mM and were generally more active against gram-positive bacteria than against gram-negative bacteria (Nobmann and others 2009).

An alternative packaging is the combination of food-packaging materials with antimicrobial substances to control microbial surface contamination of foods. For both migrating and nonmigrating antimicrobial materials, intensive contact between the food product and packaging material is required and therefore potential food applications include especially vacuum or skin-packaged products (Vermeiren and others 2002).

Edible coatings from renewable sources can function as barriers to water vapor, gases, and other solutes and also as carriers of many functional ingredients, such as antimicrobial and antioxidant agents, thus enhancing quality and extending shelf life of fresh and minimally processed fruits and vegetables (Lin and Zhao 2007). Edible coatings may also be used in processed fruits and vegetables for improving structural integrity of frozen fruits and vegetables and preventing moisture absorption and oxidation of freeze-dried fruits or vegetables (Baker and others 1994).

The materials used in these type of films include lipids, polysaccharides, and proteins. Starch (Maizura and others 2007), methylcellulose (Olivas and others 2003), hydroxypropyl cellulose (Brindle and Krochta 2008), chitosan (No and others 2007), xanthan gum (Mei and others 2002), alginate or zein (Zapata and others 2008), and soy protein (Park and others 2001) have been used for edible coatings.

Biological

Bacteriocins are peptides or small proteins with antimicrobial activity, produced by different groups of bacteria. Many lactic acid bacteria produce bacteriocins with rather broad spectra of inhibition. Bacteriocins can be added to foods in the form of concentrated preparations as food preservatives, or they can be produced *in situ* by bacteriocinogenic cultures. Several bacteriocins show additive or synergistic effects when used in combination with other antimicrobial agents, including chemical preservatives and natural phenolic compounds as well as other antimicrobial proteins. The combination of bacteriocins and physical treatments, for example high-pressure processing, also offers a good opportunity for more effective preservation of foods (Gálvez and others 2007).

The effectiveness of bacteriocins is often a function of environmental factors such as pH, temperature, food composition, structure, and food microflora (De Vuyst and Leroy 2007). A novel bacteriocin-like substance produced by *Bacillus licheniformis* P40 inhibits the activity of the soft rot bacterium *Erwinia carotovora*. This compound caused a bactericidal effect on the pathogen cells at a 30 μg/mL concentration (Cladera-Olivera and others 2006).

The application of antagonist bacteria has been suggested as a biological control for fungal spoilage. Sathe and others (2007) reported the antifungal activity of *Weissella paramesenteroides* and *Lactobacillus paracollinoides* and *L. plantarum* against a wide range of food-spoilage fungi, including *Aspergillus flavus*, *Fusarium graminearum*, *Rhizopus stolonifer*, and *Botrytis cinerea*. *Pediococcus pentosaceus* inhibited the growth of *Penicillium expansum* in apple (Rouse and others 2008). Partial characterization of the antifungal compounds indicates that their activity is likely to be because of production of antifungal peptides.

The yeasts also have been evaluated for antifungal activity. Spadaro and others (2008) reported that *Hanseniaspora uvarum*, *Rhodotorula* spp., and *Metschnikowia pulcherrima* reduced the development of *P. expansum* on apples. In this work the biocontrol effectiveness was assessed on four apple cultivars, Golden Delicious, Stark Delicious, Granny Smith, and Royal Gala. The efficacy was higher on the cv. Golden Delicious.

The applications of bacteriophages to biofilm of spoilage microorganisms have been suggested. Phage jIBB-PF7A is highly efficient in removing *Pseudomonas fluorescens* biofilms. Biomass removal due to phage activity varied between 63 and 91% depending on the biofilm age and the conditions under which the biofilm had been formed (static or dynamic conditions and with or without media renewal every 12 hr) and the phages applied (Sillankorva and others 2008).

Conclusions

The microbial spoilage processes in fruits and vegetables are associated with sensory and nutritional deterioration such as poor texture, off-flavors, and nutrient loss. Generally, the causes of such changes are associated with enzymatic activities, and sometimes microorganisms are considered to be source of these degradative enzymes. However, if there is little information concerning changes of the most important quality factors such as color, odor, and texture, information about the enzymatic deterioration of the components of nutrition is even scarcer.

On the other hand, although the use of simple sugars by microorganisms is well known, our understanding of the microbial degradation of phytochemicals is still limited. Therefore, the effect of microorganisms and microbial enzymes on the degradation of phytochemicals in fruits and vegetables is an attractive area for future research.

References

Ackland MR and Reeder JE. 1984. A rapid chemical spot test for the detection of lactic acid as an indicator of microbial spoilage in preserved foods. J Appl Bacteriol 56:415–419.

Agrios G. 1997. Plant Pathology, 4th ed. New York: Academic Press.

Annand SK and Griffiths MW. 2003. Quorum sensing and expression of virulence in *Escherichia coli* O157:H7. Int J Food Microbiol 85(1–2):1–9.

Badosa E, Trias R, Pares D, Pla M and Montesinos E. 2008. Microbiological quality of fresh fruit and vegetable products in Catalonia (Spain) using normalised plate-counting methods and real time polymerase chain reaction (QPCR). J Sci Food Agric 88(4):605–611.

Baker RA, Baldwin EA and Nisperos Carriedo MO. 1994. Edible Coatings and Films to Improve Food Quality. Lancaster, PA: Technomic, pp. 89–104.

Barras F, Van Gijsegem F and Chatterjee AK. 1994. Extracellular enzymes and pathogenesis of soft-rot *Erwinia*. Ann Rev Phytopathol 32:201–234.

Barnett JA, Payne RW and Yarrow D. 2000. Yeasts: Characteristics and Identifications, 3rd ed. Cambridge, UK: Cambridge University Press, pp. 395–401.

Bearson BL and Bearson SM. 2008. The role of the QseC quorum-sensing sensor kinase in colonization and norepinephrine-enhanced motility of *Salmonella enterica* serovar Typhimurium. Microb Pathogenesis 44(4):271–278.

Bennick MHJ, Vorstman W, Smid EJ and Gorris LGM. 1998. The influence of oxygen and carbon dioxide on the growth of prevalent Enterobacteriaceae and *Pseudomonas* species isolated from fresh and controlled-atmosphere-stored vegetables. Food Microbiol 15(5):459–469.

Brindle LP and Krochta JM. 2008. Physical properties of whey protein-hydroxypropylmethylcellulose blend edible films. J Food Sci 73(9):446–454.

Brocklehurst TF and Lund BM. 1981. Properties of pseudomonads causing spoilage of vegetables stored at low temperatures. J Appl Bacteriol 50:259–266.

Bulgarelli MA and Brackett RE. 1990. The importance of fungi in vegetables. In: Akora DK, Mukerji KG, Marth ED, editors. Handbook of Applied Mycology, Volume 3, Foods and Feeds. New York: Marcel Dekker, pp. 179–199.

Chen J. 2002. Microbial enzymes associated with fresh-cut produce. In: Lamikanra O, editor. Fresh-Cut Fruits and Vegetables: Science, Technology and Market. Boca Raton, FL: CRC Press, pp. 249–266.

Choquer M, Fournier L, Kunz C, Levis C, Pradier JM, Simon A and Viaud M. 2007. *Botrytis cinerea* virulence factors: new insights into a necrotrophic and polyphageous pathogen. FEMS Microbiol Lett 277(1):1–10.

Cladera-Olivera F, Caron GR, Motta AS, Souto AA and Brandelli A. 2006. Bacteriocin-like substance inhibits potato soft rot caused by *Erwinia carotovora*. Can J Microbiol 52(6):533–539.

Croxatto A, Chalker VJ, Lauritz J, Jass J, Hardman A, Williams P, Camara M and Milton DL. 2002. VanT, a homologue of *Vibrio harveyi* LuxR, regulates serine, metalloprotease, pigment, and biofilm production in *Vibrio anguillarum*. J Bacteriol 184:1617–1629.

De Vuyst L and Leroy F. 2007. Bacteriocins from lactic acid bacteria: production, purification, and food applications. J Mol Microbiol Biotechnol 13:194–199.

Di Cagno R, Surico RF, Paradiso A, De Angelis M, Salmon JC, Buchin S, De Gara L and Gobbetti M. 2009. Effect of autochthonous lactic acid bacteria starters on health-promoting and sensory properties of tomato juices. Int J Food Microbiol 128(3):473–483.

Dijksterhuis J and Samson RA. 2006. Zygomycetes. In: Blackburn C, editor. Food Spoilage Microorganisms. Cambridge: Woodhead, pp. 415–436.

Eberl L, Winson MK, Sternberg C, Stewart GSAB, Christiansen G, Chhabra SR, Bycroft B, Williams P, Molin S and Givskov M. 1996. Involvement of *N*-acyl-L-homoserine lactone autoinducers in controlling the multicellular behaviour of *Serratia liquefaciens*. Mol Microbiol 20:127–136.

Erkan ME and Vural A. 2008. Investigation of microbial quality of some leafy green vegetables. J Food Technol 6(6):285–288.

Farkas J. 1998. Irradiation as a method for decontaminating food. A review. Int J Food Microbiol 128(3):440–445.

Fawe A, Abou-Zaid M, Menzies JG and Bélanger RR. 1998. Silicon-mediated accumulation of flavonoid phytoalexins in cucumber. Phytopathology 88(5):396–401.

Fleming HP and Pharr DM. 1980. Mechanism for bloater formation in brined cucumbers. J Food Sci 45:1595–1600.

Francés J, Bonaterra A, Moreno MC, Cabrefiga J, Badosa E and Montesinos E. 2006. Pathogen aggressiveness and postharvest biocontrol efficiency in *Pantoea agglomerans*. Postharvest Biol Technol 39:299–307.

Franklin G, Conceicão LF, Kombrink E and Dias AC. 2009. Xanthone biosynthesis in *Hypericum perforatum* cells provides antioxidant and antimicrobial protection upon biotic stress. Phytochem 70(1):60–68.

Gálvez A, Abriouel H, López RL and Ben Omar N. 2007. Bacteriocin-based strategies for food biopreservative. Int J Food Microbiol 120(1–2):51–70.

Garg N, Churey JJ and Splittstoesser DF. 1990. Effects of processing conditions on the microflora of fresh-cut vegetables. J Food Prot 53:701–703.

Gómez-López VM, Devlieghere F, Bonduelle V and Debevere J. 2005. Intense light pulse decontamination of minimally processed vegetables and their shelf-life. Int J Food Microbiol 103(1):79–89.

Grass MH, Druet-Michaud C and Cerf O. 1994. La flore bacterienne des feuiles de salade fraiche. Sci des aliments 14(2):173–188.

Harris LJ, Farber JN, Beuchat LR, Parish ME, Suslow TV, Garrett EH and Busta FF. 2003. Outbreaks associated with fresh produce: incidence, growth, and survival of pathogens in fresh and fresh-cut produce. Compreh Rev Food Sci Food Saf 2(suppl):78–141.

Harmon SM, Kautter DA and McKee C. 1987. Spoilage of anchovy-stuffed olives by heterofermentative lactobacilli. J Food Saf 8:205–210.

Ic E, Kottapalli B, Maxim J and Pillai SD. 2007. Electron beam radiation of dried and nuts to reduce yeast and mold bioburden. J Food Prot 70(4):981–985.

Ismail FA and Afifi SA. 1976. Control of postharvest decay in fruits and vegetables by irradiation. Nahrung 20(6):585–592.

Jacques MA and Morris CE. 1995. Bacterial population dynamics and decay on leaves of different ages of ready-to-use broad leaved endive. Int J Food Sci Technol 30:221–236.

Jay JM. 1992. Modern Food Microbiology, 4th ed. New York: Chapman & Hall, pp. 13–37.

Johnston LM, Jaykus LA, Moll D, Martínez MC, Anciso J, Mora B and Moe CL. 2005. A field study of the microbiological quality of fresh produce. J Food Prot 68(9):1840–1847.

Juven BJ and Weisslowicz H. 1981. Chemical changes in tomato juices caused by lactic acid bacteria. J Food Sci 46:1543–1545.

Kombrink E and Somssich IE. 1995. Defense responses of plants to pathogens. Adv Bot Res 21:1–24.

Koseky S and Isobe S. 2007. Microbial control of fresh produce using electrolyzed water. JARQ 41(4):273–282.

Landete JM, Curiel JA, Rodríguez H, de las Rivas B and Muño R. 2008. Study of the inhibitory activity of phenolic compounds found in olive products and their degradation by *Lactobacillus plantarum* Straits. Food Chem 107(1):320–326.

Liao CH. 2006. *Pseudomonas* and related genera. In: Blackburn C, editor. Food Spoilage Microorganisms. Cambridge: Woodhead, pp. 507–540.

Liao CH, Hung HY and Chatterjee AK. 1988. An extracellular pectate lyase is the pathogenicity factor of the soft-rotting bacterium *Pseudomonas viridiflava*. Mol Plant Microbe Interac 1:199–206.

Liao CH, Revear L, Hotchkiss A and Savary B. 1999. Genetic and biochemical characterization of an exopolygalacturonase and a pectate lyase from *Yersinia enterocolitica*. Can J Microbiol 45:396–403.

Liao CH, Sullivan J, Grady J and Wong LJC. 1997. Biochemical characterization of pectate lyases produced by fluorescent pseudomonads associated with spoilage of fresh fruits and vegetables. J Appl Microbiol 83:10–16.

Lin D and Zhao Y. 2007. Innovations in the development and application of edible coatings for fresh and minimally processed fruits and vegetables. Comprehen Rev Food Sci Food Saf 6:60–75.

Macarisin D, Cohen L, Eick A, Rafael G, Belausov E, Wisniewski M and Droby S. 2007. *Penicillium digitatum* suppresses production of hydrogen peroxide in host tissue during infection of citrus fruit. Phytopathology 97(11):1491–1500.

Maizura M, Fazilah A, Norziah MH and Karim AA. 2007. Antibacterial activity and mechanical properties of partially hydrolyzed sago starch-alginate edible film containing lemongrass oil. J Food Sci 72(6):324–330.

Manzano M, Citterio B, Maifreni M, Paganessi M and Comi G. 1995. Microbial and sensory quality of vegetables for soup packaged in different atmospheres. J Food Sci Agric 67(4):521–529.

Marchetti R, Casadei MA and Guerzoni ME. 1992. Microbial population dynamic in ready-to-eat vegetable salads. Ital J Food Sci 2:97–108.

Martínez A, Díaz RV and Tapia MS. 2000. Microbial Ecology of Spoilage and Pathogenic Flora Associated to Fruits and Vegetables. In: Alzamora SM, Tapia MA and López-Malo A, editors. Minimally Processed Fruits and Vegetables, Fundamental Aspect and Applications. Maryland: Aspen, pp. 43–62.

McFeeters RF, Fleming HP and Daeschel MA. 1984. Malic acid degradation and brined cucumber bloating. J Food Sci 49:999–1002.

McKay DL and Blumberg JB. 2006. A review of the bioactivity and potential health benefits of peppermint tea (*Mentha piperita* L.). Phytother Res 20(8):619–633.

Mei Y, Zhao Y, Yang J and Furr HC. 2002. Using edible coating to enhance nutritional and sensory qualities of baby carrots. J Food Sci 67:1964–1968.

Meyer RS, Grant MA, Luedecke LO and Leung HK. 1989. Effects of pH and water activity on microbiological stability of salad dressing. J Food Prot 52:477–479.

Morales H, Marin S, Obea L, Patiño B, Doménech M, Ramos AJ and Sanchis V. 2008. Ecophysiological characterization of *Penicillium expansum* population in Lleida (Spain). Int J Food Microbiol 122(3):243–252.

Moss MO. 2008. Fungi, quality and safety issues in fresh fruits and vegetables. J Appl Microbiol 104:1239–1243.

No HK, Meyers SP, Prinyyawiwatkul W and Xu Z. 2007. Applications of chitosan for improvement of quality and shelf life of foods: a review. J Food Sci 27(5):87–100.

Nobmann P, Smith A, Dunne J, Henehan G and Bourke P. 2009. The antimicrobial efficacy and structure activity relationship of novel carbohydrate fatty acid derivatives against *Listeria* spp. and food spoilage microorganisms. Int J Food Microbiol 128(3):440–445.

Nigan P and Singh D. 1995. Enzyme and microbial systems involved in starch processing. Enzyme Microb Technol 17:770–778.

Nguyen-The C and Carlin F. 1994. The microbiology of minimally processed fresh fruits and vegetables. Crit Rev Food Sci Nutr 34:371–401.

Nguyen-The C and Carlin F. 2000. Fresh and processed vegetables. In: Lund BM, Baird-Parker TC, Gould GW, editors. The Microbiological Safety and Quality of Food. Maryland: Aspen, pp. 620–684.

Olivas GI, Rodríguez JJ and Barbosa-Cánovas GV. 2003. Edible coatings composed of methylcellulose, stearic acid, and additives to preserve quality of pear wedges. J Food Proc Preserv 27(4):299–320.

Overy DP and Frisvad JC. 2005. Mycotoxin production and postharvest storage rot of ginger (*Zingiber officinale*) by *Penicillium brevicompactum*. J Food Prot 68:607–609.

Park SK, Rhee CO, Bae DH and Hettiarachchy NS. 2001. Mechanical properties and water-vapor permeability of soy-protein films affected by calcium salts and glucono-D-lactone. J Agric Food Chem 49(5):2308–2312.

Passador L, Cook JM, Gambello MJ, Rust L and Iglewski BH. 1993. Expression of *Pseudomonas aeruginosa* virulence genes requires cell-to-cell communication. Science 260:1127–1130.

Patiño B, Medina A, Doménech M, González-Jaén MT, Jiménez M and Vázquez C. 2007. Polymerase chain reaction (PCR) identification of *Penicillium brevicompactum*, a grape contaminant and mycophenolic acid producer. Food Addit Contam 24(2):165–172.

Pitt JI, Spotts RA, Holmes RJ and Cruickshank RH. 1991. *Penicillium solitum* revived, and its role as a pathogen of pomaceous fruit. Phytopathology 81:1108–1112.

Quevillon-Cheruel S, Leulliot N, Acosta Muniz C, Vincent M, Gallay J, Argentini M, Cornu D, Boccard F, Lemaitre B and van Tilbeurgh H. 2009. EVF, a virulence factor produced by the *Drosophila pathogen Erwinia carotovora is* a S-palmitoylated protein with a new fold that binds to lipid vesicles. J Biol Chem. 284(6):3552–3562.

Rasch M, Andersen JB, Nielsen KF, Flodgaard LR, Christensen H, Givskov M and Gram L. 2005. Involvement of bacterial quorum-sensing signals in spoilage of bean sprouts. Appl Environ Microbiol 71(6):3321–3330.

Rouse S, Harnett D, Vaughan A and Van Sinderen D. 2008. Lactic acid bacteria with potential to eliminate fungal spoilage in foods. J Appl Microbiol 104(3):915–923.

Ruiz-Cruz S, Luo Y, González RJ, Yang T and González-Aguilar GA. 2006. Acidified sodium chlorite as an alternative to chlorine to control microbial growth on shredded carrots while maintaining quality. J Sci Food Agric 86(12):1887–1893.

Samish Z, Etinger-Tulczynska R and Bick M. 1963. The microflora within the tissue of fruits and vegetables. J Food Sci 28(3):259–266.

Sathe SJ, Nawani NN, Dhakephalkar and Kapadnis BP. 2007. Antifungal lactic acid bacteria with potential to prolong shelf-life of fresh vegetables. J Appl Microbiol 103:2622–2628.

Schillinger U and Holzapfel WH. 2006. Lactic acid bacteria. In: De W. Blackburn C, editor. Food Spoilage Microorganisms. Cambridge: Woodhead, pp. 541–578.

Sharma M. 2005. Transposon mutagenesis of gene involved in limonin degradation in *Pseudomonas putida*. M.C. diss., Patiala.

Sillankorva S, Neubauer P and Azeredo J. 2008. *Pseudomonas fluorescens* biofilms subjected to phage phiIBB-PF7A. BMC Biotechnol 8:79.

Sinigaglia M, Albenzio M and Corbo MR. 1999. Influence of process operations on shelf life and microbial populations of fresh cut vegetables. J Ind Microbiol Biotechnol 23:484–488.

Smittle RB and Flowers RS. 1982. Acid tolerant microorganisms involved in the spoilage of salad dressings. J Food Prot 45:977–983.

Soković MD, Vukojević J, Marin PD, Brkić DD, Vajs V and van Griensven LJ. 2009. Chemical composition of essential oils of *Thymus* and *Mentha* species and their antifungal activities. Molecules 14(1):238–249.

Spadaro D and Gullino ML. 2004. State of the art and future prospects of the biological control of postharvest fruit diseases. Int J Food Microbiol 91:185–194.

Spadaro D, Frati S, Garibaldi A and Gullino ML. 2008. Efficacy of biocontrol yeasts against *Penicillium expansum* and patulin in different cultivars of apple in postharvest. Paper presented at 16th IFOAM Organic World Congress, Modena, Italy, June 16–20, 2008. Available at: http://orgprints.org/view/projects/conference.html.

Sperandio V, Torres AG and Kaper JB. 2002. Quorum sensing *Escherichia coli* regulators B and C (QseBC): a novel two-component regulatory system involved in the regulation of flagella and motility by quorum sensing in *E. coli*. Mol Microbiol 43(3):809–821.

Stringer SC, Plowman J and Peck MW. 2007. The microbiological quality of hot water-washed broccoli florets and cut green beans. J Appl Microbiol 102:41–50.

Thomas P. 1988. Radiation preservation of foods of plant origin. Part VI. Mushrooms, tomatoes, minor fruits and vegetables, dried fruits, and nuts. Crit Rev Food Sci Nutr 26(4):313–358.

Thunberg RL, Tran TT, Bennett RW, Matthews RN and Belay N. 2002. Microbial evaluation of selected fresh produce obtained at retail markets. J Food Prot 65(4):677–682.

Toivonen PMA and DeEll JR. 2002. Physiology of fresh-cut fruits and vegetables. In: Lamikanra O, editor. Fresh-Cut Fruits and Vegetables: Science, Technology and Market. Boca Raton, FL: CRC Press, pp. 91–123.

Tournas VH. 2005. Spoilage of vegetable crops by bacteria and fungi and related health hazard. Crit Rev Microbiol 31(1):33–44.

Ukuku DO, Fan X and Kozempel MF. 2006. Effect of vacuum-steam-vacuum treatment on microbial quality of whole and fresh-cut cantaloupe. J Food Prot 69(7):1623–1629.

Van Outryve MF, Gosselem F, Joos H and Swings J. 2008. Fluorescent *Pseudomonas* isolates pathogenic on witloof chicory leaves. J Phytopathol 125(3):247–256.

Vermeiren L, Delieghere F and Debevere J. 2002. Effectiveness of some recent antimicrobial packaging concepts. Food Addit Contam 19(Suppl):163–171.

Walton JD. 1994. Deconstructing the cell wall. Plant Physiol 104:1113–1118.

Wang CY, Wang SY, Yin JJ, Parry J and Yu LL. 2007. Enhancing antioxidant, antiproliferation, and free radical scavenging activities in strawberries with essential oils. J Agric Food Chem 55(16):6527–6532.

Watada AE, Ko NP and Minott DA. 1996. Factors affecting quality of fresh-cut horticultural products. Postharvest Biol Technol 9:115–125.

Zapata PJ, Guillén F, Martínez-Romero D, Castillo S, Valero D and Serrano M. 2008. Use of alginate or zein as edible coatings to delay postharvest ripening process and to maintain tomato (*Solanum lycopersicon* Mill) quality. J Sci Food Agric 88:1287–1293.

Zhang K, Ou M, Wang W and Ling J. 2009. Effects of quorum sensing on cell viability in *Streptococcus mutans* biofilm formation. Biochem Biophys Res Commun 379(4):933–938.

Zhuang H, Barth MM and Hankinson TR. 2003. Microbial safety, quality and sensory aspects of fresh-cut fruits and vegetables. In: Novak JS, Sapers GM and Juneja VK, editors. Microbial Safety of Minimally Processed Foods. Boca Raton, FL: CRC Press, pp. 255–278.

Zhuang H, Barth MM and Hildebrand DF. 1994. Packaging influenced total chlorophyll, soluble protein, fatty acid composition and lipoxygenase activity in broccoli florets. J Food Sci 59(6):1171–1174.

Index